ADVANCES in
FOOD DEHYDRATION

Contemporary Food Engineering

Series Editor

Professor Da-Wen Sun, Director

Food Refrigeration & Computerized Food Technology
National University of Ireland, Dublin
(University College Dublin)
Dublin, Ireland
http://www.ucd.ie/sun/

Advances in Food Dehydration, *edited by Cristina Ratti* (2009)

Optimization in Food Engineering, *edited by Ferruh Erdogdu* (2009)

Optical Monitoring of Fresh and Processed Agricultural Crops, *edited by Manuela Zude* (2009)

Food Engineering Aspects of Baking Sweet Goods, *edited by Servet Gülüm Şumnu and Serpil Sahin* (2008)

Computational Fluid Dynamics in Food Processing, *edited by Da-Wen Sun* (2007)

Contemporary Food
Engineering Series
Da-Wen Sun, Series Editor

ADVANCES in FOOD DEHYDRATION

Edited by
Cristina Ratti

CRC Press
Taylor & Francis Group
Boca Raton London New York

CRC Press is an imprint of the
Taylor & Francis Group, an **informa** business

A TAYLOR & FRANCIS BOOK

CRC Press
Taylor & Francis Group
6000 Broken Sound Parkway NW, Suite 300
Boca Raton, FL 33487-2742

First issued in paperback 2019

© 2009 by Taylor & Francis Group, LLC
CRC Press is an imprint of Taylor & Francis Group, an Informa business

No claim to original U.S. Government works

ISBN-13: 978-1-4200-5252-7 (hbk)
ISBN-13: 978-0-367-38636-8 (pbk)

Library of Congress Cataloging-in-Publication Data

Advances in food dehydration / edited by Cristina Ratti.
    p. cm. -- (Contemporary food engineering ; 5)
    Includes bibliographical references and index.
    ISBN-13: 978-1-4200-5252-7
    ISBN-10: 1-4200-5252-7
    1. Food--Drying. 2. Dried food industry. I. Ratti, Cristina. II. Title. III. Series.

TP371.5.A37 2009
664'.0284--dc22
          2008020691

Visit the Taylor & Francis Web site at
http://www.taylorandfrancis.com

and the CRC Press Web site at
http://www.crcpress.com

To my patient husband and dear family

Contents

Series Preface

CONTEMPORARY FOOD ENGINEERING

Food engineering is the multidisciplinary field of applied physical sciences combined with the knowledge of product properties. Food engineers provide the technological knowledge transfer essential to the cost-effective production and commercialization of food products and services. In particular, food engineers develop and design processes and equipment in order to convert raw agricultural materials and ingredients into safe, convenient, and nutritious consumer food products. However, food engineering topics are continuously undergoing changes to meet diverse consumer demands, and the subject is being rapidly developed to reflect the market needs.

In the development of food engineering, one of the many challenges is to employ modern tools and knowledge, such as computational materials science and nanotechnology, to develop new products and processes. Simultaneously, improving food quality, safety, and security remain critical issues in food engineering study. New packaging materials and techniques are being developed to provide more protection to foods, and novel preservation technologies are emerging to enhance food security and defense. Additionally, process control and automation regularly appear among the top priorities identified in food engineering. Advanced monitoring and control systems are developed to facilitate automation and flexible food manufacturing. Furthermore, energy saving and minimization of environmental problems continue to be an important food engineering issue, and significant progresses are being made in waste management, efficient utilization of energy, and reduction of effluents and emissions in food production.

Consisting of edited books, the *Contemporary Food Engineering* book series attempts to address some of the recent developments in food engineering. Advances in classical unit operations in engineering applied to food manufacturing are covered as well as such topics as progress in the transport and storage of liquid and solid foods; heating, chilling, and freezing of foods; mass transfer in foods; chemical and biochemical aspects of food engineering and the use of kinetic analysis; dehydration, thermal processing, nonthermal processing, extrusion, liquid food concentration, membrane processes and applications of membranes in food processing; shelf-life, electronic indicators in inventory management, and sustainable technologies in food processing; and packaging, cleaning, and sanitation. The books aim at professional food scientists, academics researching food engineering problems, and graduate level students.

The editors of the books are leading engineers and scientists from many parts of the world. All the editors were asked to present their books in a manner that will address the market need and pinpoint the cutting-edge technologies in food engineering. Furthermore, all contributions are written by internationally renowned experts who have both academic and professional credentials. All authors have attempted to provide critical, comprehensive and readily accessible information on the art and science of a relevant topic in each chapter, with reference lists to be used by readers for further information. Therefore, each book can serve as an essential reference source to students and researchers in universities and research institutions.

Da-Wen Sun
Series Editor

Series Editor

Born in Southern China, Professor Da-Wen Sun is a world authority in food engineering research and education. His main research activities include cooling, drying and refrigeration processes and systems, quality and safety of food products, bioprocess simulation and optimization, and computer vision technology. Especially, his innovative studies on vacuum cooling of cooked meats, pizza quality inspection by computer vision, and edible films for shelf-life extension of fruit and vegetables have been widely reported in national and international media. Results of his work have been published in over 180 peer reviewed journal papers and more than 200 conference papers.

He received a first class BSc Honours and MSc in mechanical engineering, and a PhD in chemical engineering in China before working in various universities in Europe. He became the first Chinese national to be permanently employed in an Irish university when he was appointed college lecturer at National University of Ireland, Dublin (University College Dublin) in 1995, and was then continuously promoted in the shortest possible time to senior lecturer, associate professor and full professor. Dr. Sun is now professor of Food Biosystems Engineering and director of the Food Refrigeration and Computerized Food Technology Research Group at University College Dublin.

As a leading educator in food engineering, Sun has contributed significantly to the field of food engineering. He has trained many PhD students, who have made their own contributions to the industry and academia. He has also given lectures on advances in food engineering on a regular basis in academic institutions internationally and delivered keynote speeches at international conferences. As a recognized authority in food engineering, he has been conferred adjunct/visiting/consulting professorships from ten top universities in China including Zhejiang University, Shanghai Jiaotong University, Harbin Institute of Technology, China Agricultural University, South China University of Technology, and Jiangnan University. In recognition of his significant contribution to food engineering worldwide and for his outstanding leadership in the field, the International Commission of Agricultural Engineering (CIGR) awarded him the CIGR Merit Award in 2000 and again in 2006. The Institution of Mechanical Engineers (IMechE) based in the United Kingdom named him Food Engineer of the Year 2004. In 2008 he was awarded the CIGR

Recognition Award in honor of his distinguished achievements in the top one percent of agricultural engineering scientists in the world.

He is a fellow of the Institution of Agricultural Engineers. He has also received numerous awards for teaching and research excellence, including the President's Research Fellowship, and he twice received the President's Research Award of University College Dublin. He is a member of CIGR Executive Board and honorary vice-president of CIGR, editor-in-chief of *Food and Bioprocess Technology— An International Journal* (Springer), series editor of the *Contemporary Food Engineering* book series (CRC Press/Taylor & Francis), former editor of *Journal of Food Engineering* (Elsevier), and editorial board member for *Journal of Food Engineering* (Elsevier), *Journal of Food Process Engineering* (Blackwell), *Sensing and Instrumentation for Food Quality and Safety* (Springer) and *Czech Journal of Food Sciences*. He is also a chartered engineer registered in the U.K. Engineering Council.

Preface

Dehydration processes have been known and used in food preservation for centuries. In spite of this, drying of foodstuffs is a complex heat and mass transfer process, which consumes large amounts of energy and unfortunately does not always lead to high-quality final products. Its current understanding is still quite limited because of the following factors: (1) the complex nature of food systems together with the deep structural and physicochemical changes that foodstuffs undergo during processing, (2) the difficulty of defining quality quantitatively and developing appropriate control techniques, and (3) the lack of realistic models and simulations to represent the phenomena. Knowledge of food physicochemical, kinetics, and sorptional properties and their variation with water content and drying conditions is also required to interpret this process correctly. In addition, recently there has been a clear need to optimize natural resources to reduce energy requirements together with an increasing demand for low-cost, high-quality products. This requires practical and advanced knowledge on dehydration in the food processing industry.

Dehydration is a traditional food engineering area that has advanced significantly in the last 20 years. New methods of dehydration have been developed, new sophisticated analytical techniques have been applied to drying to allow a better understanding of the phenomena, mathematical modeling has been improved to simulate the various driers under different conditions, and so forth. This new volume in food dehydration consists of 15 chapters on traditional and novel aspects of food dehydration. Each chapter will introduce the importance of the subject, followed by general concepts and finally by the latest developments in the area.

I sincerely acknowledge the collaboration and hard work of each contributor to this volume, the help of my graduate students and research assistant, and the valuable advice and input from Dr. Tadeusz Kudra.

Editor

Cristina Ratti, who is originally from Bahia Blanca, Argentina, graduated as a chemical engineer from the National University of the South (Bahia Blanca) and began her graduate studies in food engineering at the same institution under the supervision of Enrique Rotstein and Guillermo Crapiste. She obtained a PhD in chemical engineering in 1991 with her thesis on the "Design of batch air dryers for fruit and vegetable products." Her three postdoctoral stages were undertaken at three academic institutions in collaboration with Arun Mujumdar (chemical engineering, McGill University, Canada) from 1991 to 1993, Vijaya Raghavan (bioresource engineering, McGill University) from 1993 to 1995, and Guillermo Crapiste in PLAPIQUI (Bahia Blanca) from 1995 to 1996.

In 1996 she joined the Department of Soils and Agri-Food Engineering at Laval University (Québec, Canada) as a professor of food engineering. She is presently a full professor, the coordinator of the Food Engineering Program at Laval University (since 2001), and a member of the Institute of Nutraceutical and Functional Foods (since 2003). Her research interests have always been related to food dehydration (air drying, freeze-drying, osmotic dehydration) and physicochemical and quality properties of foodstuffs related to drying. She has published numerous scientific manuscripts and contributed to the training of many graduate and undergraduate students on the science of food engineering and dehydration.

Contributors

Janusz Adamiec
Technical University of Lodz
Lodz, Wolczanska, Poland

Benu Adhikari
School of Science and Engineering
University of Ballarat
Victoria, Australia

Monica Araya-Farias
Department of Soils and Agri-Food
   Engineering
Laval University
Québec, Québec, Canada

Bhesh Bhandari
School of Land, Crop and Food
   Sciences
University of Queensland
Queensland, Australia

Silvia Blacher
Department of Applied Chemistry
University of Liège
Liège, Belgium

Catherine Bonazzi
National Institute of Agronomic
   Research
AgroParisTech
Massy, France

Bertrand Broyart
National Institute of Agronomic
   Research
AgroParisTech
Massy, France

María Elena Carrín
Planta Piloto de Ingenieria
   Quimica
Bahia Blanca, Argentina

Stefan Cenkowski
Department of Biosystems
   Engineering
University of Manitoba
Winnipeg, Manitoba, Canada

Francis Courtois
National Institute of Agronomic
   Research
AgroParisTech
Massy, France

Guillermo Héctor Crapiste
Planta Piloto de Ingenieria
   Quimica
Bahia Blanca, Argentina

Sakamon Devahastin
Department of Food Engineering
King Mongkut's University of
   Technology
Bangkok, Thailand

Maturada Jinorose
Department of Food Engineering
King Mongkut's Institute of
   Technology
Bangkok, Thailand

Tadeusz Kudra
CANMET Energy Technology Centre
Varennes, Québec, Canada

Angélique Léonard
Department of Applied Chemistry
University of Liège
Liège, Belgium

J.I. Lombraña
Department of Chemical
  Engineering
University of the Basque Country
Bilbao, Spain

Alejandro Marabi
Food Science Department
Nestlé Research Center
Lausanne, Switzerland

Valérie Orsat
Department of Bioresource
  Engineering
Macdonald Campus of McGill
  University
Ste-Anne de Bellevue, Québec,
  Canada

G.S. Vijaya Raghavan
Department of Bioresource
  Engineering
Macdonald Campus of McGill
  University
Ste-Anne de Bellevue, Québec,
  Canada

Mohammad Shafiur Rahman
Department of Food Science and
  Nutrition
Sultan Qaboos University
Al-Khod, Oman

Cristina Ratti
Department of Soils and Agri-Food
  Engineering
Laval University
Québec, Québec, Canada

S.C.S. Rocha
State University of Campinas
Faculty of Chemical Engineering
Campinas, Sao Paulo, Brazil

I. Sam Saguy
Food Science and Nutrition
  Department
Hebrew University of Jerusalem
Revohot, Israel

John Shi
Guelph Food Research Center
Agriculture and Agri-Food Canada
Guelph, Ontario, Canada

Susan D. St. George
Department of Biosystems
  Engineering
University of Manitoba
Winnipeg, Manitoba, Canada

O.P. Taranto
State University of Campinas
Faculty of Chemical Engineering
Campinas, Sao Paulo, Brazil

Sophia Jun Xue
Guelph Food Research Center
Agriculture and Agri-Food Canada
Guelph, Ontario, Canada

1 Dehydration of Foods: General Concepts

Monica Araya-Farias and Cristina Ratti

CONTENTS

1.1   INTRODUCTION

Dehydration is probably the oldest and most frequently used method of food preservation. It is currently a versatile and widespread technique in the food industry as well as a subject of continuous interest in food research. The term *dehydration* refers to the removal of moisture from a material with the primary objective of reducing microbial activity and product deterioration (Ratti, 2001). In addition to preservation, the reduced weight and bulk of dehydrated products decreases packaging, handling, and transportation costs. Furthermore, most food products are dried for improved milling or mixing characteristics in further processing. In contrast, with literally hundreds of variants actually used in drying of particulates, solids, pastes, slurries, or solutions, it provides the most diversity among food engineering unit operations (Ratti and Mujumdar, 1995). At present, instant beverage powders, dry soup mixes, spices, coffee, and ingredients used in food transformation are the major food products that are dehydrated.

    The wide variety of dehydrated foods available to the consumer and the concern to meet the highest quality specifications emphasizes the need for a better understanding of the drying process. General theoretical aspects, historical background, properties of foods in regard to drying, types of drying methods, classification of dryers, as well as quality aspects in food dehydration will be discussed in this chapter.

1.1.1   HISTORICAL BACKGROUND ON FOOD DRYING

The removal of water has been used for centuries to preserve foods. Sun drying of fruits and smoking of fish and meat are both well-known processes originated in antiquity. In fact, Persian and Chinese people dried fruits and vegetables in the sun as long as 5000 years ago. It is believed that *chuño*, which is prepared by freeze-drying native potatoes, may have been invented 2000 or 3000 years ago in the Andean highlands; it is considered as the first food product to be specially processed by humans. Sun-dried dates, figs, raisins, and apricots were supposedly developed by the aboriginal habitants of the Mediterranean Basin and Near East (Salunkhe et al., 1991; Van Arsdel and Copley, 1963). In addition to using the sun's energy for drying, various dehydration technologies were developed at the beginning of the 20th century based on the use of an artificial heat source. The name *dehydration* was given to the drying methods under controlled conditions.

The first record of artificial drying of foods appears in the 18th century. Vegetables treated in hot water were then placed in a hot stove for drying (Van Arsdel and Copley, 1963). Thereafter, development of drying technology for foodstuffs and agricultural produce was accelerated during World War I and World War II. Intensive research was conducted on vacuum drying of various foods, including meat and fish, during World War I. Other methods were investigated and developed before World War II. Drum drying was applied to whey and buttermilk, soup mixtures, and tomato powder. Spray drying was used particularly in milk products and eggs (Van Arsdel and Copley, 1963). Space exploration in the 1960s increased studies on freeze-drying. However, there was a slower pace in drying developments later.

Consumer demands for better quality, the need to economize energy, the initiation of global meetings to exchange drying information (i.e., the International Drying Symposium series), and the publication of thousands of technical papers on the subject have forced the scientific and industrial communities to advance in the multidisciplinary fields of drying technology. Numerous emerging technologies such as microwaves, heat pumps, and superheated steam drying have been considered recently in the food industry for the development of new products.

1.2 PSYCHROMETRY IN RELATION TO DRYING

1.2.1 THEORETICAL CONSIDERATIONS OF AIR–WATER MIXTURES

The dehydration process depends on understanding the relationship between the water contained in a foodstuff and the water present in the drying medium, which is usually air. The moisture contents are directly related to the chemical potentials and thus to the driving force for the dehydration process. Air–water mixtures will be discussed in this section, and the interaction and relationship between water and food will be discussed in the section dedicated to sorption equilibrium.

Air–water relationships are normally referred to as *humid air properties* or *psychrometric properties*. The term *humidification* has been used to describe the inclusion of pure liquid molecules in a gas phase. In dehydration applications, the gas is normally air and the liquid is water. In the range of temperatures and pressures used in dehydration, these mixtures behave as ideal gases (Barbosa-Canovas and Vega-Marcado, 1996; Karel and Lund, 2003a; Vega-Mercado et al., 2001). Thus, the ideal gas equations for air and water are as follows:

$$P_a V = n_a RT \tag{1.1}$$

for air and

$$P_w V = n_w RT \tag{1.2}$$

for water, where P_a and P_w are the partial pressures of air and water in the mixture, respectively; n_a and n_w are the number of moles of air and water vapor, respectively;

V is the total volume; R is the gas constant; and T is the absolute temperature. If the mixture consists only of air and water, the total pressure (P_t) is the sum of the water and air partial pressures (Dalton law):

$$P_t = P_a + P_w \tag{1.3}$$

1.2.1.1 Absolute Humidity

The mass ratio of water to dry air is known as the *absolute humidity*, which can be defined as the amount of moisture in the air at any condition (Y):

$$Y = \frac{n_w M_w}{n_a M_a} = \frac{P_w M_w}{(P_t - P_w) M_a} = 0.6207 \frac{P_w}{P_t - P_w} \tag{1.4}$$

where M_w and M_a are the molecular weights of water and air, respectively. Dry air (78% nitrogen, 21% oxygen, and 1% other gases) has an average M_a of 29 Da and water has an M_w of 18 Da.

1.2.1.2 Saturation Absolute Humidity

Equilibrium is reached when the partial pressure of water vapor in the air equals the water saturation pressure at a given temperature. Thus, the mass ratio of water to air is called the *saturation absolute humidity*. This is the maximum amount of moisture that the air can carry at that temperature, which can be expressed as

$$Y_s = 0.6207 \frac{P_{w0}}{P_t - P_{w0}} \tag{1.5}$$

where P_{w0} is the saturation vapor pressure of water.

1.2.1.3 Relative Humidity

Relative humidity (RH) is defined as the ratio of P_w to P_{w0} at the same temperature. It is a relative measure of the amount of moisture that wet air can hold at a given temperature:

$$RH = 100 \frac{P_w}{P_{w0}} \tag{1.6}$$

1.2.2 PSYCHROMETRIC OR HUMIDITY CHART

Changes in psychrometric properties are traditionally presented in a *psychrometric chart*. Many professional charts can be found in the literature (Treybal, 1980). *Psychrometric software* is the current method that is frequently used to calculate these properties (Ratti et al., 1989). A graph of the absolute air humidity (y axis) as a function of temperature (x axis) at varying degrees of saturation constitutes the main element of a psychrometric chart (Toledo, 1991). The psychrometric chart is very useful for heat and mass balance determinations involving air–water mixtures because

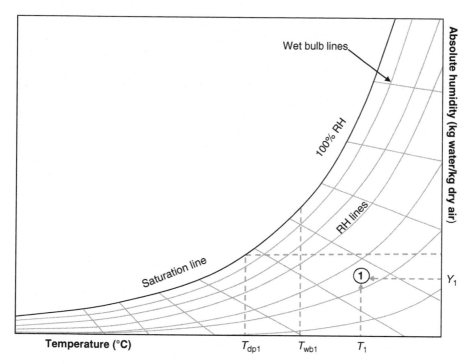

FIGURE 1.1 Schematic representation of a psychrometric chart.

these properties can be graphically represented in the chart. Dry and wet bulb temperatures, dew point temperature, absolute and relative humidity, and enthalpy and specific volume are variables commonly encountered in this chart.

Figure 1.1 is a schematic representation of a psychrometric chart. Absolute humidity is located on the vertical axis of the chart. The saturation line is identified as 100% relative humidity, and it represents the extreme left boundary of the graph. Other relative humidity levels can be identified with their respective percentage at distances to the right of the saturation line.

Dry bulb temperature is the temperature measured using an ordinary thermometer immersed in the air–vapor mixture. It is shown on the horizontal axis of the psychrometric chart in Figure 1.1. The *dew point* is the temperature at which an air–water mixture reaches saturation at constant absolute humidity. If a given air–water mixture in a closed system is cooled to a temperature just below its dew point, the water contained in the mixture starts to condense. The dew point is read in the psychrometric chart on the temperature axis at the saturation (100% relative humidity) line. *Wet bulb temperature* is the temperature determined by a thermometer placed in moving air with its bulb covered with a wet cloth. Water evaporation from the cloth cools the bulb to a temperature lower than the dry bulb temperature of the air if the air is unsaturated. This temperature difference is known as the *wet bulb depression*, a function of the relative humidity of air (Toledo, 1991). Wet bulb temperatures are read on the slanted lines of the psychrometric chart.

The enthalpy of air–water mixtures is the sum of the partial enthalpies of the constituents, and it represents the heat content of the mixture. The humid enthalpy can be defined as the enthalpy of a unit mass of dry air and its associated moisture (Mujumdar and Menon, 1995). The lines representing the enthalpy of moist air are almost parallel to the wet bulb lines on a psychrometric chart. If zero temperature (0°C) is taken as the reference point, the enthalpy of an air–water mixture is expressed as

$$H = L_0 Y + (C_{pa} + C_{pw}Y)T \tag{1.7}$$

where L_0 is the latent heat of water at 0°C (kJ/kg), C_{pa} is the specific heat of dry air (kJ/kg °C), C_{pw} is the specific heat of water vapor (kJ/kg °C), and T is the temperature (°C).

Two variables are necessary to establish a point on the psychrometric chart representing an air condition. In Figure 1.1, for instance, absolute humidity Y_1 and dry bulb temperature T_1 are required to determine the air conditions at point 1. An example of the use of the psychrometric chart in an adiabatic dehydration process, a situation commonly encountered during drying, is provided in Figure 1.2. Point a in the figure represents the ambient air condition, whereas point b is the air after being heated (before entering the dryer). Point c represents the air exit condition (after passing through the dryer), which is reached from point b by following the lines at constant enthalpy because the dryer is adiabatic. The air temperature at point c is lower than at point b and its absolute humidity is higher, because the air has lost heat and picked up moisture from the product while passing through the dryer.

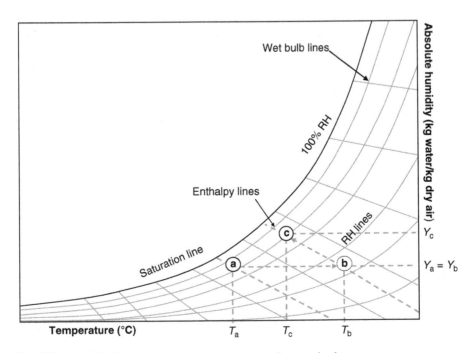

FIGURE 1.2 Adiabatic drying represented on a psychrometric chart.

Psychrometric properties change when total pressure decreases. Air relative humidity, dew point, and wet bulb temperatures decrease as a vacuum is applied in a dryer. For medium vacuum levels (i.e., vacuum drying), this variation should be taken into account in order to calculate the driving force for mass transfer.

1.3 DEHYDRATION PRINCIPLES

A wet material placed in a medium having lower water partial pressure (at the same temperature) will dehydrate until equilibrium is reached. The medium is usually air at pressures ranging from a high vacuum to atmospheric (or higher), although super-heated steam, hot oil, solvents, and solutions may also be used for this purpose (Karel and Lund, 2003a).

Food dehydration is a complex phenomenon involving simultaneous mass and energy transport in a hygroscopic and shrinking system (Mujumdar, 1997; Mujumdar and Menon, 1995; Ratti, 2001). The rate of dehydration is governed by the rate at which these processes take place. Heat transfer from the drying medium to the wet solid can be a result of convection, conduction, or radiation effects and, in some cases, a combination of them. Internal heat transfer is usually very rapid compared to external transfer; therefore, heat transfer during drying is usually controlled externally. In contrast, mass transfer depends on either the movement of moisture within the solid or the movement of water vapor from the solid surface to the bulk medium. Depending on the specific process, water removal can be limited by either heat or mass transfer or the resistances to these two transport phenomena can be coupled, that is, they depend on each other (Karel and Lund, 2003a). Table 1.1 presents a compilation of characteristics of commonly used drying methods. Note that heat and mass transfer always take place during drying whether the process is done at atmospheric or vacuum conditions or even when using different drying mediums.

Internal mass transfer is generally recognized to be the principal rate-limiting step during drying. Because of the complexity of the process, however, no generalized theory exists to explain the mechanism of internal moisture movement. The structure of food material being dried plays an important role in the mechanism of water movement within a product. As pointed out by Rizvi (2005), in liquid materials and gels, water transport is by molecular diffusion from the interior to the surface of the product, where it is removed by evaporation. In contrast, in capillary-porous materials the possible physical mechanisms are numerous and can be classified as (1) liquid movement caused by capillary and gravity forces, (2) liquid diffusion caused by a difference in concentration, (3) surface diffusion, (4) water vapor diffusion caused by partial pressure gradients, (5) water vapor flow under differences in total pressure, and (6) flow caused by an evaporation–condensation sequence (Van Arsdel and Copley, 1963). Some of these mechanisms and theories involved during drying will be explained in detail in other chapters of this book.

1.3.1 DEHYDRATION KINETICS

The drying kinetics of the product are the most important data required for the design and simulation of dryers. They represent the ease with which a product dehydrates

TABLE 1.1

Main Characteristics of Different Dehydration Methods

Operation	Drying Medium	External Driving Force	Internal Mass Transfer Mechanism	Other Characteristics	Operation Variables	Main Heat Transfer Mechanism
Hot air drying	Air at atmospheric pressure	Difference in vapor pressure	Vapor and/or liquid diffusion, capillarity, hydrodynamic flows, etc.	—	Medium: temperature, velocity, relative humidity; Product: thickness, geometry, density	Internal: conduction; External: convection
Vacuum drying	Vacuum	Difference in vapor pressure	Liquid and/or vapor diffusion	Possible boiling depending on conditions	Medium: pressure, temperature, relative humidity; Product: thickness, geometry, density	Internal: conduction; External: radiation, conduction
Superheated steam drying (SSD)	Superheated steam	Difference in vapor pressure	Vapor diffusion	Internal boiling Possible vacuum	Medium: pressure, temperature; Product: thickness, geometry, density	Internal: conduction; External: radiation, conduction
Contact drying	Heated solid	—	Vapor and/or liquid diffusion	—	Medium: surface temperature; Product: thickness, density	Internal: conduction; External: conduction
Freeze-drying	Vacuum	Difference in vapor pressure	Vapor diffusion	Sublimation Subfreezing temperatures	Medium: pressure, temperature; Product: thickness, geometry, density	Internal: conduction; External: radiation, conduction
Osmotic dehydration	Osmotic solution	Difference in osmotic pressure	Liquid diffusion, volumetric flow	—	Medium: concentration, temperature, velocity; Product: thickness, geometry	Internal: conduction; External: convection

under specific drying conditions. Drying kinetics are affected by the external conditions of the medium and the chemical and physical structure of the food (see Operation Variables, Table 1.1). In the most general case, drying a food under *constant* conditions is considered in order to obtain the kinetics curve. Under these conditions, the medium is in large excess so that its properties can be considered constant.

The moisture content of a food sample must be measured as a function of time in order to obtain the dehydration kinetics curve. The moisture content X is usually calculated on a dry basis as the ratio of the amount of water in the sample to the amount of dry solids:

$$X = \frac{m - m_s}{m_s} \qquad (1.8)$$

where m is the total mass of the sample at time t and m_s is the mass of the dry solids. The relationship between moisture content in dry and wet basis is given by

$$X_w = \frac{X}{1 + X} \qquad (1.9)$$

where X_w is the water content on a wet basis. Figure 1.3 illustrates a typical drying kinetics curve with the moisture content as a function of drying time. When a comparison of different drying kinetics curves is necessary, the ordinate of the curve is usually the ratio of the moisture content (dry basis) at time t to that at initial time X/X_0.

In Figure 1.3a, segment a–b is the start-up period, when the product comes into contact with the medium and heats up. This period could be present or not during drying of foods. Proper drying occurs during segment b–e. After long drying times, the solid reaches *equilibrium* with the medium conditions and no further drying takes place (point e, Figure 1.3a). The free moisture content X_f (i.e., the moisture that is available for drying) can thus be determined as

$$X_f = X - X_c \qquad (1.10)$$

where X_e is the equilibrium moisture content, an important food property that will be discussed later in this chapter.

The drying rate N_w (kg water/m²/h) is calculated from the kinetics curve by the following equation:

$$N_w = -\frac{m_s}{A}\left(\frac{dX}{dt}\right) \qquad (1.11)$$

where A is the area available for drying. The calculation of the derivative dX/dt in Eq. (1.11) may cause some problems in the overall determination of the drying rate. This is because a numerical derivation must be performed with the known associated errors involved in such determinations. The easiest way of estimating the numerical temporal derivative is by calculating the difference in water content at two drying times. Although simple, this method can lead to significant errors, especially in the beginning of drying where the derivative is quite steep. Numerical errors involved with this method can be decreased by increasing the number and frequency of

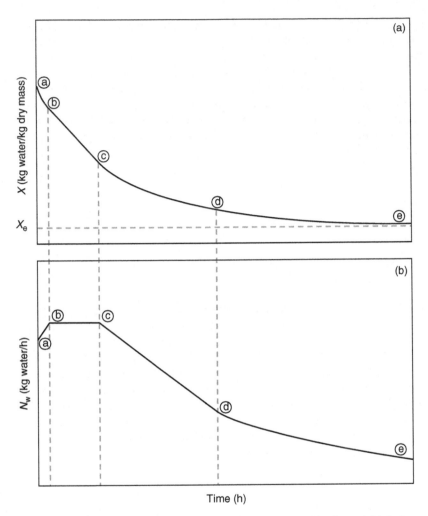

FIGURE 1.3 Drying curves as a function of drying time: (a) drying kinetics and (b) drying rate.

experimental data. A better estimation method for calculating the derivative dX/dt consists of the following two steps: a polynomial regression of X–t experimental data and an analytical derivative of the resulting polynomial. This two-step method requires trial and error in order to find the best polynomial regression for the data, and sometimes curve X–t must be divided into sectors where different fittings instead of a unique one are necessary. Although more laborious, the results obtained by this method are more precise and realistic than those from applying the difference method.

Figure 1.3b shows the drying rate as a function of time, which presents the same points as those in Figure 1.3a. Under constant conditions, the drying process can be described in a number of steps characterized by different dehydration rates. Two dehydration periods usually occur: an initial constant rate period (segment b–c) during which drying occurs as if pure water is being evaporated, and one or several falling rate periods (segments c–d and d–e) in which moisture movement is controlled by combined external–internal resistances or by purely internal resistances (Barbosa-Canovas and Vega-Mercado, 1996; Mujumdar, 1997; Toledo, 1991).

During the *constant rate period* the surface of the product is saturated with water and the rate of drying is limited by the rate at which heat is transferred to the product from air. The product temperature is usually the wet bulb temperature of the drying air. In this period, the drying rate is independent of the material surface and determined by the external conditions. Some foodstuffs do not present a constant drying rate period. However, for food materials where the liquid movement appears to be controlled by capillary and gravity forces, a constant rate period has been found. In contrast, in structured food systems where liquid movement is by diffusion, no constant rate period has been reported (Rizvi, 2005).

The constant rate period is followed by a period during which the rate of drying progressively decreases, which is called the *falling rate period*. The point at which the constant rate period ends and the falling rate period takes place is termed the point of *critical moisture content* (X_c), which is indicated by point c in Figure 1.3b. This point is represented in the drying rate curve as a sharp change of shape. At this point, the moisture content of the food is not sufficient to saturate the whole surface. In the falling rate period, the movement is controlled either by a combined external and internal mechanism or by internal mass transfer. Often during drying, one or more of these mechanisms may occur with changing relative contributions. Two or more falling rate periods could also exist in the case of hygroscopic materials.

In summary, it is evident that the drying rate of any of these periods depends on the heat and mass transfer coefficients, diffusion coefficients, the nature of the food, and the external drying conditions. Optimization of any or all of these factors would result in increased drying rates.

1.4 PHYSICAL AND THERMAL PROPERTIES

The design and optimization of any process in which the transfer of heat is involved requires knowledge of the thermal and physical properties of the materials being processed. In food drying, the product specific heat, thermal conductivity, density, and porosity are the most important thermophysical properties required for proper modeling.

1.4.1 THERMAL PROPERTIES

1.4.1.1 Thermal Conductivity

Thermal conductivity (k) is one of the most important thermophysical properties of foods that is used to estimate the rate of heat transfer by conduction in food processes such as drying, cooking, or frying (Maroulis et al., 2002; Rahman et al., 1997). It is a property that indicates how effective a material is as a heat conductor. As stated in Fourier's law for heat conduction, this constant is a proportionality factor needed for calculations of heat conduction transfer processes (Welti-Chanes et al., 2003).

Thermal conductivity depends mainly on food composition, but in some cases it can vary with the physical orientation of food compounds, as in the case of meat

TABLE 1.2
Thermal Conductivity of Five Foods

Food	Moisture X (Dry Basis)		k (W/m K)	
	Min	Max	Min	Max
Bread	0.00	0.82	0.040	3.559
Rice	0.10	0.43	0.080	0.336
Apples	0.10	19.00	0.000	2.270
Coffee	1.20	2.51	0.150	0.277
Cheese	0.50	2.94	0.340	0.548

Source: Adapted from Krokida, M.K. et al., *Int. J. Food Prop.*, 4, 111–137, 2001. With permission.

products. Water content also plays a significant role in the value of conductivity (and thus should be taken into account during drying; Cuevas and Cheyran, 1978). Data on the thermal conductivity of various food materials were reported previously (Krokida et al., 2001; Rahman, 1995). The data found in the literature present a wide variation, which is attributable to factors such as the composition or structure of the material, different experimental methods, and processing conditions (temperature, pressure, and mode of energy transfer). Table 1.2 provides a few values for the thermal conductivity of foods.

1.4.1.2   Specific Heat

Specific heat (C_p) can be defined as the amount of energy needed to increase the temperature of one unit mass of a material by $1°$. The specific heat of foods depends strongly on the water content, because water has the highest specific heat of all food compounds. Tables containing C_p data for various foodstuffs are available (ASHRAE, 1985; Rahman, 1995). Specific heat values for several foods are presented in Table 1.3.

TABLE 1.3
Specific Heat of Five Foods

Food	Moisture X_w (Wet Basis)		C_p (kJ/kg K)	
	Min	Max	Min	Max
Carrots	0.001	0.145	1.439	2.272
Eggs	0.001	0.145	1.686	2.276
Soybeans	0.000	0.379	1.576	2.342
Paddy rice	0.113	0.203	1.606	1.986
Potatoes	0.001	0.159	1.301	2.151

Source: Adapted from Rahman, S., *Food Properties Handbook*, CRC Press, Boca Raton, FL, 1995. With permission.

1.4.2 Physical Properties

The structure of food materials can be characterized by density (apparent and true), porosity, pore size distribution, specific volume, particle density shrinkage, and so on. Among these, density and porosity are the most common structural properties reported in the literature (Boukouvalas et al., 2006). Significant changes in structural properties can be observed as water is removed from the moist food during drying. These properties depend on various factors such as pretreatment, moisture content, processing method, and process conditions (Krokida and Maurolis, 2000).

1.4.2.1 Apparent or Bulk Density

Bulk density (ρ_b) applies to powdered and porous materials. It is defined as the mass of the sample (m_t) divided by its apparent volume (V_t), which includes the volume of dry solids, water, and air contained in the internal pores.

1.4.2.2 True or Particulate Density

Particulate density (ρ_p) is the density excluding all pores, and it is determined by the ratio of the sample mass to its true volume (excluding air pores).

1.4.2.3 Porosity

This property characterizes the open structure of a material. Porosity (ε) is the fraction of the empty volume (void fraction) in a sample volume, and it is usually estimated from the apparent and true density of a material (Boukouvalas et al., 2006; Krokida and Philippopoulos, 2005). Experimental data of porosity and density have been compiled by Rahman (1995) and more recently by Boukouvalas et al. (2006). Table 1.4 lists the physical properties of several foods.

1.4.3 Property Modeling

It is important to note that the product composition and its temperature change during drying. Therefore, to predict drying kinetics or to evaluate the quality changes during drying, thermophysical properties should be available as continuous functions

TABLE 1.4
Bulk Density (ρ_b) and Porosity (ε) of Five Foods

Food	Moisture X_w (Wet Basis) Min	Max	ρ_b (kg/m³) Min	Max	ε Min	Max
Bread	0.000	0.428	161	974	0.000	0.790
Apples	0.000	0.900	80	1466	0.020	0.950
Nuts	0.027	0.350	57	792	0.339	0.664
Coffee	0.099	0.238	330	514	0.450	0.540
Milk	0.046	0.980	593	1020	0.477	0.477

Source: Adapted from Boukouvalas, Ch.J. et al., *Int. J. Food Prop.*, 9, 715–746, 2006. With permission.

of moisture content and/or temperature. There are several sources that present theoretical and empirical equations to model these properties. However, because of the difficulty of using theoretical models in complex systems such as foods, empirical models are more popular and widely used in food materials. Some of these models will be presented in the following sections.

1.4.3.1 Linear Models

Water content plays a significant role in the thermophysical properties of foods because the properties of water are markedly different from those of other compounds such as proteins, fats, carbohydrates, and air. Therefore, thermophysical properties of foods are frequently modeled as a function of moisture content, commonly through a linear relationship if the moisture content is expressed on a wet basis. Moreover, thermal conductivity may also be strongly dependent on temperature and thus is often modeled as a function of both moisture content and temperature.

Sweat (1974) proposed a linear correlation for predicting the thermal conductivity of fruits and vegetables that yielded predictive data within ±15% of its experimental value. However, this model was limited to moisture content higher than 60% and the temperature and porosity effects were not considered.

Phomkong et al. (2006) developed a linear mathematical model for thermophysical properties of stone fruits. The thermal conductivity and specific heat were found to be linear functions of moisture content whereas the apparent bulk density and porosity were better represented by second-order polynomials. Specific heat and thermal conductivity were also investigated by Vagenas and Marinos-Kouris (1990) for sultana grapes and raisins, and these properties varied linearly with the moisture content (14 to 80%). According to Madamba et al. (1995), moisture content had a marked effect on the specific heat and thermal conductivity in garlic. Telis-Romero et al. (1998) used a linear function of moisture content, temperature, and composition for determination of specific heat in orange juice.

Changes of density, thermal conductivity, thermal diffusivity, and specific heat of plums during drying as a function of moisture content was investigated by Gabas et al. (2005). The apparent density of the fruits increased from 1042.9 to 1460.0 kg/m^3 and the bulk density increased from 706.6 to 897.5 kg/m^3 as the plums were dried. In contrast, the effective thermal conductivity and specific heat increased from 0.154 to 0.4 W/m K and from 1796 to 3536 J/kg K, respectively, according to an increased moisture content range of 14.2 to 80.4% (wet basis).

Shrivastava and Datta (1999) determined the specific heat and thermal conductivity of mushrooms at various levels of moisture content, temperature, and bulk density. The results of this investigation showed that changes in the moisture range (10.24 to 89.68%, wet basis) had a strong linear correlation on these properties. However, Niesteruk (1996) found that the thermal properties of fruit and vegetables vary linearly with the moisture content but not for all ranges of humidity (0 to 100%). The study noted that this effect was due to different properties of bound water in the product being dried.

1.4.3.2 Nonlinear Models

The applicability of the linear correlations mentioned earlier is sometimes limited to a reduced range of water content. Their fitting parameters also vary, depending on the type of food material. Efforts were thus carried out to develop generalized correlations to predict all food properties by means of nonlinear regression models (Rahman et al., 1997).

Lozano et al. (1979) and Mattea et al. (1989) developed nonlinear models in exponential form for correlating conductivity data of apples and pears during drying. The changes in bulk density, porosity, and bulk shrinkage during air dying of fruits and vegetables were also investigated by Lozano et al. (1980, 1983). The bulk shrinkage coefficient and porosity changes of these foods during drying were related to the moisture content by empirical equations. In addition, Kim and Bhowmik (1997) found that the bulk density of yogurt changed nonlinearly with moisture content, and Marousis and Saravacos (1990) found that the particle density data of starch materials correlated to the moisture content in a polynomial form.

Rahman et al. (1997) developed a generic correlation to predict thermal conductivity as a function of water, porosity, and temperature. The model was developed for moisture contents from 14 to 88% (wet basis), temperatures from 5 to 100°C, and porosities from 0.0 to 0.56. Maroulis et al. (2002) developed a structural generic model to predict the effective thermal conductivity of fruits and vegetables during drying. It combines a generic structural model for shrinkage and the distribution factor concept for thermal conductivity.

Table 1.5 summarizes some models found in the literature on thermophysical properties as a function of moisture content. In recent years, artificial neural network modeling has gained increasing acceptance for the estimation and prediction of food properties and processes (Mattar et al., 2004; Poonnoy et al., 2007; Sablani and Rahman, 2003).

1.5 DRYING TECHNOLOGIES

This section will introduce and compare some conventional and new (nonconventional) drying technologies and outline the principles and advantages/disadvantages of their operations. A detailed description of these processes and the associated equipment is well beyond the objective of this section.

1.5.1 SELECTION CRITERIA FOR CLASSIFICATION OF DRYERS

Drying technology has evolved markedly from the simple use of solar energy to current technology (Vega-Mercado et al., 2001). The diversity of food products generated the need for different types of dryers used in the industry. Dryers and drying processes can be classified in different ways. Kudra and Mujumdar (2002), Mujumdar and Menon (1995), and Vega-Mercado et al. (2001), have given detailed schemes for classification of industrial dryers based on numerous selection criteria. Among them, the most important criteria for evaluating a dryer system are the following:

1. Mode of operation (batch or continuous)
2. Operating pressure (vacuum, atmospheric, and high pressure)
3. Mode of heat transfer (conduction, convection, radiation, dielectric heating, and combination of different modes)

TABLE 1.5
Empirical Models Relating Food Thermophysical Properties with Moisture Content

Model	Moisture Content	Properties	Material	Ref.
$k = 0.148 + 0.00493X_w$	$X_w > 0.6$	Thermal conductivity	Strawberries, cherry tomatoes, cucumbers, onions, beets, and carrots	Sweat (1974)
$k = 0.493 - 0.359 \exp(-1.033X)$	$1.0 < X < 7.0$	Thermal conductivity	Pears	Mattea et al. (1989)
$C_p = 1.52 + 0.028X_w$	$0.4 < X_w < 0.62$	Specific heat	Garlic	Madamba et al. (1995)
$C_p = 2.89 + 0.012X_w$	$0.3 < X_w < 0.83$	Specific heat	Nectarine	Phomkong et al. (2006)
$\varepsilon = 1 - \dfrac{0.852 - 0.462 \exp(-0.66X)}{1.54 \exp(-0.051X)}$	$1.5 < X < 7.45$	Porosity	Apples	Lozano et al. (1980)
$\varepsilon = 1 - \dfrac{933.9 - 274X_w}{1485.5 + 41.8X_w - 775.6X^2}$	$0.14 < X_w < 0.8$	Porosity	Plums and prunes	Gabas et al. (2005)
$\rho_b = 0.852 - 0.462 \exp(-0.66X)$	$0 < X < 7.45$	Bulk density	Apples	Lozano et al. (1980)
$\rho_b = 933.9 - 274.4X$	$0.14 < X_w < 0.8$	Bulk density	Plums	Gabas et al. (2005)
$\rho_p = 1.54 \exp(-0.051X) - 1.15 \exp(-2.4X)$	$X < 1.5$	Particulate density	Apples	Lozano et al. (1980)
$\rho_p = 1442 + 837X - 3646X^2 + 4481X^3 - 1850X^4$	$0 < X < 1.0$	Particulate density	Starch materials	Marousis and Saravacos (1990)

X (g water/g dry mass), X_w (g water/g wet mass).

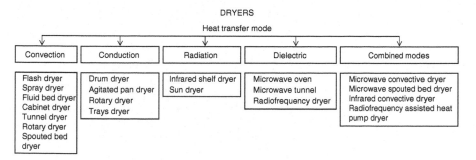

FIGURE 1.4   Classification of dryers according to heat mode transfer.

4. State of product being dried (stationary, moving, agitated, fluidized, and atomized)
5. Residence time (short = below 1 min, medium = 1–60 min, long = higher than 60 min).

The previous classification scheme gives an idea of the large number of dryers and their potential configurations. One of the most appropriate methods of classification of dryers is related to the method of heat transfer (conduction, convection, radiation, dielectric heating, etc.). Most dryers are operated at atmospheric pressure. Vacuum operations are expensive and they are recommended only for heat-sensitive products. On this basis, Figure 1.4 shows a classification of atmospheric dryers according to heat transfer mode. The list presented in the figure is brief; it is given as a simple summary of types of dryers that can be used, depending on the mode of heat transfer. As mentioned earlier, numerous other criteria can be considered for selecting a dryer (see Kudra and Mujumdar, 2002, for details). The final selection of the dryers will usually represent a compromise between the total cost, product quality, safety considerations, and convenience of installation (Mujumdar and Menon, 1995).

1.5.2   ADVANTAGES AND DISADVANTAGES

Many conventional methods are used in food drying including hot air drying, vacuum drying, drum drying, spray drying, freeze-drying, and so forth. Numerous emerging technologies have been developed recently as alternatives to more well-known methods. For multiple reasons, there is an industry preference to use conventional dryers because of simplicity of construction and operation as well as their mature status and familiarity. Dryer vendors prefer such technologies because of the low risk factor in design and scale-up (Kudra and Mujumdar, 2002). In addition, the high cost of some new technologies limits their application. New drying technologies must offer significant advantages over conventional technologies to find industrial acceptance. The selection criteria for the use of new technologies remain the same as those for conventional ones. Before a particular process is selected, different factors should be considered, including the type of product being dried (e.g., heat-sensitive materials), the desired quality of the finished product, the energy efficiency, and the cost of processing. Table 1.6 compares some

TABLE 1.6
Conventional and New Technologies Used According to the Form Feed

Drying Technologies	Feed Type	Principle or Description	Advantages	Disadvantages
Sun drying	Fruits, vegetables, meat, fish, plants, and aromatic herbs	Direct exposure of food to sunlight	Simple and low operating cost, low temperature	Long drying time, large areas of space, chemical deterioration, microbial contamination
Hot air drying	Small particulates, vegetables and fruit pieces, granular products	Food in contact with hot steam air, drying occurs by convection	Simple and continuous process, low operating cost	Nutritional quality loss, shrinkage, long drying time
Drum drying	Liquids, foods, slurry, pastelike or sludge, pulps, and suspensions	Food dried on the surface of hollow drums heated internally by steam	Continuous operation	May require modification of liquid to ensure good adhesion to the drying surface
Spray drying	Liquids foods and suspensions, instant coffee, pharmaceuticals	Liquid atomized (sprayed) into a drying chamber, where contacts hot air	Uniform and rapid drying, low cost, large-scale production	Some quality loss by thermal degradation, high temperature potential of agglomeration
Freeze-drying	Fruit and vegetables, mushrooms, instant coffee, pharmaceuticals, high value-added products	Frozen water removed from food by sublimation	High nutritional and sensorial quality, no restriction on particle size, utilization of low temperatures	Slow and expensive process, long drying time

Method	Products	Principle	Advantages	Disadvantages
Fluidized bed drying	Small particulates, diced vegetables and fruit granules, fruit juice powders, high value-added products	Material dried while suspended in an upward flowing gas stream (hot air)	High thermal efficiency, high drying rates due to good gas particle contact, low cost and maintenance	May require relatively small, uniform, and discrete particles
Spouted bed drying	Inert particles; paste and slurries; granular products like wheat, corn, oats, and cereal seeds; heat-sensitive foodstuffs	Food suspended into a spouted bed system where contacts hot air	High drying rates and low drying times, low drying temperature	Problem of scale-up and control of the cyclic pattern
Superheated steam drying (SSD)	Solid particles, value-added products, high oxygen-sensitive food and biomaterials	SSD transfers its sensible heat to the product being dried and the water to be evaporated; SSD acts as heat source and drying medium to take away the evaporated water	High drying rates, better product quality (no product oxidation), simple process control, improved drying efficiency	Appropriate insulation of the steam drying process necessary, construction of dryer above condensation temperature of SSD
Microwave (MW) and radiofrequency (RF) drying	Fruit and vegetables, high value-added products, pasta products, cookies	Food heated and dried by electromagnetic energy; MW drying frequencies from 300 to 3000 MHz; RF drying frequencies from 1 to 300 MHz	Efficiency of energy conversion, uniformity of drying, short drying times, desirable chemical physical effects, better process control	Expensive process, nonuniformity of electromagnetic field, high installation cost

types of conventional and new drying technologies that can be used according to the type of product to be dried. It also shows some of their advantages and disadvantages. For more information on other new drying technologies, see Chapters 13 and 15.

1.6   QUALITY ASPECTS OF DEHYDRATED FOODS

Quality aspects require special attention in processes dealing with foodstuffs. Unfortunately, it is well known that drying may partially or totally affect the quality of a product (Bimbenet and Lebert, 1992; Karel, 1991; Ratti, 2001).

Up until recently, R&D efforts in drying were directed toward processes and technology with the goal of extending shelf life, but little attention paid to conservation of quality attributes. Efforts were made recently to develop dried foods of high quality. This goal was attained by using innovative technologies and by improving and optimizing existing techniques. Many quality parameters can be defined for dried foods. These quality parameters can be categorized into three major groups: physical, chemical (or biochemical), and microbiological. Table 1.7 summarizes some of the quality changes that may occur in foods during drying, which include loss of aroma and flavors, changes in color and texture, and a decrease in nutritional value. Furthermore, residual enzyme and microbial activities are also important parameters that affect product quality and shelf life in a dehydrated material.

Food quality can be affected by other nondrying parameters such as pH; the composition of the food; pretreatments; and the presence of salts, oils, or solvents (Mujumdar, 1997). In contrast, analysis of processes affecting the physical, biochemical, and microbiological stability of foods is largely based on sorption isotherms of the materials involved (Tsami et al., 1990). Moreover, the removal of water during drying affects the structural and compositional characteristics of the product and can cause phase transitions.

1.6.1   WATER SORPTION ISOTHERMS

Water activity is defined as the ratio of the vapor pressure in a food material to that for pure water at the same temperature. The relationship between the equilibrium moisture content X_e and the corresponding water activity a_w at a constant temperature

TABLE 1.7
Food Quality Degradations Occurring during Drying

Physical	Chemical or Nutritional	Microbiological
Porosity changes	Enzymatic reactions	Microbial survival
Shrinkage	Lipid oxidation	Loss in activity
Change of solubility	Vitamin and protein losses	
Reduced rehydration	Browning reaction	
Hardening and cracking	Degradation of nutraceutical compounds	
Aroma and flavor loss		

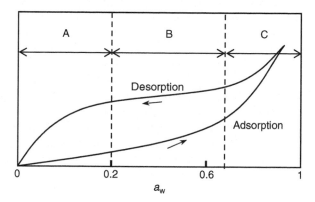

FIGURE 1.5   Water sorption isotherm of a foodstuff.

yields a water sorption isotherm. This isotherm is one of the most fundamental parameters that influence many aspects of the dehydration process and the storage stability of the dried product. For food systems, these isotherms also give useful information about the sorption mechanism and the interaction of food biopolymers with water (Chirife and Buera, 1994; Tsami et al., 1990).

The water sorption isotherms of most foods are nonlinear, generally exhibiting a sigmoidal shape when graphically represented (Figure 1.5). Isotherms in food materials are usually obtained by exposing a small sample of food to various constant relative humidity atmospheres. After equilibrium is reached, the water content of the sample is determined gravimetrically or by other methods (Karel and Lund, 2003b). Constant relative humidity is generally achieved by using salt solutions. An isotherm obtained by exposing a dry solid to increasing water activities is an adsorption isotherm; exposing a wet solid to decreasing water activities is a desorption isotherm. Adsorption isotherms are used to describe the hygroscopic behavior of a product, whereas desorption isotherms are useful for drying because the moisture content of the solid progressively decreases during the process.

Figure 1.5 shows the general shape of the sorption isotherms for a food material characterized by three different zones: A, B, and C. Zone A represents sorption of water that is strongly bound to the sites, and it is unavailable to serve as a solvent. In this region, there is essentially the formation of a monomolecular layer of water. The enthalpy of the vaporization of water in this zone is higher than that of pure water. Zone B is a zone in which the water is more loosely bound, so mainly double and multiple layers are formed. In this zone, water is held in the solid matrix by capillary condensation and multilayer adsorption. Some of the water may still be in the liquid phase, but its mobility is significantly limited because of attractive forces with the solid phase. In zone C the water is even more loosely bound and available for participation in reactions and as a solvent (Mujumdar, 1997; Mujumdar and Menon, 1995; Toledo, 1991).

The moisture content of the monomolecular layer (in zone A) is often considered as a good first approximation of the water content providing maximum stability of a dehydrated material. The best model to estimate the moisture content of monomolecular layer is described by the Brunauer–Emmett–Teller (BET) equation (Brunauer et al., 1938; Karel, 1973; Labuza, 1980):

$$\frac{a_w}{X(1-a_w)} = \frac{1}{X_mC} + \frac{C-1}{X_mC}(a_w) \qquad (1.12)$$

where X is the moisture content (g water/g dry mass); X_m is the water content corresponding to the water molecules covering the surface of particles as a monolayer; and C is a constant, which is temperature dependent. The plot of $a_w/(X(1-a_w))$ versus a_w is known as the BET plot. For most foods, this plot is linear for low water contents, and its slope and intercept can be used to determine the constants C and X_m. Determining the moisture content for the maximum shelf stability of a dehydrated material involves determining the sorption isotherm and calculating the value of X_m in Eq. (1.12) (Toledo, 1991).

Moraga et al. (2006) found that the moisture content of the monomolecular layer was similar in both whole or homogenized freeze-dried kiwi fruits, with values ranging between 0.053 and 0.061 g water/g dry solids.

1.6.1.1 Sorption Hysteresis Phenomenon

Hysteresis is related to the nature and state of the components in a food. In many cases, the process of sorbing or desorbing water may change the solid matrix and consequently affect the sorption behavior. This is represented schematically in Figure 1.5 where it is evident that the adsorption and desorption paths are different, giving a so-called hysteresis loop (Kapsalis, 1987). Thus, more moisture is retained in the desorption process compared to adsorption at a given water activity. Moisture sorption isotherms of several foods exhibit hysteresis. Some known examples in food technology include starchy foods and products with high protein or high sugar content. This hysteresis effect is also observed to a different extent in almost all hygroscopic products. Moraga et al. (2004) reported significant differences in the adsorption–desorption behavior of whole and pureed strawberries, indicating the hysteresis in the sorption pathway of both kinds of samples. Nevertheless, the hysteresis was less marked in the homogenized sample. The phenomenon of hysteresis has also been observed in other fruits such as raisins, currants, figs, prunes, apricots (Tsami et al., 1990), and bananas (Yan et al., 2008) and in starch–protein extrudates (Wlodarczyk-Stasiak and Jamroz, 2008). However, the sorption behavior of grapes did not show the phenomenon of hysteresis (Vazquez et al., 1999).

Many theories and hypotheses have been formulated to explain the phenomenon of hysteresis. One such theory is that the hysteresis of dehydrated foods can vary depending on the nature of the food, temperature, storage time, and adsorption–desorption cycles. However, no model has been found to quantitatively describe the hysteresis loop in foods (Rizvi, 2005).

1.6.1.2 Temperature Effect on Sorption Isotherms

Water activity changes with temperature. Thus, water sorption isotherms must also exhibit temperature dependence. Temperature affects the mobility of water molecules and the equilibrium between the vapor and adsorbed phases (Kapsalis, 1987). In principle, a temperature increase results in a decrease of the amount of water adsorbed. However, at high water activities ($a_w > 0.8$), foods with high sugar content exhibit opposing temperature effects (i.e., equilibrium moisture content increases with temperature) that is due to an increase in the solubility of sugars in water (Audu et al., 1978). According to Rizvi (2005), the variation of water activity with temperature can be determined by using either thermodynamic principles or the temperature terms incorporated into sorption equations. The effect of temperature on water sorption isotherms is often represented by the Clausius–Clapeyron equation:

$$\left[\frac{\partial \ln (a_w)}{\partial (1/T)} \right] = \frac{Q_s}{R} \tag{1.13}$$

where Q_s is the net isosteric heat or heat of sorption, which is defined as the difference between the total molar enthalpy change and the molar enthalpy of vaporization, and R is the universal gas constant. The Clausius–Clapeyron equation is commonly used for the calculation of the isosteric heat of sorption (Iglesias et al., 1989; Kiranoudis et al., 1993; Lim et al., 1995; Sinija and Mishra, 2008; Tsami et al., 1990).

Several studies have reported sorption data for different food materials and temperatures (Cassini et al., 2006; Iglesias et al., 1986; Kiranoudis et al., 1993; Labuza et al., 1985; Saravacos et al., 1986). A typical example of the temperature effect on the isotherms is shown in Figure 1.6 for carrot samples. The binding forces decrease

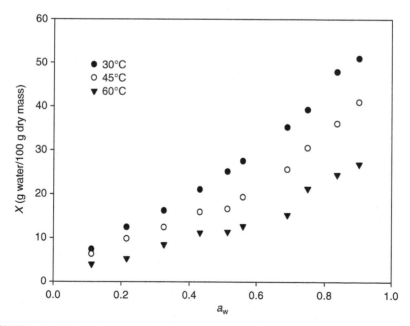

FIGURE 1.6 Effect of temperature on water sorption isotherm for carrot. (Adapted from Kiranoudis, C.T. et al., *J. Food Eng.*, 20, 55–74, 1993. With permission.)

with increasing temperature, that is, less moisture is adsorbed at higher temperatures at the same water activity or the equilibrium relative humidity. Aside from temperature, other factors that affect sorption isotherms are chemical composition, physical structure, and pretreatment of the sample, among others related to the method of handling and drying the product (Van der Berg, 1986).

1.6.1.3   Empirical and Theoretical Description of Water Sorption Isotherms

Numerous empirical and semiempirical equations have been proposed for the correlation of equilibrium moisture content with water activity. They require extensive experimental data at each temperature in order to evaluate the parameters in the model. Good compilations of sorptional data were presented by Iglesias and Chrife (1982) and Wolf et al. (1985). Useful information on the sorption and desorption enthalpies, as well as the microbiological stability of the product, can be obtained from sorptional data (Wolf et al., 1985).

One of the most relevant equations used to represent sorption data of foods is the Guggenheim–Anderson–de Boer (GAB) equation:

$$\frac{X_e}{X_m} = \frac{CKa_w}{(1 - ka_w)(1 - ka_w + Cka_w)} \tag{1.14}$$

The GAB equation is based on BET theory and generally involves three coefficients: K, C, and X_m, where C and X_m are similar to those in the BET equation and K is a third parameter that improves the fit to a wider range of moisture contents. Temperature effects can usually be determined by introducing a temperature exponential function into coefficients C and K (Kiranoudis et al., 1993).

1.6.1.4   Influence of Drying Methods

The effect of drying methods on the sorption equilibrium isotherms of food powders was demonstrated by various studies (Araya-Farias et al., 2007; Khallouffi et al., 2000; Ratti et al., 2007; Tsami et al., 1999). Tsami et al. (1999) found that freeze-dried gel adsorbed more water vapor than microwave-dried gel, which had a higher sorption capacity than vacuum and conventionally dried products. The sorption isotherms agreed with the reported shape for high sugar foodstuffs. Freeze- and vacuum-dried pectin developed the highest porosity, whereas the lowest porosity was obtained using conventional and microwave drying.

Figure 1.7 provides a comparison of sorption equilibrium isotherms at 20°C for freeze-dried or hot air dried seabuckthorn fruits and garlic samples. Both curves had the typical sigmoidal adsorption form. Freeze-dried powders showed higher water adsorption than convective dried ones, indicating that the former powders will be more hygroscopic than convective-dried powders during storage. The porous structure generated in the products during freeze-drying can certainly explain their higher water sorption. Nevertheless, the seabuckthorn freeze-dried samples were less hygroscopic than the garlic samples.

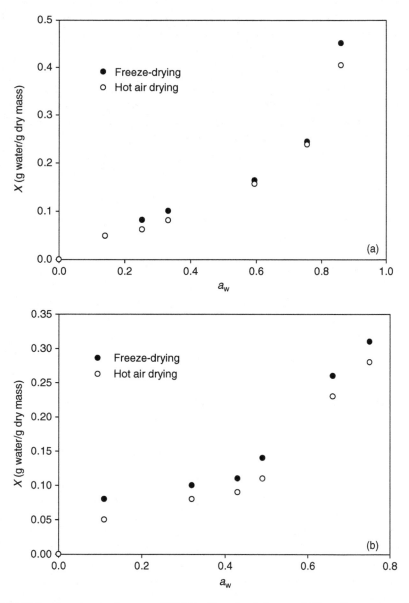

FIGURE 1.7 Sorption isotherms at 20°C for (a) seabuckthorn and (b) garlic powders. (Data from Araya-Farias et al. (2007) for seabuckthorn and from Ratti et al. (2007) for garlic.)

1.6.2 Physical Quality

Physical properties such as color, texture, density, porosity, and rehydration capacity are affected by the drying method. A tough and woody texture, slow and incomplete rehydration, and loss of the typical fresh food juiciness are the most common defects encountered during drying (Barbosa-Canovas and Vega-Mercado, 1996; Karel and

Lund, 2003a). The physicochemical basis for these changes is complex, and its understanding requires tedious lab measurements.

1.6.2.1 Color

Color is a major quality parameter in dehydrated food. During drying, color may change because of chemical or biochemical reactions. Enzymatic oxidation, Maillard reactions, caramelization, and ascorbic acid browning are some of the chemical reactions that can occur during drying and storage (Perera, 2005). Ratti (2001) provided a comparison of changes in color during air drying and freeze-drying of strawberries at different temperatures. Changes in the color of air-dried samples was significantly higher compared to freeze-dried ones. Discoloration and browning during air drying may be the result of various chemical reactions including pigment destruction (Garcia-Viguera et al., 1998; Krokida et al., 1998). For more detailed information on color measurements during drying, refer to Chapter 3.

1.6.2.2 Shrinkage, Porosity, and Bulk Density

One of the most important physical changes that foodstuffs undergo during drying is the reduction of their volume, often called shrinkage. Shrinkage is caused by structural collapse attributable to the loss of water (Prothon et al., 2003; Ratti, 1994). Changes in shape and size, loss of rehydration capacity, surface cracking, and hardening of food materials are among the most important physical phenomena associated with shrinkage (Mayor and Sereno, 2004; Prothon et al., 2003; Ratti, 1994).

Three different types of shrinkage have been reported during drying: one dimensional (i.e., when the volume change follows the direction of diffusion), isotropic (or three dimensional), and anisotropic or arbitrary. In contrast, volume changes of individual particles are normally expressed as the bulk shrinkage ratio of the sample volume at any time to the initial volume (V/V_0; Ratti and Mujumdar, 2005).

Shrinkage is rarely negligible in food systems, and it is necessary to take it into account when predicting moisture content profiles in the material undergoing dehydration (Mayor and Sereno, 2004). Modeling, design, and control of drying operations require accounting for the changes in physical dimensions of the product, moisture content, shrinkage, porosity, bulk density, and volume. For such purposes, attempts to model shrinkage, porosity, and bulk density during drying have been made by several researchers for different fruits and vegetables like apples, bananas, carrots, potatoes, and garlic. An extensive compilation of those mathematical models for shrinkage was presented by Mayor and Sereno (2004). The most common way to model shrinkage during drying is to adopt an empirical correlation between shrinkage and moisture content, including in some cases the effects of temperature and relative humidity of air. Some of the linear shrinkage models during air drying are delineated in Table 1.8.

Porosity and bulk density are important physical properties in dried foods. These two properties play an important role in rehydration of dried materials and their handling and packaging aspects (Perera, 2005). The extent of shrinkage influences the resulting changes in porosity during drying. Shrinkage and porosity changes during air drying of fruits and vegetables were investigated by Lozano et al. (1980, 1983).

TABLE 1.8
Linear Empirical Models for Shrinkage[a]

Model	Geometry	Material	Ref.
$\dfrac{V}{V_0} = k_1 X + k_2$	Cylinder	Apples	Lozano et al. (1980)
$\dfrac{V}{V_0} = k_8 + k_9 X$ for $X < X_e$	Cylinder	Apples and potatoes	Ratti (1994)
$\Delta\left(\dfrac{V}{V_0}\right) = k_{13} + (k_{14} + k_{15}RH + k_{16}T)\Delta X$	Sphere	Wheat and canola	Lang and Sokhansanj (1993)
$\dfrac{V}{V_0} = (k_{17}T + k_{18}) + (k_{19}T + k_{20})X$	Cylinder	Potatoes	McMinn and Magee (1997a)

[a] For reduced volume (current value/initial value).
Source: Adapted from Mayor, L. and Sereno, A.M., *J. Food Eng.*, 61, 373–386, 2004. With permission.

The bulk shrinkage and the porosity changes were related to moisture content by empirical correlations. It was demonstrated in apple tissue that both the overall shrinkage of the sample and the cellular shrinkage cooperate, resulting in an increase in porosity as the moisture content decreases (Lozano et al., 1980). Porosity in fruits and vegetables increases during drying, depending on the initial moisture content, composition, and size, as well as the type of drying (Saravacos, 1967).

The effects of drying methods on the shrinkage and porosity of fruits and vegetables were reported by Krokida and Maroulis (1997) and Krokida and Philippopoulus (2005). Table 1.9 provides the volume shrinkage coefficients (β') estimated for

TABLE 1.9
Effect of Drying Method on Volume Shrinkage Coefficient (β')

Material	Drying Method	β'
Apples	Convective	0.99
	Vacuum	0.96
	Microwave	1.01
	Freeze	0.34
Bananas	Convective	1.04
	Vacuum	0.90
	Microwave	1.05
	Freeze	0.43
Potatoes	Convective	1.03
	Vacuum	1.03
	Microwave	0.81
	Freeze	0.29

Source: Adapted from Krokida, M.K. and Maurolis, Z.B., *Dry. Technol.*, 15, 2441–2458, 1997. With permission.

different drying methods. The shrinkage coefficient depends on the drying method and the material. Its values approach unity for conventional, vacuum-, and microwave-dried materials, but they are much less for freeze-dried materials. Less than 10% shrinkage is normally expected in the final products after freeze-drying compared to 80 to 90% when air dried (Ratti and Mujumdar, 2005). Volumetric changes in strawberries indicated that shrinkage attributable to freeze-drying was 8% for whole fruits and less than 2% for slices (Shishehgarha et al., 2002). These results are similar to those obtained for raspberries and blackberries (Janković, 1993), where shrinkage of berries due to freeze-drying was minimal (5 to 15%) compared to 80% obtained during air drying (Ratti, 2001).

Figure 1.8 shows the effects of drying methods on the porosity of tropical fruits. Freeze-dried guava had the highest porosity, whereas heat pump dried (with normal air) had the lowest porosity. In contrast, the porosity of freeze-dried papaya was lower than that of the heat pump, and vacuum-dried samples had the largest porosity. The possible reason is a visual "puff effect" observed in vacuum-dried papaya that causes its largest porosity. Reduced pressure may have contributed to it (Hawlader et al., 2006). However, freeze-drying generally produces materials with fragile structures and higher porosity values (Krokida and Philippopoulus, 2005). In a study realized by Krokida and Maroulis (1997), the porosity of freeze-dried fruits and vegetables (80 to 90%) was always higher compared to all of the other dehydration processes that were used. The porosity of microwave potatoes and carrots followed (75%), but microwave-dried apples and bananas did not developed high porosities (60 and 25%, respectively). Vacuum-dried bananas and apples developed high porosity (70%),

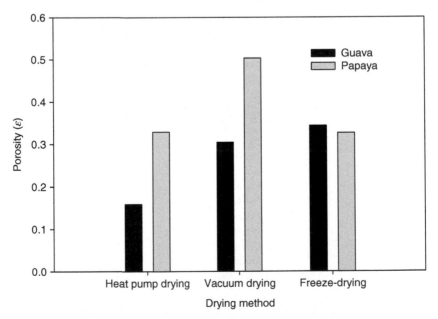

FIGURE 1.8 Effect of drying method on porosity of tropical fruits. (Adapted from Hawlader, M.N.A. et al., *Dry. Technol.*, 24, 77–87, 2006. With permission.)

but vacuum-dried carrot and potato porosity values were lower (50 and 25%, respectively). Krokida and Maroulis reported that the bulk density values during vacuum drying were significantly lower than those of conventional drying for all materials.

1.6.2.3 Rehydration Properties

As mentioned earlier, changes in the physical structure of foods take place during drying. Thus, rehydration can be considered as a measure of the injury to the material caused by drying and treatment preceding dehydration (Krokida and Philippopoulus, 2005). The rate and degree of rehydration depend on the drying conditions as well as the drying method employed. The influence of the drying temperature (30 to 60°C) on rehydration of potato samples was presented in several studies. McMinn and Magee (1997b) observed a concurrent increase in rehydration capacity with air-drying temperatures. Hawlader et al. (2006) observed that the rehydration rate of heat pump dried samples was slightly higher compared to freeze-dried and vacuum-dried samples. Similar behaviors were found in guava and papaya fruits. However, freeze-drying generally produces materials with fragile structures and higher porosity values. Products normally rehydrate faster when they are more porous (Krokida and Philippopoulus, 2005).

1.6.3 MICROBIOLOGICAL AND CHEMICAL QUALITY

Dehydrated foods are preserved because their water activity is at a level where no microbiological activity can occur and where deteriorative chemical and biochemical reaction rates are reduced to a minimum. The relationships between a_w and the rate of deteriorative reactions in food can be found extensively in the literature (Fennema, 1985). Beuchat (1983) reported minimal water activities required for growth of a number of microorganisms. Reducing a_w below 0.7 prevents microbiological spoilage. Most oxidation reactions and enzyme reactions will be inhibited as a_w decreases. However, autooxidation of lipids could take place at very low a_w values (<0.2). The maximal rate of nonenzymatic browning reactions (Maillard reaction) is achieved at intermediate a_w values (0.4 to 0.65). These reactions result in the loss of nutritive value, formation of brown pigments, as well as the formation of *off-flavors*, especially when the products are stored at high temperatures. Therefore, the browning reactions can be prevented by an appropriate treatment such as sulfur dioxide (SO_2), a decrease of the pH (2 to 3), or lowering of a_w (0.2 to 0.4) by drying. In contrast, short heat treatments such as blanching can be used to inactivate enzymes and antioxidants are useful to prevent lipid oxidation. Dried products are considered to be more stable at their monolayer moisture content (Fennema, 1985; Perera, 2005).

1.6.4 NUTRITIONAL QUALITY

Many nutritional compounds in foods such as antioxidants (carotenoids and phenolic compounds), vitamins, and essential fatty acids have recently been the subject of growing interest because of their possible role in the prevention of human diseases, including cancer and heart disease. Fruits and vegetables are essential foods in the human diet. Common dietary guidelines recommend increased intake of fruits and vegetables, which contain low levels of fat and high levels of vitamins, minerals, and fiber. They are also considered as a rich source of antioxidants. Many health benefits have been related to consumption of fruits and vegetables. However, nutritional quality can be affected by handling, processing, and packaging. Aside from physical and chemical changes, drying can also cause loss of nutritional value. The major losses of vitamins and other substances take place because of solubility in water, enzymatic oxidation, oxygen and heat sensitivity, and metal ion catalysis during processing. In addition, sugar–amine interactions (Maillard reaction) can occur during drying and storage, causing loss of nutrients. A survey of the literature reveals studies that suggest means of reducing such losses: pretreatments, proper selection of drying methods, new and innovative drying methods, and optimization of drying conditions.

1.6.4.1 Influence of Drying Methods and Pretreatments

Several studies focusing on the effect of drying on loss or retention of nutritional quality were published in the last several years. Conventional and new technologies were utilized in these studies. The effects of pretreatments as well as the operating drying conditions on nutrient loss were also considered in these works. This information was exhaustively surveyed by Sablani (2006) and Lewicki (2006). For further information on this subject, the reader is encouraged to refer to Chapter 10.

Pretreatments are often used to accelerate the drying process as well as to improve certain quality attributes of the final dried product. The pretreatments depend on the kind of material to be processed, such as liquids, pastelike foods, and solids. Table 1.10 details some important pretreatments used prior to drying according to food material. Useful details and information on those pretreatments are found in Lewicki (2006).

TABLE 1.10

Pretreatments Used prior to Drying according to Physical Form of Material[a]

Material Type	Pretreatments	
Liquid foods	Concentration, cryoconcentration, enzymatic treatments	
Pastelike foods	Whipped into foams	
	Chemical	**Physical**
Solid foods	Sulfiting	Blanching at short times and high temperatures or low temperatures and long times; utilization of hot water or hot steam
	Immersion in NaCl, CaCl$_2$, or sugars (osmotic treatment) or in acid solutions	Freezing
	Use of surfactants (ethyloleate)	High pressure
	Impregnation with biopolymers (dextrans)	Microwave

[a] According to Lewicki, P.P., *Trends Food Sci. Technol.*, 17, 153–163, 2006. With permission.

1.6.5 GLASS-TRANSITION TEMPERATURE AND QUALITY CONSIDERATIONS

The application of the knowledge of the glass transition of polymers to food systems was successful for understanding and predicting the behavior of foodstuffs during processing (Ratti, 2001). The glass-transition temperature (T_g) is defined as the temperature at which an amorphous system changes from the glassy to the rubbery state (Roos and Karel, 1991). Schenz (1995) reported that if the temperature of a product is higher than T_g, at the correspondent water content, the quality of foodstuffs can be seriously affected. The T_g is a function of moisture content as well as the concentration of solutes such as sugars. The glass transition in foods may occur during drying and can have dramatic effects on product quality (physical and chemical). It can affect the diffusion of components or reactants, resulting in the loss of aroma as well as affecting the nonenzymatic browning reactions (Mujumdar, 1997). Some studies on the relationship of nonenzymatic browning and glass transition were reported by Karmas et al. (1992) and Roos and Himberg (1994).

Shrinkage and T_g are interrelated and significant structural changes may be avoided if dehydration is carried out below the T_g. In experiments on freeze-drying of strawberries, a high percentage of strawberries collapsed at temperatures higher than the glass transition of the dry matrix (38°C). At heating temperatures higher than 50°C, the percentage of collapse exceeded 20% (Shishehgarha et al., 2002). For more detailed information on this subject, please read Chapter 2.

NOMENCLATURE

VARIABLES

A	area available for drying (m²)
a_w	water activity
C	BET constant
C_{pa}	specific heat of dry air (kJ/kg °C)
C_{pw}	specific heat of vapor water (kJ/kg °C)
C_p	specific heat (kJ/kg K)
H	enthalpy (kJ/kg dry air)
K	GAB constant
k	thermal conductivity (W/m °C)
L_0	latent heat of water at 0°C (kJ/kg)
m	wet mass (kg)
m_s	mass of dry solids (kg)
M_w	molecular weight of water (Da)
M_a	molecular weight of air (Da)
n_a	number of moles of air
n_w	number of moles of water
N_w	drying rate (kg water m²/h)
P_a	partial pressure of air (kPa)
P_w	partial pressure of water (kPa)
P_t	total pressure (kPa)

P_{w0} saturation vapor of water (kPa)
Q_s heat of sorption (kJ/kg)
R universal gas constant (J/K mol)
RH   relative humidity
T temperature (°C)
T_g glass-transition temperature (°C)
X water content on dry basis (g water/g dry mass)
X_m water content of monomolecular layer (g water/g dry mass)
X_w water content on wet basis (g water/g wet mass)
X_f free moisture content (g water/g dry mass)
X_e equilibrium moisture content (g water/g dry mass)
V volume (m³)
V_0 initial volume (m³)
V_t apparent volume (m³)
Y absolute humidity (kg water/kg dry air)
Y_s saturation absolute humidity (kg water/kg dry air)

GREEK LETTERS

β' volume shrinkage coefficient
ρ_b bulk density (g/cm³)
ρ_p particle density (g/cm³)
ε porosity

REFERENCES

Araya-Farias, M., Macaigne, O., and Ratti, C., Osmotic dehydration of seabuckthorn (*Hippophaë rhamnoides* L.) fruits, paper presented at the International Seabuckthorn Association Conference (ISA2007), Québec City, Québec, Canada, August 12–16, 2007.

ASHRAE, *ASHRAE Handbook of Fundamentals*, American Society of Heating, Refrigerating and Air-Conditioning Engineers, Inc., Atlanta, 1985.

Audu, T.O.K., Loncin, M., and Weisser, H., Sorption isotherms of sugars, *Lebensm.-Wiss. Technol.*, 11, 31–34, 1978.

Barbosa-Canovas, G. and Vega-Marcado, H., Eds., Fundamentals of air–water mixtures and ideal dryers, in *Dehydration of Foods*, Chapman & Hall, New York, 1996, pp. 9–27.

Beuchat, L.R., Influence of water activity on growth, metabolic activities and survival of yeast and molds, *J. Food Protect.*, 46, 135–141, 1983.

Bimbenet, J.J. and Lebert, A., Food drying and quality interactions, in *Drying '92*, Mujumdar, A.S., Ed., Elsevier Science, Amsterdam, 1992, pp. 42–57.

Boukouvalas, Ch.J., Krokida, M.K., Maroulis, Z.B., and Marinos-Kouris, D., Density and porosity: Literature data compilation for foodstuffs, *Int. J. Food Prop.*, 9, 715–746, 2006.

Brunauer, S., Emmett, H.P., and Teller, E., Adsorption of gases in multi-molecular layers, *J. Am. Chem. Soc.*, 60, 309–319, 1938.

Cassini, A.S., Marcsak, L.D.F., and Norena, C.P.Z., Water adsorption isotherms of texturized soy protein, *J. Food Eng.*, 77, 194–199, 2006.

Chirife, J. and Buera, M.P., Water activity, glass transition and microbial stability in concentrated semi-moist food systems, *J. Food Sci.*, 59, 921–927, 1994.

Cuevas, R. and Cheyran, M., Thermal conductivity of liquid foods: A review, *J. Food Process Eng.*, 2, 238–306, 1978.

Fennema, O.R., *Food Chemistry*, Marcel Dekker, New York, 1985.

Gabas, A.L., Marra-Junior, W.D., Telis-Romero, J., and Telis V.R.N., Changes of density, thermal conductivity, thermal diffusivity and specific heat of plums during drying, *Int. J. Food Prop.*, 8, 233–242, 2005.

Garcia-Viguera, C., Zarilla, P., Artes, F., Romer, F., Abelan, P., and Barberan, T., Colour and anthocyanin stability of red raspberry jam, *J. Food Sci. Agric.*, 78, 565–573, 1998.

Hawlader, M.N.A., Perera, C.O., Tian, M., and Yeo, K.L., Drying of guava and papaya: Impact of drying methods, *Dry. Technol.*, 24, 77–87, 2006.

Iglesias, H.A. and Chrife, J., *Handbook of Food Isotherms*, Academic Press, New York, 1982.

Iglesias, H.A., Chirife, J., and Ferro-Fortan, C., Temperature dependence of water sorption isotherms of some foods, *J. Food Sci.*, 51, 551–553, 1986.

Iglesias, H.A., Chirife, J., and Ferro-Fortan, C., On the temperature dependence of isosteric heats of water sorption in dehydrated foods, *J. Food Sci.*, 54, 1620–1631, 1989.

Janković, M., Physical properties of convectively dried and freeze-dried berrylike fruits, *Facul. Agric.*, 38, 129–135, 1993.

Kapsalis, J.G., Influence of hysteresis and temperature on moisture sorption isotherms, in *Water Activity Theory and Applications to Food*, Rockland, L.B. and Beuchat, L.R., Eds., Marcel Dekker, New York, 1987, pp. 173–207.

Karel, M., Recent research and development in the field of low moisture and intermediate-moisture foods, *Crit. Rev. Food Technol.*, 3, 329–373, 1973.

Karel, M., Physical structure and quality of dehydrated foods, in *Drying '91*, Mujumdar, A.S. and Filkova, I., Eds., Elsevier Science, Amsterdam, 1991, pp. 26–35.

Karel, M. and Lund, D.B., Dehydration, in *Physical Principles of Food Preservation*, 2nd ed., Marcel Dekker, New York, 2003a, pp. 378–460.

Karel, M. and Lund, D.B., Water activity, in *Physical Principles of Food Preservation*, 2nd ed., Marcel Dekker, New York, 2003b, pp. 117–169.

Karmas, R., Buera, M.P., and Karel, M., Effect of glass transition on rates of nonenzymatic browning in food systems, *J. Agric. Food Chem.*, 40, 873–879, 1992.

Khallouffi, S., Giasson, J., and Ratti, C., Water activity of freeze-dried berries and mushrooms, *Can. Agric. Eng.*, 41, 51–56, 2000.

Kim, S.S. and Bhowmik, S.R., Thermophysical properties of plain yogurt as functions of moisture content, *J. Food Eng.*, 32, 109–124, 1997.

Kiranoudis, C.T., Maroulis, E., Tsami, E., and Marinos-Kouris, D., Equilibrium moisture content and heat of desorption of some vegetables, *J. Food Eng.*, 20, 55–74, 1993.

Krokida, M.K. and Maroulis, Z.B., Effect of drying method and shrinkage and porosity, *Dry. Technol.*, 15, 2441–2458, 1997.

Krokida, M.K. and Maurolis, Z.B., The effect of drying method on viscoelastic behaviour of dehydrated fruits and vegetables, *Int. J. Food Sci. Technol.*, 35, 391–400, 2000.

Krokida, M.K., Panagoutou, N.M., Maroulis, Z.B., and Saravacos, G.D., Thermal conductivity: Literature data compilation for foodstuffs, *Int. J. Food Prop.*, 4, 111–137, 2001.

Krokida, M.K. and Philippopoulos, C., Rehydration of dehydrated foods, *Dry. Technol.*, 23, 799–830, 2005.

Krokida, M.K., Tsami, E., and Maroulis, Z.B., Kinetics on color changes during drying of some fruits and vegetables, *Dry. Technol.*, 16, 667–685, 1998.

Kudra, T. and Mujumdar, A.S., *Advanced Drying Technologies*, Marcel Dekker, New York, 2002.

Labuza, T.P., The effect of water activity on reaction kinetics of food deterioration, *Food Technol.*, 34, 36–41, 1980.

Labuza, T.P., Kanane, A., and Chen, J.Y., Effect of temperature on the moisture sorption isotherms and water activity shift of two dehydrated foods, *J. Food Sci.*, 50, 358–391, 1985.

Lang, W. and Sokahansanj, S., Bulk volume shrinkage during drying of wheat and canola, *J. Food Process Eng.*, 16, 305–314, 1993.

Lewicki, P.P., Design of hot air drying for better foods, *Trends Food Sci. Technol.*, 17, 153–163, 2006.

Lim, L.T., Tang, J., and Jianshan, H., Moisture sorption characteristics of freeze-dried blueberries, *J. Food Sci.*, 60, 810–814, 1995.

Lozano, J.E., Rotstein, E., and Urbicain, M.J., Thermal conductivity of apples as a function of moisture content, *J. Food Sci.*, 44, 198–199, 1979.

Lozano, J.E., Rotstein, E., and Urbicain, M.J., Total porosity and open-pore porosity in the drying of fruits, *J. Food Sci.*, 45, 1403–1407, 1980.

Lozano, J.E., Rotstein, E., and Urbicain, M.J., Shrinkage, porosity and bulk density of foodstuffs at changing moisture contents, *J. Food Sci.*, 48, 1497–1553, 1983.

Madamba, P.S., Driscoll, R.H., and Buckle, K.A., Models for the specific heat and thermal conductivity of garlic, *Dry. Technol.*, 13, 295–317, 1995.

Maroulis, Z.B., Krokida, M.K., and Rahman, M.S., A structural generic model to predict the effective thermal conductivity of fruits and vegetables during drying, *J. Food Eng.*, 52, 47–52, 2002.

Marousis, S.N. and Saravacos, G.D., Density and porosity in drying starch materials, *J. Food Sci.*, 55, 1367–1372, 1990.

Mattar, H.L., Minim, L.A., Coimbra, J.S.R., Minim, V.P.R., Sraiva, S.H., and Telis-Romero, J., Modeling thermal conductivity, specific heat, and density of milk: A neural network approach, *Int. J. Food Prop.*, 7, 531–539, 2004.

Mattea, M., Urbicain, M.J., and Rotstein, E., Effective thermal conductivity of cellular tissues during drying: Prediction by a computer assisted model, *J. Food Sci.*, 54, 194–197, 1989.

Mayor, L. and Sereno, A.M., Modelling shrinkage during convective drying of food materials: A review, *J. Food Eng.*, 61, 373–386, 2004.

McMinn, W.A.M. and Magee, T.R.A., Physical characteristics of dehydrated potatoes— Part I, *J. Food Eng.*, 33, 37–48, 1997a.

McMinn, W.A.M. and Magee, T.R.A., Physical characteristics of dehydrated potatoes— Part II, *J. Food Eng.*, 33, 49–55, 1997b.

Moraga, G., Martinez-Navarrete, N., and Chiralt, A., Water sorption isotherms and phase transitions in strawberries: Influence of pretreatment, *J. Food Eng.*, 62, 315–321, 2004.

Moraga, G., Martinez-Navarrete, N., and Chiralt, A., Water sorption isotherms and phase transitions in kiwifruit, *J. Food Eng.*, 72, 147–156, 2006.

Mujumdar, A.S., Drying fundamentals, in *Industrial Drying of Foods*, Baker, C.G.J., Ed., Blackie Academic and Professional, London, 1997, pp. 7–30.

Mujumdar, A.S. and Menon, A.S., Drying of solids: Principles, classification, and selection of dryers, in *Handbook of Industrial Drying*, 2nd ed., Vol. 1, Mujumdar, A.S., Ed., Marcel Dekker, New York, 1995, pp. 1–39.

Niesteruk, R., Changes of thermal properties of fruits and vegetables during drying, *Dry. Technol.*, 14, 415–422, 1996.

Perera, C., Selected quality attributes of dried foods, *Dry. Technol.*, 23, 717–730, 2005.

Phomkong, W., Srzednicki, G., and Driscoll, R.H., Thermophysical properties of stone fruit, *Dry. Technol.*, 24, 195–200, 2006.

Poonnoy, P., Tansakul, A., and Chinnan M., Artificial neural network modelling for temperature and moisture content prediction in tomato slices under microwave-drying, *Food Eng. Phys. Prop.*, 72, E42–E47, 2007.

Prothon, F., Ahrné, L., and Inggerd, S., Mechanism and prevention of plant tissue collapse during dehydration: A critical review, *Crit. Rev. Food Sci. Nutr.*, 43, 447–479, 2003.

Rahman, M.S., Chen, X.D., and Perera, C., An improved thermal conductivity prediction model for fruits and vegetables as a function of temperature, water content and porosity, *J. Food Eng.*, 31, 163–170, 1997.

Rahman, S., *Food Properties Handbook*, CRC Press, Boca Raton, FL, 1995.

Ratti, C., Shrinkage during drying of foodstuffs, *J. Food Eng.*, 23, 91–105, 1994.

Ratti, C., Hot air and freeze-drying of high value foods: A review, *J. Food Eng.*, 49, 311–389, 2001.

Ratti, C., Araya-Farias, M., Mendez-Lagunas, L., and Makhlouf, J., Drying of garlic (*Allium sativum*) and its effect on allicin retention, *Dry. Technol.*, 25, 349–356, 2007.

Ratti, C., Crapiste, G.H., and Rotstein, E., PSYCHR: A computer program to calculate psychrometric properties, *Dry. Technol.*, 7, 575–580, 1989.

Ratti, C. and Mujumdar, A.S., Infrared drying, in *Handbook of Industrial Drying*, 2nd ed., Vol. 1, Mujumdar, A.S., Ed., Marcel Dekker, New York, 1995, pp. 567–589.

Ratti, C. and Mujumdar A.S., Drying of fruits, in *Processing Fruits*, 2nd ed., Barret, D.M., Somogyi, L., and Ramaswamy, H., Eds. CRC Press, Boca Raton, FL, 2005, pp. 127–159.

Rizvi, S.S.H., Thermodynamic properties of foods in dehydration, in *Engineering Properties of Foods*, 3rd ed., Rao, M.A., Rizvi, S.S.H., and Datta, A.K., Eds., CRC Press, Boca Raton, FL, 2005, pp. 239–325.

Roos, Y. and Himberg, M., Nonenzymatic browning behaviour as related to glass transition of a food model at chilling temperatures, *J. Food Agric. Food Chem.*, 42, 893–898, 1994.

Roos, Y. and Karel, M., Applying state diagrams to food processing and development, *Food Technol.*, 45, 66–70, 107, 1991.

Sablani, S.S., Drying of fruits and vegetables, retention of nutritional/functional quality, *Dry. Technol.*, 24, 123–135, 2006.

Sablani, S.S. and Rahman, M.S., Using neural networks to predict thermal conductivity of food as a function of moisture content, temperature and apparent porosity, *Food Res. Int.*, 36, 617–623, 2003.

Salunkhe, D.K., Bolin, H., and Reddy, N.R., *Storage, Processing and Nutritional Quality of Fruits and Vegetables*, 2nd ed., Vol. 1, CRC Press, Boca Raton, FL, 1991.

Saravacos, G.D., Effect of drying method on the water sorption of dehydrated apple and potato, *J. Food Sci.*, 3, 81–84, 1967.

Saravacos, G., Tsiourvas, A., and Tsami, E., Effect of temperature on the water adsorption isotherms of sultana raisins, *J. Food Sci,* 51, 381–383, 1986.

Schenz, T.W., Glass transitions and product stability—An overview, *Food Hydrocolloids*, 9, 307–315, 1995.

Shishehgarha, F., Makhlouf, J., and Ratti, C., Freeze-drying characteristics of strawberries, *Dry. Technol.*, 20, 131–145, 2002.

Shrivastava, M. and Datta, A.K., Determination of specific heat and thermal conductivity of mushrooms (*Pleurotus florida*), *J. Food Eng.*, 11, 147–158, 1999.

Sinija, V.R. and Mishra, H.R., Moisture sorption isotherms and heat of sorption of instant (soluble) green tea powder and green tea granules, *J. Food Eng.*, 86, 494–500, 2008.

Sweat, V.E., Experimental values of thermal conductivity of selected fruits and vegetables, *J. Food Sci.*, 39, 1080–1083, 1974.

Telis-Romero, J., Telis, V.R.N., Gabas, A.L., and Yamashita, F., Thermophysical properties of Brazilian orange juice as affected by temperature and water content, *J. Food Eng.*, 38, 27–40, 1998.

Toledo, R.M., Dehydration, in *Fundamentals of Food Process Engineering*, 2nd ed., Van Nostrand Reinhold, New York, 1991, pp. 456–505.

Treybal, R.E., *Mass Transfer Operations*, McGraw-Hill, New York, 1980.

Tsami, E., Krokida, M.K., and Drouzas, A.E., Effect of drying method on the sorption characteristics model fruit powders, *J. Food Eng.*, 38, 381–392, 1999.

Tsami, E., Marinos-Kouris, D., and Maroulis, Z.B., Water sorption isotherms of raisins, currants, figs, prunes and apricots, *J. Food Sci.*, 55, 1594–1625, 1990.

Vagenas G.K. and Marinos-Kouris, D., Thermal properties of raisins, *J. Food Eng.*, 11, 147–158, 1990.

Van Arsdel, W. and Copley, M., *Food Dehydration*, Vol. 1, AVI Publishing, Westport, CT, 1963.

Van der Berg, C., Water activity, in *Concentration and Drying Foods*, MacCarthy, D., Ed., Elsevier, London, 1986, pp. 11–36.

Vasquez, G., Chenlo, F., Moreira, L., and Carballo, L., Desorption isotherms of muscatel and aledo grapes and the influences of pre treatments on muscatel isotherms, *J. Food Eng.*, 39, 409–414, 1999.

Vega-Mercado, H., Gongora-Nieto, M.M., and Barbosa-Canovas, G., Advances in dehydration of foods, *J. Food Eng.*, 49, 271–289, 2001.

Welti-Chanes, J., Mujica-Paz, H., Valdez-Fragoso, A., and Leon-Cruz, R., Fundamentals of mass transport, in *Transport Phenomena in Food Processing*, Welti-Chanes, J., Velez-Ruiz, J.F., and Barbosa-Canovas, G., Eds., CRC Press, Boca Raton, FL, 2003, pp. 3–22.

Wlodarczyk-Stasiak, M. and Jamroz, J., Analysis of sorption properties of starch–protein extrudates with the use of water vapour, *J. Food Eng.*, 85, 580–589, 2008.

Wolf, W., Spies, W.E.L., and Jung, G., *Sorption Isotherms and Water Activity of Food Materials*, Elsevier, New York, 1985.

Yan, Z., Sousa-Gallagher, M.J., and Oliveira, F.A.R., Sorption isotherms and moisture sorption hysteresis of the intermediate moisture content banana, *J. Food Eng.*, 86, 342–348, 2008.

2 Glass-Transition Based Approach in Drying of Foods

Bhesh Bhandari and Benu Adhikari

CONTENTS

2.1 INTRODUCTION

Drying is one of the major operations in food processing. There are several drying techniques, the most common being hot-air (fixed and fluidized bed), freeze-, and spray drying. In most drying conditions, a significant proportion of the dried product remains in an amorphous state. In some slow drying processes such as tray and fluidized bed drying, the high molecular weight carbohydrates and proteins normally remain amorphous but other low molecular weight sugars or organic acids may be partially crystallized. This crystallization can be intentional or undesirable. In the low temperature process (freeze-drying), almost all of the ingredients become amorphous because of the rapid cooling or prefreezing. Subsequent sublimation is followed by low temperature drying. The majority of spray-dried products are amorphous because this process is rapid enough to hinder the relatively slower crystallization process. In some cases pre- or postspray-drying crystallization of sugars is also practiced. In all drying processes, the degree of amorphous fractions in the powder is influenced by the processing conditions, compositions, and properties of the individual ingredients.

There are many product quality attributes that are related to the physical state (amorphous or crystalline) of the ingredients in dried products. Minor components, such as flavors, vitamins, enzymes, and microorganisms, in dried products are encapsulated within the amorphous matrix. These products become unstable or lose their quality in a certain environment defined by temperature and moisture content. Any change in the physical state of the matrix or encapsulated compounds has the propensity to affect their physicochemical characteristics. If stored above a certain temperature, the amorphous part may undergo sudden structural changes. This is a critical value, known as the glass-transition temperature (T_g), when a glassy solid structure begins to change to a "rubbery" state. This state change is responsible for the alteration of the physicochemical properties of the products, which is attributable to a sudden change in the molecular mobility of the system.

The phenomenon of glass transition and its indicator (T_g) is important for formulation, production, and quality control of products in all drying processes. It is more important in drying operations, such as spray and freeze-drying. The T_g is intimately associated with stickiness during spray drying and structural collapse during freeze-drying. The glass-transition based material science approach in dehydration and drying of sugar- and acid-rich foods and the stability of powders have been extensively illustrated by various research groups worldwide.

2.2 GLASS TRANSITION: THEORY, MEASUREMENTS, AND CORRELATIONS

Solid foods can be crystalline, amorphous (vitreous or glassy), or a mixture of the two, depending on the process conditions. An amorphous state consists of a non-aligned molecular structure. The crystalline state consists of a three-dimensional ordered array of molecules in which there is periodicity and symmetry. The differences among amorphous, vitreous, and glassy materials are less well defined (Flink, 1983). A fully amorphous material can exist in the glassy state or in the viscous, supercooled, liquid state (Roos, 1995). The term "glass" is used to show that the physical characteristics of a material in its glassy state are molecularly or atomically disordered and mechanically "brittle" and "hard" as seen in a glass in everyday life. Therefore, glassy materials have the structure of a liquid but the property of a solid. A glassy material may contain patches of a short-range molecular order that may not be detectable by normal analytical procedures (Bhandari and Hartel, 2002).

2.2.1 Glass Formation

A glass is described as any supercooled liquid with a viscosity greater than 10^{13} Pa s. A typical viscosity–temperature curve is provided in Figure 2.1. This definition accords well with the familiar meaning of the term glass, because a liquid of this viscosity is capable of supporting its own weight (Allen, 1993; White and Cakebread, 1966). The glass is thermodynamically metastable and tends to convert to the crystalline form if the molecules attain adequate mobility. An essential prerequisite for glass formation from a solution is that the cooling rate or the removal of the dispersed phase (such as water) should be sufficiently fast to preclude nucleation and crystal growth (Hemminga et al., 1993).

2.2.2 Prediction of Glass-Transition Temperature of Food Model Mixtures

One of the important areas of immediate interest is how to predict the T_g of a mixture (or a complex food system) from the known T_g values of its components. The mixtures can be pure compound and water or a mixture of solid compounds and water. Some equations that have been used widely in foods will be briefly described.

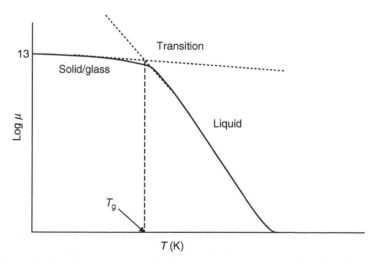

FIGURE 2.1 Viscosity–temperature curve for a typical material. The viscosity is on the order of 10^{13} to 10^{14} Pa s of a glass and the transition from solid to liquid is not sharp. The intersection between the extrapolated glass and liquid lines is normally taken as the glass-transition temperature.

2.2.2.1 Gordon–Taylor (GT) Equation

The GT equation (Gordon and Taylor, 1952) was originally developed for polymer blends. It was based on expansion coefficients ($\gamma = dV/dT$) and the assumption of volume additivity:

$$T_g = \frac{\Delta\gamma_1 x_1 T_{g1} + \Delta\gamma_2 x_2 T_{g2}}{\Delta\gamma_1 x_1 + \Delta\gamma_2 x_2} \tag{2.1}$$

where $\Delta\gamma_i$ ($i = 1$ or 2) is the difference between the expansion coefficients of liquid and glass of component i. The GT equation has been rewritten in the following form (Jouppila and Roos, 1994):

$$T_g = \frac{x_1 T_{g1} + Kx_2 T_{g2}}{x_1 + Kx_2} \tag{2.2}$$

For an anhydrous system, subscript 1 is component 1 and subscript 2 is component 2 (the one with a lower T_g). For an aqueous system, subscript 1 is solid component 1 (or a dry mixture of various solid components), subscript 2 is water, and x is either the mass or mole fraction. The empirical constant K can be obtained by a least squares method, as shown in Eq. (2.3):

$$\frac{d}{dK} \sum_{i=1}^{n} \left(\frac{x_{1i} T_{g1} + Kx_{2i} T_{g2}}{x_{1i} + Kx_{2i}} - T_{gmi} \right)^2 = 0 \tag{2.3}$$

where n is the number of data points; T_{gmi} is the measured T_g of model mixture i; and x_{1i} and x_{2i} are the fractions of components 1 and 2, respectively, of model mixture i.

The GT equation has been applied to polymer blends such as sucrose–maltodextrin (Roos and Karel, 1991a), glucose–fructose (Arvanitoyannis et al., 1993), corn syrup solids or maltodextrins (Gabarra and Hartel, 1998), sucrose–poly(vinylpyrrolidone) (Shamblin et al., 1998); and to solute–diluent systems such as sucrose–water and fructose–water (Roos, 1993; Roos and Karel, 1991a, 1991b).

2.2.2.2 Couchman–Karasz Equation

Couchman (1978) and Couchman and Karasz (1978) derived the relation between the glass-transition temperature and composition in a compatible binary mixture using classical thermodynamic theory. Below is a brief introduction to their derivation based on the entropy of the systems. They started by expressing the mixed system entropy S as a function of component entropy S_i ($i = 1$ and 2):

$$S = x_1 S_1 + x_2 S_2 + \Delta S_m \tag{2.4}$$

where the entropy of component i at a given temperature T is the summation of the entropy at the glass-transition temperature (S_i^o) and the entropy difference, $S_i = S_i^o + (S_i - S_i^o) = S_i^o + \Delta S_i$. It is known that

$$Q = mC_p \frac{dT}{dt}, \qquad C_{pi} = \frac{dQ}{dt} = dS_i \left(\frac{T}{dT}\right)$$

Therefore,

$$\Delta S_i = \int_{T_{gi}}^{T} dS_i = \int_{T_{gi}}^{T} \frac{C_{pi}}{T} dT = \int_{T_{gi}}^{T} C_{pi} \, d \ln T \qquad (2.5)$$

At the T_g of the mixture, $T = T_g$, so the entropy S for the glassy state is identical to that for the liquid state. Thus, by subtraction of the entropy between the liquid and glassy states of the form in Eq. (2.4), the following equation was obtained:

$$x_1 \int_{T_{g1}}^{T_g} (C_{pL1} - C_{pG1}) d \ln T + x_2 \int_{T_{g2}}^{T_g} (C_{pL2} - C_{pG2}) d \ln T + \Delta S_{mL} - \Delta S_{mG} = 0 \qquad (2.6)$$

Assuming that the excess entropy at the glassy state (ΔS_{mG}) is not far from that of the liquid state (ΔS_{mL}), together with the temperature independence of ($C_{pLi} - C_{pGi}$) through the T_g region, and an approximation of $\ln(1 + y) = y$, Eq. (2.6) becomes

$$T_g = \frac{x_1 \Delta C_{p1} T_{g1} + x_2 \Delta C_{p2} T_{g2}}{x_1 \Delta C_{p1} + x_2 \Delta C_{p2}} \qquad (2.7)$$

where $\Delta C_{pi} = (C_{pLi} - C_{pGi})$ is the change in heat capacity of component i between its liquidlike and glassy states, which can be measured by differential scanning calorimetry. Equation (2.7) is derived for a purely conformational system. It is identical to the GT equation when $K = \Delta C_{p2}/\Delta C_{p1}$. The Couchman–Karasz equation has been applied successfully for some polymer blends, especially for solid–solid mixtures because the difference between the T_{gi} values of the components is rather small (Tenbrinke et al., 1983). For a solid–liquid system, this difference can be high (e.g., sucrose $T_g = 65°C$ and water $T_g = -135°C$), so the assumption of temperature independence of ($C_{pLi} - C_{pGi}$) is no longer valid. Tenbrinke et al. (1983) discussed the errors of Eq. (2.7) for cross-linked polymers and explained that the interaction between the cross-links and liquid was significant.

2.2.2.3 Couchman–Karasz Equation for Ternary Systems

This equation is an expansion of Eq. (2.7), where in an aqueous system subscripts 1, 2, and 3 refer to components 1, 2, and water, respectively. For polymer blends of three components, subscript 3 refers to the component with the lowest T_g. This distinction is for convenience because water always has the lowest T_g among the components. The application of Eq. (2.8) to a glucose–fructose–water mixture was carried out by Arvanitoyannis et al. (1993). Equation (2.8) can be rewritten in the following form by division of the numerator and denominator for ΔC_{p1}:

$$T_g = \frac{x_1 \Delta C_{p1} T_{g1} + x_2 \Delta C_{p2} T_{g2} + x_3 \Delta C_{p3} T_{g3}}{x_1 \Delta C_{p1} + x_2 \Delta C_{p2} + x_3 \Delta C_{p3}} \tag{2.8}$$

$$T_g = \frac{x_1 T_{g1} + x_2 K_{12} T_{g2} + x_3 K_{13} T_{g3}}{x_1 + x_2 K_{12} + x_3 K_{13}} \tag{2.9}$$

Constants K_{12} and K_{13} can be estimated using the Simha–Boyer rule (Simha and Boyer 1962), where ρ is the density of each component:

$$K_{12} = \rho_1 T_{g1}/\rho_2 T_{g2} \quad \text{and} \quad K_{13} = \rho_1 T_{g1}/\rho_3 T_{g3} \tag{2.10}$$

Equation (2.9) is also considered to be a GT equation for ternary components. When $K_{12} = K_{13} = 1$, Eq. (2.9) becomes a linear equation. Application of Eq. (2.9) using constants K_{12} and K_{13} in Eq. (2.10) was carried out by Cantor (2000) for ternary hydrocarbon blends. The equation was found to work well for this system.

2.2.2.4   Truong et al.'s Analytical Equation

Truong et al. (2002) developed an analytical equation to determine the mixture T_g of a ternary system containing two sugars and one water. This equation was developed by assuming that all of the binary mixtures of the components follow the GT equation and that the third-order interaction between the components can be neglected. The final form of the equation for prediction of the T_g for a ternary mixture is described in Eq. (2.11):

$$T_{g123} = \frac{K_{12} x_2}{K_{12} x_2 + x_1} \frac{(1 - x_3) T_{g2} + K_{23} x_3 T_{g3}}{1 - x_3 + K_{23} x_3} + \frac{x_1}{K_{12} x_2 + x_1} \frac{(1 - x_3) T_{g1} + K_{13} x_3 T_{g3}}{1 - x_3 + K_{13} x_3} \tag{2.11}$$

where x_1, x_2, and x_3 are the mass fractions of components 1, 2, and 3, respectively; and K_{12}, K_{13}, and K_{23} are the K values in the GT equation for binary mixtures of components 1–2, 1–3, and 2–3, respectively. A least squares method is used to determine these constants separately from data of binary mixtures. The interaction factors in this equation are incorporated through the binary mixtures. The quantity $(1 - x_3)$ in Eq. (2.11) is the total fraction of component 1 and 2 in the mixture.

2.2.3   Application of Glass-Transition Temperature

The understanding of the T_g of wet and dried products is essential to predict and optimize the conditions for drying and subsequent storage (Bhandari and Howes, 1999; Roos and Karel, 1991a). Because of very limited molecular mobility, a food product in the glassy state is presumed not to deteriorate rapidly during storage.

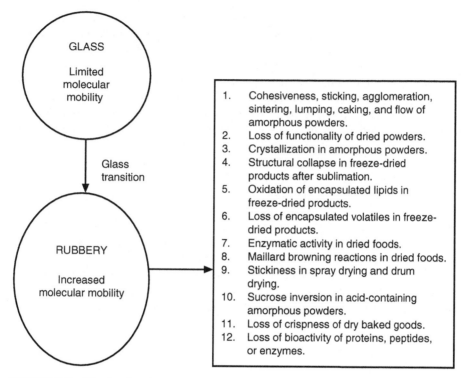

FIGURE 2.2 Physicochemical changes likely to occur during processing and/or storage at a temperature related to the T_g of foods.

In the T_g region, there is not only a sudden change in thermal and mechanical properties of the system but also a sudden change in the rates of translational molecular diffusion (Franks, 1985). As the product temperature exceeds the T_g, amorphous materials enter the rubbery state and the decreasing viscosity induces flow and deformation. Many thermophysical and physicochemical properties of food products that are related to the glass-transition temperature are summarized in Figure 2.2 (Bhandari and Howes, 1999; Levine and Slade, 1988).

2.3 SPRAY DRYING

Spray drying is a widely used dehydration technique for the transformation of feed from a fluid state into a dry particulate form by spraying it into a hot drying medium (Masters, 1994a). Spraying or atomization produces 10 to 500 μm droplets. Because the surface–volume ratio of those droplets is very large, the drying process concludes within a few seconds. Spray dryers are mechanically simple, operationally hygienic, and automation friendly. These units use inexpensive means to produce dry air, which makes them less expensive and thus a better choice compared with many competing dehydration techniques such as freeze-, vacuum-, and microwave drying (Bonazzi et al., 1996; King et al., 1984; Ranz and Marshall, 1952). Spray drying is a well-established and broadly used dehydration technique for producing food and pharmaceutical powders, including delicate flavor encapsulated foods. Furthermore, spray drying is ideally suited to heat-sensitive food materials because the product temperature usually remains well below 100°C, the exposure time is on the order of seconds, and the product is already in a solid state in which the reaction rate is slow.

The change in the physical state of a drying droplet during spray drying is illustrated in Figure 2.3 (Bhandari, Datta, and Howes, 1997). The liquid feed is atomized into swarms of droplets. The small droplets will dry completely within a short distance from the atomizer, the medium-sized droplets will get caught in a circulating eddy, and the relatively large droplets with sufficient momentum avoid being entrained in this eddy and reach the dryer wall (Langrish and Fletcher, 2001). The rapid evaporation of water and the subsequent concentration of solutes in the droplets transform them into syrup and finally to glassy particulates (Bhandari and Howes, 1999). If the evaporation process is sufficiently rapid so that the particle surface is glassy before reaching the wall and the bottom of the dryer, the sticky problem can be avoided. However, the particles containing a higher amount of low molecular weight sugars and organic acids remain thermoplastic (because of their low T_g) even if they are sufficiently dry and have sufficiently low moisture content. This is the reason why the particulates containing high sugar and organic acid remain as soft and sticky particles even at the end of the drying process (Bhandari et al., 1993).

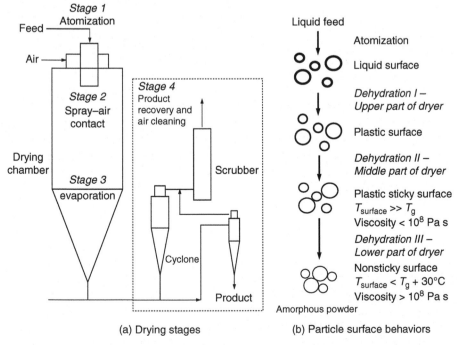

(a) Drying stages   (b) Particle surface behaviors

FIGURE 2.3 Schematic presentation of physical changes on the drop surface during spray drying. Dehydration I, II, and III represent three arbitrary drying stages along the drying history; $T_{surface}$ is the surface temperature of a drying droplet or particle. The particle is in a completely glassy form when it is fully dried.

2.3.1 PROBLEMS ASSOCIATED WITH SPRAY DRYING OF SUGAR- AND ACID-RICH FOODS

The foods to be spray dried can be subjectively classified into two broad groups: nonsticky and sticky. In general, nonsticky materials can be dried using a simple drier design and the final products remain free flowing. Materials such as skim milk, malto-dextrins, gums, and proteins belong to this group. The glass-transition temperature of these materials is high, and the particles are in a solid state at the lower part of the spray dryer. Sticky materials are difficult to dry under normal spray-drying condi-tions (Bhandari, Datta, and Howes, 1997). Natural sugar-rich and acid-rich foods such as fruit juices, vegetable extracts, and honey belong to this group. The sticky behavior of sugar-rich and acid-rich materials is attributed to low molecular weight sugars such as fructose, glucose, and sucrose and organic acids such as citric, malic, and tartaric acid, which constitute up to 90% of the solids (Dolinsky et al., 2000). During spray drying, they form lumps or syrup and resist converting into powder. They stick to the dryer wall or roof when the trajectory and recirculation allows them to reach it. Even if they do not collide at the dryer chamber, they ultimately come in contact with each other at the bottom of the dryer at the cyclone and conveying ducts and form unwanted agglomerates. Their high hygroscopicity causes them to covert to a plastic mass that is manifested as stickiness, even when the outlet temperature of the air is maintained as low as 50°C. The resultant wall deposits cause lower product yields, operational problems such as frequent plant shutdowns and cleanings, product handling difficul-ties, and fire hazards (Papadakis and Bahu, 1992).

2.3.1.1 Approaches to Remedy the Sticky Problem

Two approaches are commonly employed to overcome the sticky problem. The first approach, the *process-based approach*, manipulates the process parameters and equipment designs such as introduction of cold air at the bottom of the drying chamber (Lazar et al., 1956), control of the wall temperature (Bhandari et al., 1993), frequent scraping or sweeping of the inner wall of the dryer chamber to remove wall depositions (Karatas and Esin, 1990), and low temperature low humidity dryers (Hayashi, 1989). The second approach, the *material science based approach*, introduces high molecular weight additives (carriers) such as maltodextrin into the solution (Bhandari, Datta, Cooks, et al., 1997; Bhandari et al., 1993; Gupta, 1978) to raise the overall T_g of the mixture. The introduction of additives may inevitably alter the strength of the natural flavor, color, and taste of the resulting powders. This may affect consumer acceptance of the final product, particularly if added in large amounts.

There are various fundamental issues that require attention to be able to understand and make use of the materials science based method to overcome the stickiness. First, the phenomenon that causes stickiness has to be identified and interpreted. Second, an objectively measured indicator has to be found that can be used to optimize the drying process and amount of additives. In this aspect materials science provides us with the fundamentals of the glass transition and its indicator, the T_g. The glass transition is basic to all sugar- and acid-rich foods and the T_g can be measured objectively. The glass transition and T_g can be used not only to overcome the stickiness problem but also to design and produce composite powders from sugar- and acid-rich materials that are stable at the given handling and storage conditions.

2.3.1.1.1 Process-Based Approach to Remedy Stickiness

2.3.1.1.1.1 Low Temperature and Humidity of Drying Air (Birs Dryer)

In an attempt to dry the product at low temperature (30°C) and humidity (3% RH) to avoid stickiness, a mammoth 15 × 60 m (diameter × height) drying tower was developed (Hayashi, 1989; Mizrahi et al., 1967). The principle was based on drying below the stickiness temperature condition of the dried product. In this method, even if the drying was successful, the powders that were obtained were still sticky (Mizrahi et al., 1967). This is not an economically feasible system.

2.3.1.1.1.2 Cooling of the Dryer Chamber Wall

As early as 1968, a cool wall spray dryer was used by Robe et al. (1968) to produce tomato juice powder. Air was drawn through a hollow jacket surrounding the drying chamber, maintaining a wall temperature of between 38 and 50°C. This range of temperatures was well below the sticky point temperature (60°C) of tomatoes at 2% moisture content as measured by Lazar et al. (1956). Brennan et al. (1971) conducted experiments with orange juice by cooling the dryer wall. Similarly, Bhandari et al. (1993) used a LEAFLASH spray dryer with a temperature controlled chamber wall. Both of these studies found some improvement in the product yield (reduced stickiness) compared to conventional drying methods. However, this method did not improve the performance of the dryers to the required level because the cold chamber wall cools the surrounding air while simultaneously producing localized high relative humidity regions close to the wall. This increases the surface moisture content of the powders (Bhandari, Datta, and Howes, 1997). This type of dryer in association with a fluidized bed at the bottom is used commercially for drying of tomato juice (Masters, 1994b). The actual success of such a drier is still questionable.

2.3.1.1.1.3 Introduction of Cooling Air to the Bottom Dryer Chamber

Lazar et al. (1956) conducted experiments in which cooling air was introduced to the lower part of the dryer chamber to cool the powder. This technique can result in decreases in particle temperature and stickiness. The recovery was still low at 60 to 65%.

2.3.1.1.1.4 Secondary Air near the Wall

To avoid deposition of the powder on the dryer walls, secondary air was inserted near the walls through wall slots to maintain motion in the particles and prevent powder settling onto the walls (Masters, 1994b). In this technique, the droplets are retained in the central part of the chamber until moisture evaporation has been completed. This method is useful for hygiene requirements. However, stickiness can occur at the bottom part of the dryer if the surrounding air temperature is higher than the sticky point temperature of the particles.

2.3.1.1.1.5 Using a Ceiling Air Disperser

For a concurrent spray dryer with a rotary atomizer, a ceiling air disperser with rotation air can be used to eliminate the area of semiwet deposits on the top and side walls (Masters, 1994a). However, the stickiness of the powder deposited on the bottom walls remains unresolved. This is more of a localized problem as a result of nonuniform airflow or droplet trajectory.

2.3.1.1.1.6   Partial Crystallization of Sugar

This method is normally applied for relatively pure components such as lactose, sucrose, or whey. A partially crystallized sugar solution is produced before spray drying (Hansen and Rasmussen, 1977; Rheinlander, 1982). The particles may deposit initially on the dryer walls, but the layers can loosen from the walls because of partial crystallization within the particles. This method produces a nonhygroscopic powder in semicrystalline form but does not solve the stickiness problem for the amorphous fraction in the powder. This method to dry problematic fruit juices has not been reported (Bhandari, Datta, Cooks, et al., 1997), which is probably attributable to the inability to crystallize the main sugars (glucose or fructose) of fruit juices.

2.3.1.1.1.7   Recirculation of Fines

Recirculation of fine particles recovered from the cyclone into the atomization zone can cause agglomeration and produce a less sticky product (Bhandari, Datta, and Howes, 1997). A selected amount of dried particles can be recycled to the feed inlet to mix with the cloud of atomized product, and these particles may consist of recycled material (Woodruff et al., 1971). In the case of sucrose, the final products contain about 45 to 85% crystalline material (Woodruff et al., 1971). This high crystallinity also assists in minimizing stickiness. The particle size range is high (75 to 1000 μm), which is due to agglomeration. The use of this method requires a high recycled portion and a large volume of air in order to attain good contact between the recycled particles and atomized droplets. Therefore, it is not economical (Hansen and Rasmussen, 1977). There is no extension of this method to drying of fruit juices (Masters, 1994a).

2.3.1.1.1.8   Three-Stage Drying Technique

A three-stage drying system with an integrated belt (Filtermat) is reported to be the most suitable technique for fruit juices (Masters, 1994a; Rheinlander, 1982). The product is initially dried in a wide tower for a short time (7 to 8 s) to a moisture content of 10 to 20% (first stage) and then falls to a polyester moving belt for further drying (second stage). After drying in the first and second stages, the powder is conveyed to the third drying stage where the drying is completed by means of low temperature drying air. The final product contains bigger agglomerates that must be passed through a sifting system to break them into the desired particle size.

2.3.1.1.2   Approaches Based on Material Properties

The composition of the feed in this method is altered by introducing carriers in order to increase the T_g of the product. Consequently, the sticky point temperature of the products is increased. Carriers are normally high molecular weight polymers such as maltodextrins or starch ($T_g = 100$ to $243°C$) or dried milk powder. The other high molecular weight carriers such as carboxymethyl cellulose, gum arabic, and pectin have also been used without success. This is because those carriers do not contribute significantly to the increase in T_g of the products, because of the limited amount of them that can be added due to the high viscosity of the solution limiting atomization. Thus, the T_g of the product is the main factor governing the success of the spray-drying process. In general, the addition of maltodextrin has become a common practice because of its neutral color and taste and relative inexpensiveness. Many types of fruit juices have been tried such as orange, black currant, cherry, tomato, coconut, and tamarind. The fraction of the carrier is normally in the range of 40 to 45 wt% juice solids, but up to 70 wt% juice solids have been reported (Masters, 1994a).

Brennan et al. (1971) found that the addition of 44 wt% glucose syrup in relation to orange juice solids increased the sticky point temperature of the product to 44°C (at a moisture content of 2% dry basis) and improved the recovery dramatically compared to nonadditives. However, wall deposition of 10 to 60% was still encountered depending on the drying air temperature. The wall deposition was minimal when the wall temperature was controlled below the sticky point temperature (Brennan et al., 1971). The stickiness point temperature was still low enough to be accepted as a practical solution in drying of food products.

2.3.1.1.2.1   Empirical Relation between Stickiness and the Drying Aids Required in Spray Drying

For a given geometry of the spray dryer, the improvement in the drying performance when a carrier is added to the feed can be related to the T_g of the feed. Busin et al. (1996) conducted spray-drying experiments for different sugars (sucrose, glucose, fructose) using maltodextrin with a different dextrose equivalent (DE) as the drying aid and found that the recovery of the powder was proportional to the T_g of the mixture. For a 50% recovery of powder of a sucrose–maltodextrin mixture (total powder collected in the cyclone and dryer walls), the drying aids required were 16 and 18% for DE 6 ($T_g = 164°C$) and DE 21 ($T_g = 135.5°C$), respectively. It was interesting that the final T_{gmix} of the powder in the two cases was the same (75.5°C). This means that, for a recovery of 50% of the sucrose–carrier mixture, the amount of carrier can be calculated using the GT equation if the T_g of the carrier is known. However, for drying of glucose and fructose using the same drying aids as for sucrose, the T_{gmix} values of the final product were 52.5 and ~51°C, respectively. There was a difference between the T_{gmix} for glucose and fructose compared to sucrose even if the drying condition was the same (inlet and outlet air temperatures were 180 and 100°C, respectively). This may be due to the difference in drying characteristics among sugars that results in different final moisture content of the product that affects the T_{gmix} values. Thus, the drying performance (recovery, stickiness) in spray drying of sugar-rich foods is governed not only by the T_g value but also by the drying characteristic of the feed. This drying characteristic depends on the inlet drying conditions and the interaction between the drying air and the atomization. The interaction, in turn, is a function of the construction of the dryer and the mode of feed–air contact, for example, concurrent, countercurrent, or mix flow type.

As a simple practical solution, Bhandari et al. (1997a) developed a drying index model to determine the drying aid required for a 50% recovery of a food model mixture:

$$\sum_{i=1}^{n} I_i x_i = Y_I \qquad (2.12)$$

where I_i and x_i are the respective drying index and weight fraction of component i in the mixture and Y_I is the indicator for the degree of successful drying that is equal to 1 for 50% recovery. In the drying experiment using inlet and outlet air temperatures of

150 and 65°C, respectively, the drying indices of sucrose, glucose, fructose, and citric acid were 0.85, 0.51, 0.27, and −0.40, respectively (Bhandari, Datta, and Howes, 1997), where $I = 1.6$ was arbitrarily set for the index of maltodextrin DE 6. The model was validated for honey and pineapple juices. In addition, the drying index was linearly related to the T_g of the mixture. This is useful for the determination of the drying aid required for 50% recovery (total powder collected in the cyclone and dryer chamber) in spray drying for any food model.

2.3.1.1.2.2 Control Based on T_g

The T_g is recognized as a fundamental parameter to explain stickiness (Bhandari and Howes, 1999; Roos and Karel, 1991a) in sugar- and acid-rich foods. It is a fundamental property of amorphous materials and polymers. When a glassy material is subjected to moisture and heat and subsequently approaches the phase transition, the viscosity of the system decreases dramatically from 10^{12-14} to 10^{6-8} Pa s. The lowered viscosity is unable to support the glassy microstructure, giving way to stickiness of the dried materials (Flink, 1983; Wallack and King, 1988).

The material undergoing drying can be a mixture of low or high molecular weight solids and water. The T_g values of starches, proteins, and gums (high molecular weight materials) are higher than those of sugars, organic acids, and polyols (low molecular weight materials). The T_g of water is the lowest at around −135°C. Therefore, the influence of water in depressing the T_g of the mixture is significant especially in the early stages of drying. The T_g of materials increases as the drying progresses because of the removal of water. The evolution of the glass-transition temperature of a material that is difficult to dry with or without the addition of additives (such as maltodextrin) is illustrated in Figure 2.4. In this case, the addition of maltodextrin raises the overall T_g of the mixture. This results in a nonsticky powder at the exit of the drier. Note in Figure 2.4 that the particle temperature should be lower than the sticky temperature (curve) of the product to achieve the powder in a nonsticky state.

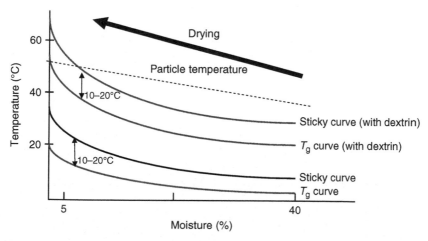

FIGURE 2.4 Illustration of the sticky point temperature versus T_g of a drying product, with or without the addition of maltodextrin to raise the overall T_g of the mixture.

The critical viscosity of stickiness is reached at temperatures 10 to 20°C above the measured T_g value of the material, according to reports. For skim milk powders, Hennigs et al. (2001) and Ozmen and Langrish (2002) found that the differences between the stickiness temperature and T_g were 23.3 and 14 to 22°C, respectively. Chuy and Labuza (1994) reported that the differential temperature was 30 to 50°C for dairy-based infant formula powders. The stickiness curve at 10 to 20°C above the glass-transition curve is depicted schematically in Figure 2.4. Flow dynamics within the dryer can also influence the stickiness property of the drop. Particles that are tangentially directed toward the dryer wall can impact and stick easily onto the wall even at their glass-transition temperature. In contrast, the inertia of mobile particles and the kinetic energy of the moving air can loosen the contact point of the particles from the wall, which eventually results in an increase in the sticky temperature. Overall, the dominating effects of the film-forming properties of the material along with the dynamic conditions in the drier can locally elevate the stickiness temperature above the glass-transition temperature (Bhandari and Hartel, 2005).

Adhikari et al. (2003, 2004) demonstrated experimentally that the surface stickiness of a drying droplet depends on the T_g based on the surface composition rather than the average or bulk composition. They showed that a maltodextrin drop exhibited peak stickiness at a mean moisture of around 50% (wet basis) and the drop surface became completely nonsticky at 40% (wet basis) while drying at 63°C and 2.5% relative humidity. Further, the average T_g of the drop at those moisture levels were −95.9 and −81.1°C, respectively, and the drop surface was completely nonsticky even when the drop temperature was about 138°C higher than the glass-transition temperature (Adhikari et al., 2005). These findings cannot be explained based on the classical sticky point concept that states that the sticky point temperature is about 10 to 20°C higher than the glass-transition temperature. This disparity can be explained by accepting that there is a strong moisture gradient in a drying drop, so the composition of the surface layer (and hence the T_g value) can be quite different than the T_g value based on the average moisture content. In some cases, the migration of surface active proteins on the surface of the powder (water–air interface) during atomization has influenced the stickiness property of the drying material. Despite these facts and the difficulty in measuring the composition of the particle surface layer and the surface layer T_g experimentally, the glass-transition mechanism is the most successful one to explain and quantify the droplet and particle stickiness.

We should also remember that the occurrence of stickiness of a very viscous material is a time-dependent phenomenon. Even if such a material cannot show stickiness in the drier, it can manifest it during storage. The time dependency is a consequence of the slow viscous flow of the material in between the particles in contact (discussed further in Section 2.7.1).

The trajectory of the drying air and the geometry of the dryer are also important factors. Because of the design principle of the spray dryer, many air circulation zones exist within the dryer chamber, especially at the corners and the intersection between the cylindrical and conical parts. These air circulation zones entrap the particles. This leads to a longer residence time and a higher probability of particle–particle contact and/or particle–wall contacts (Oakley and Bahu, 1991). Both of these increase the risk of stickiness. The trajectory of air and particles can be predicted by the computational fluid dynamic (CFD) technique (Oakley and Bahu, 1991; Kieviet, Raaij, De Moor, et al., 1997; Kieviet, Raaij, and Kerkhof, 1997; Straatsma et al., 1999a, 1999b). The CFD modeling suggests that the most likely areas of wall deposition are the annular area of the roof, corresponding to the small recirculation eddy (medium-sized droplets) and a region below the atomizer where large particles are likely to deposit (Langrish and Fletcher, 2001).

2.4 FREEZE-DRYING AND STRUCTURAL COLLAPSE

Freezing-drying involves the removal of frozen water by sublimation followed by the removal of unfrozen water by vacuum drying. During freezing, depending on the composition of the food material, 65 to 90% of the initial water is in the frozen state and the remaining 10 to 35% of the water is in the unfrozen (sorbed) state. The rate and amount of ice formation during freezing is dependent on the composition and viscosity of the unfrozen fraction. If the water concentration in the product is low, all of the water remains in the unfrozen state. Water remained unfrozen in partially dried strawberries when the water activity of the materials was below 0.75 (Roos, 1987) and water in honey (moisture $\approx$ 16% w/w, $a_w \approx 0.6$) was unfreezable (Sopade et al., 2003).

The stickiness temperature (T_{sticky}) in the context of freeze-drying is termed the collapse temperature (T_c). Therefore, the stickiness during spray drying is analogous to the collapse of the drying material during freeze-drying. Understanding the collapse of the structure of a material undergoing freeze-drying is also important for its subsequent storage and handling. Collapse is defined as a time-, temperature-, and moisture-dependent viscous flow that results in a loss of structure and a reduction in sample volume (Tsouroflis et al., 1976). This is particularly evident at the later stage of freeze-drying when the majority of the ice is sublimed and the temperature of the drying front is higher than the critical temperature. This is because the highly porous solid matrix of the freeze-dried material can no longer support its structure against gravity (Bellows and King, 1973; MacKenzie, 1975).

Collapse progresses in two stages. In the first stage, collapse is localized to the concentration of collapsed pores in certain areas. In the second stage, collapse of the entire structure takes place. The collapse temperature is normally the temperature at which the entire structure is collapsed because of the viscous flow of the solid matrix. The viscosity during the initiation of collapse (μ_c) can be estimated using the Williams–Landel–Ferry (WLF) equation. The collapse temperatures of food products in the dry and liquid states are presented in Table 2.1. To and Flink (1978a, 1978b) reported that the collapse of dehydrated carbohydrates and the T_g of polymers were related phenomena. As shown in Table 2.2, the T_c is equivalent to T_g', which is the glass-transition temperature of maximally concentrated nonfrozen liquid in the product undergoing drying.

2.5 HOT-AIR DRYING

There are few published reports available on the consequences of the T_g on hot-air drying. Nevertheless, the knowledge gained in other types of drying can be logically

TABLE 2.1

Collapse Temperature (T_c) of Freeze-Dried Anhydrous Food Powders and Fruit Juices during Freeze-Drying

Materials	T_c (°C)
Lactose powder	101.1
Maltose powder	96.1
Sucrose powder	55.6
Maltodextrin DE 10 powder	248.9
Maltodextrin DE 15 powder	232.2
Maltodextrin DE 20 powder	232.2
Maltodextrin DE 25 powder	204.4
Orange juice/maltose (100 : 10) powder	65.6
Xylose, 25%	−49
Orange juice, 23%	−24
Grapefruit juice, 16%	−30.5
Lemon juice, 9%	−36.5
Apple juice, 22%	−41.5
Pineapple juice, 10%	−41.5
Prune extract, 20%	−35.0
Coffee extract, 25%	−20
Sodium citrate, 10%	−40
Sodium ascorbate, 10%	−37

Source: Compiled data from Tsourouflis, S. et al., *J. Sci. Food Agric.*, 27, 509–519, 1976; Bellows, R. and King, C.J., *AIChE Symp. Ser.*, 69, 33–41, 1973; and Fonseca, F. et al., *Biotechnol. Prog.*, 20, 229–238, 2004. With permission.

applied in this process (Ratti, 2001). Collapse during freeze-drying is analogous to shrinkage during hot-air drying. Because the products during hot-air drying are dried above the T_g, shrinkage or collapse is inevitable during hot-air drying. Hot-air drying (fixed or fluidized bed) is normally employed to dry solid foods. Glass-transition related problems exist during the drying of sugar-rich products such as fruit leathers and other products that are extremely sticky. These products appear wet even at very low moisture content. The leathery texture of these products even at lower moisture content is due to the plasticizing effect of low molecular weight sugars that depress the T_g. Drying of other fruits may pose problems, particularly if there is an excessive disruption of skins or tissues, resulting in leaching of sugars out onto the surface.

Sticky products will not fluidize well in a fluidized bed system. This problem may also be encountered while dehydrating thick pasty products and cut fruit pieces. The same problem will exist during the coating of the product. The coating material is normally sprayed in a liquid state. If it is sprayed in an aqueous solution, the coating should be followed by drying. If the coating material has a very low T_g, it will cause sticking of the coated particulates during the water removal in a fluidized bed. The stickiness will be very serious toward the later stage of drying when the viscosity of the coating material will increase substantially at low moisture.

TABLE 2.2

Glass-Transition Temperature of Maximally Concentrated Solutions (T_g') and Collapse Temperature (T_c) of Aqueous Solutions

Solution (w/w)	T_g' (°C)	T_c (°C)
Sucrose, 10%	−32 (−32.9)[a]	−32 (−32)[b]
Maltose, 10%	−30 (−29.5)[c]	−30 (−30 to −35)[b]
Maltodextrin DE 5–8, 10%	−9 (−9.5)	−7 (−10)[b]
Sodium glutamate, 10%	−47	−46 (−50)[b]
Sodium citrate, 10%	−43 (−41)[d]	−40
Sodium ascorbate, 10%	−37	−37
Sorbitol, 10%	−45 (−46.1)[a]	−46 (−45)[b]
Sucrose 5% + maltodextrin (DE 5–8) 5%	−24	−19 (−20)[b]
Sucrose 5% + sorbitol 5%	−41	−38 (−40)[b]

[a] Compiled data from Her, L.M. and Neil, S.L., *Pharmaceut. Res.*, 11, 54–59, 1994. With permission.
[b] Compiled data from Fonseca, F. et al., *Biotechnol. Prog.*, 20, 229–238, 2004; and MacKenzie, A.P., in *Freeze-Drying and Advance Food Technology*, Academic Press, New York, 1975, pp. 277–307. With permission.
[c] Compiled data from Levine, H. and Slade, L., *Cryo-Letters*, 9, 21–63, 1988. With permission.
[d] Compiled data from Chang, B.S. and Randall, C.S., *Cryobiology*, 29, 632–656, 1992. With permission.

The shrinking phenomenon is dependent upon the physical pliability and microstructure of the material that should be influenced by the T_g and other factors. Karathanos et al. (1993) demonstrated that continuous shrinkage occurred in celery during drying. This was attributed to the drying air temperature that was well above the T_g of celery undergoing drying. In high sugar products, if the dehydration temperature is higher than the T_g, the product can be soft during drying (Del Valle et al., 1998; Riva et al., 2005). Upon cooling they will then harden as the product temperature drops below the T_g. A simple relationship may not be found between the T_g and shrinkage because of the drying temperature being normally well above the T_g at any time, the influence of the tissue structure of the biological materials, the contribution of nonsoluble materials on the structure, and the large gradient of moisture in the drying materials. The shrinkage, however, can continue even after drying if the dehydrated product is still in the rubbery state (which is the case in many dried fruits and vegetables). A reduced rate of shrinkage at a later stage of drying of many food products is certainly related to the increased viscosity of the matrix, when the material approaches the T_g.

2.6 INSTANTIZATION

Agglomeration followed by drying is a common process in food powder industries. Agglomerate is composed of two or more individual primary particles, which are bonded together by submicron particles, viscous amorphous materials between the

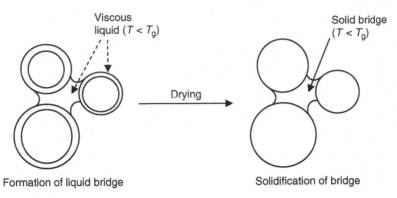

FIGURE 2.5 Agglomeration and drying of particulates as related to the surface T_g.

particles, or a binder. Agglomerates tend to offer an instantizing property to the powder. They disperse and dissolve quickly into solvents such as water, milk, tea, and so forth during reconstitution. The reconstitution performance of the agglomerated powder is enhanced because the open porous structure of the agglomerate allows solvent to penetrate and disperse throughout, forcing the agglomerate to sink faster compared to individual particles. Agglomeration of the particles also provides a dustless, free flowing powder that facilitates handling.

Agglomeration is achieved by applying heat, rewetting, or a combination of both. If the dried product with relatively higher moisture content is warmed to above its glass transition, its surface becomes viscous and can form a bridge between adjacent particles that results in an agglomeration. The viscosity should be low enough to cause virtually instantaneous sticking of the particles. The agglomeration generally involves sticking of the powder particles followed by drying. The viscous liquid material responsible for sticking the particles is converted to a solid during drying (Figure 2.5).

Few direct studies have been published on this process in relation to T_g. However, the glass-transition concept can provide a better understanding and optimize desired and undesired agglomeration of amorphous powders. The glass-transition temperature and its dependence on moisture content is a material property, which allows an estimation of the mechanical properties of food products at varying moisture and temperature conditions. Moisture–temperature and time combinations for agglomeration of raw materials and its mechanism can be defined (Palzer, 2005). It is also possible to predict parameters for the blockage of a fluidized bed that is sometimes a problem in the powder industry.

2.7 OTHER GLASS-TRANSITION RELATED PHENOMENA OF DEHYDRATED FOODS

2.7.1 CAKING

The mechanism of caking is the same as agglomeration. The caking of an amorphous powder is a time-dependent phenomenon. The rate of caking attributable to the viscous flow property of the material is a function of $T - T_g$, and the kinetics of caking can be determined by using a WLF type relationship (Aguilera et al., 1995):

$$\log_{10}\left(\frac{t}{t_g}\right) = \frac{-C_1(T - T_g)}{C_2 + (T - T_g)} \tag{2.13}$$

where t is the time of caking, t_g is the time to cake at the T_g, C_1 and C_2 are constants (17.44 and 51.6 K, respectively), and T is the temperature (K). The WLF relationship is also applicable to predict the crystallization of the glass as a function of temperature. Note that crystallization is one of the causes of caking. The WLF equation is also used to predict the viscosity of the material above its glass-transition temperature [by replacing t with μ in the right-hand side of Eq. (2.13)].

Downton et al. (1982) studied the following well-known Frenkel mechanistic model [Eq. (2.14)] for estimating the μ_c (Pa s) that causes caking:

$$\mu_c = \frac{\kappa\gamma\tau}{KD} \tag{2.14}$$

where κ is the dimensionless proportionality constant of order unity; τ is the contact time (s); γ is the surface tension, which for an interstitial concentrate $= 0.07 \, \text{N/m}$; K is the fraction of the particle diameter required as the bridge width for a sufficiently strong interparticle bond, usually 0.01 to 0.001; and D is the particle diameter (m).

It is important to note that the caking event is time dependent whereas sticking is a relatively instantaneous phenomenon (say, $\tau = 1$ s). Longer contact times increase the tendency toward sticking and caking, when all other things are equal. Thus, the dried product recovered during drying with a relatively free flowing property could also cake in a collection or packaging container over a period of time if the surface viscosity is still relatively low because of higher temperature or moisture levels (Bhandari, Datta, and Howes, 1997). For this reason, the dried product needs to be cooled immediately to an appropriate temperature before packaging. The interesting consequence of this is that the viscosity (and thus temperature) requirements for storage are much more stringent than those for processing; however, it is still possible to estimate critical values for both. Many powders can be a mixture of amorphous and crystalline materials. The proportion of the crystalline fraction (which is the noncaking structure) will influence the overall caking behavior of the powder (Fitzpatrick et al., 2007).

Caking is strongly influenced by compression that decreases the distance between the particulates (Lloyd et al., 1996). The final shrinkage of the product under compaction increases when increasing $T - T_g$. The storage of sugar-rich dried products like fruit pieces or whole fruits such as sultanas in a bulk bin can cause clumping and the formation of a hard block. This may be caused by several factors. The free sugar can leach out onto the surface because of skin damage. Storing this product at low temperature can cause the viscosity of highly concentrated sugar syrup to increase substantially, which results in cementing the pieces together into a solid mass. There could also be a partial crystallization of sugars and coalescence. This phenomenon can be more prominent when the products are packed into a big mass that reduces the interparticulate distance and can also cause further leaching of the syrup from internal tissues at the surface. To minimize this problem, the product is normally surface coated with oil to avoid direct contact between sticky surfaces or further dehydration (which will elevate the T_g).

2.7.2 DIFFUSION AND REACTIONS IN A DRIED GLASSY MATRIX

The glass-transition temperature is directly related not only to the molecular mobility of the matrix but also to water and other components entrapped in the matrix. Volatile components in a food system have very limited mobility in a glassy matrix. Flavors encapsulated in carbohydrate and/or protein glasses are retained on drying and subsequent storage until released by the addition of water or an increase in temperature (Karel et al., 1993; Karel and Flink, 1983; King, 1988; Shimada et al., 1991). Above the glass-transition temperature of the matrix, diffusion of the volatiles through the matrix primarily occurs through its pores (Levi and Karel, 1995; Whorton and Reineccius, 1995). Volatile diffusivity is greatly increased when the temperature exceeds the T_g and continues to increase with an increase in temperature (Roos and Karel, 1991a). The diffusivity of oxygen and mobility of the volatile compounds in the matrix promotes an oxidation reaction. Shimada et al. (1991) found that methyl linoleate encapsulated in a lactose–gelatin matrix does not oxidize below the T_g of the matrix, but it oxidizes above it with oxidation proceeding more rapidly as ΔT $(T - T_g)$ increases. Labrousse et al. (1992) studied the oxidation of lipids (encapsulated in sugars) at different storage temperatures and times. They found that there was no oxidation of encapsulated lipid below the T_g of the matrix. The oxidation increased when the storage temperature exceeded the T_g of the matrix. The retention of diacetyl in spray-dried lactose and skim milk systems was also related to ΔT (Senoussi et al., 1995). Karmas et al. (1992) and Buera and Karel (1993) found an increased rate of nonenzymic reactions in dried vegetables stored above the T_g.

In an earlier study, To and Flink (1978a,b) found that the loss of volatiles was very high at collapse temperatures during freeze-drying or during storage of the dried product. Gejlhansen and Flink (1977) reported that the oxygen absorption into a freeze-dried lipid–carbohydrate system was increased significantly at the collapse temperature, which is related to the T_g.

Relating the reactions on the basis of the glass-transition temperature has become a controversy in recent days, particularly in complex food systems. A Maillard reaction occurred in dry pasta systems (Fogliano et al., 1999) below their glass-transition temperature. The creation of small mobile molecules because of deamidation of glutamine was attributed to this reaction even below the T_g. Schebor et al. (1999) also reported a Maillard reaction below the glass-transition temperature in anhydrous milk powder systems. In this case, the authors suspected that other types of localized mobility such as glass relaxation, intermixing of reactants increasing the proximity of the molecules in concentrated state, and localized mobility near the pores might have existed because there were no other components present in the milk. However, the milk still contained some amount of fat that would have been mobile and the protein that can be a part of the fat globule membrane will still have some mobility at the interface of the solid–oil. It is not clear though if such factors can influence aminocarbonyl condensation. Any preexisting smaller molecules (amino compounds) can still have translational motion within the matrix. Dissolution of some protein or peptides in lactose cannot be ruled out as causing this effect (localized β-relaxation can accelerate the condensation of amino and carbonyl groups). In an early study, Kamman and Labuza (1985) also found that the Maillard reaction was accelerated by the presence of liquid-phase oil in a powder starch mix containing glucose and glutamate. Their assumption was that the oil may act as a solvent for the reactants. However, they did not explain how a hydrophobic oil can solubilize these polar reactants. This is an important area that needs further investigation (Bhandari, 2007).

2.7.3 Crispness and Plasticity

Crispness and brittleness are some of the important attributes of dried products. Dried products are normally brittle. In normal dried products, an increased amount of low molecular weight sugars and water reduces the T_g of the product to below ambient temperature and the textural attributes are lost (Figure 2.6). Some semidried products are leathery or rubbery because the requirement is to make them chewy rather than brittle or hard. Examples are dried fruits, fruit bars, fruit leathers, and so forth. Water acts a plasticizer in such products. Addition of low molecular weight sugars and polyols as humectants can maintain the plastic or chewy texture of intermediate dried product even at relatively low moisture content by lowering the T_g of the product. The common humectants used in food are glucose, fructose, sucrose, invert sugar, and glycerol. Loss of crispness due to lowering of the T_g of dried extruded products was also reported (Labuza and Katz, 1981; Nikolaidis and Labuza, 1996; Roos, 1995).

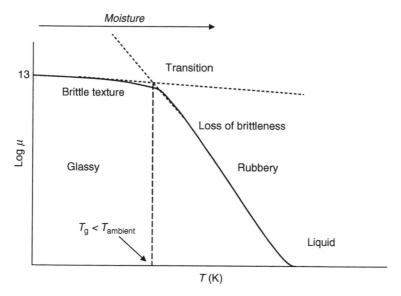

FIGURE 2.6 Illustration of the loss of crispness of the product during the glass–rubber transition (either by moisture adsorption or increased temperature).

NOMENCLATURE

VARIABLES

C_1 and C_2	constants
C_p	heat capacity
$\Delta C_{pi} = (C_{pLi} - C_{pGi})$	change in heat capacity of component i between its liquidlike and glassy states
D	particle diameter (m)
I_i	drying index
K	fraction of particle diameter required as the bridge width for a sufficiently strong interparticle bond
n	number of data points
S	mixed system entropy
S_i	component entropy
ΔS_{mG}	excess entropy at the glassy state
ΔS_{mL}	excess entropy at the liquid state
t	time of caking
t_g	time to cake at glass-transition temperature
T	temperature (K)
ΔT	change in temperature (K)
T_c	collapse temperature
T_g	glass-transition temperature
T_g'	glass-transition temperature of maximally concentrated nonfrozen liquid in the product undergoing drying
T_{g123}	glass-transition temperature of a ternary mixture
T_{gmi}	measured glass-transition temperature of model mixture i
T_{sticky}	sticky point temperature
$T_{surface}$	surface temperature of a drying droplet or particle
x	mass or mole fraction
x_{1i}, x_{2i}	fractions of components 1 and 2, respectively, of model mixture i
Y_I	indicator for the degree of successful drying

GREEK LETTERS

γ	surface tension
$\Delta \gamma_i$	difference between the expansion coefficients of liquid and glass of component i
κ	dimensionless proportionality constant of order unity
μ_c	viscosity during the initiation of collapse (Pa s)
ρ	density
τ	contact time (s)

REFERENCES

Adhikari, B., Howes, T., Bhandari, B.R., and Truong, V., In situ characterization of stickiness of sugar-rich foods using a linear actuator driven stickiness testing device, *J. Food Eng.*, 58, 11–22, 2003.

Adhikari, B., Howes, T., Bhandari, B.R., and Truong, V., Effect of addition of maltodextrin on drying kinetics and stickiness of sugar and acid-rich foods during convective drying: Experiments and modelling, *J. Food Eng.*, 62, 53–68, 2004.

Adhikari, B., Howes, T., Lecomte, D., and Bhandari, B.R., A glass transition temperature approach for the prediction of the surface stickiness of a drying droplet during spray drying, *Powder Technol.*, 149, 168–179, 2005.

Aguilera, J.M., Valle, J.M., and Karel, M., Amorphous food powders, *Trends Food Sci. Technol.*, 6, 149–155, 1995.

Allen, G., A history of the glassy state, in *The Glassy State in Foods*, Blanshard, J.M.V. and Lillford, P.J., Eds., Nottingham University Press, Nottingham, U.K., 1993, pp. 1–12.

Arvanitoyannis, I., Blanshard, J.M.V., Ablett, S., Izzard, M.J., and Lillford, P.J., Calorimetric of the glass transition occurring in aqueous glucose: Fructose solutions, *J. Sci. Food Agric.*, 63, 177–188, 1993.

Bellows, R. and King, C.J., Product collapse during freeze drying of liquid foods, *AIChE Symp. Ser.*, 69, 33–41, 1973.

Bhandari, B.R., Beyond water: Water-like functions of other biological compounds in water-less system, paper presented at the 10th International Symposium on Properties of Water, ISOPOW 2007, Bangkok, September 2–7, 2007.

Bhandari, B.R., Datta, N., Cooks, R., Howes, T., and Rigby, S., A semi-empirical approach to optimize the quantity of drying aids required to spray dry sugar rich foods, *Dry. Technol.*, 15, 2509–2525, 1997.

Bhandari, B.R., Datta, N., and Howes, T., Problems associated with spray drying of sugar-rich foods, *Dry. Technol.*, 15, 671–684, 1997.

Bhandari, B.R. and Hartel, R.W., Co-crystallization of sucrose at high concentration in the presence of glucose and fructose, *J. Food Sci.*, 67, 1797–1802, 2002.

Bhandari, B.R. and Hartel, R.W. Phase transitions during food powder production and powder stability, in *Encapsulated and Powdered Foods*. Chapter 11, Onwulata, C., Ed., Taylor and Francis, New York, 2005, pp. 261–291.

Bhandari, B.R. and Howes, T., Implication of glass transition for the drying and stability of foods, *J. Food Eng.*, 40, 71–79, 1999.

Bhandari, B.R., Senoussi, A., Demoulin, E.D., and Lebert, A., Spray drying of concentrated fruit juices, *Dry. Technol.*, 11, 33–41, 1993.

Bonazzi, C., Demoulin, E., Wack, A.L.R., Berk, Z., Bimbenet, J.J., Courtois, F., Trystram, G., and Vasseur, J., Food drying and dewatering, *Dry. Technol.*, 14, 2135–2170, 1996.

Brennan. J. G., Herrera, J., and Jowitt, R., A study of some of the factors affecting the spray drying of concentrated orange juice, on a laboratory scale, *Food Technol.*, 6, 295–307, 1971.

Buera, M.D. and Karel, M., Application of the WLF equation to describe the combined effects of moisture and temperature on nonenzymatic browning rates in food systems, *J. Food Proc. Preserv.*, 17, 31–45, 1993.

Busin, L., Buisson, P., and Bimbenet, J.J., Notion of glass transition applied to the spray-drying of carbohydrate solutions, *Sci. Aliment.*, 16, 443–459, 1996.

Cantor, A.S., Glass transition temperatures of hydrocarbon blends: Adhesives measured by differential scanning calorimetry and dynamic mechanical analysis, *J. Appl. Polym. Sci.*, 77, 826–832, 2000.

Chang, B.S. and Randall, C.S., Use of subambient thermal analysis to optimize protein lyophilization, *Cryobiology*, 29, 632–656, 1992.

Chuy, L. and Labuza, T.P., Caking and stickiness of dairy based food powders as related to glass transition, *J. Food Sci.*, 59, 43–46, 1994.

Couchman, P.R., Compositional variation of the glass transition temperature. Application of the thermodynamic theory to compatible polymer blends, *Macromolecules*, 11, 1156–1161, 1978.

Couchman, P.R. and Karasz, F.E., A classical thermodynamic discussion of the effect of composition on the glass transition temperature, *Macromolecules*, 11, 117–119, 1978.

Del Valle, J.M., Cuadros, T.R.M., and Aguilera, J.M., Glass transitions and shrinkage during drying and storage of osmosed apple pieces, *Food Res. Int.*, 31, 191–204, 1998.

Dolinsky, A., Maletskaya, K., and Snezhkin, Y., Fruit and vegetable powders production technology on the bases of spray and convective drying methods, *Dry. Technol.*, 18, 747–758, 2000.

Downton, G.E., Flores-Luna, J.L., and King, C.J., Mechanism of stickiness in hygroscopic, amorphous powders, *Ind. Eng. Chem. Fundam.*, 21, 447–451, 1982.

Fitzpatrick, J.J., Hodnett, M., Twomey, M., Cerqueira, P.S.M., O'Flynn, J., and Roos, Y.H., Glass transition and the flowability and caking of powders containing amorphous lactose, *Powder Technol.*, 178, 119–128, 2007.

Flink, J.M., Structure and structure transitions in dried carbohydrate materials, in *Physical Properties of Foods*, Peleg, M. and Bagley, E.B., Eds., AVI, Westport, CT, 1983, pp. 473–521.

Fogliano, V., Monti, M.S., Musella, T., Randazzo, G., and Ritieni, A., Formation of coloured Maillard reaction products in a gluten–glucose model system, *Food Chem.*, 66, 293–299, 1999.

Fonseca, F., Passot, S., Cunin, O., and Marin, M., Collapse temperature of freeze-dried *Lactobacillus bulgaricus* suspensions and protective media, *Biotechnol. Prog.*, 20, 229–238, 2004.

Franks, F., *Biophysics and Biochemistry at Low Temperatures*, Cambridge University Press, Cambridge, U.K., 1985.

Gabarra, P. and Hartel, R.W., Corn syrup solids and their saccharide fractions affect crystallization of amorphous sucrose, *J. Food Sci.*, 63, 523–528, 1998.

Gejlhansen, F. and Flink, J.M., Freeze-dried carbohydrate containing oil-in-water emulsions—Microstructure and fat distribution, *J. Food Sci.*, 42, 1049–1055, 1977.

Gordon, M. and Taylor, J.S., Ideal copolymers and the second-order transitions of synthetic rubbers. I. Non-crystalline copolymers, *J. Appl. Chem.*, 2, 493–500, 1952.

Gupta, A.S., Spray drying of orange juice, U.S. Patent 4,112,130, 1978.

Hayashi, H., Drying technologies of foods—Their history and future, *Dry. Technol.*, 7, 315–369, 1989.

Hansen, O.R. and Rasmussen, S., Method of evaporation and spray drying of a sucrose solution, U.S. Patent 4,099,982 (to A/s Niro Atomizer, Soborg, DK), 1977, pp. 1–8.

Hemminga, M.A., Roozen, M.J.G.W., and Walstra, P., Molecular motions and the glassy state, in *The Glassy State in Foods*, Blanshard, J.M.V. and Lillford, P.J., Eds., Nottingham University Press, Nottingham, U.K., 1993, pp. 157–171.

Hennigs, C., Kockel, T.K., and Langrish, T.A.G., New measurements of the sticky behavior of skim milk powder, *Dry. Technol.*, 19, 471–484, 2001.

Her, L.M. and Neil, S.L., Measurement of glass transition temperatures of freeze-concentrated solutes by differential scanning calorimetry, *Pharm. Res.*, 11, 54–59, 1994.

Jouppila, K. and Roos, Y.H., Glass transitions and crystallization in milk powders, *J. Dairy Sci.*, 77, 2907–2915, 1994.

Kamman, J.F. and Labuza, T.P., A comparison of the effect of oil versus plasticized vegetable shortening on rates of glucose utilization in nonenzymatic browning, *J. Food Proc. Preserv.*, 9, 217–222, 1985.

Karatas, S. and Esin, A., A laboratory scrapped surface drying chamber for spray drying of tomato paste, *Lebensm.-Wiss.-Technol.*, 23, 354–357, 1990.

Karathanos, V., Angelea, S., and Karel, M., Collapse of structure during drying of celery, *Dry. Technol.*, 11, 1005–1023, 1993.

Karel, M., Buera, M.P., and Roos, Y.H., Effects of glass transition on processing and storage, in *The Glassy State in Foods*, Blanshard, J.M.V. and Lillford, P.J., Eds., Nottingham University Press, Nottingham, U.K., 1993, pp. 13–34.

Karel, M. and Flink, J.M., Some recent developments in food dehydration research, in *Advances in Drying*, Vol. 2, Mujumdar, A.S., Ed., Hemisphere Publishing, New York, 1983, pp. 103–154.

Karmas, R., Buera, M.P., and Karel, M., Effect of glass-transition on rates of nonenzymatic browning in food systems, *J. Agric. Food Chem.*, 40, 873–879, 1992.

Kieviet, F., Raaij, J.V., De Moor, P.P.E.A., and Kerkhof, P.J.A.M., Measurement and modelling of the air flow pattern in a pilot-plant spray dryer, *Trans. IChemE*, 75, 321–328, 1997.

Kieviet, F., Raaij, J.V., and Kerkhof, P.J.A.M., A device for measuring temperature and humidity in a spray dryer chamber, *Trans. IChemE.*, 75, 329–333, 1997.

King, C.J., Spray drying of food liquids, and volatiles retention, in *Preconcentration and Drying of Food Materials*, Bruin, S., Ed., Elsevier Science, Eindhoven, 1988, pp. 147–162.

King, C.J., Kieckbusch, T.G., and Greenwald C.G., Food quality factors in spray drying, in *Advances in Spray Drying*, Vol. 3, Mujumdar, A.S., Ed., Hemisphere Publishing, New York, 1984, pp. 71–120.

Labrousse, S., Karel, M., and Roos, Y.H., Collapse and crystallization in amorphous matrices with encapsulated compounds, *Sci. Aliment.*, 12, 757–769, 1992.

Labuza, T.P. and Katz E.E., Structure evaluation of 4 dry crisp snack foods by scanning electron-microscopy, *J. Food Proc. Preserv.*, 5, 119–127, 1981.

Langrish, T.A.G. and Fletcher, D.F., Spray drying of food ingredients and applications of CFD in spray drying, *Chem. Eng. Process.*, 40, 345–354, 2001.

Lazar, M.E., Brown, A.H., Smith, G.S., Wong, F.F., and Lindquist, F.E., Experimental production of tomato powder by spray drying, *Food Technol.*, 3, 129–134, 1956.

Levi, G. and Karel, M., Volumetric shrinkage (collapse) in freeze-dried carbohydrates above their glass transition temperature, *Food Res. Int.*, 28, 145–151, 1995.

Levine, H. and Slade, L., Principles of "cryostabilization" technology from structure/property relationship of carbohydrate/water systems—A review, *Cryo-Letters*, 9, 21–63, 1988.

Lloyd, R.J., Chen, X.D., and Hargreaves, J.B., Glass transition and caking of spray-dried lactose, *Int. J. Food Sci. Technol.*, 31, 305–311, 1996.

MacKenzie, A.P., Collapse during freeze-drying: Qualitative and quantitative aspects, in *Freeze-Drying and Advanced Food Technology*, Goldblith, S.A., Rey, L., and Rothmayr, W.W., Eds., Academic Press, New York, 1975, pp. 277–307.

Masters, K., *Spray Drying Handbook*, Longman Scientific & Technical, New York, 1994a.

Masters, K., Scale-up of spray dryers, *Dry. Technol.*, 12, 235–257, 1994b.

Mizrahi, S., Berk, Z., and Cogan, U., Isolated soybean protein as a banana spray-drying aid, *Cereal Sci. Today*, 12, 322–325, 1967.

Nikolaidis, A. and Labuza, T.P., Glass transition state diagram of a baked cracker and its relationship to gluten, *J. Food Sci.*, 61, 803–806, 1996.

Oakley, D.E. and Bahu, R.E., Spray/gas mixing behaviour within spray dryers, in *Drying '91*, Mujumdar, A.S. & Filkova, I., Eds., Elsevier Science, Amsterdam, 1991, pp. 303–313.

Ozmen, L. and Langrish, T.A.G., Comparison of glass transition temperature and sticky point temperature for skim milk powder, *Dry. Technol.*, 20, 1177–1192, 2002.

Palzer, S., The effect of glass transition on the desired and undesired agglomeration of amorphous food powders, *Chem. Eng. Sci.*, 60, 3959–3968, 2005.

Papadakis, S.E. and Bahu, R.E., Sticky issue of drying, *Dry. Technol.*, 10, 817–837, 1992.

Ranz, W.E. and Marshall, W.R. Jr., Evaporation from drops, *Am. Inst. Chem. Eng. J.*, 48, 141–146 and 173–180, 1952.

Ratti, C., Hot air and freeze-drying of high-value foods: A review, *J. Food Eng.*, 49, 311–319, 2001.

Rheinlander, P., Drying of hydrolyzed whey, *Nordeuro. Mej.-Tidsskrift.*, 48, 121–126, 1982.

Riva, M., Campolongo, S., Leva, A.A., Maestrelli, A., and Torreggiani, D., Structure–property relationships in osmo-air dehydrated apricot cubes, *Food Res. Int.*, 38, 533–542, 2005.

Robe, K., Malvick, A., and Heid, J.L., First US installation is applicable to spray drying other heat-sensitive, hygroscopic food products, *Food Process. Market.*, 29, 48, 1968.

Roos, Y. and Karel, M., Plasticizing effect of water on thermal behavior and crystallization of amorphous food materials, *J. Food Sci.*, 56, 38–43, 1991a.

Roos, Y. and Karel, M., Phase transitions of amorphous sucrose and frozen sucrose solutions, *J. Food Sci.*, 56, 266–267, 1991b.

Roos, Y.H., Effect of moisture on the thermal behaviour of strawberries study using differential scanning colorimetry, *J. Food Sci.*, 52, 146–149, 1987.

Roos, Y.H., Melting and glass transitions of low molecular weight carbohydrates, *Carbohydr. Res.*, 238, 39–48, 1993.

Roos, Y.H., *Phase Transitions in Foods*, Academic Press, San Diego, CA, 1995.

Schebor, C., Buera, M.P., Karel M., and Chirife, J., Color formation due to non-enzymatic browning in amorphous, glassy, anhydrous, model systems, *Food Chem.*, 65, 427–432, 1999.

Senoussi, A., Dumoulin, E., and Berk, Z., Retention of diacetyl in milk during spray drying and storage, *J. Food Sci.*, 60, 894–905, 1995.

Shamblin, S.L., Taylor, L.S., and Zografi, G., Mixing behavior of colyophilized binary systems, *J. Pharm. Sci.*, 87, 694–701, 1998.

Shimada, Y., Roos, Y.H., and Karel, M., Oxidation of methyl linoleate encapsulated in amorphous lactose-based food model, *J. Agric. Food Chem.*, 39, 637–641, 1991.

Simha, R. and Boyer, R.F., On a general relation involving the glass temperature and coefficients of expansion of polymers, *J. Chem. Phys.*, 37, 1003–1007, 1962.

Sopade, P.A., Halley, P., Bhandari, B., D'Arcy, B., Doebler, C., and Caffin, N., Application of the Williams–Landel–Ferry model to the viscosity–temperature relationship of Australian honeys, *J. Food Eng.*, 56, 67–75, 2003.

Straatsma, J., Houwelingen, G.v., Steenbergen, A.E., and De Jong, P., Spray drying of food products: 1. Simulation model, *J. Food Eng.*, 42, 67–72, 1999a.

Straatsma, J., Houwelingen, G.v., Steenbergen, A.E., and De Jong, P., Spray drying of food products: 2. Prediction of insolubility index, *J. Food Eng.*, 42, 73–77, 1999b.

Tenbrinke, G., Karasz, F.E., and Ellis, T.S., Depression of glass transition temperatures of polymer networks by diluents, *Macromolecules*, 16, 244–249, 1983.

To, E.C. and Flink, J.M., Collapse, a structural transition in freeze dried carbohydrates. I. Evaluation of analytical methods, *J. Food Technol.*, 13, 551–565, 1978a.

To, E.C. and Flink, J.M., Collapse, a structural transition in freeze dried carbohydrates. II. Effect of solute composition, *J. Food Technol.*, 13, 567–581, 1978b.

Truong, V., Bhandari, B.R, Howes, T., and Adhikari, B., Analytical model for the prediction of glass transition temperature of food systems, in *Amorphous Foods and Pharmaceutical Systems*, Levine, H., Ed., RSC Publishing, London, 2002, pp. 31–47.

Tsourouflis, S., Flink, J.M., and Karel, M., Loss of structure in freeze-dried carbohydrate solutions: Effect of temperature, moisture content and composition, *J. Sci. Food Agric.*, 27, 509–519, 1976.

Wallack, D.A., and King, C.J., Sticking and agglomeration of hygroscopic, amorphous carbohydrate and food powders, *Biotechnol. Prog.*, 4, 31–35, 1988.

White, G.W. and Cakebread, S.H., The glassy state in certain sugar-containing food products, *J. Food Technol.*, 1, 73–82, 1966.

Whorton, C. and Reineccius, G.A., Evaluation of the mechanisms associated with the release of the encapsulated flavor materials from maltodextrin matrices, in *Encapsulation and Controlled Release of Food Ingredients, ACS Symposium Series 590*, Risch, S.J. and Reineccius, G.A., Eds., American Chemical Society, Washington, DC, 1995, pp. 142–159.

Woodruff, E., Woodbine, M., Andersen, V., and Hackettstown, N., Sugar drying method, U.S. Patent 3,706,599 (to W.R. Grace & Co.), 1971, pp. 1–8.

3 Application of Image Analysis in Food Drying

Maturada Jinorose, Sakamon Devahastin,
Silvia Blacher, and Angélique Léonard

CONTENTS

3.1 INTRODUCTION

An image is a two-dimensional (2-D) or three-dimensional (3-D) reproduction of a form of an object that can be captured by such optical devices as cameras and microscopes. Although the human visual system is very fast in creating and capturing images and very good in making qualitative judgments about image features, it is not suitable for making a quantitative measurement. In order to quantify image data, a biological imaging system such as human eyes requires assistance from image analysis tools.

Image analysis is a means of obtaining meaningful information from an image, mainly from a digital image (Plataniotis and Venetsanopoulos, 2000). The task can be as simple as measuring the color (Louka et al., 2004; Mendoza et al., 2006) of a product to something more sophisticated such as analysis of food microstructure (Acevedo et al., 2008; Devaux et al., 2006; Kerdpiboon et al., 2007; Ramos et al., 2004; Varela et al., 2008). Although basic tools such as a ruler, perimeter, and spot densitometer can be used to analyze image data, the use of basic tools is a tedious and time-consuming task. Therefore, computers (digital image processing) are used to decrease the data analysis time.

Computer image analysis is indispensable for the analysis of large amounts of data, especially for tasks that require complex computation or extraction of quantitative or qualitative information from 2-D images as in computer vision or 3-D images as in medical imaging applications (Jennane et al., 2007; Louka et al., 2004). The use of computers to perform digital image processing allows a much wider range of algorithms to be applied to analyze the input data and can avoid problems such as the build-up of noise and signal distortion during processing (Imaging Research Inc., 2007). It should be noted, however, that although computers are useful when the task is too impractical to perform by hand, many important image analysis tools such as edge detectors are inspired by human visual perception models, so human analysts still cannot be completely replaced by computers.

A variety of image analysis systems are currently available from dozens of manufacturers; some of them are designed for specific applications and no programming is required. The reduction in hardware costs allows the system to be built for more general use in food research and applications.

There are currently a number of techniques and algorithms, which involve advanced mathematics such as Fourier and wavelet transforms (Antonelli et al., 2004; Fernández et al., 2005), probability and statistics (Dan et al., 2007; Ros et al., 1995), or discrete topology and geometry (Chevalier et al., 2000) that can be applied to food research applications. However, this chapter will focus only on selected techniques that have proved useful in food drying applications. These techniques range from the use of a simple computer vision system (CVS) to scanning electron microscopy (SEM) to more advanced techniques such as magnetic resonance imaging (MRI), combined image and fractal analysis, and x-ray microtomography. The coverage here is certainly not inclusive, and the choices of examples discussed are sometimes arbitrary. However, these examples should nevertheless serve to illustrate the possible use of image analysis in food drying research and quality control. Indeed, the use of image analysis in food drying is still at an early stage and more efforts are certainly needed to utilize the advantages of the technique to a higher level.

3.2 APPLICATION OF IMAGE ANALYSIS IN FOOD DRYING

3.2.1 Color Evaluation

Because the visual (external) appearance of food is one of the major factors in determining product quality and consumer acceptance, as these are often the only direct information consumers receive when buying a product, the ability to assess the external appearance (e.g., size, shape, and color) of a product qualitatively and quantitatively is of great importance (Fernández et al., 2005; Locht, 2007; Yan et al., 2008). Visual inspection has to be performed with consumer acceptance that is often based on external features in mind. Until recently, this inspection has been performed mostly by human eyes. Thus, it has been qualitative rather than quantitative because it relies more on feeling (subjective) than on scientific (objective) perspective.

Among the various properties, color is an important quality parameter that plays a major role in the assessment of external (or apparent) quality of food because it correlates well with other physical, chemical, and sensorial indicators of product quality (Fernández et al., 2005; Louka et al., 2004; Mendoza et al., 2006). During drying the color of food can change significantly because of such reactions as enzymatic browning, as in the case of dehydrated fruits, Maillard reactions, or other oxidation reactions (Acevedo et al., 2008; Fernández et al., 2005). Browning discoloration of a product lowers its quality value and is often accompanied by textural changes, off-flavors, decreased solubility, and loss of nutritional value.

Perception of color is a result of an interaction among a light source (daylight, tungsten, etc.), the object (colorants, surface texture, etc.), and the observer (human eye sensitivity, experience, etc.; HunterLab, 2007). Thus, in the assessment of dried (or drying) food color the difficulty lies in measuring the color of a heterogeneous product because color changes of food during drying are not homogeneous (Fernández et al., 2005; Louka et al., 2004). The first and simplest method of color measurement is the sensory perception by a group of trained people, which is imprecise and subjective. In addition, this method cannot describe color in the form of color space, which is a system for describing color numerically.

To build an instrument that can quantify color perception, the three elements of a visual observing situation must be quantified. The source must be quantified by a selection of illuminant, the object quantified by measuring the reflectance or transmission curve, and the observer quantified by CIE Standard Observer functions (HunterLab, 2007). The conventional color specification instruments (such as a colorimeter), which usually provide readings in CIEXYZ, RGB, or LAB color space, use filters in combination with a light source and detector to spectrally emulate the standard observing functions of the eyes and can give direct evaluations of X, Y, and Z. However, because the normal viewing area of these instruments is only 2 to 5 cm^2, these instruments are inappropriate for discriminating the overall color changes of a whole piece of a heterogeneous product. Therefore, many locations must be measured to obtain a representative color profile (Fernández et al., 2005; HunterLab, 2007; Mendoza et al., 2006).

The use of the CVS computerized image analysis technique for quality inspection of food is now an established technique because of its simplicity and nondestructive and time-saving nature. The system offers an alternative for measuring uneven coloration and it can be applied to the measurement of other attributes of food such as size and shape, especially when dealing with irregular surfaces. It can also be easily incorporated into any on-line processing tasks. Calibrated average color measurements and other appearance features using CVS have been demonstrated to closely correlate with those from visual assessment. Changes in average color and color patterns of whole foods during dehydration have been effectively determined by this technique (Acevedo et al., 2008; Fernández et al., 2005).

An image acquisition system usually consists of four basic components: illuminant (lighting system), camera, hardware, and software. Vision systems require the use of a proper light source in order to avoid glitter and to obtain sharp contrasts at the border of the sample image. Usually, most systems use one of the CIE standard illuminants representing white light: illuminant *A* represents incandescent lamp, *C* represents average daylight, and *F2* represents cool white fluorescent. However, the most commonly used illuminant is D_{65}, which represents noon daylight (HunterLab, 2007). It should be noted, however, that *F7*, a typical fluorescent illuminant, which is sometimes called D_{65} daylight, does not give the visual result that matches CIE D_{65} because the spectral power distributions are not the same as those of CIE illuminant D_{65}. Proper calibration should be performed if matching of measured color to a standard one is desired. Choosing an incorrect light may cause metamerism to happen; a common case is that two color samples appear to match when viewed under a particular light but do not match when illuminated by different lights.

The setup of a camera and illuminant (viewing environment) is critical for image acquisition and for obtaining meaningful and reproducible data. The angle between the axis of a camera as well as between the axis of a light source and the axis of a sample being observed is usually 45° in order to capture the reflected light, which occurs at that angle (Figure 3.1). Circumferential illumination via the use of a

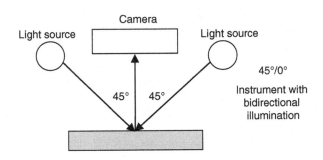

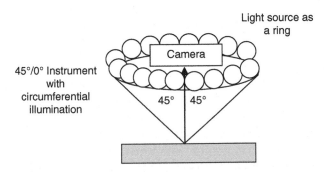

FIGURE 3.1 Setup of a camera and illuminant with 45° reflection.

number of lights set around the sample as a ring at 45° is also recommended because it helps reduce the effect of the sample surface by providing more even lighting across the surface than using only mono- or bidirectional illumination (see also Figure 3.1). The surface texture of a sample also has an important role in color measurement. Although the sample may have the same color, a shiny surface looks darker and more saturated than a matte one. Thus, for dark glossy products, diffuse or 8° geometry or "sphere" geometry is more suitable because a sphere coating (a substance with an excellent capacity to diffuse light such as titanium dioxide) helps diffuse the light; the sample is then illuminated diffusely from all directions.

Digital and video cameras, ranging from a consumer digital camera with one charge-coupled device (CCD) to a high-end CCD camera with three CCDs (one each for red, green, and blue light) are generally powerful enough for color studies because the image analysis systems for color measurements have been developed in response to low spatial resolution. The majority of color digital still cameras being used for technical applications employ a single array of light-sensitive elements on a CCD chip, with a filter array that allows some elements to see red, some green, and some blue; the relative intensities could be manually or automatically adjusted by a function called "white balance." Such cameras devote one half of the elements to green sensitivity and one quarter each to red and blue, emulating human vision, which is most sensitive to the green portion of the spectrum. A digital color image is represented in RGB form with three components per pixel, conventionally stored using 8 bits per color component, and electronically combined to produce a digital color picture (Russ, 2005). It is obvious that signals generated by a CCD are device dependent; each camera has its own color sensor characteristics and produces different responses for the same image when it is displayed through hardware such as a graphics card and monitor. This may not be too serious for gray-scale imaging, but it is for color imaging, especially when colorimetric precision is important.

Another important issue is gamma correction. Gamma is the standard for device encoding to match the eye nonlinearity. It helps minimize transmission noise in the dark areas. The ITU-R BT.709 standard, in particular, defines the transformation of CIEXYZ tristimulus values into a target RGB monitor space (ICC, 2007). The resulting image is obviously not an exact match of the original scene but instead is a preferred reproduction of the original scene, which is consistent with the limitations of a monitor. Different computer platforms like Macintosh and PC use different gamma values.

To obtain a device-independent color of a food sample using a digital or video camera, it is necessary to correlate the camera RGB signals and CIEXYZ tristimulus values using a power function with a suitable gamma value, which is applied to each red, green, and blue signal. This process is known as camera characterization. The most common technique for camera characterization consists of presenting the camera with a series of color patches in a standardized reference chart with known CIEXYZ values and recording the average RGB signals for each patch. Polynomial fitting techniques can then be applied to interpolate the data over the full range and to generate inverse transformations (Mendoza et al., 2006).

The choice of background color also plays an important role in image processing. White background is easily removed when performing image analysis via the use of a simple thresholding technique. However, if the background is too bright, the color of the sample may be affected because of the chroma of the detectors in a camera (Louka et al., 2004). A black background is thus suitable for general applications as it helps reduce the shadow effect when analyzing an image. However, if a sample is a transmit object, black backing will reduce its color gamut. When the product color is very bright, it is more suitable to use a white backing for image analysis. A suitable backing medium should not be fluorescent; high brightness white ceramic tiles or other materials whose surfaces are coated with similar colorants are recommended. Mendoza et al. (2006) measured the color of bananas and found that the lightness (L^* value) of bananas was significantly altered when the color of the background changed from black to white.

In addition, the image should be taken in a closed environment (e.g., inside a light box) to avoid external light and reflections (a black body absorbs light at all frequencies, hence no reflections). The camera should be set to manual mode with no zoom and no flash. The image should then be stored in a noncompressed or lossless compression format such as TIFF, BMP, or RAW, but not in a lossless JPG, which is not a lossless compression. Generally, a digital camera saves an image in a JPG format by default. Switching the saving format to TIFF or RAW improves the quality of an image. Nevertheless, the differences in the analysis results may be difficult to observe if the image is analyzed manually with a photo editing program. Note that loading the image into a graphics program and resaving it in JPG format will degrade the image even without editing it. If a video camera is used, the still image should be acquired with video editing or frame grabbing software.

Image analysis can be performed manually by common photo editing software such as Adobe Photoshop™ (Adobe System, Inc., San Jose, CA) or free graphic editor software such as the GNU Image Manipulation Program (GIMP; The GIMP Team, 2007), which is one of the most widely used bitmap editors in the printing and graphics industries. The easiest way to measure color is to use a wand selection tool to select the area to measure the average color. If using only the wand selection to separate the object of interest from the background is difficult, converting the color image to a gray-scale image by selecting the binary process through thresholding should be performed before using the wand selection. The RGB values can be converted to any color system such as the CIELAB color model, which is widely used in food research and quality control, using mathematical formulas described in any computer image analysis textbook (e.g., Plataniotis and Venetsanopoulos, 2000). Free Java-based image analysis software such as ImageJ (National Institutes of Health, 2007) or free image processing and analysis software such as UTHSCSA ImageTool (University of Texas Health Science Center, 2007) may be helpful for more accurate results.

MATLAB® (MathWorks, Inc., Natick, MA) software is widely used to analyze images in food research. Segmentation of a sample image from the background by the algorithms within MATLAB usually involves the following steps:

1. Convert a color image into a gray-scale image using the function *rgb2gray*.
2. Apply a threshold and perform background subtraction using an edge detection technique based on such operators as Laplacian-of-Gauss or Sobel

to obtain a binary image, where "0" (black) means background and "1" (white) means object.

3. If necessary, close small noisy holes within the object of interest using such function as *medfilt2*.
4. Remove all objects surrounding the sample.
5. Overlap the contour of the binary image on the original color image and extract the sample from the background (Mendoza et al., 2006).

However, thresholding a color image is not an easy task. There is no simple generalization for the thresholding technique; in other words, an appropriate threshold is dependent on the image itself. For more information, refer to the Help section of MATLAB.

Selection of an appropriate color space to measure the color of food is important because this may cause the result to vary. Louka et al. (2004) compared the color results of dried fish (cod filet) obtained from image analysis to those obtained from sensory analysis performed by a trained jury. Four different color systems (RGB, CIELAB, HLC, and Aclc2) were used to define the color of the fish. Although the RGB system was unsuitable in this case, the instrumental measurement of color was accurate, reliable, and reproducible. The relative errors of all measured parameters remained low; the highest error belonged to chroma values in CIELAB, with its variance close to 6%. Although chroma was less precise than the entire CIELAB space, its variance was able to statistically distinguish the results from different drying methods: hot-air drying (HAD), vacuum drying, freeze-drying, a novel DDS (a process of dehydration by successive pressure drop), and DIC (a process of dehydration by controlled instantaneous pressure drop). For continuous production control, however, the use of CIELAB color space is recommended because chroma variance cannot be used to study the color of a sample based on a single image and requires that repeated measurements be made. Acevedo et al. (2008) also confirmed that CIELAB is more suitable for measuring food color.

Mendoza et al. (2006) studied the effect of the curved surface of a yellow sheet on the measured sheet color. They examined the color profile variation of the yellow sheet when it was flat and rolled. They noted that the curvature of the surface affected the efficiency of the system. RGB was most affected but the lightness (L^* and V) of CIELAB and the hue–saturation–value (HSV) model also seemed to be affected by the curvature of the sample as well. This is because color is based on the surface reflectance of an object and the intensity of the light source shining on it. Later, when analyzing the color of samples with curved surfaces (yellow and green bananas), it was revealed that CIELAB was more appropriate for color representation of glossy surfaces. This is because the color profiles were less affected by the degree of curvature, shadows, and glossiness of the surfaces when comparing with the HSV color system, as shown in Figure 3.2.

A CVS could be used to measure food color instead of a colorimeter as demonstrated by Mendoza et al. (2006). Figure 3.3 shows a relationship between the color scale values of the 125 Pantone® color sheets obtained from a CVS and those obtained from a colorimeter. The plot of CIEXYZ values obtained from the two measuring methods exhibited a linear relationship with a reasonable fit ($R^2 > 0.97$).

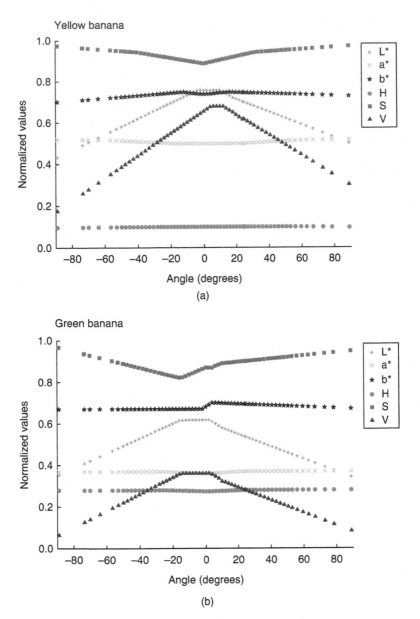

FIGURE 3.2 Color profiles expressed in CIELAB and HSV color scales for (a) yellow and (b) green bananas. (From Mendoza, F. et al., *Postharvest Biol. Technol.*, 41, 285–295, 2006. With permission.)

Fernández et al. (2005) also mentioned that the color changes of apple tissue browning during drying as measured by image analysis and by a hand-held colorimeter gave high average correlation coefficients in terms of the CIELAB color model.

Martynenko (2006) applied CVS via the use of a portable CCD camera with an IEEE-1396 interface and LabView™ 7.0 and IMAQ™ 6.1 software (National

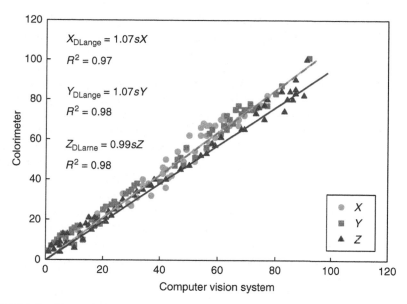

FIGURE 3.3 Plots of color scale values acquired by CVS against values measured by a colorimeter on 125 Pantone color sheets. (From Mendoza, F. et al., *Postharvest Biol. Technol.*, 41, 285–295, 2006. With permission.)

Instruments Corp., Austin, TX) to monitor the drying process for ginseng. A ginseng sample was continuously monitored by extracting the green plane from the RGB color space; this was followed by thresholding and pixel counting. The sample color was monitored continuously as color intensity in the hue–saturation–intensity (HSI) color space. The correlations between various image attributes (i.e., area, color, and texture) and the physical parameters of drying (i.e., moisture content) were also determined through the use of CVS. Color measurements demonstrated a high correlation between the product quality and the drying conditions. A CVS therefore has the potential to be a means to control a drying process on a real-time basis.

Image analysis can also be used to inspect browning reactions to attain results that agree with other techniques including the ones involving chemical analysis. Acevedo et al. (2008) used image analysis to study the effects of pretreatments and processes (blanching, freezing, and drying), which affected the sample (apple disks) structure at macroscopic and microscopic levels. The results showed that image analysis provided reproducible values for browning changes without invasion of the sample. The results were also in good agreement with those of White and Bell (1999) who analyzed brown pigment formation via a Maillard reaction in low moisture model food systems.

3.2.2 Size and Shape Evaluation

Shrinkage is one of the major physical changes during drying because it modifies the shape and dimension (size) of food. It is usually expressed in terms of the ratio between the volume of the sample before and after drying; the volume can be measured either by the Archimedes principle or by any displacement techniques. Shrinkage can also be expressed as a function of the change of selected dimensions of a sample.

The shrinkage, shape, and size of a food sample have been quantified mostly by direct measurement with such instruments as a vernier caliper, digital caliper, micrometer, or planimeter. However, the use of computer vision and image analysis to quantify shrinkage and deformation of food products has been increasing during the past decade because these techniques provide an alternative to manual inspection, have several advantages, and can be integrated directly with other on-line processing tasks.

The basic idea in the evaluation of shrinkage, shape, and size is the use of shape descriptors to represent the shape and size of a sample. One advantage of using image analysis to evaluate shrinkage is the ease of comparing the sample before and after drying. Software such as Image-Pro™ (Media Cybernetics, Inc., Bethesda, MD) provides ready to use geometrical parameters to identify objects, thus simplifying the task significantly.

As an illustrative example, Fernández et al. (2005) developed a method based on image analysis to determine the effect of drying on various physical properties of apple disks. A pattern recognition program was developed in MATLAB using Fourier descriptors, which represent the shape of the samples in terms of the 1-D or 2-D signal, as a shape representation method. This program analyzed all of the sample images, matching them with the prototype image and deciding to which class (degree of dryness) the images belonged. Although it was difficult to identify classes after an extended period of drying because the external images of the samples remained almost unchanged, 95% of the samples were correctly classified. The advantage of this method is that it represents the quality factors perceived by the consumers better than using other process parameters (i.e., rate of drying and moisture content), which have limited significance from the standpoint of product quality.

3.2.3 EVALUATION OF INTERNAL CHARACTERISTICS

During drying a food sample not only undergoes external changes but also obviously suffers internal changes. It is indeed these internal changes that affect or lead to many apparent changes of the sample (e.g., shrinkage, deformation, and textural changes). The ability to assess various internal changes of a product undergoing drying is thus of interest. There are a number of methods that can be used to characterize the internal changes of food, ranging from the use of CVS to various microscopic methods such as MRI as well as other techniques, for example, combined image and fractal analysis or x-ray microtomography, which will be discussed in later sections of the chapter. Note that the following brief review covers only selected illustrative examples of the use of image analysis in characterizing various internal changes of food. No attempt is made to discuss the fundamentals or principles behind each technique. The reader is referred to some excellent resources on image analysis (e.g., Russ, 2005) and microstructural principles (e.g., Aguilera and Stanley, 1999) for more details.

Techniques such as light microscopy (LM), SEM, or confocal laser scanning microscopy (CLSM) are widely used to study the microstructure of a food sample. Images obtained from a microscope are generally converted into a number of matrixes (RGB) with the help of computer software. The matrixes can then be modified to obtain better images through such processes as contrasting, darkening, and thresholding before an analysis be made.

Alamilla-Beltrán et al. (2005) described morphological changes, in terms of crust (skin) formation as well as deformation, of maltodextrin particles during spray drying using LM and after spray drying using SEM. The fractal dimension (FD) of the projected perimeter of the particles (as observed by a scanning electron microscope) was also calculated. During spray drying the samples were sampled out via the use of a specially designed sampler at various vertical locations along the height of the dryer. The calculated FD supported findings related to morphology–temperature relations, which are important to predict particle properties such as density, size distribution, moisture content, and strength.

Varela et al. (2008) applied image analysis to quantify fracture pattern and microstructural changes of almonds (*Prunus amygdalus* L.) during roasting. Image analysis was used to characterize the fracture pattern of almonds roasted under different conditions through the measurement of the particle size distribution after compression (to the point of sample fracture), and LM and SEM were used to observe microstructural changes of the samples during roasting. The microstructure of the samples was then related to the sample compression and fracture behavior. A typical digital camera was used to capture images of fractured samples and Image Pro-Plus™ 4.5 imaging software was used to analyze the obtained images. The software was also used to analyze the LM images of the tissue sections obtained from the center of the samples. The proposed image analysis methods successfully quantified the features at both macroscopic and microscopic levels related to the fracture of roasted almonds.

Ghosh et al. (2007) applied MRI to study moisture profiles in single rewetted wheat kernels during drying, which is a rather difficult task using more traditional determination techniques. A Magnex (Magnex Scientific Ltd., Yarnton, U.K.) 11.7-T (500-MHz) superconducting vertical-bore magnet was used, which was equipped with a Magnex SGRAD 123/72/S72 mm capable of producing a maximum gradient strength of 550 mT/m. An Avance DRX console was interfaced to the magnet to record the magnetic resonance images via ParaVision™ software (Bruker BioSpin, Billerica, MA). The MRI probe and coil (Helmholtz configuration: 7-mm inner diameter) were built by the authors. A schematic diagram of the experimental setup is provided in Figure 3.4.

The obtained 2-D images were processed as 8-bit images in a $128 \times 4 \times 8$ pixel matrix with a field of view of $1.28 \times 1.28 \times 0.4$ cm, resulting in a pixel resolution of $100 \times 200 \times 500$ μm. One-dimensional spectral data were obtained with a spectral width of 50 kHz. Image analysis (image acquiring and visualization, normalization, pixel selection, and thresholding) was performed with ImageJ and MATLAB. Nitrogen was used as the drying medium instead of hot air because nitrogen is non-corrosive to the instruments. A calibration curve of the magnetic resonance image intensity versus the moisture content of the grains was obtained using nuclear

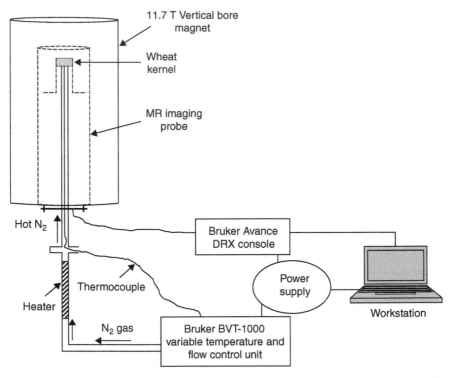

FIGURE 3.4 A schematic diagram of MRI experimental setup with the dryer assembly. (From Ghosh, P.K. et al., *Biosyst. Eng.*, 97, 189–199, 2007. With permission.)

magnetic resonance (NMR) spectra of wheat at different known moisture contents. The calibration curve was needed because the magnetic resonance signal intensity was not always directly proportional to the actual moisture content at different locations in an image of a wheat kernel.

The magnetic resonance images showed the variation in the internal distribution of moisture both before and during drying. Moisture was concentrated mainly in the embryo region and the embryo signal intensity remained high, even after some period of drying at all test temperatures (30, 40, and 50°C). The results also showed that moisture reduction was dependent on the moisture content of the kernels as well as their compositions and regions (e.g., embryo, endosperm, and pericarp). The real-time conditions obtained from these measurements form a basis for the development of a more accurate grain drying models describing transport phenomena in individual kernels. Indeed, recent MRI results showed that the simple moisture receding front assumption may not be realistic because no sharp front was observed in MRI studies (Chen, 2006).

In addition to MRI and standard SEM or CLSM, microstructural analysis can be performed using other optical techniques such as environmental SEM (ESEM). The use of ESEM also allows the microstructure analysis to be done in situ and on a real-time basis (Chen, 2006).

3.3 APPLICATION OF COMBINED IMAGE AND FRACTAL ANALYSIS IN FOOD DRYING

3.3.1 INTRODUCTION

During drying the physical properties of food change mainly because of the loss of its moisture. Attempts have been made to characterize these physical changes in terms of such parameters as the changes in volume, area, and shape (Khraisheh et al., 1997; Ochoa et al., 2002a,b; Panyawong and Devahastin, 2007; Ratti, 1994). However, as mentioned in the previous section, these external changes are caused by internal changes within a sample, which are directly related to the structure of the drying material. Among many methods that could be used to describe the structural and physical changes of materials, optical instruments such as light microscopes and scanning electron microscopes are common (Aguilera and Stanley, 1999). However, the microstructural images that are obtained cannot be easily quantified without the use of other appropriate evaluation techniques.

Because of the usefulness of image analysis, some researchers have applied this technique in combination with a technique called fractal analysis to quantify the various changes of food products undergoing different thermal and nonthermal processing. Fractal analysis can be used to describe the surface, structure, and morphology as well as the mechanical properties of foods. This combined technique (fractal and image analysis) has been used successfully to characterize ruggedness changes of tapped agglomerated food powders (Barletta and Canovas, 1993), ruggedness of potatoes after deep fat frying (Rubnov and Saguy, 1997), as well as food surfaces and microstructural changes of many raw food materials (Quevedo et al., 2002), among others.

Fractals or fractal geometry refers to the geometry of irregular or fragmented shapes of common natural materials such as shorelines, clouds, and plants (Mandelbrot, 1983). The FD contains three characteristics, which are described as self-similar (a shape that has no characteristic length because the shape does not change with the change of the observation scale; Rahman, 1997), described by recurrent dependencies (not by a mathematical formula), and containing a dimension that is not an integer (Dziuba et al., 1999). Fractals can be used to describe the geometry of complex particles and to quantify the ruggedness of particles.

The FD can be calculated with many techniques, and each technique yields different values from the same image (Rahman, 1997). It consists of dimensionless units, and their value depends on the images being studied. The value is in the range of 1 to 2 if the image has two dimensions, or the value is 2 to 3 if the image has three dimensions. Many studies have applied different techniques to calculate FDs from the same images (Pedreschi and Aguilera, 2000; Quevedo et al., 2002; Rahman, 1997). They found that the same images yielded different FD values when different FD calculation techniques were used. Therefore, FD interpretation without physical understanding could be misleading.

The box counting technique is one of the most popular algorithms used to calculate the FD of signals and images (Turner et al., 1998). This technique applies the size–measure relationships to calculate the FD. Cubic boxes of different sizes (r) are mounted on an image and the number of boxes intercepted (N_r) with the object in the image for each iteration of the r value is then counted. The FD is determined from the slope of the least squares linear regression of the logarithmic plot of N_r and $1/r$:

$$FD = \frac{\log (N_r)}{\log (1/r)} \tag{3.1}$$

The sizes of the boxes used to obtain the FD vary, depending on the application of interest.

3.3.2 APPLICATIONS IN FOOD DRYING

Very few attempts have been made to apply fractal analysis to food drying. Among the early attempts, Chanola et al. (2003) applied fractal analysis to the distribution of the surface temperature of a model food (mixture of glucose syrup and agar gel) in order to monitor the drying kinetics of the sample. In addition, the evolution of the FD of the images of the sample surface was also monitored. Based on the fractal analysis of the surface temperature distributions, different periods of drying could be identified. It was also observed that the FD of the images of the sample surface increased with the drying time because of an increase in surface irregularity as drying proceeded. Furthermore, higher air temperatures and velocities led to higher FD values of the surfaces. However, fewer attempts have been made to use fractal analysis to monitor physical and/or microstructural changes of food products.

Kerdpiboon and Devahastin (2007a,b) and Kerdpiboon et al. (2007) made the first attempts to apply fractal analysis to characterize the microstructure of food materials undergoing drying and to correlate the microstructural changes of foods with their apparent physical changes (e.g., volume shrinkage). In their studies the FD of the microstructural images of model food materials (carrot and potato) was calculated using the box counting method. Prior to being able to calculate the FD, however, several steps were necessary and these are briefly outlined here. For the detailed procedures, refer to Kerdpiboon and Devahastin (2007a,b) and Kerdpiboon et al. (2007).

The process of microstructural imaging consists of fixation, in which the sample is preserved with formaldehyde prior to passing through the remaining steps. The objective of the fixation step is to immobilize cellular components to ensure that the structure and tissue shown in the microstructural images reflects the living state of the sample as closely as possible. After fixation, the sample is soaked with flowing distilled water before removing the remaining moisture within the sample cells by flushing the sample with a series of isopropyl alcohol solutions. Isopropyl alcohol within the sample is subsequently removed by flushing the sample with absolute xylene. Finally, the pores of the sample are replaced with paraffin (Aguilera and Stanley, 1999; Humason, 1979).

The treated sample is then embedded in paraffin wax. Each embedded sample is sectioned with a microtome cutter, and the sliced sample is dried and then fixed on a glass slide. The finished slide is dyed to highlight the cell walls. Finally, microstructural images are obtained using a light microscope. A light microscopic image of the sample is transformed from an RGB format to a black and white format before a calculation of the FD is performed.

The FD of a black and white image is calculated using the box counting method (Quevedo et al., 2002). The boxes are composed of 4, 5, 10, 13, 26, 65, 130, and 260 pixels. The threshold values are between 0.5 and 0.8, and the value of the threshold used for each image is chosen as the lowest value that can yield the clearest image of the cell walls of the sample. Kerdpiboon and Devahastin (2007a,b) and Kerdpiboon et al. (2007) performed FD calculations using MATLAB (version 6.5).

The normalized change of the FD was reported (Kerdpiboon and Devahastin, 2007a,b; Kerdpiboon et al., 2007) to compare the changing values of the FD of the sample undergoing drying:

$$\Delta\overline{FD} = \frac{\Delta FD}{FD_0} \tag{3.2}$$

$$\Delta FD = FD_t - FD_0 \tag{3.3}$$

where FD_0 and FD_t are the FDs of the fresh sample and the sample at any instant (time) during drying, respectively.

Kerdpiboon and Devahastin (2007a,b) utilized two different drying methods, HAD, and low-pressure superheated steam drying (LPSSD), to dry the sample (carrot cube). Drying was performed at 60, 70, and 80°C and in the case of LPSSD at an absolute pressure of 7 kPa. For each drying experiment, the drying process was carried out up to a predetermined sampling time; that particular experiment was terminated at that time. A new experimental run was then performed up to the next predetermined sampling time. These steps were repeated until a complete drying curve was obtained. At the end of each experimental run, the sample was taken out to determine the moisture content, shrinkage (percentage change of the volume of the sample compared with its original volume), and microstructural changes following the steps described earlier.

Images from a light microscope (10× magnification level) were used for FD calculation. The FD of a fresh carrot (Figure 3.5a) was approximately 1.75, but the FD of the sample increased upon drying. Initially, the cell walls of the carrot were round. After a period of time, however, the moisture within the cells started to migrate to the

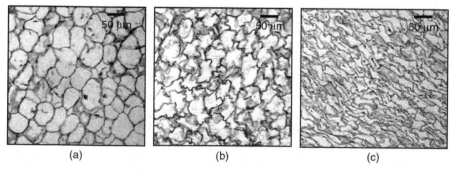

(a)	(b)	(c)

FIGURE 3.5 Microstructure of a carrot cube undergoing HAD at 60°C and 0.5 m/s after (a) 0, (b) 150, and (c) 300 min.

surface. As a result, moisture gradients started to develop, leading to internal stresses and shrinkage of the cells (Figure 3.5b and c). The ruggedness of the cell walls increased, leading to an increase in the FD of the cells, as expected.

The rates of change of the FD of carrot cubes undergoing HAD (Figure 3.6a and b) and LPSSD (Figure 3.6c) were monitored. These rates were divided into roughly two periods, consisting of linear and nonlinear rates of change. The drying time that identified the end point of the period where the FD rates of change were in a linear fashion was also the same as the drying time that divided the drying curves (not shown here) into the first and second falling rate periods, which divided the physical changes (in terms of the percentage of shrinkage) into rapid increase and slow increase periods, respectively. However, the microstructural changes of the cell walls of the carrots undergoing different drying techniques and conditions were quite different (Figure 3.7). Carrots that were dried using HAD suffered more cell structure deformation than with LPSSD.

To compare (and hence to generalize) the results, Kerdpiboon et al. (2007) captured light microscopic images of potatoes at a 10× magnification level. Each image represented both cell walls and starch granules. However, because it was not easy to describe the changes of the starch granules of potato cells through LM, the starch granules were eliminated from every image. Thus, only the cell walls of the potato were displayed within the images. Figure 3.8 represents the steps in treating the images starting from an original image (Figure 3.8a), deleting the starch granules image via the use of MATLAB (version 6.5; Figure 3.8b), and converting the image without starch granules into a black and white format (Figure 3.8c). Some starch granules that could not be deleted via the use of the program could be deleted manually to allow better analysis of the image (Aguilera and Stanley, 1999).

The FD of fresh potato was approximately 1.61 and increased as the drying time increased. For example, the FD of a potato cube undergoing drying at 60°C and an air velocity of 1 m/s was approximately 1.70 and 1.73 at 180 and 420 min, respectively. The FD of the potato cube did not change much compared to that of the carrot cube undergoing HAD under the same conditions; the carrot FD varied between 1.75 and 1.91 (Kerdpiboon and Devahastin, 2007a,b).

The normalized changes of the FD could be used to monitor the physical changes of carrots and potatoes during drying. For example, at an $\overline{\Delta FD}$ of approximately 0.06 the percentage of shrinkage of carrots was around 70 to 80% for all HAD (Figure 3.9) and LPSSD cases (Figure 3.10). This type of relationship was also observed at other $\overline{\Delta FD}$ and percentage of shrinkage values. These results are supported by the results presented in Figure 3.11, which illustrates the microstructure of the samples at an $\overline{\Delta FD}$ of 0.06 when drying with HAD at 60°C, 1 m/s, and 150 min; 70°C, 1 m/s, and 120 min; and 80°C, 1 m/s, and 90 min as well as with LPSSD at 60°C, 7 kPa, and 210 min; 70°C, 7 kPa, and 150 min; and 80°C, 7 kPa, and 90 min. All samples displayed similar levels of shrinkage and had the same $\overline{\Delta FD}$ values. The deviations in the percentage of shrinkage might have arisen from the errors that occurred during the collection of the experimental data.

In comparison, the normalized changes of the FD were also used to monitor the percentage of shrinkage of potatoes undergoing HAD at different conditions

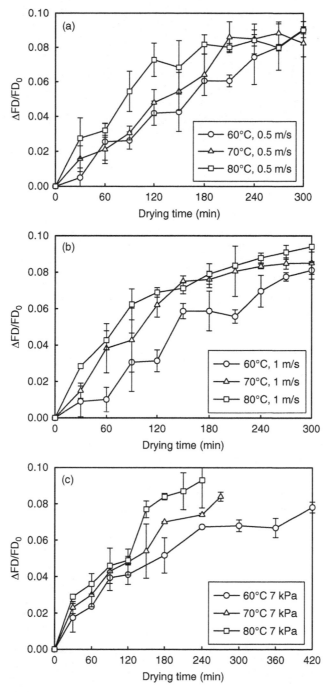

FIGURE 3.6 The $\Delta FD/FD_0$ of a carrot cube undergoing HAD at velocities of (a) 0.5 and (b) 1 m/s and (c) LPSSD at 7 kPa.

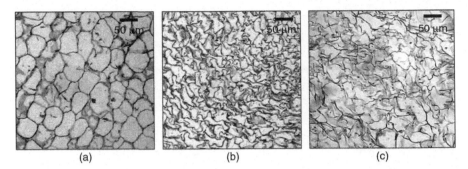

FIGURE 3.7 Microstructure of (a) fresh carrot cube and carrot cube dried until reaching equilibrium moisture content using (b) HAD and (c) LPSSD.

(Kerdpiboon et al., 2007). For instance, an $\overline{\Delta FD}$ value of ~0.06 again refers to around 70 to 80% shrinkage at all conditions of HAD. This therefore proved that the ΔFD could be used to monitor shrinkage (deformation) of both potatoes and carrots undergoing HAD. Thus, it has the potential of being an indicator for describing deformation of other products undergoing drying as well. Further investigation is needed to confirm this hypothesis.

3.4 APPLICATION OF X-RAY MICROTOMOGRAPHY IN FOOD DRYING

3.4.1 INTRODUCTION

X-ray tomography (from the Greek *tomê*, section and *graphein*, to describe) is a nondestructive imaging technique that allows access to the internal structure of the investigated object. The technique is based upon the local variation of the x-ray attenuation coefficient of the object located along the x-ray path. Initially, x-ray computed tomography was developed in the field of medicine to allow nondestructive examination of slices of the human body (Cormack, 1963). Since the first system,

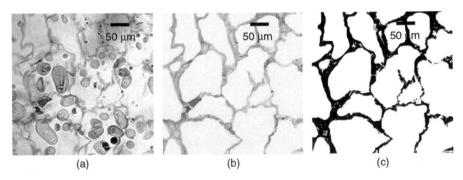

FIGURE 3.8 Image processing of a potato cube: (a) original image, (b) image after deletion of starch granules, and (c) black and white image.

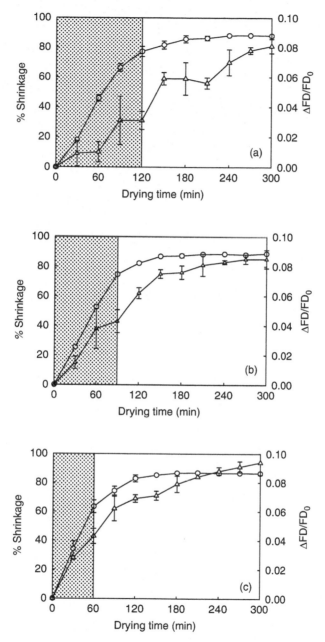

FIGURE 3.9   Relationship between the (○) percentage of shrinkage and (△) ΔFD/FD$_0$ of a carrot cube undergoing HAD at a velocity of 1 m/s and temperatures of (a) 60, (b) 70, and (c) 80°C.

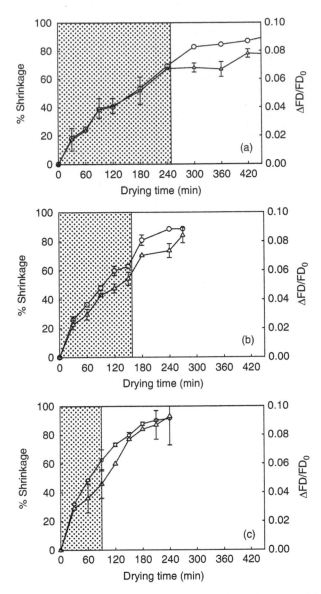

FIGURE 3.10 Relationship between the (○) percentage of shrinkage and (△) ΔFD/FD$_0$ of a carrot cube undergoing LPSSD at temperatures of (a) 60, (b) 70, and (c) 80°C.

built by the EMI Company in 1971 following Hounsfield's design (Petrik et al., 2006), the technique has been successfully applied in several domains such as geology (Ueta et al., 2000) and hydrogeology (Wildenschild et al., 2002). In particular, it has been extensively applied in chemical engineering during the past decade to determine the phase saturation spatial distribution within a process vessel (Marchot et al., 1999; Toye et al., 1998).

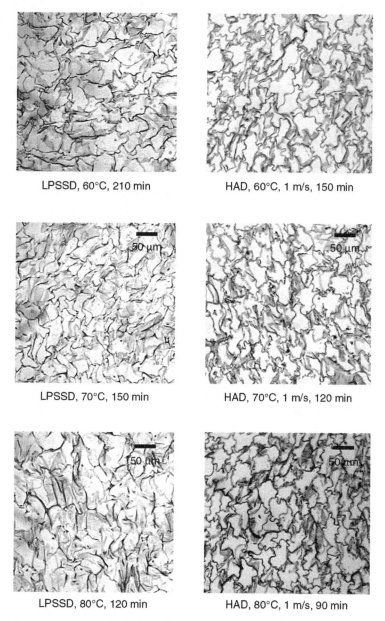

LPSSD, 60°C, 210 min HAD, 60°C, 1 m/s, 150 min

LPSSD, 70°C, 150 min HAD, 70°C, 1 m/s, 120 min

LPSSD, 80°C, 120 min HAD, 80°C, 1 m/s, 90 min

FIGURE 3.11 Microstructural images of a carrot cube at $\Delta FD/FD_0 = 0.06$.

Through the rapid development of microfocus x-ray sources and high-resolution CCD detectors, the commercialization of microtomographs with improved resolution (on the order of a few microns) has been realized, especially during the last 10 years. Their main components are the x-ray source (usually of cone–beam type), rotating stage, and detector (Figure 3.12a). At present, however, no x-ray microtomography

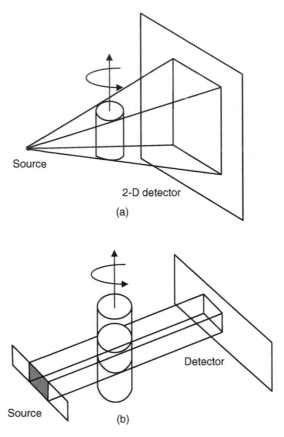

FIGURE 3.12　Schematic views of (a) an x-ray microtomograph with cone–beam geometry and (b) synchrotron microtomography.

computed system has been integrated directly with laboratory drying equipment. The follow-up of any drying process thus requires an interruption of the drying process to scan the sample being dried.

For high-resolution (on the order of ≤1 μm) applications, an x-ray microtomographic investigation can be realized with the use of synchrotron facilities. The x-ray beams produced in a synchrotron are parallel (Figure 3.12b), monochromatic, and highly coherent. The main advantage of a monochromatic beam is the absence of beam hardening artifacts, which results from preferential absorption of low energy x-rays in the periphery of an object. The high spatial coherence of the beam produces a phase contrast that can be exploited to detect material interfaces. For drying related problems, the major drawback of synchrotron x-ray microtomography is the small sample size that can be scanned. These systems are thus more suitable for the characterization of the microstructure of a dried material than for the follow-up of the drying process itself.

X-rays have the same wave–particle duality characteristics attributed to visible light, that is, they can be described either as an electromagnetic wave or a particle or photon with energy proportional to frequency. For x-ray microtomography systems, the source usually covers energies between 20 and 150 keV (1 eV corresponds to 1.602×10^{-19} J). For the sake of brevity, the physical principles of x-rays are not included here. Kak and Slaney (1988) and Brooks and Di Chiro (1976) provide detailed physical principles of x-rays and x-ray tomography.

3.4.2 Image Reconstruction

Once the transmission data, called projections, have been recorded at different angular positions, a reconstruction algorithm is used to produce 2-D cross-section images of an object. The gray value of each pixel of the reconstructed image corresponds to the linear attenuation coefficients (related to linear densities).

In the case of a planar parallel x-ray beam, a stabilized and discretized version of the inverse radon transform known as the filtered backprojection algorithm (Brooks and Di Chiro, 1976; Kak and Slaney, 1988) is used for image reconstruction. Parallel geometry, as encountered within synchrotron facilities, leads to a shorter reconstruction time. In the cone–beam sources usually found in x-ray microtomographs, 3-D reconstruction cone–beam algorithms such as Feldkamp's (1984) are employed. The resolution of the reconstructed images mainly depends on the source spot size, the source–object–detector distance, and the definition of the detector.

Beam hardening artifacts can be observed when using polychromatic sources. Because of preferential absorption of low energy x-rays in the periphery of an object, the mean energy of the beam increases and leads to a decrease in the attenuation coefficient. For a homogeneous object, the border of the reconstructed image will appear denser than its center, giving rise to the so-called cupping effect. In addition, strakes and flares can appear when the object contains hard elements or when strong variations of the attenuation occur within the sample.

3.4.3 X-Ray Microtomography and Food Drying

3.4.3.1 Shrinkage Determination

Determining the level (and pattern) of shrinkage is of much importance to food drying because the information on product shrinkage is necessary for many subsequent calculations (e.g., calculation of drying rates) and for modeling purposes. Shrinkage is also directly related to the acceptance of the product by consumers. Except for some studies employing image analysis (Abud-Archila et al., 1999; Fernandez et al., 2005; Lewicki and Piechnik, 1996; Mayor et al., 2005; Mulet et al., 2000; Yan et al., 2008), most of the published works refer to the use of a caliper (Moreira et al., 2002; Ratti, 1994) or volume displacement methods (Devahastin et al., 2004; Ochoa et al., 2002a,b; Panyawong and Devahastin, 2007) to determine shrinkage of food samples during drying. With its accuracy and its nondestructive nature, x-ray tomography coupled with image analysis constitutes an alternative for an accurate 3-D determination of shrinkage (Lewicki and Witrowa, 1992; Lozano et al., 1983; Ratti, 1994). To follow shrinkage, however, it is still necessary to interrupt the drying process at different sampling periods to take the sample out of the dryer for microtomographic analysis. However, these repeated interruptions have proved to have a negligible effect on the drying kinetics (Léonard et al., 2002). The volume of the sample can then be obtained by applying image analysis algorithms on the 2-D or 3-D reconstructed images (Léonard et al., 2004).

3.4.3.2   Crack Detection and Quantification

When drying food materials, the final quality of the products sometimes relies on the absence of cracks or product breakage. This is especially true in rice (Abud-Archila et al., 1999; Zhang et al., 2005), pasta (Andrieu and Stamatopoulos, 1986a; Andrieu et al., 1988; Antognelli, 1980; Sannino et al., 2005), wheat grains (Wozniak et al., 1999), or noodles (Inazu et al., 2005). Therefore, the detection and quantification of cracks during drying is essential for quality criteria estimation or model validation. x-ray microtomography can be efficiently used to follow crack formation (Job et al., 2006; Léonard et al., 2004) and to help optimize the drying process. Following the interrupted drying methodology described for shrinkage measurement, it is possible to determine the drying time and the location corresponding to crack onset. A simple adaptation of the image analysis allows detection and quantification of the crack extent from the same set of 2-D reconstructed images (Léonard et al., 2003).

3.4.3.3   Internal Moisture Profile Determination

In general, the determination of internal moisture profiles, which develop in a material during drying, is rather complicated, especially when steep gradients are present at the external sample surface. Moisture profiles within the drying material have been mainly determined either by destructive techniques such as slicing (Andrieu and Stamatopoulos, 1986b; Litchfield and Okos, 1992) or by a sophisticated nondestructive method such as NMR or MRI (Frias et al., 2002; Rongsheng et al., 1992; Ruan et al., 1991; Ruiz-Cabrera et al., 2004). These technical difficulties have limited possibility to validate many recently developed advanced mathematical models because direct comparison between actual and predicted moisture profiles is almost impossible (Waananen et al., 1993). In contrast, using x-ray microtomography together with an appropriate calibration method makes it possible to accurately determine internal moisture profiles, provided that the attenuation properties of the solid matrix and those of moisture are sufficiently different (Léonard et al., 2005). This measurement can be performed simultaneously with the determination of shrinkage or cracks, because the set of reconstructed images is the same.

3.4.3.4   Microstructure Characterization

One of the most interesting applications of x-ray microtomography in the context of food drying is to characterize the microstructure of a sample after drying. Although SEM is traditionally used to analyze the microstructure of dried food materials (Aguilera and Stanley, 1999), the technique does not give reliable information on the total pore volume and pore size distribution of the sample. Indeed, because only small parts can be investigated at a time, measurements must be repeated in order to give statistically relevant results. Moreover, SEM fails to describe the whole 3-D morphology because only 2-D information is obtained.

In contrast, there are several advantages of using x-ray microtomography to characterize the microstructure of dried products. The technique does not require any specific sample preparation and is nondestructive as opposed to SEM, which requires cutting of a sample after impregnation. X-ray microtomography also yields a 3-D characterization of the internal structure of the sample versus the 2-D structure obtained by SEM. The technique is particularly suitable to materials exhibiting larger pore diameters (d_p), for which classical measurement methods such as nitrogen adsorption (2 nm $< d_p <$ 50 nm) or mercury porosimetry (7.5 nm $< d_p <$ 150 μm) cannot be applied.

An illustrative example of the use of x-ray microtomography in microstructural evaluation of a dried food product was provided by Léonard et al. (2008). They applied the technique to evaluate the effects of different drying methods (LPSSD and vacuum drying), as well as far-infrared radiation (FIR) to assist these drying processes, on the microstructure of banana slices.

Projections (assembling of transmitted beams) were recorded for several angular positions by rotating the sample between 0 and 180°; the rotation step was fixed at 0.4°. Then, a backprojection algorithm was used to reconstruct 2-D or 3-D images, depending on the method used. In the 2-D images, each pixel had a gray-level value corresponding to the local attenuation coefficient. A Skyscan-1172 high-resolution desktop micro-CT system (Skyscan, Kontich, Belgium) was used in Léonard et al.'s (2008) work. The cone–beam source was operated at 60 kV and 167 μA. The detector was a 2-D, 1048 × 2000 pixel, 16-bit x-ray camera. The source–object–camera distance was adjusted to produce images with a 15 μm pixel size. Because of the sample height, three successive subscans had to be performed, each corresponding to one third of the slice height. For each angular position, a radiograph of the sample (instead of a 1-D projection of a cross section) was recorded by a 2-D camera.

Figure 3.13 shows a typical radiograph obtained after the three subscans were linked together. The Feldkamp backprojection algorithm was used to reconstruct 2-D images of the cross sections. For each banana slice, about 200 cross sections, separated by 150 μm, were reconstructed. Figure 3.14a and b show typical gray-level cross sections obtained for two vertical positions in the sample.

Three-dimensional images of the samples were built by stacking the 200 cross sections. The 3-D image was preliminary segmented by assigning a value of 1 to all pixels whose intensity was below a given gray tone value and 0 to the others, which implies fixing a threshold on the 3-D gray-level image. This threshold was determined through the use of an automatic threshold based on the entropy of the histogram (Sahoo et al., 1988) calculated for each 2-D cross section. After thresholding, some small black holes were still present in the image and were removed by applying a closing filter (Soille, 1999). Figure 3.14c and d show the result of the segmentation process.

Porosity, which is defined as the fraction of voxels of the image that belong to the pores, could be evaluated from the 3-D processed binary images. Because the 3-D images of the banana slices presented a continuous and rather disordered pore structure in which it was not possible to assign a precise geometry to each pore, a standard granulometry measurement could not be applied. To quantify the larger pores, the

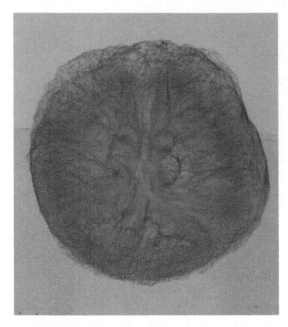

FIGURE 3.13   Radiograph of a dried banana slice.

opening size distribution (Soille, 1999), which allows assigning a size to both continuous and individual particles, was calculated. When an opening transformation was performed on a binary image with a structuring element of size λ, the image was replaced by an envelope of all structuring elements inscribed in its objects. Spheres of increasing radii λ (approximated by octahedra) were used. As the size of the sphere increased, larger parts of the objects were removed by the opening transformation. Opening could thus be considered as equivalent to a physical sieving process.

After the microtomographic and image analysis steps are performed, the microstructural and/or pore size and pore size distribution information can be obtained. As an illustrative example, Figure 3.15a and b show the histograms comparing the pore size distribution of the samples obtained at 80 and 90°C with LPSSD and LPSSD-FIR, respectively. These figures clearly show that the pore sizes were nonnormally distributed but pores with sizes lower than 100 μm prevailed. The use of FIR as well as a higher drying temperature resulted in a shift of the distribution toward larger pore sizes for both drying techniques.

3.5   CLOSING REMARKS

Because of the many advantages of image analysis, it has been increasingly used in food research and applications. More recently, the technique has also proved useful in food drying research and applications. Image analysis, through the use of CVSs, has been successfully applied to determine the changes in color, shape, and size of different foods undergoing drying. More advanced techniques such as SEM or MRI have been useful for the microstructural analysis of foods undergoing drying.

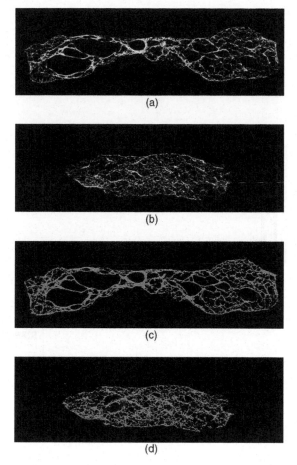

FIGURE 3.14   (a, b) Gray-level cross sections and (c, d) the corresponding binary images after thresholding.

Other techniques including the combination of image and fractal analysis and x-ray microtomography have also provided more accurate and comprehensive microstructural results than the more conventional microscopic methods in some cases. We hope that this chapter has provided some ideas on how different image analysis techniques can be applied to food drying research and applications and that the reader will find more ways to utilize the benefits of these techniques to a higher level.

NOMENCLATURE

VARIABLES

d_p	pore diameter
N_r	number of boxes
r	size of box

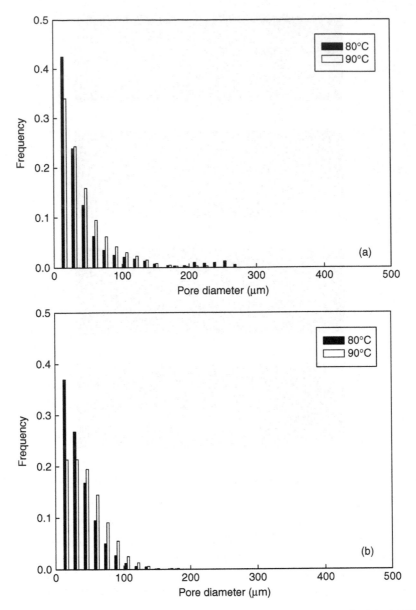

FIGURE 3.15 Pore size distributions of the samples dried at 80 and 90°C with (a) LPSSD and (b) LPSSD-FIR.

GREEK LETTERS

λ   size of structuring element in a binary image

SUBSCRIPTS

0   initial or fresh sample
t   time instant

ABBREVIATIONS

CCD	charge-coupled device
CLSM	confocal laser scanning microscopy
CVS	computer vision system
ESEM	environmental scanning electron microscopy
FD	fractal dimension
ΔF̅D̅	normalized change of fractal dimension
FIR	far-infrared radiation
HAD	hot-air drying
HSI	hue–saturation–intensity
HSV	hue–saturation–value
LM	light microscopy
LPSSD	low-pressure superheated steam drying
LPSSD-FIR	low-pressure superheated steam drying with far-infrared radiation
MRI	magnetic resonance imaging
NMR	nuclear magnetic resonance
SEM	scanning electron microscopy

REFERENCES

Abud-Archila, M., Bonazzi, C., Aichoune, H., and Heyd, B., Image analysis of shrinkage and cracking of rice during convective drying, in *Récents Progrès en Génie des Procédés*, Bimbenet, J.J., Ed., Lavoisier Technique et Documentation, Paris, 1999, pp. 411–418.

Acevedo, N.C., Briones, V., Buera, P., and Aguilera, J.M., Microstructure affects the rate of chemical, physical and color changes during storage of dried apple discs, *J. Food Eng.*, 85, 222–231, 2008.

Aguilera, J.M. and Stanley, D.W., *Microstructural Principles of Food Processing and Engineering*, Aspen Publishers, Gaithersburg, MD, 1999.

Alamilla-Beltrán, L., Chanona-Pérez, J., Jiménez-Aparicio, A., and Gutiérrez-López, G., Description of morphological changes of particles along spray drying, *J. Food Eng.*, 67, 179–184, 2005.

Andrieu, J. and Stamatopoulos, A., Durum wheat pasta drying kinetics, *Lebensm.-Wiss. Technol.*, 19, 448–456, 1986a.

Andrieu, J. and Stamatopoulos, A., Moisture and heat transfer modelling during durum wheat pasta drying, in *Drying '86*, Mujumdar, A.S., Ed., Hemisphere Publishing, Washington, DC, pp. 492–498, 1986b.

Andrieu, J., Boivin, M., and Stamatopoulos, A., Heat and mass transfer modeling during pasta drying, application to crack formation risk prediction, in *Handbook of Preconcentration and Drying of Food Materials*, Bruin, S., Ed., Elsevier Science, Amsterdam, 1988, pp. 183–192.

Antognelli, C., The manufacture and applications of pasta as a food and as a food ingredient: A review, *J. Food Technol.*, 15, 125–145, 1980.

Antonelli, A., Cocchi, M., Fava, P., Foca, G., Franchini, F.C., Manzini, D., and Ulrici, A., Automated evaluation of food colour by means of multivariate image analysis coupled to a wavelet-based classification algorithm, *Anal. Chim. Acta*, 515, 3–13, 2004.

Barletta, B.J. and Canovas, V.B., An attrition index to assess fines formation and particle size reduction in tapped agglomerated food powders, *Powder Technol.*, 77, 89–93, 1993.

Brooks, R.A. and Di Chiro, G., Principles of computer assisted tomography, *Phys. Med. Biol.*, 21, 689–732, 1976.

Chanola, P.J.J., Alamilla, B.L., Farrera, R.R.R., Quevedo, R., Aguilera, J.M., and Gutierrez, L.G.F., Description of the convective air drying of a food model by means of the fractal theory, *Food Sci. Technol. Int.*, 9, 207–213, 2003.

Chen, X.D., Guest editorial, *Dry. Technol.*, 24, 121–122, 2006.

Chevalier, D., Le Bail, A., and Ghoul, M., Freezing and ice crystals formed in a cylindrical food model: Part I. Freezing at atmospheric pressure, *J. Food Eng.*, 46, 277–285, 2000.

Cormack, A.M., Representation of a function by its line integrals, with some radiological applications, *J. Appl. Phys.*, 34, 2722–2727, 1963.

Dan, H., Azuma, T., and Kohyama, K., Characterization of spatiotemporal stress distribution during food fracture by image texture analysis methods, *J. Food Eng.*, 81, 429–436, 2007.

Devahastin, S., Suvarnakuta, P., Soponronnarit, S., and Mujumdar, A.S., A comparative study of low-pressure superheated steam and vacuum drying of a heat-sensitive material, *Dry. Technol.*, 22, 1845–1867, 2004.

Devaux, M.-F., Taralova, L., Levy-Vehel, J., Bonnin, E., Thibault, J.-F., and Guillon, F., Contribution of image analysis to the description of enzymatic degradation kinetics for particulate food material, *J. Food Eng.*, 77, 1096–1107, 2006.

Dziuba, J., Babuchowski, A., Smoczynski, M., and Smietana, Z., Fractal analysis of caseinate structure, *Int. Dairy J.*, 9, 287–292, 1999.

Feldkamp, L.A., Practical cone–beam algorithm, *J. Opt. Soc. Am.*, 1, 612–619, 1984.

Fernández, L., Castillero, C., and Aguilera, J.M., An application of image analysis to dehydration of apple discs, *J. Food Eng.*, 67, 185–193, 2005.

Frias, J.M., Foucat, L., Bimbenet, J.J., and Bonazzi, C., Modeling of moisture profiles in paddy rice during drying mapped with magnetic resonance imaging, *Chem. Eng. J.*, 86, 173–178, 2002.

Ghosh, P.K., Jayas, D.S., Gruwel, M.L.H., and White, N.D.G., A magnetic resonance imaging study of wheat drying kinetics, *Biosyst. Eng.*, 97, 189–199, 2007.

Humason, G.L., *Animal Tissue Techniques*, 4th ed., W.H. Freeman, San Francisco, CA, 1979.

HunterLab, http://www.hunterlab.com, November 2007.

ICC, http://www.color.org, November 2007.

Imaging Research Inc., http://www.imagingresearch.com, November 2007.

Inazu, T., Iwasaki, K., and Furuta, T., Stress and crack prediction during drying of Japanese noodle (udon), *Int. J. Food Sci. Technol.*, 40, 621–630, 2005.

Jennane, R., Harba, R., Lemineur, G., Bretteil, S., Estrade, A., and Benhamou, C.L., Estimation of the 3D self-similarity parameter of trabecular bone from its 2D projection, *Med. Image Anal.*, 11, 91–98, 2007.

Job, N., Sabatier, F., Pirard, J.P., Crine, M., and Léonard, A., Towards the production of carbon xerogel monoliths by optimizing convective drying conditions, *Carbon*, 44, 2534–2542, 2006.

Kak, A.C. and Slaney, M., *Principles of Computerized Tomographic Imaging*, IEEE Press, New York, 1988.

Kerdpiboon, S. and Devahastin, S., Fractal characterization of some physical properties of a food product under various drying conditions, *Dry. Technol.*, 25, 135–146, 2007a.

Kerdpiboon, S. and Devahastin, S., Erratum to fractal characterization of some physical properties of a food product under various drying conditions, *Dry. Technol.*, 25, 1127, 2007b.

Kerdpiboon, S., Devahastin, S., and Kerr, W.L., Comparative fractal characterization of physical changes of different food products during drying, *J. Food Eng.*, 83, 570–580, 2007.

Khraisheh, M.A.M., Cooper, T.J.R., and Magee, T.R.A., Shrinkage characteristics of potatoes dehydrated under combined microwave and convective air conditions, *Dry. Technol.*, 15, 1003–1022, 1997.

Léonard, A., Blacher, S., Marchot, P., and Crine, M., Use of x-ray microtomography to follow the convective heat drying of wastewater sludges, *Dry. Technol.*, 20, 1053–1069, 2002.

Léonard, A., Blacher, S., Marchot, P., Pirard, J.P., and Crine, M., Image analysis of x-ray microtomograms of soft materials during convective drying, *J. Microsc.*, 212, 197–204, 2003.

Léonard, A., Blacher, S., Marchot, P., Pirard, J.P., and Crine, M., Measurement of shrinkage and cracks associated to convective drying of soft materials by x-ray microtomography, *Dry. Technol.*, 22, 1695–1708, 2004.

Léonard, A., Blacher, S., Marchot, P., Pirard, J.P., and Crine, M., Moisture profiles determination during convective drying using x-ray microtomography, *Can. J. Chem. Eng.*, 83, 127–131, 2005.

Léonard, A., Blacher, S., Nimmol, C., and Devahastin, S., Effect of far-infrared radiation assisted drying on microstructure of banana slice: An illustrative use of x-ray microtomography in microstructural evaluation of a food product, *J. Food Eng.*, 85, 154–162, 2008.

Lewicki, P.P. and Piechnik, H., Computer image analysis of shrinkage during food dehydration, paper presented at the 10th International Drying Symposium, Krakow, Poland, 1996.

Lewicki, P.P. and Witrowa, D., Heat and mass transfer in externally controlled drying of vegetables, in *Drying '92*, Mujumdar, A.S., Ed., Elsevier, Amsterdam, 1992, pp. 884–891.

Litchfield, J.B. and Okos, M.R., Moisture diffusivity in pasta during drying, *J. Food Eng.*, 17, 117–142, 1992.

Locht, http://www.machinevisiononline.org/public/articles/articles.cfm?cat=50 (November 2007).

Louka, N., Juhel, F., Fazilleau, V., and Loonis, P., A novel colorimetry analysis used to compare different drying fish processes, *Food Control*, 15, 327–334, 2004.

Lozano, J.E., Rotstein, E., and Urbicain, M.J., Shrinkage, porosity and bulk density of foodstuffs at changing moisture contents, *J. Food Sci.*, 48, 1497–1553, 1983.

Mandelbrot, B.M., *The Fractal Geometry of Nature*, Freeman Press, New York, 1983.

Marchot, P., Toye, D., Crine, M., Pelsser, A.M., and L'Homme, G., Investigation of liquid maldistribution in packed columns by x-ray tomography, *Chem. Eng. Res. Design*, 77, 511–518, 1999.

Martynenko, A.I., Computer-vision system for control of drying processes, *Dry. Technol.*, 24, 879–888, 2006.

Mayor, L., Silva, M.A., and Sereno, A.M., Microstructural changes during drying of apple slices, *Dry. Technol.*, 23, 2261–2276, 2005.

Mendoza, F., Dejmek, P., and Aguilera, J.M., Calibrated color measurements of agricultural foods using image analysis, *Postharvest Biol. Technol.*, 41, 285–295, 2006.

Moreira, R., Figueiro, A., and Sereno, A.A., Shrinkage of apple disks during drying by warm air convection and freeze drying, *Dry. Technol.*, 18, 279–294, 2002.

Mulet, A., Garcia-Reverter, J., Bon, J., and Berna, A., Effect of shape on potato and cauliflower shrinkage during drying, *Dry. Technol.*, 18, 1201–1219, 2000.

National Institutes of Health, http://rsb.info.nih.gov/ij, November 2007.

Ochoa, M.R., Kesseler, A.G., Pirone, B.N., Marquez, C.A., and De Michelis, A., Shrinkage during convective drying of whole rose hip fruits (*Rosa rubiginosa* L.), *Lebensm.-Wiss. Technol.*, 35, 400–406, 2002a.

Ochoa, M.R., Kesseler, A.G., Pirone, B.N., Marquez, C.A., and De Michelis, A., Volume and area shrinkage during dehydration of whole sour cherry fruits (*Prunus cerasus*) during dehydration, *Dry. Technol.*, 20, 147–156, 2002b.

Panyawong, S. and Devahastin, S., Determination of deformation of a food product undergoing different drying methods and condition via evolution of a shape factor, *J. Food Eng.*, 78, 151–161, 2007.

Pedreschi, F. and Aguilera, J.M., Characterization of food surfaces using scale-sensitive fractal analysis, *J. Food Process Eng.*, 23, 127–143, 2000.

Petrik, V., Apok, V., Britton, J.A., Bell, B.A., and Papadopoulos, M.C., Godfrey Hounsfield and the dawn of computed tomography, *Neurosurgery*, 58, 780–786, 2006.

Plataniotis, K.N. and Venetsanopoulos, A.N., *Color Image Processing and Application*, Springer, Berlin, 2000.

Quevedo, R., Carlos, L.-G., Aguilera, J.M., and Cadoche, L. Description of food surfaces and microstructural changes using fractal image texture analysis. *J. Food Eng.* 53, 361–371, 2002.

Rahman, M.S., Physical meaning and interpretation of fractal dimensions of fine particles measured by different methods, *J. Food Eng.*, 32, 447–456, 1997.

Ramos, I.N., Silva, C.L., Sereno, A.M., and Aguilera, J.M., Quantification of microstructural changes during first stage air drying of grape tissue, *J. Food Eng.*, 62, 159–164, 2004.

Ratti, C., Shrinkage during drying of foodstuffs, *J. Food Eng.*, 23, 91–95, 1994.

Rongsheng, R., Litchfield, J.B., and Eckhoff, S.R., Simultaneous and non-destructive measurement of transient moisture profiles and structural changes in corn kernels during steeping using microscopic nuclear magnetic resonance imaging, *Cereal Chem.*, 69, 600–606, 1992.

Ros, F., Guillaume, S., Rabatel, G., and Sevila, F., Recognition of overlapping particles in granular product images using statistics and neural networks, *Food Control*, 6, 37–43, 1995.

Ruan, R., Schmidt, S.J., Schmidt, A.R., and Litchfield, J.B., Non-destructive measurement of transient moisture profiles and moisture diffusion coefficient in a potato during drying and absorption by NMR imaging, *J. Food Process Eng.*, 14, 297–313, 1991.

Rubnov, M. and Saguy, I.S., Fractal analysis and crust water diffusivity of a restructured potato product during deep-fat frying, *J. Food Sci.*, 62, 135–154, 1997.

Ruiz-Cabrera, M.A., Gou, P., Foucat, L., Renou, J.P., and Daudin, J.D., Water transfer analysis in pork meat supported by NMR imaging, *Meat Sci.*, 67, 169–178, 2004.

Russ, J.C., *Image Analysis of Food Microstructure*, CRC Press, Boca Raton, FL, 2005.

Sahoo, P.K., Soltani, S., Wong, K.C., and Chen, Y.C., A survey of thresholding techniques, *Comput. Vision Graph. Image Process.*, 41, 233–260, 1988.

Sannino, A., Capone, S., Siciliano, P., Ficarella, A., Vasanelli, L., and Maffezzoli, A., Monitoring the drying process of lasagna pasta through a novel sensing device-based method, *J. Food Eng.*, 69, 51–59, 2005.

Soille, P., *Morphological Image Analysis—Principles and Applications*, Springer, Berlin, 1999.

The GIMP Team, http://www.gimp.org, November 2007.

Toye, D., Marchot, P., Crine, M., Pelsser, A.-M., and L'Homme, G., Local measurements of void fraction and liquid holdup in packed columns using x-ray computed tomography, *Chem. Eng. Process.*, 37, 511–520, 1998.

Turner, M.J., Blackledge, J.M., and Andrews, P.R., *Fractal Geometry in Digital Imaging*, Academic Press, New York, 1998.

Ueta, K., Tani, K., and Kato, T., Computerized x-ray tomography analysis of three-dimensional fault geometries in basement-induced wrench faulting, *Eng. Geol.*, 56, 197–210, 2000.

University of Texas Health Science Center, http://ddsdx.uthscsa.edu/dig, November 2007.

Varela, P., Aguilera, J.M., and Fiszman, S., Quantification of fracture properties and micro-structural features of roasted *Marcona* almonds by image analysis, *Food Sci. Technol.*, 41, 10–17, 2008.

Waananen, K.M., Litchfield, J.B., and Okos, M.R., Classification of drying models for porous solids, *Dry. Technol.*, 11, 1–40, 1993.

White, K.L. and Bell, L.N., Glucose loss and Maillard browning in solids as affected by porosity and collapse, *J. Food Sci.*, 64, 1010–1014, 1999.

Wildenschild, D., Vaz, C.M.P., Rivers, M.L., Rikard, D., and Christensen, B.S.B., Using x-ray computed tomography in hydrology: Systems, resolutions, and limitations, *J. Hydrol.*, 267, 285–297, 2002.

Wozniak, W., Niewczas, J., and Kudra, T., Internal damage vs. mechanical properties of microwave-dried wheat grains, *Int. Agrophys.*, 13, 259–268, 1999.

Yan, Z., Sousa-Gallagher, M.J., and Oliveira, F.A., Shrinkage and porosity of banana, pineapple and mango slices during air-drying, *J. Food Eng.*, 84, 430–440, 2008.

Zhang, Q., Yang, W., and Sun, Z., Mechanical properties of sound and fissured rice kernels and their implications for rice breakage, *J. Food Eng.*, 68, 65–72, 2005.

4 Dehydration and Microstructure

Mohammad Shafiur Rahman

CONTENTS

4.1   INTRODUCTION

Structure is the manner or form of building of food components. In addition to natural structures, man-made structured foods use assembly or structuring processes to build product microstructure from the microscopic level details of each component (such as water, starch, sugars, proteins, lipids, and salts) or cells up to the molecular level. As food materials undergo different treatments or processes, their microstructure may be preserved or destroyed and develop useful processed products. Drying methods create or destroy the microstructure of foods by specific distribution of component phases. This chapter reveals the food microstructural parameters, porosity, and characteristics of pores by defining the parameters and discussing their measurement methods, mechanisms involved, prediction methods, and relevance to the quality of dried food products.

4.2   FOOD MICROSTRUCTURE PARAMETERS

4.2.1   POROSITY

Porosity indicates the volume fraction of void space or air in a material, which is defined as

$$\text{porosity} = \frac{\text{air or void volume}}{\text{total volume}} \tag{4.1}$$

Different forms of porosity are also used in food process calculations and food product characterization (Rahman, 1995, 2005). These are defined in the following sections.

4.2.1.1   Open Pore Porosity

The open pore porosity (ε_{op}) is the volume fraction of pores connected to the exterior boundary of a material:

$$\varepsilon_{op} = \frac{\text{volume of open pore}}{\text{total volume of material}} = 1 - \frac{\rho_a}{\rho_p} \tag{4.2}$$

where and ρ_a and ρ_p are the apparent and particle density (kg/m³), respectively. There are two types of open pores: one type is connected to the exterior boundary only (noninterconnected), and the other type is connected to the other open pores and to the exterior geometric boundary (interconnected). Interconnected pores are accessible on both ends, but noninterconnected pores are accessible from only one end.

4.2.1.2   Closed Pore Porosity

The closed pore porosity (ε_{cp}) is the volume fraction of pores closed inside the material, which are not connected to the exterior boundary of the material. They are also termed dead-end or blind pores and can be defined as

$$\varepsilon_{cp} = \frac{\text{volume of closed pores}}{\text{total volume of material}} = 1 - \frac{\rho_p}{\rho_m} \qquad (4.3)$$

where ρ_m is the material density (kg/m^3). Inaccessible closed pores are enclosed within a solid or liquid phase and in many instances behave as part of the solid.

4.2.1.3   Apparent Porosity

The apparent porosity (ε_a) is the volume fraction of total air or void space inside the material boundary ($\varepsilon_a = \varepsilon_{op} + \varepsilon_{cp}$). It is defined as

$$\varepsilon_a = \frac{\text{volume of all pores}}{\text{total volume of material}} = 1 - \frac{\rho_a}{\rho_m} \qquad (4.4)$$

4.2.1.4   Bulk Porosity

The bulk porosity (ε_B) is the volume fraction of voids outside the boundary of individual particles when packed or stacked as bulk:

$$\varepsilon_B = \frac{\text{volume of voids outside material boundary}}{\text{total bulk volume of stacked materials}} = 1 - \frac{\rho_B}{\rho_a} \qquad (4.5)$$

where ρ_B is the bulk density (kg/m^3).

4.2.1.5   Bulk Particle Porosity

The bulk particle porosity (ε_{BP}) is the volume fraction of voids outside the individual particle and open pore when packed or stacked as bulk ($\varepsilon_{BP} = \varepsilon_B + \varepsilon_{op}$):

$$\varepsilon_{BP} = 1 - \frac{\rho_B}{\rho_p} \qquad (4.6)$$

4.2.1.6   Total Porosity

The total porosity (ε_T) is the total volume fraction of air or void space (i.e., inside and outside of the materials) when the material is packed or stacked as bulk.

$$\varepsilon_T = \varepsilon_a + \varepsilon_B = \varepsilon_{op} + \varepsilon_{cp} + \varepsilon_B = \varepsilon_{cp} + \varepsilon_{BP} \qquad (4.7)$$

4.2.1.7   Pore Size Distribution

Porosity is a global characteristics and it does not characterize a pore's size, shape, orientation, and population. When pore or particle size and population varies, size distribution is most commonly used to characterize this variation.

4.2.2   CHARACTERISTICS OF PORES

In addition to porosity and pore size distribution, other characteristics are used to identify the types of pores, such as the shape, thickness of walls, and building structure of pores. Microscopic as well as other instruments are used to explore these characterisitics.

4.3   MEASUREMENT METHODS

4.3.1   POROSITY

Porosity can be measured by a direct method or an optical microscopic method followed by image analysis and estimated from density data.

4.3.1.1   Direct Method

In the direct method the bulk volume of a piece of porous material is measured after compacting the material in order to destroy all of its voids and then measure the volume. The porosity can be determined from the difference of the two volumes that are measured. This method can be applied if the material is very soft and there is no repulsive or attractive force present between the surfaces of solid particles.

4.3.1.2   Optical Microscopic Method

In the optical microscopic method the porosity can be determined from the microscopic view of a random section of the porous medium. This method is reliable if the sectional [two-dimensional (2-D)] porosity is same as the three-dimensional (3-D) porosity. Image analysis techniques are used to estimate quantitative results. A novel technique was also proposed for measuring the 3-D structure using a microsliced image processing system (Sagara, 2003). This system constructs the 3-D image based on the imaged data of exposed cross sections obtained by multislicing a sample with a minimum thickness of 1 μm. Then the internal structure is displayed through computer simulation.

4.3.1.3   Density Method

Porosity can also be estimated from the densities of the materials from Eqs. (4.2) to (4.6). Alternatively, pore volume can be measured directly by liquid or gas displacement methods. Interconnected pores can be identified using one-dimensional (1-D) to 3-D directions. In the case of a cube, a 1-D system can be achieved by coating all five sides of the sample before measurement. If all open pores are interconnected, even when the fluid is introduced in one direction, flow takes place in 3-D.

4.3.1.4   X-Ray Microtomography (XMT)

Low-resolution x-ray tomography or computerized axial tomography is widely used in medicine to image body tissues in a noninvasive manner. This technology, which originated in the 1970s, has a conventional resolution of ~100μm, which is insufficient to explore the microstructure of many types of foam. The use of very high energy x-rays from high intensity synchrotrons or other sources allows a resolution of 1μm and enables the use of XMT (Coker et al., 1996). Two types of XMT are common: one based on the absorbance of x-rays and the other based on the phase shift (or phase contrast) of incident x-rays produced by an object (Falcone et al., 2004; Snigirev et al., 1995). Noninvasive imaging technology XMT can be used to visualize and measure microstructural features. It allows noninvasive imaging of sample cross sections at various depths and facilitates accurate and hitherto impossible measurements of features like true cell size distribution (bimodal), average diameter, open wall area fraction, cell wall thickness, presence or absence of an interconnected network, and void fraction (Trater et al., 2005). All of these features are impossible to ascertain using destructive, 2-D imaging techniques like light microscopy and scanning electron microscopy (SEM). Another problem in SEM or optical imaging is obtaining adequate contrast between the air and solid phases, in which the lighting and angle of illumination play an important role. The XMT technique was used for 3-D visualization of the microstructural features of expanded extruded products (Trater et al., 2005) and bread (Falcone et al., 2004).

4.3.1.5 Mercury Porosimetry

In addition to porosity, the pore's characteristics and size distribution can be determined by mercury porosimetry. Modern mercury porosimetry can achieve pressures in excess of 414 MPa, which translates into a pore size of 0.003 nm. This technique involves constructing an isotherm of either pressure or intrusion volume. The total open pores, characteristics of pores, and pore size distribution can be derived from these isotherms.

4.3.1.5.1 Intrusion and Extrusion Curves
The intrusion curve denotes the volume change with increasing pressure, and the extrusion curve indicates the volume change with decreasing pressure (Figure 4.1).

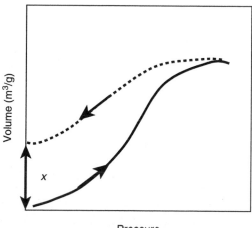

FIGURE 4.1 Intrusion and extrusion curves of mercury porosimetry showing hysteresis.

Mercury intrusion and extrusion results were compared in order to assess capillary hysteresis, which can be used to explore the characteristics of pores, such as shape and structure. Most of the porosimetry curves exhibit hysteresis. The difference in the curvatures of the intrusion and extrusion curves indicates that the path followed by the extrusion curve is not the same as the intrusion path. At a given pressure the volume indicated on the extrusion curve is lower than that on the intrusion curve, and the difference in volume is the entrapped volume. The entrapped mercury volume is indicated by x in Figure 4.1. The intrusion–extrusion cycle does not close when the initial pressure is reached after extrusion, indicating that some mercury is permanently entrapped by the sample, thereby preventing the loop from closing. Often at the completion of an intrusion–extrusion cycle mercury slowly continues to extrude, sometimes for hours (Lowell and Shields, 1984). The maximum hysteresis volume (x) found for fresh apples was 0.135cm^3/g (Rahman et al., 2005). This indicated the amount of mercury entrapped in the sample after a complete extrusion cycle and the complexity of the pore network. The total volume of entrapped mercury was 0.135cm^3/g, which was 85% of the total mercury intruded. The shape of the intrusion and extrusion curves for dried apple samples at 20 and 30h are different from those for the fresh apples. A completely different extrusion curve was observed for the dried sample after 30h. These graphs created by Rahman et al. (2005) indicated that the pore structure is changed in the course of drying in addition to the porosity. Bread and cookie samples showed capillary hysteresis (Hicsasmaz and Clayton, 1992). About 95% of the intruded mercury was entrapped in bread samples, whereas 80% mercury was entrapped in cookies. This indicated that both bread and cookie contain pores where narrow pore segments are followed by segments, which are larger in diameter. The reason that mercury entrapment is lower in cookies than in bread is that bread is a highly expanded product containing very large pores connected by narrow necks. This phenomenon (narrow neck pores) was also supported by measurement of the cellular structure by using SEM. Another cause could be due to the filtered out of fine particles from the solid suspension tend to form agglomerates in the pores during deposition.

4.3.1.5.2 Pore Size Distribution Curves

Pore size distribution curves are characterized by different methods. In addition to the intrusion–extrusion curve, other plots are used to characterize the pores:

- Derivative of the cumulative volume curve (dV/dP) versus pressure (P) or diameter
- Volume distribution (dV/dr) versus P or diameter
- Pore size distribution function (f_v) versus P or pore diameter

Here, dV is the volume of intruded mercury in the sample (considered to be exactly equal to the volume of pores) and P is the pressure (Pa). The term dV/dP is the first derivative (slope) of the volume versus pressure or diameter data. It is used in subsequent calculations of the volume distribution function, area distribution function, and pore number fraction. The pore size distribution is the computed relation between the pore radius and pore volume, assuming cylindrical pores (Lowell and Shields, 1984). It can be defined as

$$f_v = \left(\frac{P}{r}\right) \times \left(\frac{dV}{dP}\right) \tag{4.8}$$

where f_v is the pore size distribution function (m³/m kg) and r is the pore radius (μm). This function uses the slope times P/r; thus, it becomes increasingly susceptible to minor fluctuations in volume or pressure data or to slight changes in slope. These slight variations in slope can result from actual discontinuities in the pore size distribution, that is, some pore sizes may be absent or present in lesser volume than pores immediately adjacent. This means that pore size distributions are usually not smooth and continuous functions (Rahman et al., 2002).

Figure 4.2 shows pore size distribution curves of air-dried, vacuum-dried, and freeze-dried tuna meat (Rahman et al., 2002). A pore size distribution curve is usually characterized by the number, size, and shape of the peaks (Rahman and Sablani, 2003).

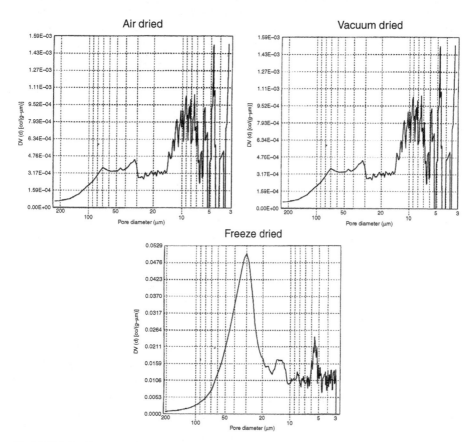

FIGURE 4.2 Pore size distribution cures of air-dried, vacuum-dried, and freeze-dried tuna meat (D_v in the diagram should read as f_v). (From Rahman, M.S. et al., *J. Food Eng.*, 53, 301–313, 2002. With permission.)

Although pore size distribution curves are presented in the literature, there is very little information available on how this graph should be analyzed in order to explore the characteristics of pores. The large peak in the figure indicates that most of the pores exist in that size range, and the wider peak indicates that a relatively larger segment of pores is followed by smaller ones. The sharp peak indicates the extent of similar size pores; the higher the height is, the more pores at this size. A freeze-dried abalone sample dried at −15°C showed only one large peak at 10 µm. Increased the freeze-drying temperature moved the peaks slightly lower than 10 µm (Rahman and Sablani, 2003). Karathanos et al. (1996) found three peaks for freeze-dried carrots: two sharp peaks at 0.2 and 1.1 µm, and one shorter and wider peak at 21 µm. Two discrete peaks were found for freeze-dried potatoes, cabbage, and apples. Completely different types of distribution curves were observed for freeze-dried garlic samples when the samples were dried at different shelf temperatures (Sablani et al., 2007). Rahman et al. (2002) determined the pore size distribution graphs for tuna meat dried by different methods and found more complex curves. They pointed that simply counting the peaks did not characterize the entire curve. Moreover, the lines also showed significant ruggedness compared to the data for fruits and vegetables provided by Karathanos et al. (1996). The ruggedness in the air-dried sample was to such an extent that it was very difficult to count the number of peaks. The freeze-dried sample showed three major peaks (5.5, 13, and 30 µm) and there were many other small peaks, making the curve rugged at the end of the graphs (at lower pore radius). Fresh apples showed two sharp peaks at 5.8 and 3.6 µm (Rahman et al., 2005). The second peak was so sharp that it was just a line. An apple sample air dried for 20 h showed only one sharp peak at 3.6 µm; the sample dried for 30 h showed three peaks at 10, 5.8, and 3.6 µm. The peak at 10 µm was much wider and shorter, which may have been due to the formation of cracks or channels during longer drying. In addition, all of the curves were extremely skewed to the right (toward a lower pore diameter). The sharp peak indicated that most of the pores exist in that size range, and the wider peak indicated that a relatively larger segment of pores is followed by smaller ones. For Amioca starch containing 11% moisture, two peaks were found when using low-pressure mercury porosimetry (one at 6 to 8 µm and one at ~1 to 3.5 µm) and another peak was seen from high-pressure mercury porosimetry at very small pore sizes of 3 nm (Karathanos and Saravacos, 1993). Xiong et al. (1991) studied the pore size distribution curves of regular pasta and puffed pasta and found that regular pasta showed a horizontal line with no peak whereas puffed pasta showed a large peak at 20 µm. Extruded regular pasta with mixtures of gluten and starch was very dense and homogeneous, but the puffed pasta samples were porous. When the gluten–starch ratio was 1, they found two peaks: one at 0.02 µm and one at 60 µm. Similarly, when the gluten–starch ratio was 3, there were two peaks: 0.006 and 11 µm. However, the peaks for a gluten–starch ratio of 3 were very sharp and high. This is attributable to the cross-linking of protein through stronger covalent and hydrogen bonds during extrusion. Fractal analysis was also used to characterize the pores (Rahman et al., 2005). Pfeifer and Avnir (1983) derived a scaling equation for the pore size distribution:

$$\frac{dV}{dD} \propto D^{2-\delta} \tag{4.9}$$

where D is the pore diameter (m), V is the pore volume (m³), and δ is the fractal dimension. In terms of applied pressure, the scaling equation is

$$\frac{dV}{dP} \propto P^{\delta-4} \qquad (4.10)$$

The fractal dimension can be estimated from the slope of the plot of $\log(dV/dP)$ versus $\log P$. Rahman et al. (2005) measured the fractal dimensions of fresh and dried apples. Fractal dimensions of >2 and ~3 indicate the characteristics of pores in fresh and dried apples, respectively. Dimensions increased with increasing drying time, indicating the formation of micropores on the surface during air drying. Rahman (1997) determined the fractal dimensions of native, gelatinized, and ethanol deformed starches, which were 3.09, 3.10, and 2.45, respectively. Lower values indicated the removal of micropores within the starch particles by ethanol modification.

4.3.1.5.3 Pore Size
Mean and median pore diameters are usually calculated. The mean pore diameter is the weighted arithmetic average of the distribution function, and the median pore diameter is the diameter at which an equal volume of mercury is introduced into larger and smaller pores than the median. Table 4.1 shows the pore diameter with mercury porosimetry. The pore radius for strawberries was the largest (60.8 μm), followed by apples (38.3 μm) and pears (28.7 μm) according to Khalloufi and Ratti (2003). They also found that porosimetry gave a lower average pore diameter compared to the microscopic techniques.

4.3.1.6 Gas Adsorption Method

In the gas adsorption method the sample is cooled, usually to cryogenic temperatures, and then exposed to an inert adsorptive gas, typically nitrogen, at a series of precisely controlled pressures. As the pressure increases a number of the gas molecules are adsorbed on the surface of the sample. This volume of adsorbed gas is measured using sensitive pressure transducers. As the pressure increases, more and more gas is adsorbed. The very small micropores (<20 Å) are filled first, followed by the free surface and the rest of the pores. Each pressure corresponds to a pore size, and the volume of these pores can be obtained from the volume of gas adsorbed. A desorption isotherm as well as hysteresis can be derived by collecting data as the pressure is reduced.

4.4 CREATION AND COLLAPSE OF MICROSTRUCTURE BY DRYING METHODS

4.4.1 MECHANISMS

4.4.1.1 Mechanisms of Collapse

Collapse is a result of a decrease in porosity, and it is caused by shrinkage of the product. The factors affecting formation of pores can be grouped as intrinsic and

TABLE 4.1
Pore Size (Diameter) of Different Dried Foods

Material	Drying Method	X_w	ρ_a (kg/m³)	ρ_m (kg/m³)	ε_a	Mercury Porosimetry		Microscopy		Ref.
						Range (μm)	Average (μm)	Range (μm)	Average (μm)	
Abalone	Freeze-drying (PT = 15°C)	—	664	—	0.664	—	8.8 (0.2)	—	—	Rahman and Sablani (2003)
Abalone	Freeze-drying (PT = −5°C)	—	715	—	0.715	—	9.3 (0.6)	—	—	Rahman and Sablani (2003)
Abalone	Freeze-drying (PT = −20°C)	—	737	—	0.737	—	7.6 (0.4)	—	—	Rahman and Sablani (2003)
Apple	Air drying (80°C, 0h)	0.860	849	1091	0.222	3.6–231	12.1 (2.7)	—	—	Rahman et al. (2005)
Apple	Air drying (80°C, 0h)	0.304	660	1352	0.512	3.6–214	12.2 (1.7)	—	—	Rahman et al. (2005)
Apple	Air drying (80°C, 0h)	0.103	560	1536	0.613	3.6–240	15.0 (1.0)	—	—	Rahman et al. (2005)
Apple[a]	Freeze-drying	<0.013	145	1478	0.902	—	76.6 (3.0)	46–416	146.4 (27.2)	Khalloufi and Ratti (2003)
Garlic	Freeze-drying (PT = −5°C)	0.090	469	1534	0.690	4.0–249	12.0 (2.0)	—	—	Sablani et al. (2007)

Material	Drying method									Reference
Garlic	Freeze-drying (PT=-15°C)	0.081	440	1517	0.710	4.0–227	12.0 (1.0)	—	—	—	Sablani et al. (2007)
Garlic	Freeze-drying (PT=-25°C)	0.061	431	1504	0.710	4.0–224	12.0 (1.0)	—	—	—	Sablani et al. (2007)
Pear[b]	Freeze-drying	<0.013	162	1049	0.845	—	57.4 (8.4)	44–240	96.0 (20.6)	—	Khalloufi and Ratti (2003)
Pear[c]	Freeze-drying	<0.013	140	1148	0.878	—	75.4 (8.4)	32–224	103.2 (10.4)	—	Khalloufi and Ratti (2003)
Pear[d]	Freeze-drying	<0.013	187	844	0.779	—	25.8	—	—	—	Khalloufi and Ratti (2003)
Strawberry[a]	Freeze-drying	<0.013	117	910	0.898	—	121.6 (4.6)	82–378	183.8 (20.4)	—	Khalloufi and Ratti (2003)
Strawberry[d]	Freeze-drying	<0.013	300	1476	0.797	—	121.0	—	—	—	Khalloufi and Ratti (2003)
Tuna meat	Air drying (70°C)	0.114	960	1255	0.240	3.3–229	33.9 (3.1)	—	—	—	Rahman et al. (2002)
Tuna meat	Vacuum drying (70°C)	0.077	709	1309	0.460	3.3–241	30.1 (4.6)	—	—	—	Rahman et al. (2002)
Tuna meat	Freeze-drying (PT = -20°C)	0.086	317	1259	0.760	3.3–220	27.2 (3.6)	—	—	—	Rahman et al. (2002)

[a] Frozen at -27 and -17°C (hot plate temperature = 20–70°C).
[b] Frozen at -27°C (hot plate temperature = 20–70°C).
[c] Frozen at -17°C (hot plate temperature = 20–70°C).
[d] Collapsed sample.

extrinsic factors. The extrinsic factors are temperature, pressure, relative humidity, gas atmosphere, air circulation, and electromagnetic radiation applied in the process. The intrinsic factors are chemical composition, inclusion of volatile components (such as alcohol and carbon dioxide), and initial structure before processing (Rahman, 2001, 2004). Genskow (1990) and Achanta and Okos (1995) mentioned several mechanisms that affect the degree of collapse or shrinkage and formation of pores. An understanding of these mechanisms aids in achieving the desired shrinkage or collapse in the products. The following physical mechanisms play an important role in the control of shrinkage or collapse (Rahman and Perera, 1999):

1. Surface tension: considers collapse in terms of the capillary suction created by a receding liquid meniscus
2. Plasticization: considers collapse in terms of the plasticizing effect of solvents on various polymer solutes
3. Electrical charge effects: considers collapse in terms of van der Waals electrostatic forces
4. Mechanism of moisture movement in the process
5. Gravitational effects

The rate at which shrinkage occurs is related to the viscoelastic properties of a matrix. The higher viscosity of the mixtures caused the lower rate of shrinkage and vice versa (Achanta and Okos, 1996).

4.4.1.2 Glass-Transition Concept

Levine and Slade first applied the concept of glass transition to identify or explain the physicochemical changes in foods during processing and storage (Levine and Slade, 1986; Slade and Levine, 1988, 1991). The glass-transition theory is one of the concepts that have been proposed to explain the process of shrinkage, collapse, fissuring, and cracking during drying (Cnossen and Siebenmorgen, 2000; Karathanos et al., 1993, 1996; Krokida et al., 1998; Rahman, 2001). The hypothesis indicates that there is significant shrinkage during drying only if the temperature of the drying or processing is higher than the glass transition of the material at that particular moisture content (Achanta and Okos, 1996).

4.4.1.2.1 Support for Glass-Transition Concept

The methods of freeze-drying and hot-air drying can be compared based on this theory. In freeze-drying, with the drying temperature below or close to the maximally freeze-concentrated glass-transition temperature (T_g', independent of solids content) or T_g (glass transition as a function of solids content), the material is in the glassy state. Hence, the shrinkage is negligible. As a result, the final product is very porous. In contrast, in hot-air drying with the drying temperature above T_g' or T_g, the material is in the rubbery state and substantial shrinkage occurs causing a lower level of pores. During the initial stage of freeze-drying, the composition of the freeze-concentrated phase surrounding the ice dictates the T_g'. In the initial or early stage of drying, the T_g' is very relevant and the vacuum must be sufficient to ensure that sublimation is occurring. At the end of the initial stage of drying the pore size and the porosity are dictated by the ice crystal size, if collapse of the wall of the matrix that surrounded the ice crystal does not occur. Conversely, the secondary stage of drying refers to removal of water from the unfrozen phase. After sublimation is completed, the sample begins to warm up to the shelf temperature. At this stage, the T_g of the matrix is related to the collapse and no longer to T_g'. This is attributable to the $T_g > T_g'$ (T_g increases from T_g' as the concentration of solids increases during the process of drying). Karel et al. (1994) performed freeze-drying under high vacuum (0.53 Pa) and reduced vacuum conditions (90.64 and 209.28 Pa) to obtain varying initial sample temperatures that were −55, −45, and −28°C. Collapse was determined by measuring the apparent shrinkage before and after freeze-drying of apples, potatoes, and celery. Samples dried at −55°C showed no shrinkage (more pores) but shrinkage increased with the increase of the drying temperature, justifying the glass-transition concept.

4.4.1.2.2 Evidence against Glass-Transition Concepts

Recent experimental results indicated that the glass-transition concept is not valid for freeze-drying of all types of biological materials, indicating the need for the incorporation of other concepts (Sablani and Rahman, 2002). Thus, a unified approach needs to be used. In freeze-drying the pore formation in food materials showed two distinct trends when shelf temperatures were maintained at a constant level between −45 and 15°C (Sablani and Rahman, 2002). The materials in group I (i.e., abalone, potatoes, and brown dates) showed a decreasing trend, whereas those in group II (i.e., apples and yellow dates) showed an increasing trend in pore formation. This may be due to the structural affects of the materials. However, none of the research measured the actual temperature and moisture history of the sample passing through freeze-drying. The temperature and moisture history of the sample during freeze-drying could provide more fundamental knowledge to explain the real process of pore formation or collapse in freeze-drying. In many cases during convection air drying, the observations related to collapse are just the *opposite of the glass-transition* concept (Del Valle et al., 1998; Rahman et al., 2005; Ratti, 1994; Wang and Brennan, 1995). The mechanism proposed for this was the concept of case hardening (Achanta and Okos, 1996; Rahman et al., 2005; Ratti, 1994). Bai et al. (2002) studied surface structural changes in apple rings during heat-pump drying with controlled temperature and humidity. Electron microscopy showed tissue collapse and pore formation. Case hardening occurred in the surface of the dried tissue when the apple slices were dried at 40 to 45 and 60 to 65°C, and in the extreme case (at 60 to 65°C) cracks formed on the surface. Low case hardening was observed in samples dried at 20 to 25°C.

Wang and Brennan (1995) utilized microscopy to study the structural changes in potatoes during air drying (40 and 70°C). They found that shrinkage occurred first at the surface and then gradually moved to the bottom with an increase in drying time. The cell walls became elongated. The degree of shrinkage at a low drying temperature (40°C) was greater (i.e., less porosity) than that at high temperature (70°C). At a low drying rate (i.e., at low temperature) the moisture content at the center of a piece was never much higher than at the surface, internal stresses were minimized, the material shrunk fully into a solid core, and shrinkage was uniform.

At higher drying rates (i.e., higher temperature) the surface moisture decreased very fast so that the surface became stiff, the outer layers of the material became rigid, and the final volumes were fixed early in the drying. This caused the case hardening phenomenon and limited subsequent shrinkage, thus increasing pore formation. As the drying proceeded, the tissues split and rupture internally, forming an open structure, and cracks were formed in the inner structure. When the interior finally dried and shrunk, the internal stresses pulled the tissue apart. Thus, the increase of surface case hardening was increased with the increase in drying temperature; at the extreme case, cracks were formed on the surface and/or inside.

In case hardening the permeability and integrity of the crust play a role in maintaining the internal pressure inside the material boundary. Internal pressure always tries to puff the product by creating a force to the crust. During air drying stresses are formed that are due to nonuniform shrinkage resulting from nonuniform moisture and/or temperature distributions. This may lead to stress crack formation, when stresses exceed a critical level. Crack formation is a complex process influenced interactively by heat and moisture transfer, physical properties, and operational conditions (Kowalski and Rybicki, 1996; Liu et al., 1997). Liu et al. (1997) identified the relative humidity and temperature of the air as the most influential parameters that need to be controlled to eliminate the formation of cracks. Sannino et al. (2005) used a DVS-1000 system to study the drying process of lasagna pasta at controlled humidity and temperature with a sensing device to measure the electrical conductivity of pasta during the drying process. An anomalous diffusion mechanism was observed, typical of the formation of a glassy shell on the surface of the pasta slice during drying at low relative humidity, which inhibits fast diffusion from the rubbery internal portion. Internal stresses at the interface of the glassy–rubbery surfaces are responsible for crack formation and propagation and thus lasagna sample and breakage. Accurate control of the sample water activity and external environment humidity needs to be maintained to avoid stress generation in the sample and crack formation. The glass-transition concept cannot explain the effects of crust, case hardening, cracks, and internal pressure. Vacuum drying of tuna meat produced higher porosity compared to air drying when both samples were dried at 70°C (Rahman et al., 2002). The porosity of dehydrated products is increased as the vacuum pressure is decreased, which means shrinkage can be prevented by controlling the pressure (Krokida and Maroulis, 2000). Microwaves create a massive vaporization situation and cause puffing (Pere et al., 1998). This indicates that, in addition to the temperature effect, environment pressure can also affect pore formation, and this effect cannot be explained by only a single glass-transition concept. Similarly, in the case of extrusion, the higher the processing temperature is above 100°C, the higher the porosity, which is contrary to the glass-transition concept (Ali et al., 1996). This is attributable to the rapid vaporization of the water vapor at the exit of the die. Rahman et al. (2005) also found much higher pore formation with 105°C convection air drying compared to 50 or 80°C.

4.4.1.3 Rahman Hypothesis

After analyzing the experimental results from the literature, Rahman (2001) stated that the glass-transition theory does not hold true for all products or processes. Other concepts, such as surface tension, pore pressure, structure, environment pressure, and mechanisms of moisture transport, also play important roles in explaining the formation of pores. Rahman (2001) hypothesized that because capillary force is the main force responsible for collapse, counterbalancing this force causes the formation of pores and lower shrinkage. The counterbalancing forces are due to generation of internal pressure that is caused by vaporization of water or other solvents, variations in the moisture transport mechanism, and pressure outside the material. Other factors could be the strength of the solid matrix (i.e., ice formation; case hardening; surface crack formation; the permeability of water through the crust; changes in the tertiary and quaternary structures of the polymers; the presence or absence of crystalline, amorphous, and viscoelastic natures of solids; matrix reinforcement; and residence time). Capillary force is related to the degree of water saturation of the porous matrix. In apples with an initial porosity of 0.20, a final porosity of 0.56 was achieved during air drying for the dried sample. However, the final porosity of dried samples was reduced to around 0.30 (nearly half) if the initial porosity of the apple matrix was reduced to 0.0 to 0.023 by saturation with a vacuum infiltration of water before air drying (Bengtsson et al., 2003). Hussain et al. (2002) also identified that the prediction accuracy improved significantly when the initial porosity was included in the generic prediction model of porosity. This indicated that the initial air phase (related to pore or matrix pressure) has significant effects on the subsequent pore formation in the matrix.

4.4.2 PREDICTION

There are negligible theoretical methods to predict the porosity in foods during processing. Lozano et al. (1980) proposed the geometric model of Rotstein and Cornish (1978) to predict the porosity of fruits and vegetables. These authors mentioned that the change in porosity is the result of two processes: the shrinkage of the overall dimensions and the shrinkage of the cells themselves. The model is based on the truncated spheres forming a cube. On the basis of geometry considerations, the porosity at full turgor and dried conditions were proposed. However, it is difficult to estimate two proposed geometric parameters for real foods. The formation of pores in foods during drying can be grouped into two generic types: with (Figure 4.3) and without an inversion point (Figure 4.4). Figure 4.3a shows that during drying initially the pores are collapsed and reach a critical value, and a further decrease in moisture causes the formation of pores until complete drying. An opposite condition exists in Figure 4.3b. Figure 4.4 reveals that the level of pores is increased or decreased as a function of the moisture content (Rahman, 2000, 2001). Usually the porosity of foods increases but the moisture content decreases. The final porosity depends on the material's structure as well as the drying conditions. Dried apples develop high porosity while the shrinkage phenomenon seems to be intense. Bananas and carrots seem to shrink intensely and develop lower porosity. Most of the porosity is predicted from the density data or from empirical correlations of the porosity and moisture content. Mainly empirical correlations are used to correlate the porosity. Rahman et al. (1996) developed the following correlations for open and closed pores in calamari meat during air drying up to zero moisture content as

$$\varepsilon_{\text{op}} = 0.079 - 0.164(X_w/X_w^0) + 0.099(X_w/X_w^0)^2 \qquad (4.11)$$

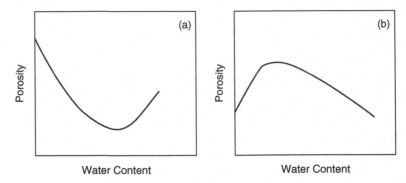

FIGURE 4.3 Change of porosity as a function of water content for inversion point. (From Rahman, M.S., *Dry. Technol.*, 19, 1–13, 2001. With permission.)

$$\varepsilon_{cp} = 0.068 - 0.216(X_w/X_w^0) + 0.138(X_w/X_w^0)^2 \qquad (4.12)$$

where X_w and X_w^0 are the moisture content at any time and the initial moisture content (wet basis, fraction) before drying, respectively. Rahman (1991) developed the apparent porosity of squid mantle meat during air drying up to zero moisture content as

$$\varepsilon_a = 0.109 - 0.219(X_w/X_w^0) + 0.099(X_w/X_w^0)^2 \qquad (4.13)$$

Lozano et al. (1980) ascertained the correlation for the open pore porosity of apples during air drying as (X_w = 0.89 to 0.0)

$$\varepsilon_{op} = 1 - \frac{852.0 - 462.0[\exp(-0.66M_w)]}{1540.0[\exp(-0.051M_w)] + \{-1150.0[\exp(-2.4M_w)]\}} \qquad (4.14)$$

where M_w is the moisture content at any time (dry basis, fraction). They found a peak at low moisture content (Figure 4.3b). Ali et al. (1996) studied the expansion

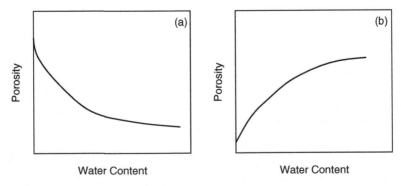

FIGURE 4.4 Change of porosity as a function of water content for no inversion point. (From Rahman, M.S., *Dry. Technol.*, 19, 1–13, 2001. With permission.)

characteristics of extruded yellow corn grits in a single-screw extruder with various combinations of barrel temperatures (100 to 200°C) and screw speeds (80 to 200 r/min). They observed that the open and apparent pore volumes increased with the increase of temperature and screw speed when the initial moisture content during extrusion was 0.64 (wet basis). Ali et al. (1996) developed the multiple regression of porosity as a function of the extrusion temperature and speed. Hussain et al. (2002) developed generic porosity prediction models using regression analysis and a hybrid neural network technique. They considered 286 data points of foods during convection air drying. For the temperature and water content as two inputs, the best regression model was the following:

$$\varepsilon_a = 0.5X_w^2 - 0.8X_w - 2.0 \times 10^{-4}t^2 + 0.02t - 0.15 \qquad (4.15)$$

where t is the temperature (°C). This equation gave an average error of $34 \pm 4\%$. However, inclusion of the initial porosity in the input variables resulted an average error of $24 \pm 4\%$. In this case the regression equation was

$$\varepsilon_a = 0.5X_w^2 - 0.8X_w - 2.0 \times 10^{-3}t^2 + 0.02t - 0.15(1 - \varepsilon_a^0) \qquad (4.16)$$

where ε_a^0 is the apparent porosity before drying. The type of product was then included with the other three inputs by assigning a numerical value and then the error was reduced to $14 \pm 4\%$. The regression equation resulted in

$$\varepsilon_a = 0.5X_w^2 - 0.8X_w - 2.0 \times 10^{-3}t^2 + 0.02t - 0.05(1 - \varepsilon_a^0)F \qquad (4.17)$$

A number was given to each type of product: 1 for sugar-based products, 2 for starch-based products, and 3 for other products. Hussain et al. (2002) also found that the error could be further reduced to <1% by applying a hybrid neural network technique. Although this approach was not based on a theoretical foundation, this empirical approach could be considered as generic to some extent and easy to use. Rahman et al. (1996) mentioned that further detailed studies beyond the empirical correlations are necessary to understand the physicochemical nature of the interactions of component phases and the collapse and formation of pores in food materials during processing.

4.4.3 THEORETICAL PREDICTIONS

The theoretical prediction models are mainly based on the conservation of mass and volume principle. However, further improved generic models need to be developed that could be valid in real situations.

4.4.3.1 Density and Shrinkage

Most of the density prediction models in the literature are empirical in nature. Fundamental models exist that are based on the conversion of both mass and volume.

Thus, a number of authors have proposed models of this kind. When food is treated as an ideal mixture (where the mixing process conserves both mass and volume), the density of food can be predicted using fundamental models (Mannapperuma and Singh, 1989). Rahman (1991) proposed a model using mass and volume conservation principles and a number of new terms to account for the interaction of the phases and formation of the air phase during processing. Food materials can be considered as multiphase systems (i.e., gas–liquid–solid systems). When the mixing process conserves both mass and volume, then the density of the multiphase system can be written as

$$\frac{1}{\rho_T} = \sum_{i=1}^{N} \frac{X_i}{(\rho_T)_i} = \xi \tag{4.18}$$

where $(\rho_T)_i$ and ρ_T are the true density of component i and composite mixture, respectively; X_i is the mass fraction of component i; and N is the total number of components present in the mixtures. Miles et al. (1983) and Choi and Okos (1985) proposed the above equation for predicting the density of food materials. However, this equation has limited uses in the cases where there is no air phase present and no excess volume due to the interaction between the phases. Rahman (1991) extended the theoretical model, introducing the pore volume and interaction term into the above equation. The apparent density of a composite mixture can be divided into three parts based on the conservation law for mass and volume:

apparent volume = actual volume of pure component phases + volume of pores or
air phase + excess volume due to interaction of phases

The excess volume can be positive or negative, depending on the physicochemical nature of the process of concern, whereas the porosity is always positive. Equation (4.18) can be written as

$$\frac{1}{\rho_a} = \sum_{i=1}^{N} \frac{X_i}{(\rho_T)_i} + V_a + V_{ex} \tag{4.19}$$

where V_a is the specific apparent volume of the void or air phase (m³/kg) and V_{ex} is the specific excess volume (m³/kg). The apparent porosity, apparent volume, and excess volume fraction (ε_{ex}) are defined as

$$\varepsilon_a = \frac{V_v}{V_a}, \quad V_a = \frac{1}{\rho_a}, \quad \varepsilon_{ex} = \frac{V_{ex}}{V_a} \tag{4.20}$$

Equation (4.19) can be transformed to Eq. (4.21) using the apparent porosity, apparent volume, and excess volume fraction:

$$\frac{1}{\rho_a} = \frac{\xi}{1 - \varepsilon_{ex} - \varepsilon_a} \tag{4.21}$$

When the apparent porosity and excess volume fraction are negligible, Eq. (4.19) is reduced to the following:

$$\frac{1}{\rho_a} = \frac{1}{\rho_m} = \frac{1}{\rho_T} = \xi \tag{4.22}$$

The apparent porosity (or total air volume fraction) and excess volume fraction can be calculated from the experimental density data as

$$\varepsilon_a = 1 - \frac{\rho_a}{\rho_m} \quad \text{and} \quad \varepsilon_{ex} = 1 - \frac{\rho_m}{\rho_T} \tag{4.23}$$

The apparent shrinkage (S_a) can be written as

$$S_a = \frac{V_a}{V_a^0} = \frac{\xi}{\xi^0} \times \frac{(1 - \varepsilon_{ex} - \varepsilon_a)}{(1 - \varepsilon_{ex}^0 - \varepsilon_a^0)} \tag{4.24}$$

In the case of organic liquid–liquid mixtures, a number of authors confirmed that there was a deviation in total volume, which might be positive or negative. Positive means an increase and negative means a decrease in the net specific volume. Excess volume and excess enthalpy of binary mixtures of liquids are also called excess properties of mixtures. The excess volumes of binary mixtures showed a range of behaviors: in some cases there was a single or double peak in the plot (excess volume vs. mass fraction of one component) and in some cases there were positive and negative lobes in the plots. For example, up to a mass fraction of 0.5 the excess volume is positive and from 0.5 to 1.0 the excess volume is negative. The excess properties and porosity are usually not easy to correlate in complex systems such as foods (Rahman et al., 1996).

4.4.3.2 Porosity Prediction

Two concepts are applied to develop porosity prediction models: one is based on the concept of the shrinkage–expansion coefficient, and one is based on the critical moisture contents when material is transforming into different characteristics.

4.4.3.2.1 Rahman Model
Rahman (2003) developed a theoretical model to predict the porosity based on the conservation of mass and volume principle and defining a shrinkage–expansion coefficient. A model for the ideal condition was derived, and then it was extended to a nonideal condition.

4.4.3.2.1.1 Ideal Condition
A theoretical model (ideal condition) to predict the porosity in foods during drying was developed based on the conservation of mass and volume principle and assuming that the volume of pores formed is equal to the volume of water removed during drying (Rahman, 2003). In this case there is no shrinkage (or collapse) or expansion in the initial volume of the material. If no pores exist in the material before drying

(i.e., initially), the volume fraction of pores in the material at any water level can be expressed as

$$\varepsilon'_a = \frac{\alpha}{\alpha + \beta} \quad \text{if } \varepsilon^0_a = 0 \tag{4.25}$$

where

$$\psi = \frac{1 - X^0_w}{1 - X_w}, \quad \alpha = \frac{1 - \psi}{\rho_w}, \quad \beta = \frac{\psi}{\rho_m}$$

Here, ψ is the mass of the sample at any moisture content per unit initial mass before processing (kg/kg), α is the volume of water removed per unit mass (m³/kg), and β is the volume of water remaining and solids per unit mass (m³/kg). The material density of a multicomponent mixture can be estimated from the following equation:

$$\frac{1}{\rho_m} = \frac{1}{\rho_t} = \frac{X_w}{\rho_w} + \sum_{i=1}^{n} \frac{X_i}{\rho_i} \tag{4.26}$$

where X_i and ρ_i are the mass fraction and density of other components except water, respectively. The mass fraction of other components except water can be estimated from the mass fraction of water:

$$X_i = \frac{X^0_i}{\psi} \tag{4.27}$$

For a two component mixture, if the excess volume due to the interaction between water and solids is zero, the material density can be expressed as

$$\frac{1}{\rho_m} = \frac{1}{\rho_T} = \frac{X_w}{\rho_w} + \frac{X_s}{\rho_s} \tag{4.28}$$

where ρ_s is the substance density for dry solids at zero moisture content (kg/m³). For a porous material, the ideal porosity can be written as

$$\varepsilon_a = \varepsilon^0_a + (1 - \varepsilon^0_a)\varepsilon'_a \tag{4.29}$$

As expected, the ideal model may not be valid in many practical cases. Constant porosity and no shrinkage are often stated as key assumptions in the model for dryer design. Although the Eq. (4.29) does not predict the real situation, it can be used for ideal conditions. In addition, it could be used as an initial estimation when experimental results are missing in the case of nonideality and to assess which processes of pore formation are closer to the ideal condition.

4.4.3.2.1.2 Nonideal Conditions
The ideal model is then extended for nonideal conditions when there is shrinkage, collapse, or expansion by defining a shrinkage–expansion coefficient as

$$\varepsilon''_a = \frac{\phi \alpha}{\phi \alpha + \beta} \tag{4.30}$$

where ϕ is the shrinkage–expansion coefficient for pore formation or collapse and the porosity can be calculated as (Rahman, 2003)

$$\varepsilon_a = \varepsilon_a^0 + (1 - \varepsilon_a^0)\varepsilon_a'' \qquad (4.31)$$

The shrinkage–expansion coefficient can be estimated using Eqs. (4.30) and (4.31) and applying the initial porosity and actual measured porosity using the values of α and β:

$$\phi = \frac{\beta(\varepsilon_a^0 - \varepsilon_a)}{\alpha(\varepsilon_a - 1)} \qquad (4.32)$$

When the value of ϕ is 1, the nonideal model is transformed to the ideal conditions, that is, the volume of water loss is equal to the pores formed. A ϕ value of 0 indicates complete collapse of all pores, that is, no pore formation occurs. Values of ϕ that are >1 show there is an expansion of the material's boundary during the process. Values of ϕ that are <1 indicate that there is shrinkage of the material by collapse of the initial air filled pores or collapse of water filled pores. When the moisture content of the product is very close to the initial moisture content (e.g., only 2% change), this model may provide an unrealistic prediction. Future work needs to be targeted on estimating the values of ϕ for a wide variation of the material's characteristics and processing conditions.

4.4.3.2.2 Sarma–Heldman Concept

Sarma and Heldman (cited by Heldman, 2001) proposed a concept of the visual structure of food particles as a function of the moisture content. There are three critical moisture ranges: $0 < M_w \leq 0.20$, $0.20 \leq M_w \leq 0.60$, and $M_w \geq 0.60$. Based on observations for starch and casein particles, six moisture content ranges were defined by Heldman (2001). However, all of the critical moisture contents varied with product types and their compositions, as well as the process used to manufacture the food product. Thus, the generic concept transforms into an empirical nature unless critical conditions can be predicted from fundamental theoretical concepts.

4.5 QUALITY CREATION AND DETERIORATION BY MICROSTRUCTURE

Pores or bubbles occur in diverse solid food products, such as cakes; breads; wafers; breakfast cereals; and puffed, extruded, and dried foods. The issue of pores in food products is generic. Many foods possess complex microstructures. The safety and quality of foods rely on the ability to control *microbial growth* in each microscopic location. Variations in microenvironmental regions can affect colony formation as a direct result of a restricted supply of nutrients or oxygen to the growing cells (Robins and Wilson, 1994). The effect of oxygen on lipid oxidation is also closely related to the product porosity. Freeze-dried tuna meat was found to be more susceptible to oxygen because of its high porosity and exposed surface area compared to air-dried and vacuum-dried samples (Rahman et al., 2002). The characteristics of pores affect the flow of gas, thus controlling many processes and the safety of the products. Pore size distribution affects the mass transfer in a product.

Microstructure determines the mechanical properties, and thus the texture of extruded food products. Many solid food materials are brittle and fragile, especially when dry. The puffiness of foods also depends on the porosity. Specific processing can be applied to attain specific porosity for better texture and mouth feel of food. Pores have affects on the sensory properties of foods. The perception of food fracturability involves receptors sensitive to vibration. Vibrations are likely to be produced when food is fractured because of the cellular nature of most fracturable foods (Vickers and Bourne, 1976). Stiff-walled cells are filled with air in low moisture foods and with liquid in high moisture foods. When sufficient force is applied during crushing by bite, there is a serial bursting of the cell walls. Thus, vibratory sensations by sound are involved when cellular foods are crushed (Vickers and Bourne, 1976). A broad frequency band of sound with irregularly varying amplitude was produced by crushing low and high moisture foods, which is identified as being crisp (Christensen, 1984). Air bubbles enhance the creamy perception, thus low fat products can be developed. The texture, density, wettability, rehydration capacity, and mechanical properties of dried food products directly depend on drying process conditions. For example, freeze-drying typically results in biological products with a porous crust and superior rehydration capacity, whereas hot-air drying results in a dense product with an impermeable crust. Depending on the end use, these properties may be desirable or undesirable. If a long bowl life is required for a cereal product, a crust product that prevents moisture reabsorption may be preferred. If a product with good rehydration capacity is required, a slow drying process, such as freeze-drying, may be preferred (Achanta and Okos, 1995; Rahman and Perera, 1999). Raw materials may end up with completely different levels of shrinkage or collapse, depending on the type of process applied. For example, drying at 5°C results in a porous product with good rehydration capacity whereas drying at 80°C results in a dense product with poor rehydration capacity (Karathanos et al., 1993). Products baked at a low temperature have a different texture than those baked at a higher temperature. The low-temperature product has a crumbly texture, and the high-temperature product has a crispy crust.

Porosity is also important to estimate other properties, such as moisture diffusivity, hardness, and thermal conductivity. Fat bloom is a physical defect, which appears during chocolate storage. The grayish or white appearance results from the formation of large fat crystals on the chocolate surface. It seems to be related to the polymorphic forms of butter and recrystallization of solutes. A higher level of crystallization could generate pores and microfractures in the matrix and release part of the liquid phase. Loisel et al. (1997) analyzed the structure of dark chocolate samples by mercury porosimetry to determine whether the formation of fat bloom was related to the presence of pores. They found the presence of a porous matrix partly filled with liquid cocoa butter fractions in dark chocolate. This matrix is likely a network of cavities closed by multiple walls made by intermigrating fat crystals impregnated with liquid fat (Loisel et al., 1997).

4.6   CONCLUSION

Different types of porosity can be defined to characterize pores. In addition to porosity and pore size distribution, other characteristics are used to identify the types of pores, such as shape, thickness of pore wall, and structure of pores. Porosity is mainly measured by density, mercury porosimetry, and microscopic techniques. Different mechanisms of the creation and collapse of microstructures are being proposed. Most generic hypotheses posit that capillary force is the main force responsible for collapse, so counterbalancing this force causes the formation of pores and lower shrinkage. Porosity is mainly predicted by empirical correlations because the theoretical models are far from the reality of being applied as a generic approach.

NOMENCLATURE

VARIABLES

D	pore diameter (m)
dV/dr	volume distribution
f_v	pore size distribution function (m³/m kg)
M_w	moisture content at any time (dry basis, fraction)
N	total number of components present in the mixtures
P	pressure (Pa)
r	pore radius (μm)
S_a	apparent shrinkage
t	temperature (°C)
T_g	glass-transition temperature
T_g'	maximally freeze-concentrated glass-transition temperature
V	pore volume (m³)
V_a	specific volume of void or air phase (m³/kg)
V_{ex}	specific excess volume (m³/kg)
X_i	mass fraction of component i
X_w	moisture content at any time
X_w^0	initial moisture content (wet basis, fraction) before drying

GREEK LETTERS

α	volume of water removed per unit mass (m³/kg)
β	volume of water remaining and solids per unit mass (m³/kg)
δ	fractal dimension
ε_a	apparent pore porosity ($\varepsilon_a = \varepsilon_{op} + \varepsilon_{cp}$)
ε_a^0	apparent porosity before drying
ε_B	bulk porosity
ε_{BP}	bulk particle porosity ($\varepsilon_{BP} = \varepsilon_B + \varepsilon_{op}$)
ε_{cp}	closed pore porosity
ε_{op}	open pore porosity
ε_T	total porosity
ϕ	shrinkage–expansion coefficient
ψ	mass of sample at any moisture content per unit initial mass before processing (kg/kg)
ρ_a	apparent density (kg/m³)

ρ_B bulk density (kg/m³)
ρ_i density of other components except water
ρ_m material density (kg/m³)
ρ_s substance density for dry solids at zero moisture content (kg/m³)
ρ_T true density of composite mixture (kg/m³)
$(\rho_T)_i$ true density of component i (kg/m³)
ξ true specific volume (m³/kg)

REFERENCES

Achanta, S. and Okos, M.R., Impact of drying on the biological product quality, in *Food Preservation by Moisture Control. Fundamentals and Applications*, Barbosa-Canovas, G.V. and Welti-Chanes, J., Eds., Technomic Publishing, Lancaster, PA, 1995, p. 637.

Achanta, S. and Okos, M.R., Predicting the quality of dehydrated foods and biopolymers—Research needs and opportunities, *Dry. Technol.*, 14, 1329–1368, 1996.

Ali, Y., Hanna, M.A., and Chinnaswamy, R., Expansion characteristics of extruded corn grits, *Food Sci. Technol.*, 29, 702–707, 1996.

Bai, Y., Rahman, M.S., Perera, C.O., Smith, B., and Melton, L.D., Structural changes in apple rings during convection air-drying with controlled temperature and humidity, *J. Agric. Food Chem.*, 50, 3179–3185, 2002.

Bengtsson, G.B., Rahman, M.S., Stanley, R.A., and Perera, C.O., Apple rings as a model for fruit drying behavior: Effect of surfactant and reduced osmolality reveal biological mechanisms, *J. Food Sci.*, 68, 563–566, 2003.

Choi, Y. and Okos, M.R., Effects of temperature and composition on the thermal properties of foods, in *Food Engineering and Process Applications. Transport Phenomena*, Le Maguer, M. and Jelen, P., Eds., Elsevier Applied Science, London, Vol. 1, 1985.

Christensen, C.M., Food texture perception, *Adv. Food Res.*, 29, 159–199, 1984.

Cnossen, A.G. and Siebenmorgen, T.J., The glass transition temperature concept in rice drying and tempering: Effect on milling quality, *Trans. ASAE*, 43, 1661–1667, 2000.

Coker, D.A., Torquato, S., and Dunsmuir, J.H., Morphology and physical properties of Fontainebleau sandstone via a tomographic analysis, *J. Geophys. Res.*, 101, 17497–17506, 1996.

Del Valle, J.M., Cuadros, T.R.M., and Aguilera, J.M., Glass transitions and shrinkage during drying and storage of osmosed apple pieces, *Food Res. Int.*, 31, 191–204, 1998.

Falcone, P.M., Baiano, A., Zanini, F., Mancini, L., Tromba, G., and Montanari, F., A novel approach to the study of bread porous structure: Phase-contrast x-ray microtomography, *J. Food Sci.*, 69, FEP38–FEP43, 2004.

Genskow, L.R., Consideration in drying consumer products, in *Drying '89*, Mujumdar, A.S. and Roques, M., Eds., Hemisphere Publishing, New York, 1990.

Heldman, D.R., Prediction models for thermophysical properties of foods, in *Food Processing Operations Modeling. Design and Analysis*, Irudayaraj, J., Ed., Marcel Dekker, New York, 2001, pp. 1–23.

Hicsasmaz, Z. and Clayton, J.T., Characterization of the pore structure of starch based food materials, *Food Struct.*, 11, 115–132, 1992.

Hussain, M.A., Rahman, M.S., and Ng, C.W., Prediction of pores formation (porosity) in foods during drying: Generic models by the use of hybrid neural network, *J. Food Eng.*, 51, 239–248, 2002.

Karathanos, V., Anglea, S., and Karel, M., Collapse of structure during drying of celery, *Dry. Technol.*, 11, 1005–1023, 1993.

Karathanos, V.T., Kanellopoulos, N.K., Belessiotis, V.G., Development of porous structure during air drying of agricultural plant products, *J. Food Eng.*, 29, 167–183, 1996.

Karathanos, V.T., Saravacos, G.D., Porosity and pore size distribution of starch materials, *J. Food Eng.*, 18, 259–280, 1993.

Karel, M., Anglea, S., Buera, P., Karmas, R., Levi, G., and Roos, Y., Stability-related transitions of amorphous foods. *Thermochimica Acta.*, 246, 249–269, 1994.

Khalloufi, S. and Ratti, C., Quality determination of freeze-dried foods as explained by their glass transition temperature and internal structure, *J. Food Sci.*, 68, 892–903, 2003.

Kowalski, S.J. and Rybicki, A., Drying induced stresses and their control, paper presented at the 10th International Drying Symposium (Drying '96), Krakow, Poland, 1996.

Krokida, M.K., Karathanos, V.T., and Maroulis, Z.B., Effect of freeze-drying conditions on shrinkage and porosity of dehydrated agricultural products, *J. Food Eng.*, 35, 369–380, 1998.

Krokida, M. and Maroulis, Z., Quality changes during drying of food materials, in *Drying Technology in Agricultural and Food Sciences*, Mujumdar, A.S., Ed., Science Publishers, Enfield, NH, 2000, pp. 61–106.

Levine, H. and Slade, L., A polymer physico-chemical approach to the study of commercial starch hydrolysis products (SHPs), *Carbohydr. Polym.*, 6, 213–244, 1986.

Liu, H., Zhou, L., and Hayakawa, K., Sensitivity analysis for hygrostress crack formation in cylindrical food during drying, *J. Food Sci.*, 62, 447–450, 1997.

Loisel, C., Lecq, G., Ponchel, G., Keller, G., and Ollivon, M., Fat bloom and chocolate structure studied by mercury porosimetry, *J. Food Sci.*, 62, 781–788, 1997.

Lowell, S. and Shields, J.E., *Powder Surface Area and Porosity*, 2nd ed., Chapman & Hall, London, 1984.

Lozano, J.E., Rotstein, E., and Urbicain, M.J., Total porosity and open-pore porosity in the drying of fruits, *J. Food Sci.*, 45, 1403–1407, 1980.

Mannapperuma, J.D. and Singh, R.P., A computer-aided method for the prediction of properties and freezing/thawing times of foods, *J. Food Eng.*, 9, 275–304, 1989.

Miles, C.A., Beek, G.V., and Veerkamp, C.H., Calculation of thermophysical properties of foods, in *Thermophysical Properties of Foods*, Jowitt, R., Escher, F., Hallstrom, B., Meffert, H.F.T., Spiess, W.E.L., and Vos, G., Eds., Applied Science Publishers, London, 1983, pp. 269–312.

Pere, C., Rodier, E., and Schwartzentruber, J., Effects of the structure of a porous material on drying kinetics in a microwave vacuum laboratory scale dryer, paper presented at the 11th International Drying Symposium (IDS '98), 1998.

Pfeifer, P. and Avnir, D., Chemistry in noninteger dimensions between two and three. I. Fractal theory of heterogenous surfaces, *J. Chem Phys.*, 79, 3558–3565, 1983.

Rahman, M.S., Thermophysical Properties of Seafoods, Ph.D. thesis, University of New South Wales, Sydney, 1991.

Rahman, M.S., *Food Properties Handbook*, CRC Press, Boca Raton, FL, 1995.

Rahman, M.S., Physical meaning and interpretation of fractal dimensions of fine particles measured by different methods, *J. Food Eng.*, 32, 447–456, 1997.

Rahman, M.S., Mechanism of pore formation in foods during drying: Present status, in *Proceedings of the Eighth International Congress on Engineering and Food (ICEF-8)*, Welti-Chanes, J., Barbosa-Canovas, G.V., and Aguilera, J.M., Eds., Technomic Publishing, Lancaster, PA, 2000, pp. 1111–1116.

Rahman, M.S., Toward prediction of porosity in foods during drying: A brief review. *Dry. Technol.*, 19, 1–13, 2001.

Rahman, M.S., A theoretical model to predict the formation of pores in foods during drying, *Int. J. Food Prop.*, 6, 61–72, 2003.

Rahman, M.S., Prediction of pores in foods during processing: From regression to knowledge development from data mining (KDD) approach, paper presented at the International Conference on Engineering and Food (ICEF-9), Montpellier, France, March 7–11, 2004.

Rahman, M.S., Mass–volume–area-related properties of foods, in *Engineering Properties of Foods*, Rao, M.A., Rizvi, S.S.H., and Datta, A.K., Eds., CRC Press, Boca Raton, FL, 2005, pp. 1–39.

Rahman, M.S., Al-Amri, O.S., and Al-Bulushi, I.M., Pores and physico-chemical characteristics of dried tuna produced by different methods of drying, *J. Food Eng.*, 53, 301–313, 2002.

Rahman, M.S., Al-Zakwani, I., and Guizani, N., Pore formation in apple during air-drying as a function of temperature: Porosity and pore-size distribution, *J. Sci. Food Agric.*, 85, 979–989, 2005.

Rahman, M.S. and Perera, C.O., Drying and food preservation, in *Handbook of Food Preservation*, Rahman, M.S., Ed., Marcel Dekker, New York, 1999, pp. 173–216.

Rahman, M.S., Perera, C.O., Chen, X.D., Driscoll, R.H., and Potluri, P.L., Density, shrinkage and porosity of calamari mantle meat during air drying in a cabinet dryer as a function of water content, *J. Food Eng.*, 30, 135–145, 1996.

Rahman, M.S. and Sablani, S.S., Structural characteristics of freeze-dried abalone: Porosimetry and puncture, *Trans. IChemE*, 81C, 309–315, 2003.

Ratti, C., Shrinkage during drying of foodstuffs, *J. Food Eng.*, 23, 91–105, 1994.

Robins, M.M. and Wilson, P.D.G., Food structure and microbial growth, *Trends Food Sci. Technol.*, 5, 291–293, 1994.

Rotstein, A. and Cornish, A.R.H., Prediction of the sorption equilibrium relationship for the drying of foodstuffs, *AIChE J.*, 24, 966, 1978.

Sablani, S. and Rahman, M.S., Pore formation in selected foods as a function of shelf temperature during freeze drying. *Dry. Technol.*, 20, 1379–1391, 2002.

Sablani, S.S., Rahman, M.S., Al-Kuseibi, M.K., Al-Habsi, N.A., Al-Belushi, R.H., Al-Marhubi, I., and Al-Amri, I.S., Influence of shelf temperature on pore formation in garlic during freeze-drying, *J. Food Eng.*, 80, 68–79, 2007.

Sannino, A., Capone, S., Siciliano, P., Ficarella, A., Vasanelli, L., and Maffezzolo, A., Monitoring the drying process of lasagna pasta through a novel sensing device-based method, *J. Food Eng.*, 69, 51–59, 2005.

Slade, L. and Levine, H., Non-equilibrium behavior of small carbohydrate–water systems, *Pure Appl. Chem.*, 60, 1841–1864, 1988.

Slade, L. and Levine, H., A food polymer science approach to structure property relationships in aqueous food systems: Non-equilibrium behavior of carbohydrate–water systems, in *Water Relationships in Food*, Levine, H. and Slade, L., Eds., Plenum Press, New York, 1991, pp. 29–101.

Snigirev, A., Snigireva, I., Kohn, V., Kuznetsov, S., and Schelokov, I., On the possibilities of x-ray phase contrast microimaging by coherent high-energy synchrotron radiation, *Rev. Sci. Instrum.*, 66, 5486–5492, 1995.

Trater, A.M., Alavi, S., and Rizvi, S.S.H., Use of non-invasive x-ray microtomography for characterizing microstructure of extruded biopolymer foams, *Food Res. Int.*, 38, 709–719, 2005.

Vickers, Z. and Bourne, M.C., A psychoacoustical theory of crispness, *J. Food Sci.*, 41, 1158–1164, 1976.

Wang, N. and Brennan, J.G., Changes in structure, density and porosity of potato during dehydration, *J. Food Eng.*, 24, 61–76, 1995.

Xiong, X., Narsimhan, G., and Okos, M.R., Effect of comparison and pore structure on binding energy and effective diffusivity of moisture in porous food, *J. Food Eng.*, 15, 187–208, 1991.

5 Convective Drying of Foods

María Elena Carrín and Guillermo Héctor Crapiste

CONTENTS

5.1 INTRODUCTION

Preserving food and agricultural products via air drying has been utilized worldwide for centuries. At present, there is still increasing demand for high-quality shelf-stable dried food products, especially fruits and vegetables. The design of drying equipment and the selection of dryer operating conditions for food products must be done on the basis of raw material characteristics, quality requirements for the final dry product, and economic analysis.

Convective drying of foods is a combined heat and mass transfer operation. Its modeling and optimization require understanding of the transport mechanisms inside and between the solid food and the drying air, as well as knowledge of the thermophysical, equilibrium, and transport properties of both systems (Crapiste, 2000; Crapiste and Rotstein, 1997).

5.2 BASIC CONCEPTS RELATED TO CONVECTIVE DRYING

Calculations in convective drying are based on the air and material properties. The transfer of energy mainly depends on air and food temperatures, air-flow rate, and exposed area of food material. Internal transfer of moisture is governed by the nature of the food, including composition and structure, temperature, pressure, and particularly moisture content. The transfer of moisture from solid to air mainly depends on the surface water activity, air humidity, air-flow rate, exposed area of the food material, and pressure.

5.2.1 BASIC CONCEPTS AND PROPERTIES ASSOCIATED WITH WATER VAPOR AND AIR MIXTURES

Humidity (H) and *relative humidity* (RH) are calculated according to the following:

$$H = \frac{P_w}{P - P_w} \frac{M_w}{M_g} \tag{5.1}$$

where M_w is the molecular weight of the moisture vapor, M_g is the molecular weight of dry air (gas), P is the total pressure, and P_w is the partial pressure of water vapor.

When the partial pressure of the vapor in the gas phase equals the vapor pressure of the liquid at the temperature of the system (T), the gas is saturated. The relative humidity is a measure of moisture saturation. It is defined as the ratio of the partial pressure of water vapor in a gaseous mixture with air to the saturated vapor pressure of water at a given temperature. The relative humidity is expressed as a percentage and is calculated in the following manner:

$$RH = \frac{P_w}{P_w^0} \times 100 \tag{5.2}$$

where P_w^0 is the saturated vapor pressure.

The *dry bulb temperature* (T_{db}) is the temperature of the air as measured by a thermometer freely exposed to the air but shielded from radiation and moisture.

The *wet bulb temperature* (T_{wb}) is measured by a gas passing rapidly over a wet thermometer bulb. It is used along with dry bulb temperature to measure the relative humidity of a gas.

The relationships between air and water vapor and the psychrometric properties of moist air are commonly found in the form of psychrometric tables and charts

(Keey, 1978; Mujumdar, 1987; Perry and Green, 1999), computer programs (Ratti et al., 1989), or calculating equations (ASAE, 1982). For more detailed information on psychrometric charts, please refer to Chapter 1.

5.2.2 TERMINOLOGY AND CONCEPTS ASSOCIATED WITH DRYING OF FOOD MATERIALS

The *moisture content* of a material is the weight of water per weight of wet solid (wet basis, X_w) or the weight of water per weight of dry solid (dry basis, X). They are related in the following manner:

$$X_w = \frac{X}{X + 1} \quad \text{and} \quad X = \frac{X_w}{1 - X_w} \tag{5.3}$$

The *moisture ratio* (MR) is the moisture content of a material during drying. It is usually expressed in a dimensionless form as

$$\text{MR} = \frac{X - X_e}{X_0 - X_e} \tag{5.4}$$

where X is the moisture content at any time t, X_e is the equilibrium moisture content, and X_0 is the initial moisture content.

Water activity (a_w) is an index of the availability of water for chemical reactions and microbial growth. It can be defined by the following equation:

$$a_w = \frac{P_w}{P_w^0} = \frac{\text{RH}}{100} \tag{5.5}$$

Moisture content can be classified according to its availability in the food matrix.

1. *Bound moisture*: Bound moisture is the amount of water tightly bound to the food matrix, mainly by physical adsorption on active sites of hydrophilic macromolecular materials such as proteins and polysaccharides, with properties significantly different from those of bulk water.
2. *Free moisture content*: Free moisture content is the amount of water mechanically entrapped in the void spaces of the system. Free water is not in the same thermodynamic state as liquid water because energy is required to overcome the capillary forces. Furthermore, free water may contain chemicals, especially dissolved sugars, acids, and salts, altering the drying characteristics.

An important value is the equilibrium moisture content, which is the moisture content of a product in equilibrium with the surrounding air at given temperature and humidity conditions. Theoretically, it is the minimum moisture content to which a material can be dried under these conditions. A plot of the equilibrium moisture content versus the relative humidity or water activity at constant temperature, which is called sorption isotherm, is used to illustrate the degree of water interactions with foods. The value of the equilibrium moisture content for some solids depends on the

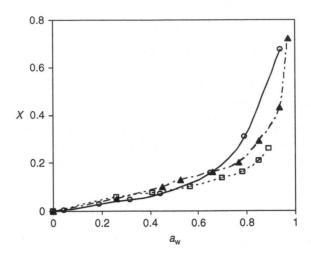

FIGURE 5.1 Desorption isotherms for typical food products at 40°C: (o) apple (Román et al., 1982), (□) potato (Crapiste and Rotstein, 1986), and (▲) garlic (Pezzutti and Crapiste, 1997).

direction from which equilibrium is approached, and the desorption equilibrium is of particular interest for drying calculations.

Although several models have been proposed to estimate the equilibrium moisture content of food materials, experiments are still needed to determine the correct values because water binding is affected not only by temperature but also by composition and structure. Theoretical predictions and empirical models are not generalized. Iglesias and Chirife (1982) provide water sorption parameters for many foods. Figure 5.1 presents sorption isotherms for typical food products.

The equilibrium moisture content for biological materials generally increases rapidly with a relative humidity above ~60 to 80% because of capillary and dissolution effects. The equilibrium moisture content of food materials high in soluble solids (apples in Figure 5.1) increases slowly at low water activity and shows a steep rise at high relative humidities. At low relative humidities, the equilibrium moisture content is greater for food materials with a high content in high molecular weight polymers, such as potatoes, which have more active sites for sorption, and lower for food materials high in soluble solids.

For more information on sorption isotherms and their models, please refer to Chapter 1.

5.3 CONVECTIVE DRYING KINETICS

Drying kinetics refer to the changes in the average material moisture content and temperature with time. They are needed to calculate the amount of evaporated moisture, drying time, energy consumption, and product quality. The change of material moisture content and temperature is usually controlled by heat and mass transfer among the solid surface, the surroundings, and the inside of the drying material.

A drying process is well illustrated in diagrams based on the material moisture content and drying time (drying curve), the drying rate and drying time (drying rate curve), and the material temperature and drying time (temperature curve). The process can also be represented for similar curves by replacing the drying time with the moisture ratio.

The *drying rate* is defined as the amount of moisture removed from the dried material in unit time per unit of drying surface. It is calculated from the time derivative of the moisture content, as will be shown later.

Representative curves of the drying process are constructed from data obtained under laboratory conditions by measuring the mass and temperature changes of the food sample with time. Experiments are carried out at the steady state of external conditions (constant air temperature, velocity, and humidity) using hot air as the drying agent.

An illustration of a typical drying process for convective drying with constant external conditions is presented in Figure 5.2 for the case when the initial product temperature is lower than the air wet bulb temperature. The entire process can be divided into three characteristic periods:

1. Initial drying period (line A to B): The material is heated while surface evaporation occurs.
2. First drying period (line B to C): The moisture content decreases linearly with time, so the drying rate that is proportional to the slope of this line is constant. This behavior, called the *constant drying rate period*, takes place until critical point C is reached.
3. Second drying period (line C to E): The MR functionality with time becomes a curve at the critical point that approaches the equilibrium moisture content X_e asymptotically. This period is called the *falling drying rate period*.

The explanation for the drying curve shapes shown in Figure 5.2 must be focused on heat and mass transfer phenomena. Initially, the solid surface is totally covered by a thin liquid layer that behaves as unbound moisture. Thus, the drying rate is usually independent of the solid and is essentially the same as the evaporation rate from a free liquid surface under the same air conditions. However, the roughness of the solid surface and shrinkage may affect the drying rate during this period.

In the initial drying period, the solid (including its surface covered with a liquid layer) has a lower temperature than the equilibrium temperature T_{wb}. The drying rate in the range between points A and B in Figure 5.2 increases with the solid temperature until the surface temperature reaches the T_{wb} value (corresponding to the line B to C). An opposite effect, cooling of the solid, can be observed when the initial temperature is higher than T_{wb}. The initial drying period is usually very short and can be neglected in practice.

No constant drying rate period appears during the drying of hygroscopic solids or when the process is controlled by internal mass transport.

Temperature curves are important in the development of drying techniques because the quality of the dry material depends considerably on the product temperature during the drying process. In the constant drying rate period (B to C in

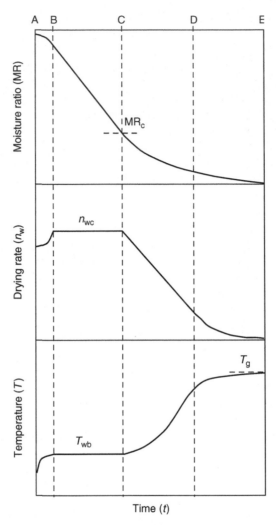

FIGURE 5.2 Typical behavior for convective drying at constant external conditions; MR_c, critical moisture ratio; n_{wc}, constant drying rate; T_g, gas temperature; T_{wb}, wet bulb temperature.

Figure 5.2) the material temperature usually equals the wet bulb temperature, allowing the use of higher temperatures for the drying air.

The falling drying rate period begins when the moisture ratio has reached the critical value. The critical moisture point greatly depends on the material properties and external conditions. High drying rates will raise the critical point, and low drying rates will decrease it (Okos et al., 1992). The complex and nonuniform structure of dried porous materials creates difficulties in calculating the critical moisture content, which must be determined by drying experiments performed under conditions similar to those used in industrial practice. Then, when the drying rate is represented as a function of the moisture content in a figure, the lower X value fitting the initial horizontal line corresponds to the critical moisture content.

At this point the amount of moisture at the drying surface begins to decrease gradually. Thus, the vapor pressure above the material surface also decreases and the drying rate decreases. Depending on the drying conditions and material properties, internal and external resistance can be important. At the beginning of this period the entire surface is no longer wetted, and the wetted area continually decreases until the surface is completely dry at point D (Figure 5.2). The second falling rate period begins at this point when the surface is completely dry. The plane of evaporation slowly recedes from the surface.

As the moisture content decreases, the internal resistance for mass transfer increases and may become the prevailing step while the product temperature approaches the dry bulb temperature. The moisture content asymptotically reaches the equilibrium value at the relative humidity and temperature of the air.

The shape of the falling drying rate period curve depends on the type of the material being dried and on the mass transfer conditions. In some cases no sharp discontinuity occurs at point D (Figure 5.2), and the change from partially wetted to completely dry conditions at the surface is so gradual that no sharp change is detectable. Strumillo and Kudra (1986) presented six types of drying rate curves versus moisture content in the falling drying period distinguished for the material characteristics: a straight line or concave downward curve for capillary–porous bodies with large specific evaporation surfaces, a concave upward curve for capillary–porous bodies with small specific evaporation surfaces, and more complex curves for colloidal–capillary–porous bodies.

In convective drying, moisture evaporation from the exposed surface of the food material occurs because moisture moves from the inside of the solid to the surface. Geankoplis (1993) briefly reviewed the principles of two of the theories explaining the various types of falling rate curves: liquid diffusion and capillary movement in porous solids. Liquid diffusion occurs when there is a concentration difference between the inside of the solid and the surface. This mechanism of moisture transport is usually found in nonporous solids. Conversely, when granular and porous materials are being dried, unbound or free moisture moves through the capillaries and voids of the solids because of a difference in capillary pressure. At the beginning of the falling rate period (point C in Figure 5.2) the water moves to the surface by capillary action, but the water surface layer starts to recede as drying proceeds. As the water is continuously removed and replaced by air, a point is reached where there is insufficient water left to maintain continuous films across the pores and the rate of drying suddenly decreases, starting the second falling rate period (point D). Then, diffusion of water vapor in the pores may become the prevailing mechanisms of mass transfer.

Several other mechanisms of internal mass transfer have been proposed for drying: *surface diffusion* of water, suggesting that molecules hop from one adsorption site to another; *thermal diffusion*, which is due to a temperature gradient; or *vapor mass transfer* (Knudsen diffusion, effusion, hydrodynamic flow, slip flow, etc.). More than one mechanism may contribute to the total flow, and the contribution of different mechanisms may change as the drying process proceeds (Bruin and

Luyben, 1980). At temperatures approaching and exceeding the boiling point of water, rapid vapor generation may produce significant total pressure gradients in addition to partial vapor pressure gradients. Total pressure-driven flow may occur in moderate temperature vacuum drying, high temperature convective and contact drying, and superheated steam drying.

Crapiste et al. (1988) developed a complete theory of drying for cellular material based on water activity as the driving force. They concluded that most of these mechanisms of water migration can be lumped together into a diffusion-like equation.

The drying rate curve in the second falling rate period in fine porous solids may conform to the diffusion law and the curve is concave upward. The drying rate curve in this period is often straight for very porous solids with large pores, and the diffusion equations do not apply. In drying of many food materials, the movement of water in the falling rate period has been represented by a diffusion theory. The effective diffusivity (D_{eff}) is usually lower than that of the first falling rate period although the temperature increases to T_{db}, because vapor diffusion is the prevailing mechanism in this stage.

When more than one driving force is present in a process, cross-effects can occur. Development of drying models accounting for cross-effects between different driving forces was performed using irreversible thermodynamics theory. Okos et al. (1992) summarized the results of different studies utilizing the irreversible thermodynamics framework to model heat and mass transfer in a porous medium and drying of porous food material. They concluded that cross-effects are negligible compared to direct transport mechanisms.

5.4   DRYING TIME CALCULATIONS

The drying rate (n_w) can be calculated from

$$n_w = -\frac{m_s}{A_s}\frac{dX}{dt} = -\frac{\rho_s}{a_v}\frac{dX}{dt} \tag{5.6}$$

where m_s is the mass of the dry solid, ρ_s is the density of the dry solid, A_s is the external area exposed to drying, and a_v is the area per unit volume.

Drying times should be determined separately for the constant and falling rate drying periods because of the different characters of the drying rate curves.

5.4.1   CONSTANT DRYING RATE PERIOD (I)

The drying rate in this period is constant and depends on the convective heat (h) and mass (k_g) transfer coefficients between the air and the solid surface. Thus, this period is strongly dependent on the drying agent velocity.

During the constant rate period, the temperature and moisture profiles inside the product are approximately flat because the moisture movement rate within the solid is sufficient to keep the surface saturated. Therefore, the drying rate is controlled by the rate of heat transfer to the surface that supplies the latent heat for evaporation.

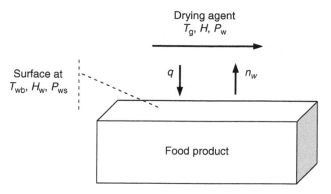

FIGURE 5.3   Scheme of heat and mass transfer in constant rate drying.

By considering only heat transfer to the solid surface by convection from hot air and moisture transfer from the surface to the air (Figure 5.3), it is possible to obtain the following equation:

$$\frac{dX}{dt} = -\frac{k_g A_s (P_{ws} - P_w)}{m_s} = -\frac{hA_s (T_g - T_{wb})}{m_s \Delta H_v} \tag{5.7}$$

where P_{ws} and P_w represent the water vapor partial pressure at the surface of the material and at the air, respectively; T_g is the gas temperature; T_{wb} is the surface material temperature; and ΔH_v is the heat of vaporization of water at the drying temperature. The vapor pressure at the product surface can be evaluated from the sorption isotherm according to Eq. (5.5).

The heat transfer coefficient (h) can be predicted using correlations given in the literature in terms of dimensionless numbers (Table 5.1). Mass transfer coefficients can be derived from direct correlations or from the better-known heat transfer coefficients using mass energy transfer analogies (Crapiste and Rotstein, 1997).

During the constant drying rate period, the rate of evaporation between points B and C (Figure 5.2) will be constant and equal to n_{wc}. In classical drying theory it is assumed that n_{wc} is equal to the evaporation rate from a free surface. Nevertheless, Strumillo and Kudra (1986) referred to more thorough observations showing that the value of n_{wc} depends on the type of dried material. They found that it is usually up to 30% lower than the values obtained in the process of liquid evaporation from a free surface.

Although the rate of drying n_w can be calculated using the heat transfer equation or the mass transfer equation, it is more reliable to use the heat transfer equation, because an error in determining the interface temperature T_{wb} affects the driving force $(T - T_{wb})$ much less than it affects the difference $(P_{ws} - P_w)$ (Geankoplis, 1993).

To estimate the drying time during the constant rate period, integration of Eq. (5.7) led to

$$t_I = \int_0^{t_1} dt = \frac{m_s \Delta H_v}{A_s h (T_g - T_{wb})}(X_0 - X_c) = \frac{m_s}{A_s k_g (P_{ws} - P_w)}(X_0 - X_c) \tag{5.8}$$

TABLE 5.1

Correlations for Heat Transfer Coefficients

Correlation	Conditions	Ref.
	Single Particles	
$Nu = 0.664\ Re^{0.5}\ Pr^{1/3}$	Parallel to flat sheet, $10^3 < Re < 3 \times 10^5$	Perry and Green (1999)
$Nu = 0.0366\ Re^{0.8}\ Pr^{1/3}$	Parallel to flat sheet, $Re > 3 \times 10^5$	Geankoplis (1993)
$Nu = 0.683\ Re^{0.466}\ Pr^{1/3}$	Cylinder, $40 < Re < 4 \times 10^3$	Perry and Green (1999)
$Nu = 2.0 + 0.6\ Re^{0.5}\ Pr^{1/3}$	Single sphere, $Re < 7 \times 10^4$	Perry and Green (1999)
$Nu = 0.036\ Re^{0.8}\ Pr^{1/3}$	Drying, $Re > 1.5 \times 10^5$	Treybal (1980)
$Nu = 0.249\ Re^{0.64}$	Drying food particles	Ratti and Crapiste (1995)
	Particle Beds	
$Nu = 1.95\ Re^{0.49}\ Pr^{1/3}$	Packed bed, $Re < 350$	Perry and Green (1999)
$Nu = 1.064\ Re^{0.59}\ Pr^{1/3}$	Packed bed, $Re > 350$	Perry and Green (1999)
$Nu = 2.0 + 1.1\ Re^{0.6}\ Pr^{1/3}$	Packed bed, steady and unsteady state	Wakao et al. (1979)
$Nu = 0.33\ Re^{0.6}$	Rotary dryers	Bakker-Arkema (1986)
$Nu = 0.0133\ Re^{1.6}$	Fluidized beds, $Re < 80$	Strumillo and Kudra (1986)
$Nu = 0.316\ Re^{0.8}$	Fluidized beds, $80 < Re < 500$	Strumillo and Kudra (1986)
$Nu = 2.0 + a\ Re^{0.6}Pr^{1/3}$	Spray dryers, $a = 0.5\text{--}0.6$	Masters (1979)

$Nu = hL_c/k$, $Re = GL_c/\mu$, $Pr = c_p\mu/k$, where h is the heat transfer coefficient, L_c is the characteristic dimension of the material, k is the gas thermal conductivity, G is the gas mass flux, μ is the gas viscosity, and c_p is the heat capacity of air.

where X_c is the critical moisture content. It is important to take into account that the thickness of the dried material is present in Eqs. (5.7) and (5.8) through the relation m_s/A_s ($=\rho_s/a_v$).

5.4.2 FALLING DRYING RATE PERIOD (II)

The shape of the drying rate curve in this period depends mainly on the type of material. The drying conditions used in the constant drying rate period can produce physical modifications on the material that influence moisture transport properties. As the drying rate decreases, the influence of external conditions becomes less pronounced.

5.4.2.1 Graphical Resolution

The drying time in the falling drying rate period can be calculated from Eq. (5.6) in its integrated form,

$$t_{II} = \int_{t_1}^{t} dt = -\frac{m_s}{A_s} \int_{X_c}^{X_f} \frac{dX}{n_w} \tag{5.9}$$

where X_f is the final material moisture content. However, in this period the drying rate is variable and cannot be eliminated from the integral.

For any shape of the falling rate drying curve, Eq. (5.9) can be graphically integrated by plotting $1/n_w$ versus X and determining the area under the curve. The drying time of the falling drying rate period can be obtained by substituting the values of the integral obtained in Eq. (5.9).

5.4.2.2 Theoretical Models

As stated, in many instances the moisture transfer rate in the falling rate period is governed by the internal movement of moisture via diffusion or via capillary movement.

5.4.2.2.1 Diffusive Models
When liquid or vapor diffusion of moisture controls the rate of drying in the falling rate period, the equations for diffusion can be used:

$$\frac{\partial X}{\partial t} = \nabla_z (D_{\text{eff}} \nabla_z X) \tag{5.10}$$

where D_{eff} represents an effective transport coefficient and z represents the spatial coordinate.

Drying curves for several foods have been fitted by diffusion equations. This type of diffusion is often characteristic of relatively slow drying in nongranular materials such as gelatin, and in the later stages of drying of bound water in foods, starches, and other hydrophilic solids.

During diffusion drying, the resistance to mass transfer of water vapor from the surface is usually very small, and the diffusion in the solid controls the rate of drying. However, in some cases both internal and external resistance should be considered.

Crank (1967) gave analytical solutions of Fick's law [Eq. (5.10)] with different boundary conditions, which are valid for one-dimensional transport, regular geometries, constant diffusivity, and uniform initial moisture distribution (X_0 uniform at initial time $t = t_0$). The solutions of Fick's law for internal mass transfer control are the following:

1. For a sphere of radius R:

$$\text{MR} = \frac{6}{\pi^2} \sum_{n=1}^{\infty} \frac{1}{n^2} \exp\left(-\pi^2 n^2 \frac{D_{\text{eff}}(t - t_0)}{R^2} \right) \tag{5.11}$$

Examples of products modeled as spheres include agricultural products (Alves-Filho and Rumsey, 1985), carrots (Doymaz, 2004), and parboiled wheat (Mohapatra and Rao, 2005).

2. For a slab of thickness L and moisture loss occurring from both sides:

$$\text{MR} = \frac{8}{\pi^2} \sum_{n=0}^{\infty} \frac{1}{(2n + 1)^2} \exp\left(-(2n + 1)^2 \pi^2 \frac{D_{\text{eff}}(t - t_0)}{L^2} \right) \tag{5.12}$$

Foods modeled using a slab geometry include apples (Wang et al., 2007), pumpkins (Doymaz, 2007), potatoes (Hassini et al., 2007), bananas (Nguyen and Price, 2007), and yogurt (Hayaloglu et al., 2007).

3. For a cylinder of radius R:

$$\text{MR} = \frac{8}{R^2} \sum_{n=1}^{\infty} \frac{1}{(\beta_n)^2} \exp(-D_{\text{eff}} \beta_n^2 (t - t_0))$$

(5.13)

where β_n^2 are the Bessel function roots of the first kind and zero order. Pasta (Andrieu and Stamatopoulos, 1986) and garlic (Sharma and Prasad, 2004) were modeled using the solution of Fick's law for a cylinder.

At sufficiently large times, only the leading term in the series expansion [Eqs. (5.11) to (5.13)] needs to be taken. The falling period time of drying for a slab, for example, can be obtained from Eq. (5.12) using only the first term as follows:

$$t - t_0 = \frac{4L^2}{\pi^2 D_{\text{eff}}} \ln\left(\frac{8}{\pi^2} \text{MR}\right)$$

(5.14)

This equation states that when internal diffusion controls the process, the drying time varies directly as the square of the thickness. For more exact calculations, Eq. (5.10) must be solved by numerical methods.

Diffusion coefficients are typically determined from the slope of the linear segments when experimental drying data are plotted in terms of ln MR versus time. Experiments should be carried out under drying conditions, particularly with air velocity, in which the external resistance to mass transfer is negligible.

Various research groups have made efforts to include other important parameters in the diffusion model. Rotstein et al. (1974) studied the effects of the sample shape and size on the diffusivity constant. They concluded that the cross-sectional area should be incorporated into an analytical model to describe diffusivity. Steffe and Singh (1980) used a more complex model to incorporate shape effects with constant diffusivity and without heat transfer and shrinkage effects.

Equations (5.11) to (5.13) can be used to predict the drying curves only for small changes in moisture, because the effective diffusivity changes with moisture and temperature. However, these equations are widely employed with average effective diffusivity values as a good approximation to represent the drying curves of most food products.

The dependence of the effective diffusivity on temperature and moisture is generally described by an Arrhenius type equation as follows:

$$D_{\text{eff}} = D_0(X) \exp\left(-\frac{E_a(X)}{R_c T}\right)$$

(5.15)

where E_a is the activation energy that may depend on the type of solid and the moisture content, D_0 is a preexponential term, and T is the absolute temperature.

TABLE 5.2

Effective Water Diffusivities (D_{eff}) and Activation Energies (E_a) for Food Products

Product	T (°C)	$D_{eff} \times 10^{10}$ (m²/s)	E_a (kJ/mol)	Ref.
Apple pomace	75–105	11.48–21.95 (first period)	23.951	Wang et al. (2007)
		68.05–12.82 (second period)	25.492	
Banana	30–70	1.3–7.8	39.8	Nguyen and Price (2007)
Black tea	80–120	0.114–0.298	406.02	Panchariya et al. (2002)
Broccoli	50–75	131.85–182.56		Mrkić et al. (2007)
Carrot	50–70	13.04–24.18	28.36	Doymaz (2004)
Carrot (core)	50–70	8.64–13.73		Srikiatden and Roberts (2006)
	40–70	6.42–14.7	24.78	
Carrot (cortex)	50–70	8.32–11.53		Srikiatden and Roberts (2006)
	40–70	6.68–11.83	16.53	
Garlic	45–75	0.34–0.59	16.92	Pezzutti and Crapiste (1997)
Mint leaves	35–60	30.7–194.1	62.96	Doymaz (2006)
Parboiled wheat	40–60	1.218–2.861	37.013	Mohapatra and Rao (2005)
Pasta (regular)	44–71	0.166–0.358	21.757	Xiong et al. (1991)
Potato	50–70	5.94–9.69		Srikiatden and Roberts (2006)
	40–70	4.68–10.2	23.61	
Pumpkin	50–60	3.88–9.38	78.93	Doymaz (2007)
Yogurt	40–50	9.5–13	26.07	Hayaloglu et al. (2007)

Table 5.2 provides the effective diffusivity and activation energy data reported for various food products as an example of the great applicability of the diffusion model found in the literature. Bruin and Luyben (1980) and Okos et al. (1992) conducted reviews of these parameters for several foods at different moisture contents and external conditions.

Xiong et al. (1991) developed a model based on the postulation that the decrease in the effective diffusivity at lower moisture contents is a result of a decrease in the availability of water molecules for diffusion.

5.4.2.2.1.1 Simplified Models

Simplified equations for diffusion at a sufficiently long period of processing can be rewritten as

$$MR = A \exp(-Kt) \tag{5.16}$$

where K is a drying rate constant. This equation has been applied to represent experimental data on drying of grains (Brooker et al., 1974) and tropical fruits (Nguyen and Price, 2007). Krokida et al. (2004) reviewed the drying experimental data for cereal products, fish, fruits, legumes, nuts, vegetables, and model foods. They obtained an empirical equation to represent the functionality of drying rate constant K with the temperature, air velocity, water activity, and particle diameter.

Kiranoudis et al. (1992) applied this phenomenological mass transfer model to represent the drying behavior of vegetables such as potatoes, onions, carrots, and green peppers. They proposed that the drying constant varied with the material characteristic dimension, air temperature, humidity, and velocity. Kiranoudis and colleagues compared the behavior of this phenomenological model with mechanistic mass transfer, including moisture diffusion in the solid phase toward its external surface plus vaporization and convective transfer of the vapor into the air stream. They found that the mechanistic mass transfer improved the representation of the experimental moisture content.

5.4.2.2.1.2 Capillary Flow Model

Depending on the pore sizes of granular materials, water can flow by capillary action rather than by diffusion from regions of high concentrations to regions of low concentrations. Capillary forces provide the driving force for moving the water through the pores of the material to the drying surface.

Geankoplis (1993) derived an equation for the drying time when the flow is by capillary movement in the falling rate period, assuming that rate of drying n_w varies linearly with X through the origin:

$$t = \frac{L\rho_s(X_c - X_e)}{n_{wc}} \ln \frac{(X_c - X_e)}{(X - X_e)} = \frac{L\rho_s \Delta H_v(X_c - X_e)}{h(T - T_w)} \ln \frac{(X_c - X_e)}{(X - X_e)} \qquad (5.17)$$

where L is the thickness of the material.

Equation (5.17) states that when capillary flow controls the falling rate period, the drying time varies directly with the thickness and depends on the gas velocity, temperature, and humidity, contrary to the diffusion process.

Moyers and Baldwin (1999) explained how to elucidate what the mechanism of drying in the falling rate period is. The experimental drying data expressed as $\ln[(X - X_e)/(X_c - X_e)]$ are plotted versus time. If a straight line is obtained, then diffusion or capillary flow occurs. If the relation for capillary flow applies, the slope of this line is related to Eq. (5.17), which contains constant drying rate n_{wc}. The value of n_{wc} is calculated from the measured slope of the line. If it agrees with the experimental value in the constant drying period, the moisture movement is by capillary flow. If the values of n_{wc} do not agree, the moisture movement is by diffusion.

These forms of calculations allow the determination of the drying time necessary to obtain the desired final moisture content by adding the duration of the constant drying rate period (t_I) to the corresponding falling drying rate period (t_{II}).

5.4.3 GENERALIZED DRYING CURVES

One possible mass transfer model is the use of a characteristic drying curve. This approach assumes that at each volume-averaged free moisture content value there is a corresponding specific drying rate relative to the unhindered drying rate in the first drying period that is independent of external drying conditions. Van Meel (1958) developed the characteristic drying curve method to obtain a normalized drying curve for each material while avoiding the necessity of descriptions of the heterogeneous structure and various compositions throughout the food samples. This method is based on a normalized evaporation rate f, a function of a normalized moisture content, which is characteristic for each product being dried and independent of the external drying conditions (Crapiste and Rotstein, 1997), so that

$$\frac{n_w}{n_{wc}} = f\left(\frac{X - X_e}{X_c - X_e}\right) \tag{5.18}$$

Thus, the drying curve is normalized to pass through the point (1, 1) at the critical point of transition in drying behavior and the point (0, 0) at equilibrium. This representation is attractive because it leads to a simple lumped parameter expression for the drying rate:

$$n_w = f n_{wc} = f k_g (P_{ws} - P_w) \tag{5.19}$$

In particular, the concept of a characteristic drying curve states that the shape of the drying rate curve for a given material is unique and independent of the gas temperature, humidity, and velocity. Drying rate curves for the same material under different operating conditions should be geometrically similar, according to this hypothesis. Equation (5.19) has been used extensively as the basis for understanding the behavior of industrial drying plants because of its simplicity and the separation of the parameters that influence the drying process: the material itself (f) and the design of the dryer, and the process conditions (k_g, P_{ws}, and P_w).

Function f is obtained by plotting the normalized drying rate (n_w / n_{wc}) versus MR according to Eq. (5.18) (Figure 5.4). The falling rate period is then represented with a concave curve. A single linear characteristic drying curve appears in the limit of the slow drying of thick, fairly impervious materials. Because the constant drying rate does not appear clearly for most biological materials, the short period where the drying rate is maximum is often considered as the first period. Fornell et al. (1980) suggested using theoretical calculations for free evaporation to evaluate n_{wc} when the constant drying period is absent.

Keey (1972, 1978, 1992) and Keey and Suzuki (1974) discussed fundamentals and applications for which a characteristic curve might apply, using a simplified analysis based on an evaporative front receding through a porous mass. Their analysis shows that there is a unique curve when the material is thin and the effective moisture diffusivity is high. Characteristic drying curves might then be expected for small, microporous particles dried individually. The concept worked well for modest ranges of air temperature, humidity, and velocity. There is a sufficient body of data to suggest that a characteristic drying curve may be found to describe the drying of discrete particles over a range of conditions that normally exist within a commercial dryer. This method has been used to represent the drying behavior of several food materials, such as milk powder (Langrish and Kockel, 2001), bananas (Jannot et al., 2004), and spirulina (Desmorieux and Decaen, 2005). Keey (1992) reviewed some agricultural products, and Laws and Parry (1983) provided an example of the

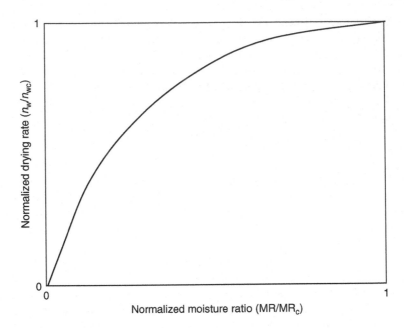

FIGURE 5.4 Typical characteristic drying curves for food convective drying.

application of a characteristic drying curve to other agricultural products. However, Baini and Langrish (2007) found that the concept of the characteristic drying curve method was not applicable to describe the intermittent drying of bananas.

Ratti and Crapiste (1992) extended those formulations to shrinking food systems and found the following expression for n_w:

$$n_w = \frac{k_g(P_{ws} - P_w)}{1 + (\theta/X_0)\,Bi_m} \tag{5.20}$$

where θ is the generalized drying parameter only as a function of water content and Bi_m is the mass transfer Biot number. On this basis, the drying process can be modeled as a coupled mass and heat transfer mechanism by using Eq. (5.20) as follows:

$$\frac{dT}{dt} = \frac{hA}{m_s(1+X)C_p}\left[(T_g - T) - \frac{\Delta H_s(a_w P_w^0 - P_w)}{(h/k_g)(1 + Bi_m(\theta/X_0))}\right] \tag{5.21}$$

$$\frac{dX}{dt} = \frac{Ak_g}{m_s(1 + Bi_m(\theta/X_0))} \tag{5.22}$$

Predictions of this model compared properly with experimental drying data for potatoes, apples, and carrots (Ratti and Crapiste, 1992). This model was later used for accurate prediction of the air-drying kinetics of grapes, white mushrooms, and bananas under diverse drying conditions (Fontaine and Ratti, 1999).

5.4.4 OTHER METHODS

Strumillo and Kudra (1986) described a method of calculating the total drying time including both the constant and falling drying rate periods, based on a single kinetic equation. The principle of this method is based on the generalized mass transfer equation written for the drying process in the form

$$\frac{dX}{dt} = K(X_0 - X)(X - X_e) \tag{5.23}$$

where K is assumed to be constant in the first and second drying periods and the expression $(X_0 - X)(X - X_e)$ represents the driving force of drying. Drying rate coefficient K includes the structural parameters as well as thermal and thermodynamic properties of the drying materials. Integration of Eq. (5.23) from the initial equilibrium moisture content X_{0e}, where the temperature at the material surface is equal to T_{wb}, to the final moisture content X gives the following relationship:

$$t = \frac{z}{K(X_0 - X_e)} \quad \text{with} \quad z = \ln\frac{(X_0 - X)(X_{0e} - X_e)}{(X_0 - X_{0e})(X - X_e)} \tag{5.24}$$

The drying curves depicted in coordinates t–z give a straight line with the slope related to the drying rate coefficient and the intercept on the z axis equal to X_{0e}.

The mass transfer coefficient for the falling drying rate period can be expressed as

$$k_g = \frac{m_s}{A_s} K(X_0 - X_e) \tag{5.25}$$

Thus, if the mass transfer coefficient is known, K can be obtained from Eq. (5.25) and the drying time can be calculated on the basis of Eq. (5.24) without any experimental drying curve.

Among semitheoretical and empirical thin layer drying models, Table 5.3 lists other equations that have been used widely in drying of food products.

5.5 SELECTIVE DRYING OF FOODS

In industrial manufacturing of many products the moisture to be removed does not consist of one component but is of a mixture of two or more solvents. Multicomponent moisture removal has been applied to a broad spectrum of pharmaceutical products and aroma retention in food drying. Desolventization of humid food products or by-products after solvent extraction is another case of multicomponent drying. Because most liquid mixtures are thermodynamically nonideal liquids, this nonideal behavior has to be considered in process simulation and design. In some cases the liquid mixture is only partially miscible, as occurs in water and essential oils mixtures, and some multicomponent systems may form two liquid phases.

TABLE 5.3
Other Mathematical Models for Representing Drying Curves

Model Name	Model MR	Ref.
Lewis	$\exp(-kt)$	Wang et al. (2007)
Page	$\exp(-kt^n)$	Doymaz (2004)
Modified Page	$\exp(-kt)^n$	Wang et al. (2007)
Modified Page equation II	$\exp(-c(t/L_c^2)^n)$	Wang et al. (2007)
Henderson and Pabis	$a \exp(-kt)$	Doymaz (2004)
Logarithmic	$a \exp(-kt) + c$	Togrul and Pehlivan (2002)
Two-term model	$a \exp(-k_0 t) + b \exp(-k_1 t)$	Henderson (1974)
Approximation of diffusion	$a \exp(-kt) + (1 - a) \exp(-kat)$	Apkinar (2006)
Wang and Singh	$1 + at + bt^2$	Wang et al. (2007)
Quadratic equation	$at^n + bt^{n-1} + \cdots + d$	Srinivasa et al. (2004)
Simplified Fick's diffusion	$a \exp(-c(t/L_c^2))$	Wang et al. (2007)
Midilli–Kucuk	$a \exp(-kt^n) + bt$	Apkinar (2006)

The favorable removal of one component, in which one component of the moisture is removed preferentially during drying of a product, is called *selective drying*. A well-known example is aroma retention in the drying of foodstuffs where it is important to remove only the water and some aroma compounds should remain inside the material.

As illustrated, the drying process of single component moisture is divided into periods of constant and falling drying rates, each period being the result of interactions between diffusion or convection in the gas and liquid or vapor diffusion and capillary movement in the solid. The direction and intensity of each process are governed by the appropriate driving force resulting from the difference between the equilibrium value of the partial pressure of moisture at the interface and its value in the bulk of the gas or a difference in capillary potential between regions of the solid. In multicomponent moisture removal, each of these processes is also complicated by the interrelationship of the individual components of the mixture as well as the sorption behavior of each solvent in the solid. The gas mass transfer, the solid mass transfer, and the multicomponent vapor–liquid equilibrium govern the selectivity of drying. Some particular issues are observed in the process of multicomponent moisture drying, that is, the wet bulb temperature changes during the process or the initial composition of moisture can influence what component evaporates preferentially.

Pakowski (1992) presented a detailed description of theoretical models for multicomponent moisture drying processes and concluded that the major problem is that the selectivity of drying is not only dependent on the identity and initial composition of the liquid system but also on the characteristics of the solid and the method of drying.

Thijssen (1971, 1979) was one of the first who reported the selectivity of drying. He studied aroma retention during the drying of foodstuffs and spray dried a solution of water, alcohol, and sugar, finding water selective diffusion.

Thurner and Schlünder (1986) investigated the drying characteristics of a porous material wetted with a binary liquid mixture. They demonstrated the dependence of the selectivity and drying rate on the drying conditions and the properties of the porous material. Quite often the drying process is nonselective because of the controlling solid or liquid mass transfer. However, at the early stages of drying, selectivity might be expected, because the gas mass transfer is in control at the beginning. This so-called initial selectivity can be extended to the whole drying process by intermittent drying (Heimann et al., 1986).

Sorptional equilibria of solvent mixtures in foodstuffs become key data in selective drying. Carrín et al. (2004) determined and modeled binary equilibrium sorption isotherms of water–ethanol over different food-related solids, reporting nonadditive equilibrium behavior.

The drying behavior of a solid material wetted with a ternary miscible liquid mixture was studied experimentally and modeled (Martinez and Setterwall, 1991; Vidaurre and Martinez, 1997). The influence of the transport of gas and liquid in the pores of the material and of the properties of the material itself was not investigated by these researchers.

Steinbeck and Schlünder (1998) examined the evaporation of a ternary liquid mixture, which forms two liquid phases, with respect to drying. They first conducted experimental and theoretical investigations of the evaporation of a ternary partially miscible liquid mixture. The selectivity of the process is controlled by the vapor–liquid equilibrium and the gas mass transfer. They carried on the work with studies on the influence of a liquid moisture that forms two liquid phases during the drying process. Steinbeck (1999) investigated the influence of two-phase moisture on the drying characteristics of a nonhygroscopic capillary porous material. He reported that the drying rate was not significantly influenced by the two-phase moisture and the selectivity could be shifted by the temperature of the drying air and by the pore size of the porous solid.

5.6   LATEST ADVANCES IN CONVECTIVE FOOD DRYING

The air temperature, humidity, and velocity in convective drying have a significant effect not only on the drying kinetics but also on the quality of food products. Most food materials are heat sensitive, so drying at relatively high temperatures implies structural, organoleptic, and nutritional changes during dehydration (Crapiste, 2000). The most important product quality parameters such as color (pigments, nonenzymatic browning) and nutritive value (antioxidants, vitamins) are often manifested by a progressive loss with increasing temperature. The dependence of the reaction rate constants on temperature implies that a low temperature drying process would result in less quality degradation. A longer constant drying rate period increases the nutrient retention because, due to evaporative cooling, the product is at a lower temperature (Chou and Chua, 2001). As technology advances, more options will be available to improve product quality by controlling the drying conditions.

The necessity of improving the drying characteristics of the food product during convective drying has caused the derivation of hybrid drying technologies, such as heat pump drying (HPD) or a combination of different drying equipment.

5.6.1 TIME VARIABLE DRYING SCHEMES

By employing time varying temperature and/or air-flow profiles during drying of food products it is possible to reduce the degradation of quality among another advantages. Several investigations have been carried out to study different time-dependent drying schemes with different foods: dry aeration of corn in a thin layer (Sabbah et al., 1972), intermittent drying of corn (Zhang and Litchfield, 1991), intermittent drying of peanuts with air-flow interruption (Troger and Butler, 1980), intermittent drying of maize in a bin dryer (Harnoy and Radajewski, 1982), sinusoidal heating of wheat in a fluidized bed (Giowacka and Malczewski, 1986), intermittent drying of carrot pieces in a vibrated bed batch (Pan et al., 1999), and drying of guava pieces under isothermal as well as time varying air temperatures (Chua, Chow, et al., 2000).

Devahastin and Mujumdar (1999) studied a mathematical model demonstrating the feasibility and advantages of operating a dryer by varying the temperature of the inlet drying air and reducing the drying time by up to 30%.

Chua, Mujumdar, Chou, Ho, et al. (2000) and Chua, Mujumdar, Chou, Howlader, et al. (2000) studied the effect of nonuniform temperature drying on color changes of food products. They found that color changes can be reduced by using different temperature profiles. Products with a high sugar content favor a time varying profile with a low starting temperature. Products with a low sugar content such as potatoes allow the use of higher temperature profiles to yield higher drying rates without any pronounced change in the overall color change. Prescribing the appropriate cyclic temperature variation schemes, Chua and colleagues showed that the reductions in the overall color change for potatoes and bananas were 87 and 67%, respectively.

Another time varying drying scheme is through the use of a pressure-regulatory system with an operating pressure range from above vacuum to 1 atm. The operation may be worked at lower pressure continuously at a fixed level, intermittently, or following a prescribed cyclic pattern, depending principally on the drying kinetics of the product and its thermal properties. Maache-Rezzoug et al. (2001) used a pressure-swing drying mechanism, consisting of dehydration by successive decompressions, for food products requiring the production of homogeneous thin sheets. The recommended process involves cycles during which the material is placed in desiccated air at a given pressure and then subjected to an instantaneous (200 ms) pressure drop to a vacuum (7 to 90 kPa), repeating this procedure until the desired moisture is obtained. The drying time savings using this pressure-swing drying process could be as high as 480 and 700 min in comparison to vacuum and hot-air drying systems.

The principal characteristics of time-dependent drying found in these studies are thermal energy savings, shorter effective drying time, higher drying rates, lower product surface temperature, and higher product quality (improved color and nutrient retention, reduced shrinkage, cracking, and brittleness).

5.6.2 HEAT PUMP DRYING (HPD)

Heat-sensitive high value products are often freeze-dried, which is an extremely expensive drying process. A heat pump can be coupled with the drying unit to increase the air temperature of the process and reduce energy consumption. Integration of the heat pump and the dryer is known as a heat pump dryer or heat pump assisted dryer. There has been great interest in the HPD system when low-temperature drying and well-controlled drying conditions are required to enhance the quality of food products. Their ability to convert the latent heat of condensation into sensible heat at the hot gas condenser makes them unique in drying applications.

In addition to the higher energy efficiency obtained with HPD compared with conventional hot-air drying, the dry product quality may be improved because of the special temperature and humidity conditions obtained during the process. The higher volatile retention in heat-pump dried samples is probably attributable to their reduced degradation when lower drying temperatures are employed compared to the higher temperatures of commercial dryers. Because HPD is conducted in a closed chamber, any compound that volatilizes will remain within the drying chamber, building its partial pressure up within the chamber, thus retarding further volatilization from the product (Perera and Rahman, 1997). Table 5.4 presents a summary of recent work on HPD of selected food products. A more detailed description about HPD is provided in Chapter 13.

TABLE 5.4
Examples of Recent Works Conducted on HPD of Food Products

Application and Results	Ref.
Herbs (Jew's mallow, spearmint, and parsley): maximum dryer productivity was obtained at 55°C, air velocity of 2.7 m/s, and dryer surface load of 28 kg/m^2.	Fatouh et al. (2006)
Apple, guava, and potato (modified environment with nitrogen and carbon dioxide): better physical properties were obtained at 45°C and 10% RH.	Hawlader et al. (2006)
Papaya and mango glace (effects of initial moisture content, cubic size, and effective diffusion coefficient of products on the optimum conditions, modeled and optimized): the physical properties of the product significantly affect the optimum air-flow rate and evaporator bypass air ratio.	Teeboonma et al. (2003)
Chicken meat (two multistage drying technologies): superheated steam drying + HPD was the most suitable method.	Nathakaranakule et al. (2007)
Potato slice: optimization of a two-stage evaporator system	Ho et al. (2001)
Agricultural and marine products: scheduling drying conditions let the quality of the products be improved.	Chou et al. (1998)
Banana: HPD resulted was suitable and cheap for drying high moisture materials.	Prasertsan and Saen-saby (1998), Prasertsan et al. (1997)
Onion slices: shorter processing time implied energy savings and better product quality in comparison to a conventional hot air system.	Rossi et al. (1992)
Fish: Dried products were high quality. Product properties (porosity, rehydration rates, strength, texture, and color) were regulated by programming temperature.	Strømmen and Krammer (1994)

When only a marginal amount of convection air is needed to evaporate moisture, as happens during drying of foods with large falling rate periods, the drying chambers can be operated in sequence as a multiple dryer system. The air from the heat pump can be directed sequentially to two or more chambers or can be divided up according to a preprogrammed schedule to two or more drying chambers. This makes it possible to dry the same or different products at the same time. Thus, the heat pump can be operated at a near optimal level at all times (Chou and Chua, 2001).

In addition to the advantages of HPD, the principal advantages of operating a dryer with multiple drying chambers are focused on cost: improved energy efficiency with proper channeling of conditioned air to chambers and reduced capital cost in equipment and floor space requirements.

5.6.3 SUPERHEATED STEAM DRYING UNDER VACUUM

An interesting alternative to using hot air is low-pressure superheated steam (LPSS). Because heat-sensitive foods are damaged at the saturation temperature of superheated steam corresponding to atmospheric or higher pressures, one possible way to avoid or moderate that problem is to operate the dryer at reduced pressure. Several groups found that it results in significantly less damage to the dried food material because it works at lower temperatures and there is no oxygen present. In general, the resulting product keeps its original color and shape and shows a highly porous structure that makes it easily rehydrated. From an economical point of view, LPSS requires less energy than hot-air drying for the same duty.

Elustondo et al. (2001) conducted experimental and theoretical studies applying LPSS to dry shrimp, bananas, apples, potatoes, and cassava slices. They developed a semiempirical mathematical model to calculate the drying rate of a given product assuming that the water removal is carried out by evaporation in a moving boundary that allows the vapor to flow through the dry layer that is built as drying proceeds. Despite its simplicity, the model adequately predicted the drying kinetics of the tested materials.

The use of LPSS as the drying agent during the drying of aromatic leaves rendered a product with higher retention of the original volatile compounds and lower modification of the composition percentage. The LPSS drying technique allows the recovery of a substantial amount of the aroma compounds withdrawn from the product, increasing the economic advantages of the new method (Barbieri et al., 2004).

A more detailed description of superheated steam drying is provided in Chapter 15.

5.6.4 MODIFIED FLUIDIZED BED DRYING (FBD)

This method has been used in many applications of granular solid foods with size particles in the range of 50 to 2000 μm, when uniformity in product drying is required. The FBD technique is most suitable for temperature-sensitive matter applications (Chaplin and Pugsley, 2005; Walde et al., 2006). Although high heat and mass transfer rates between the air and solid are obtained in FBD, avoiding product overheating, high power consumption, high potential for attrition, and low flexibility are some of the disadvantages (Mujumdar and Devahastin, 1999).

Modified fluidized bed dryers have been studied to dry different products and to look for more advantages: FBD plus HPD for heat-sensitive food materials (Alves-Filho and Strømmen, 1996; Strømmen and Jonassen, 1996), vibrofluidized bed drying for reducing gas velocities (Cabral et al., 2007; Goyel and Shah, 1992), FBD plus inert particles as energy carriers (Hatamipour and Mowla, 2003), pulsed FBD of beans for lowering pumping requirements (Nitz and Pereira Taranto, 2007), and a cross-flow fluidized bed dryer for moisture reduction and quality of paddy (Soponronnarit et al., 1995, 1999).

5.6.5   OPTIMIZATION AND ON-LINE CONTROL STRATEGIES

Drying process optimization based on reduction of quality degradation is difficult. Modern food technology makes it imperative that solutions be found that will allow optimization of complex processes with respect to complex quality factors (Karel, 1988).

Important advances in the knowledge of convective drying of foods have been made during the last decade, which allow better modeling and simulation of the process. Computer-assisted numerical simulation allows the prediction of the drying behavior of complex food systems under different conditions and the analysis of the influence of design and operating variables on process efficiency. By combining simulation and optimization techniques, with constraints on some quality parameters, it is possible to obtain optimum operating conditions, including time variable schemes. These new developments are essential to improve the drying of foods, obtaining higher quality products with more efficient energy use.

Real-time process control strategies have been investigated to minimize quality degradation of the product. Chou et al. (1997) and Chua, Mujumdar, Chou, Ho, et al. (2000) used the strategy of measuring the drying product temperature and, based on these real-time values and predefined constraints, a controller tuned the temperature of the drying air to minimize the degradation of dried product quality.

NOMENCLATURE

VARIABLES AND ABBREVIATIONS

a, b, c, d	coefficients (Table 5.3)
a_v	external solid area exposed to drying per unit volume (m^{-1})
a_w	water activity
A	coefficient
A_s	external solid area exposed to drying (m^2)
Bi_m	mass transfer Biot number
C_p	heat capacity (J/kg K)
D_{eff}	effective moisture diffusivity (m^2/s)
D_0	preexponential coefficient (m^2/s)
E_a	activation energy (kJ/mol)
f	normalized evaporation rate
G	gas mass flux $(kg/s\ m^2)$
h	heat transfer coefficient $(W/m^2\ K)$

H	air humidity (kg moisture vapor/kg dry air)
k	coefficient (Table 5.3)
k	gas thermal conductivity (W/m K)
k_g	external mass transfer coefficient (kg/m²/s Pa)
K	drying rate constant (s⁻¹)
L	thickness (m)
L_c	characteristic dimension of material (m)
m_s	mass of dry solid (kg)
M_g	molecular weight of dry air (kg/mol)
M_w	molecular weight of moisture vapor (kg/mol)
MR	moisture ratio
MR_c	critical moisture ratio
n	coefficient (Table 5.3)
n_w	drying rate (kg/m²/s)
n_{wc}	constant drying rate (kg/m²/s)
Nu	Nusselt number
P_w	moisture vapor partial pressure at air (Pa)
P_w^0	saturated moisture vapor pressure (Pa)
P_{ws}	moisture vapor partial pressure at material surface (Pa)
P	total pressure (Pa)
Pr	Prandtl number
q	heat flow (W)
R	sphere or cylinder radius (m)
R_c	universal gas constant (kJ/mol K)
Re	Reynolds number
RH	relative humidity
t	time (s)
t_I	duration of the constant drying rate period (s)
t_{II}	duration of the falling drying rate period (s)
t_0	initial time (s)
T	absolute temperature (K)
T_{db}	dry bulb temperature (K)
T_g	gas temperature (K)
T_{wb}	wet bulb temperature (K)
X	material moisture content (dry basis) (kg water/kg dry solid)
X_c	critical moisture content (kg water/kg dry solid)
X_e	equilibrium moisture content (kg water/kg dry solid)
X_f	final moisture content (kg water/kg dry solid)
X_0	initial moisture content (kg water/kg dry solid)
X_w	material moisture content (wet basis) (kg water/kg material)
z	spatial coordinate (m)

GREEKS

μ	gas viscosity (Pa s)
ΔH_v	latent heat of vaporization (J/kg)
β_n	Bessel function roots of the first kind and zero order

ρ_s density of dry solid (kg/m^3)
θ generalized drying parameter

REFERENCES

Alves-Filho, O. and Rumsey, T.R., Thin layer drying and rewetting models to predict moisture diffusion in spherical agricultural products, in *Drying '85*, Mujumdar, A.S., Ed., Hemisphere Publishing, New York, 1985, pp. 434–437.

Alves-Filho, O. and Strømmen, I., Performance and improvements in heat pump dryers, in *Drying '96*, Strumillo, C. and Pakowski, Z., Eds., Lodz Technical University, Lodz, Poland, 1996, pp. 405–415.

Andrieu, J. and Stamatopoulos, A., Durum wheat pasta drying kinetics, *Lebensm.-Wiss. Technol.*, 19, 448, 1986.

Apkinar, E.K., Determination of suitable thin layer drying curve model for some vegetables and fruits, *J. Food Eng.*, 73, 75–84, 2006.

ASAE, *Agricultural Engineers Yearbook*, American Society of Agricultural Engineers, St. Joseph, MI, 1982.

Baini, R. and Langrish, T.A.G., Choosing an appropriate drying model for intermittent and continuous drying of bananas, *J. Food Eng.*, 79, 330–343, 2007.

Bakker-Arkema, F.W., Heat and mass transfer aspects and modeling of dryers—A critical review, in *Concentration and Drying of Foods*, MacCarthy, D., Ed., Elsevier, London, 1986, pp. 165–202.

Barbieri, S., Elustondo, M., and Urbicain, M., Retention of aroma compounds in basil dried with low pressure superheated steam, *J. Food Eng.*, 65, 109–115, 2004.

Brooker, D.B., Baker-Arkema, F.W., and Hall, C.W., *Drying Cereal Grains*, AVI Publishing, Westport, CT, 1974.

Bruin, S. and Luyben, K., Drying of food materials: A review of recent developments, in *Advances in Drying*, Vol. 1, Mujumdar, A.S., Ed., Hemisphere Publishing, New York, 1980, pp. 155–215.

Cabral, R.A., Telis-Romero, J., Telis, V., Gabas, A., and Finser, J., Effect of apparent viscosity on fluidized bed drying process parameters of guava pulp, *J. Food Eng.*, 80, 1096–1106, 2007.

Carrín, M.E., Ratti, C., and Crapiste, G., Vapor–solid equilibrium of ethanol–water mixtures in food-related solids, paper presented at ICEF9, 2004.

Chaplin, G. and Pugsley, T., Application of electrical capacitance tomography to the fluidized bed drying of pharmaceutical granule, *Chem. Eng. Sci.*, 60, 7022–7033, 2005.

Chou, S.K. and Chua, K.J., New hybrid drying technologies for heat sensitive foodstuffs, *Trends Food Sci. Technol.*, 12, 359–369, 2001.

Chou, S.K., Chua, K.J., Hawlader, M.N., and Ho, J.C., A two-stage heat pump dryer for better heat recovery and product quality, *J. Inst. Eng.*, 38, 8–14, 1998.

Chou, S.K., Hawlader, M.N., and Chua, K.J., Identification of the receding evaporation front in convective food drying, *Dry. Technol.*, 15, 1353–1376, 1997.

Chua, K.J., Chou, S.K., Ho, J.C., Mujumdar, A.S., and Hawlader, M.N., Cyclic air temperature drying of guava pieces: Effects on moisture and ascorbic acid contents, *T. I. Chem. Eng. Lond.*, 78, 28–72, 2000.

Chua, K.J., Mujumdar, A.S., Chou, S.K., Hawlader, M.N., and Ho, J.C., Convective drying of banana, guava and potato pieces: Effect of cyclical variations of air temperature on convective drying kinetics and color change, *Dry. Technol.*, 18, 907–936, 2000.

Chua, K.J., Mujumdar, A.S., Chou, S.K., Ho, J.C., and Hawlader, M.N., Heat pump drying systems: Principles, applications and potential, in *Developments in Drying*, Vol. 2, Mujumdar, A.S. and Suvachittanont, S., Eds., Kasetsart University Press, Bangkok, Thailand, 2000, pp. 95–134.

Crank, J., *The Mathematics of Diffusion*, Oxford University Press, London, 1967.

Crapiste, G.H., Simulation of drying rates and quality changes during the dehydration of food-stuffs, in *Trends in Food Engineering*, Lozano, J.E., Añón, C., Parada-Arias, E., and Barbosa-Cánovas, G.V., Eds., Technomic Publishing, Lancaster, PA, 2000, pp. 135–148.

Crapiste, G.H. and Rotstein, E., Sorptional equilibrium at changing temperatures, in *Drying of Solids: Recent International Developments*, Mujumdar, A.S., Ed., John Wiley & Sons, New York, 1986, pp. 41–45.

Crapiste, G.H. and Rotstein, E., Design and performance evaluation of dryers, in *Handbook of Food Engineering Practice*, Valentas, K.J., Rotstein, E., and Singh, R.P., Eds., CRC Press, New York, 1997, pp. 125–166.

Crapiste, G.H., Whitaker, S., and Rotstein, E., Drying of cellular material. I. A mass transfer theory. II. Experimental and numerical results, *Chem. Eng. Sci.*, 43, 2919–2936, 1988.

Desmorieux, H. and Decaen, N., Convective drying of spirulina in thin layer, *J. Food Eng.*, 66, 497–503, 2005.

Devahastin, S. and Mujumdar, A.S., Batch drying of grains in a well-mixed dryer—Effect of continuous and stepwise change in drying air temperature, *Trans. ASAE*, 42, 421–425, 1999.

Doymaz, I., Convective air drying characteristics of thin layer carrots, *J. Food Eng.*, 61, 359–364, 2004.

Doymaz, I., Thin-layer drying behaviour of mint leaves, *J. Food Eng.*, 74, 370–375, 2006.

Doymaz, I., The kinetics of forced convective air-drying of pumpkin slices, *J. Food Eng.*, 79, 243–248, 2007.

Elustondo, D., Elustondo, M., and Urbicain, M., Mathematical modeling of moisture evaporation from foodstuffs exposed to subatmospheric pressure superheated steam, *J. Food Eng.*, 49, 15–24, 2001.

Fatouh, M., Metwally, M.N., Helali, A.B., and Shedid, M.H., Herb drying using a heat pump dryer, *Energ. Convers. Manage.*, 47, 2629–2643, 2006.

Fontaine, J. and Ratti, C., Lumped-parameter approach for prediction of drying kinetics in foods, *J. Food Process Eng.*, 22, 287–305, 1999.

Fornell, A., Bimbenet, J.J., and Amin, Y., Experimental study and modelization for air drying of vegetable products, *Lebensm.-Wiss. Technol.*, 14, 96–100, 1980.

Geankoplis, C.J., *Transport Processes and Unit Operations*, 3rd ed., Prentice Hall, Englewood Cliffs, NJ, 1993.

Giowacka, M. and Malczewski, J., Oscillating temperature drying, in *Drying of Solids: Recent International Developments*, Mujumdar, A.S., Ed., John Wiley & Sons, New York, 1986, pp. 77–83.

Goyel, S.K. and Shah, R.M., Drying of desiccated coconut on vibrating fluidised bed dryer, in *Drying '92*, Mujumdar, A.S., Ed., Elsevier Science, Amsterdam, 1992, pp. 815–826.

Harnoy, A. and Radajewski, W., Optimization of grain drying with rest periods, *J. Agric. Eng. Res.*, 27, 291–307, 1982.

Hassini, L., Azzouz, S., Peczalski, R., and Beighith, A., Estimation of potato moisture diffusivity from convective drying kinetics with correction for shrinkage, *J. Food Eng.*, 79, 47–56, 2007.

Hatamipour, M.S. and Mowla, D., Correlations for shrinkage, density and diffusivity for drying of maize and green peas in a fluidized bed with energy carrier, *J. Food Eng.*, 59, 221–227, 2003.

Hawlader, M.N.A., Perera, C.O., and Tian, M., Properties of modified atmosphere heat pump dried foods, *J. Food Eng.*, 74, 392–401, 2006.

Hayaloglu, A.A., Karabulut, I., Alpaslan, M., and Kelbaliyev, G., Mathematical modeling of drying characteristics of strained yoghurt in a convective type tray-dryer, *J. Food Eng.*, 78, 109–117, 2007.

Heimann, F., Thurner, F., and Schlünder, E.U., Intermittent drying of porous materials containing binary mixtures, *Chem. Eng. Process.*, 20, 167–174, 1986.

Henderson, S.M., Progress in developing the thin layer drying equation, *Trans. ASAE*, 17, 1167–1172, 1974.

Ho, J.C., Chou, S.K., Mujumdar, A.S., Hawlader, M.N., and Chua, K.J., An optimization framework for drying of heat-sensitive products, *Appl. Therm. Eng.*, 21, 1779–1798, 2001.

Iglesias, H.A. and Chirife, J., *Handbook of Food Isotherms*, Academic Press, New York, 1982.

Jannot, Y., Talla, A., Nganhou, J., and Puiggali, J.-R., Modeling of banana convective drying by the drying characteristics curve (DCC) method, *Dry. Technol.*, 22, 1949–1968, 2004.

Karel, M., Optimizing the heat sensitive materials in concentration and drying, in *Preconcentration and Drying of Food Materials*, Briun, S., Ed., Elsevier Science, Amsterdam, 1988, pp. 217–234.

Keey, R.B., *Drying: Principles and Practice*, Pergamon Press, Oxford, U.K., 1972.

Keey, R.B., *Introduction to Industrial Drying Operations*, Pergamon Press, Oxford, U.K., 1978.

Keey, R.B., *Drying of Loose and Particulate Materials*, Hemisphere Publishing, New York, 1992.

Keey, R.B. and Suzuki, M., On the characteristic drying curve, *Int. J. Heat Mass Transfer*, 17, 1455–1464, 1974.

Kiranoudis, C.T., Maroulis, Z.B., and Marinos-Kouris, D., Model selection in air drying of foods, in *Drying '92*, Mujumdar, A.S., Ed., Elsevier, Amsterdam, 1992, pp. 785–793.

Krokida, M.K., Foundoukidis, E., and Maroulis, Z., Drying constant: Literature data compilation for foodstuffs, *J. Food Eng.*, 61, 321–330, 2004.

Langrish, T.A.G. and Kockel, T.K., The assessment of a characteristic drying curve for milk powder for use in computational fluid dynamics modeling, *Chem. Eng. J.*, 84(2), 69–74, 2001.

Laws, N. and Parry, J.L., Mathematical modeling of heat and mass transfer in agricultural grain drying, *Proc. Roy. Soc. Lond. A Mater.*, 385(1788), 169–187, 1983.

Maache-Rezzoug, Z., Rezzoug, S.A., and Allaf, K., Development of a new drying process particularly adapted to thermosensitive products: Dehydration by successive pressure drops application to the drying of collagen gel, *Dry. Technol.*, 19, 961–974, 2001.

Martinez, J. and Setterwall, F., Gas-phase controlled convective drying of solids wetted with multicomponent liquid mixtures, *Chem. Eng. Sci.*, 46, 2235–2252, 1991.

Masters, K., *Spray Drying Handbook*, George Godwin Ltd., London, 1979.

Mohapatra, D. and Rao, P.S., A thin layer drying model of parboiled wheat, *J. Food Eng.*, 66, 513–518, 2005.

Moyers, C.G. and Baldwin, G.W., Psychrometry, evaporative cooling, and solid drying, in *Perry's Chemical Engineers Handbook*, 7th ed., Perry, R.H., Green, D.W., and Maloney, J.O., Eds., McGraw-Hill, New York, 1999.

Mrkić, V., Ukrainczyk, M., and Tripalo, B., Applicability of moisture transfer Bi-Di correlation for convective drying of broccoli, *J. Food Eng.*, 79, 640–646, 2007.

Mujumdar, A.S., *Handbook of Industrial Drying*, Marcel Dekker, New York, 1987.

Mujumdar, A.S. and Devahastin, S., Fluidized bed drying, in *Developments in Drying: Food Dehydration*, Vol. 1, Mujumdar, A.S. and Suvachittanont, S., Eds., Kasetsart University Press, Bangkok, Thailand, 1999, pp. 59–111.

Nathakaranakule, A., Kraiwanichkul, W., and Soponronnarit, S., Comparative study of different combined superheated-steam drying techniques for chicken meat, *J. Food Eng.*, 80, 1023–1030, 2007.

Nguyen, M.H. and Price, W.E., Air-drying of banana: Influence of experimental parameters, slab thickness, banana maturity and harvesting season, *J. Food Eng.*, 79, 200–207, 2007.

Nitz, M. and Pereira Taranto, O., Drying of beans in a pulsed fluid bed dryer: Drying kinetics, fluid-dynamic study and comparisons with conventional fluidization, *J. Food Eng.*, 80, 249–256, 2007.

Okos, M.R., Narsimhan, G., Singh, R.K., and Weitnauer, A.C., Food dehydration, in *Handbook of Food Engineering*, Heldman, D.R. and Lund, D.B., Eds., Marcel Dekker, New York, 1992, pp. 437–562.

Pakowski, Z., Drying of solids containing multicomponent moisture, in *Advances in Drying*, Vol. 5, Mujumdar, A.S., Ed., Hemisphere Publishing, Washington, DC, 1992, pp. 145–202.

Pan, Y.K., Zhao, L.J., Dong, Z.X., Mujumdar, A.S., and Kudra, T., Intermittent drying of carrots: Effect on product quality, *Dry. Technol.*, 17, 2323–2340, 1999.

Panchariya, P.C., Popovic, D., and Sharma, A.L., Thin-layer modeling of black tea drying process, *J. Food Eng.*, 52, 349–357, 2002.

Perera, C.O. and Rahman, M.S., Heat pump dehumidifier drying of foods, *Trends Food Sci. Technol.*, 8, 75–79, 1997.

Perry, R.H. and Green, D., *Perry's Chemical Engineers Handbook*, 6th ed., McGraw-Hill, New York, 1999.

Pezzutti, A. and Crapiste, G.H., Sorptional equilibrium and drying characteristics of garlic, *J. Food Eng.*, 31, 113–123, 1997.

Prasertsan, S. and Saen-saby, P., Heat pump drying of agricultural materials, *Dry. Technol.*, 16, 235–250, 1998.

Prasertsan, S., Saen-saby, P., Prateepchaikul, G., and Ngamsritrakul, P., Heat pump dryer part 3: Experiment verification of the simulation, *Int. J. Energ. Res.*, 21, 1–20, 1997.

Ratti, C. and Crapiste, G.H., A generalized drying curve for shrinking food materials, in *Drying '92*, Mujumdar, A.S., Ed., Elsevier, New York, 1992, pp. 864–873.

Ratti, C. and Crapiste, G.H., Determination of heat transfer coefficients during drying of foodstuffs, *J. Food Process Eng.*, 18, 41–53, 1995.

Ratti, C., Crapiste, G.H., and Rotstein, E., PSYCHR: A computer program to calculate psychrometric properties, *Dry. Technol.*, 7, 575–580, 1989.

Román, G.N., Urbicain, M.J., and Rotstein, E., Moisture equilibrium in apples at several temperatures: Experimental data and theoretical considerations, *J. Food Sci.*, 47, 1484–1488, 1982.

Rossi, S.J., Neues, I.C., and Kicokbusch, T.G., Thermodynamics and energetic evaluation of a heat pump applied to drying of vegetables, in *Drying '92*, Mujumdar, A.S., Ed., Elsevier, Amsterdam, 1992, pp. 1475–1483.

Rotstein, E., Laura, P.A., and Cemborain, M.E., Analytical prediction of drying performance in unconventional shapes, *J. Food Sci.*, 39, 627, 1974.

Sabbah, M.A., Foster, G.H., Hauge, C.G., and Peart, R.M., Effect of tempering after drying on cooling shelled corn, *Trans. ASAE*, 15, 763–765, 1972.

Sharma, G.P. and Prasad, S., Effective moisture diffusivity of garlic cloves undergoing microwave-convective drying, *J. Food Eng.*, 65, 609–617, 2004.

Soponronnarit, S., Wetchacama, S., Swasdisevi, T., and Poomsa-ad, N., Managing moist paddy by drying, tempering and ambient air ventilation, *Dry. Technol.*, 17, 335–344, 1999.

Soponronnarit, S., Yapha, M., and Prachayawarakorn, S., Cross-flow fluidized bed paddy: Prototype and commercialization, *Dry. Technol.*, 13, 2207–2216, 1995.

Srikiatden, J. and Roberts, J.S., Measuring moisture diffusivity of potato and carrot (core and cortex) during convective hot air and isothermal drying, *J. Food Eng.*, 74, 143–152, 2006.

Srinivasa, P.C., Ramesh, M.N., Kumar, K.R., and Tharanathan, R.N., Properties of chitosan films prepared under different drying conditions, *J. Food Eng.*, 63, 79–85, 2004.

Steffe, J.F. and Singh, R.P., Liquid diffusivity of rough rice components, *Trans. ASAE*, 23, 767–774, 1980.

Steinbeck, M.J., Convective drying of porous material containing a partially miscible mixture, *Chem. Eng. Process.*, 38, 487–502, 1999.

Steinbeck, M.J. and Schlünder, E.U., Convective drying of porous material containing a partially miscible mixture, *Chem. Eng. Process.*, 37, 79–88, 1998.

Strømmen, I. and Jonassen, O., Performance tests of a new 2-stage counter-current heat pump fluidized bed dryer, in *Drying '96*, Strumillo, C., Pakowski, Z., and Mujumdar, A.S., Eds., Elsevier, Amsterdam, 1996, pp. 563–568.

Strømmen, I. and Kramer, K., New applications of heat pumps in drying process, *Dry. Technol.*, 12, 889–901, 1994.

Strumillo, C. and Kudra, T., Drying kinetics, in *Drying: Principles, Applications and Design*, Hughes, F., Ed., Gordon and Breach Science Publishers, Geneva, 1986, pp. 67–98.

Teeboonma, U., Tiansuwan, J., and Soponronnarit, S., Optimization of heat pump fruit dryers, *J. Food Eng.*, 59, 369–377, 2003.

Thijssen, H.A.C., Flavor retention in drying preconcentrated food liquids, *J. Appl. Chem. Biotechnol.*, 21, 372–377, 1971.

Thijssen, H.A.C., Optimization of process conditions during drying with regard to quality factors, *Lebensm.-Wiss. Technol.*, 12, 308–317, 1979.

Thurner, F. and Schlünder, E.U., Progress towards understanding the drying of porous material wetted with binary mixtures, *Chem. Eng. Process.*, 20, 2–25, 1986.

Togrul, I.T. and Pehlivan, D., Mathematical modeling of solar drying of apricots in thin layers, *J. Food Eng.*, 55, 209–216, 2002.

Treybal, R.E., *Mass-Transfer Operations*, 3rd ed., McGraw-Hill, New York, 1980.

Troger, J.M. and Butler, J.L., Drying peanuts with intermittent airflow, *Trans. ASAE*, 23, 197–199, 1980.

Van Meel, D.A., Adiabatic convection batch drying with recirculation of air, *Chem. Eng. Sci.*, 9, 36–44, 1958.

Vidaurre, M. and Martinez, J., Continuous drying of a solid wetted with ternary mixtures, *AIChE J.*, 43, 681–692, 1997.

Wakao, N., Kaguei, S., and Funazkri, T., Effect of fluid dispersion coefficients on particle-to-fluid heat transfer coefficients in packed beds, *Chem. Eng. Sci.*, 34, 325–336, 1979.

Walde, S.G., Velu, V., Jyothirmayi, T., and Math, R.G., Effects of pretreatments and drying methods on dehydration of mushroom, *J. Food Eng.*, 74, 108–115, 2006.

Wang, Z., Sun, J., Liao, X., Chen, F., Zhao, G., Wu, J., and Hu, X., Mathematical modeling on hot air drying of thin layer apple pomace, *Food Res. Int.*, 4, 39–46, 2007.

Xiong, X., Narsimhan, G., and Okos, M.R., Effect of composition and pore structure on binding energy and effective diffusivity of moisture in porous food, *J. Food Eng.*, 15, 187–208, 1991.

Zhang, D. and Litchfield, J.B., An optimization of intermittent corn drying in a laboratory scale thin layer dryer, *Dry. Technol.*, 9, 383–395, 1991.

6 Advances in Spouted Bed Drying of Foods

S.C.S. Rocha and O.P. Taranto

CONTENTS

6.1 THE SPOUTED BED

The dynamic fluid–solid system developed by Mathur and Gishler in 1954, which they named spouted bed, has a wide range of applications for processing coarse particles, such as grains and tablets, which are not easily treated in a fluidized bed. The equipment consists of a cylindrical column with a conical base fitted with an orifice through which the spouting fluid is injected.

High fluid flow rates create a flow of ascending particles as a central channel inside the bed of solids. After reaching the top of the bed, these particles form a fountain directed to the outer part of the bed surface and fall into the annular region between the channel and the bed walls. The particles flow down the base of the vessel where they reach the central channel air flow, thus allowing a continuous and cycling systematic movement of particles in a unique hydrodynamic system that is called the spouted bed (Mathur and Epstein, 1974).

Figure 6.1 illustrates the three distinct regions that constitute the spouted bed: the spout, the annulus, and the fountain.

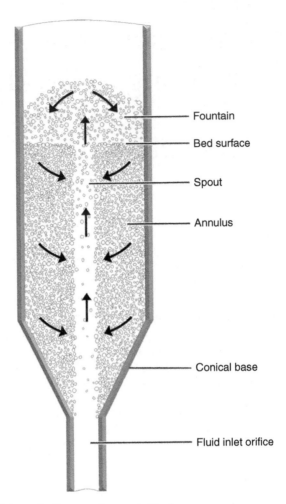

FIGURE 6.1   Schematic diagram of a spouted bed.

The central channel in this system is called the spout and the peripheral region is known as the annulus. The term fountain is used to determine the mushroom form above the annulus, where the particles move in a decelerated regime, falling into the annulus.

In the spout region the particles move in the same direction as the gas flow, as in a pneumatic conveyor, because of the high velocity of the gas. It is characterized as a diluted-phase region with high porosity.

In the annular region the particles move countercurrent to the gas. It is a dense region with low porosity.

Mathur and Epstein (1974) refer to spouting as a visible phenomenon that occurs for a determined range of fluid velocity, which is a function of the particle diameter and the height and the geometry of the vessel.

From the standpoint of gas–solid contact, the spouted bed is considered to be a combination of a moving bed and a fluidized bed, because it promotes good mixing and circulation of particles.

The main characteristics of the spouted bed are a high rate of solids circulation, intense mixing of particles, intimate fluid–particle contact, and high rates of mass and heat transfer.

The transition mechanism from packed to spouted bed can be verified in the fluid dynamics curves, that is, the pressure drop versus the fluid flow rate. With an increase in fluid flow rate, the particles in the vicinity of the gas entry orifice move and form a cavity surrounded by a compact layer of solids, causing the bed pressure drop to increase. The pressure drop increases continuously up to a maximum value Δp_{max}. From this point on, the effect of internal spouting becomes more significant than the solid layer that surrounds the cavity and the pressure drop begins to diminish as the fluid flow rate is raised, down to the point of minimum spout (Δp_{ms} and Q_{ms}), from which a small rise in the fluid flow rate results in the establishment of the spout. Then, the bed pressure drop (Δp_s) remains constant and any rise in the fluid flow rate only causes the fountain height to increase. Therefore, the important fluid dynamic parameters of the spouted bed regime are Δp_{max}, Δp_{ms}, Q_{ms}, and Δp_s. Because of the instability caused by the rupture of the jet stream in the bed, the point of minimum spout is not easily reproducible. Thus, it is determined using the fluid dynamics curve obtained with the inverse process, that is, slowly diminishing the gas flow rate from the steady spout regime down to the collapse of the spout.

In addition to the experimental determination, several empirical correlations are available to obtain these parameters (Mathur and Epstein, 1974; Strumillo and Kudra, 1986).

The spouted bed technique was initially developed for drying wheat, but it was later utilized with several other types of solid particles. This made it possible to check process efficiency. It is also currently utilized in the processes of particle coating, granulation, heat transfer, and so forth.

The following limitations have restricted the application of the spouted bed on a large scale: high head loss before attaining the steady spout regime, gas flow limited by the demands of the spout stability rather than by necessities of heat and mass transfer during drying, narrow operation range, limitation of a maximum particle bed height for steady spouting (H_{max}), and difficulties in scaling up.

The main limitations are the existence of a maximum bed height for steady spouting and the inability to scale up for vessels with a diameter >1 m. Some modifications have been proposed to overcome the restrictions in the use of the spouted bed. Attempts are reported for conical spouted beds (Oliveira and Passos, 1997; Reyes and Massarani, 1993; Reyes et al., 1996, 1997) and for a jet spouted bed with high porosity (Markowski and Kaminski, 1983; Tia et al., 1995). In addition, the introduction of a draft tube at some distance above the inlet nozzle and different designs, like the two-dimensional spouted bed and the slot spouted bed, are modifications on the conventional spouted bed that eliminate the former limitation (Kalwar and Raghavan, 1992, 1993; Kalwar et al., 1991; Mujumdar, 1989).

The introduction of a draft tube promotes the separation between the spout and the annular region, avoiding the continuous reentry of particles into the spout region.

Thus, the circulation of particles begins at lower pressure drops, the bed height and diameter can be substantially altered, and lower gas flow rates are required. However, some disadvantages of the draft tube are notable, such as the reduction in quality of particle mixing, the possibility of particles becoming stuck at the beginning or end of the process, and lower mass and heat transfer rates.

The spouted bed dryer (SBD) can operate in a batch or a continuous regime. The batch operation consists of loading the bed with wet particles and then injecting heated air until the desired particle moisture content is reached.

Figure 6.2 is a schematic diagram of an SBD with a draft tube in continuous operation. The flow of particles in the annular region can be considered as plug flow, promoting a uniform distribution of residence time in the dryer.

It is important to note that the SBD can be used for solids having constant and/ or decreasing drying rates. Utilizing an inert material in order to obtain and maintain

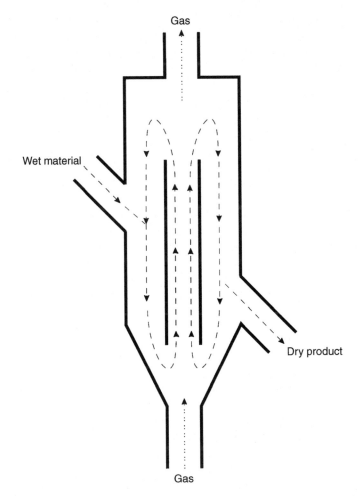

FIGURE 6.2 Schematic diagram of a continuous spouted bed dryer with draft tube.

the fluid dynamic regime, the SBD has been successfully applied for drying pastes and suspensions.

Generally, the appropriate particles for obtaining the spouted bed regime have an average diameter >1 mm. The Geldart diagram (Geldart, 1986) classifies different types of particulate materials based on the size and density of the particles, which is very useful for the choice of appropriate equipment to process a specific material. For the SBD, particles in group D, called spoutable, are the most appropriate, that is, particles in group D are appropriate for the process of steady spouting and do not present good fluidization quality (coarse particles promote channeling and dead zones in fluidized beds). Examples of particles belonging to group D are corn, oats, beans, rice, paddy, and some seeds.

The application of the SBD in food drying is not only restricted to the mentioned materials, but it can also be economically and efficiently applied to the drying of pastes and suspensions.

6.2   DRYING OF PASTES AND SUSPENSIONS IN SPOUTED BEDS WITH INERTS

The technique of drying pastes and suspensions on inert solids was developed in the former USSR and was industrially applied to dry chemicals and biological materials. The technique is suitable for drying a large variety of pastes and suspensions, producing fine and homogeneous powders, and can compete with the spray drying technique. However, it did not arouse interest in other countries until the 1980s. Since then, R&D in this area has become important in many countries of the Americas, Europe, and Asia.

Spouted bed systems (dryers, granulators, and coaters) with throughputs from 200 g to 3000 kg/h can be found on the market today (Glatt, 2007).

Figure 6.3 illustrates a simple schematic diagram of a spouted bed utilized for the drying of pastes and suspensions. The paste or suspension is fed in batches or continuously over the bed of inert particles already moving in a steady spouting regime. The drying gas is the spouting gas itself. The following mechanisms take place in this type of drying. The suspension fed into the bed coats the particles of the inert material with a thin film. As the film becomes fragile, because of the inter-particle collisions inside the bed, it fractures and releases the inert particles being carried by the drying gas stream. This powder is then collected by a cyclone or other separation equipment installed in the upper part of the dryer.

Some studies refer to making use of this technique for drying food pastes and extracts: tomato paste (Kachan and Chiapetta, 1988); artichoke vegetable extracts, *guarana*, and *cascara sagrada* (Ré and Freire, 1988); nonhomogenized low-fat milk (Martinez et al., 1993); low-fat and whole milk, orange and carrot juices, and coffee (Martinez et al., 1995); beets (Lima et al., 1998); acerola pulp (Alsina et al., 1996); tropical fruit pulps (Medeiros et al., 2002); vegetable starch (Benali and Amazouz, 2006); and mango pulp (Cunha et al., 2006), among others.

The process of deposition, drying, and removal of the film occurs only if the movement of the spout is not jeopardized by agglomeration of particles.

As a rule of thumb, the removing rate must always be greater than or equal to the suspension feed rate. Thus, two factors govern the dryer performance: heat and mass

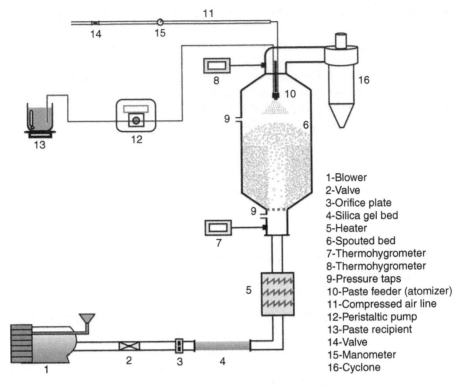

1-Blower
2-Valve
3-Orifice plate
4-Silica gel bed
5-Heater
6-Spouted bed
7-Thermohygrometer
8-Thermohygrometer
9-Pressure taps
10-Paste feeder (atomizer)
11-Compressed air line
12-Peristaltic pump
13-Paste recipient
14-Valve
15-Manometer
16-Cyclone

FIGURE 6.3   Schematic diagram of a spouted bed dryer of pastes.

transfer and the friability of the film adhered to the particle surface. One of these two factors can be limiting when process stability is concerned, because it depends upon the drying rate as well as the mechanical properties of the film, which can continue to grow if it is strongly attached to the particle. Surface properties, such as superficial energy and wettability, are also important for the definition of particle coating and suspension drying processes. The adhesion characteristics define the tendency of the particle–suspension pair to go in a coating or drying process. Therefore, the appropriate choice of inert particle is strongly related to the performance of the paste or suspension drying.

High-density polyethylene (HDPE), low-density PE (LDPE), polypropylene (PP), polystyrene (PS), Sangel®, Teflon, glass, and silica, among others, are possible inert particles only if they have diameters that define them as spoutable (Geldart, 1986). All of these materials had already been tested as inert in the spouted bed. Their performance is a function of the paste or suspension inert surface and physical properties.

LDPE has already shown some problems when pastes with high lipid concentration are processed. The inert particles become breakable soon after the drying of such materials.

An important consideration for the inert particles used in food industries is that the effects of attrition due to the intense motion of the particles may cause contamination of the powder produced, especially after having been used for a long time. Thus, the possibility of contamination and its hazard to health must be critically analyzed.

The analysis of the drying performance using this technique is based on two approaches. The first approach is the evaluation of the efficiency of powder production (η_{po}), defined in Eq. (6.1) as the ratio between the mass of powder (m_{po}) collected in the cyclone and the mass of paste fed into the bed (m_{paste}), and the evaporation rate (W_{ev}) given in Eq. (6.2), both expressed on a dry basis (db):

$$\eta_{po} = \frac{m_{po} \times (1 - X_{po}) \times 100}{m_{paste} \times (1 - X_{paste})} \tag{6.1}$$

$$W_{ev} = \dot{m}_{ar} \times (Y_o - Y_0) \tag{6.2}$$

where X_{po} and X_{paste} are the moisture content of the powder and paste, respectively; $\dot{m}_{ar}$ is the mass flow rate of air; and Y_o and Y_0 are the outlet and initial absolute humidity, respectively.

The second approach refers to the paste or suspension effects on the hydrodynamic parameters of the spouted bed, which is of great relevance to efficient spouted bed drying.

Modifications of the fluid dynamic parameters by adding liquid to the inert particles have been evaluated (Medeiros et al., 2002; Passos et al., 1997; Patel et al., 1986; Schneider and Bridgwater, 1993; Spitzner Neto and Freire, 1997). Patel et al. (1986) reported a reduction in the value of the minimum spout flow rate because of the presence of liquid or paste by adding water and glycerol to the bed. Lima and Alsina (1992) also came to this conclusion when analyzing the drying of the pulp of a tropical fruit, *umbu* (*Spondias tuberosa*). The decrease in the minimum spouting flow rate was attributed to a decrease in the number of particles in the spout region. The increase in moisture content of the particles in the annulus region caused a decrease in the circulation velocity and fewer particles entered the spout region, resulting in a more dilute spout region. Lima and Alsina (1992) also observed an increase in the stable spout pressure drop with the increase in amount of pulp in the bed.

An important work on this topic is by Schneider and Bridgwater (1993), who analyzed the dynamic stability of the spouted bed with the addition of glycerol. The results were somehow different from those of Lima and Alsina (1992) and Patel et al. (1986). The stable spout pressure drop decreased with the increasing amount of liquid in the bed and with higher liquid viscosity. The minimum spouting flow rate initially increased and then decreased as a function of the addition of liquid to the bed.

Variations in the hydrodynamic behavior of the spouted bed that are due to continuous water feeding into the bed are also evidenced in the study by Spitzner Neto and Freire (1997). The assumptions for variations in the fluid dynamic parameters are based on the increase in the adhesion forces between the particles produced by the presence of liquid in the bed and on the shear stress forces caused by the liquid viscosity, in agreement with the study conducted by Passos et al. (1997).

Although the results presented in different works are apparently contradictory concerning the fluid dynamic parameters, all of them account for the influence of the presence of the liquid, paste, or suspension on the spouted bed hydrodynamics. In some cases, this presence compromises the spout stability, causing it to collapse.

The main conclusions concerning the effects of the introduction of paste on the bed dynamics are as follows:

1. A critical paste feed rate exists (above this critical value, interparticle cohesion forces caused by liquid bridges are significant).
2. The cohesion forces depend on the paste and inert properties.
3. If the cohesion forces can be neglected, the minimum spouting flow rate decreases with the feeding of paste (fine paste layers on the inert promotes particle slipping, enhancing the spout regime).

The change in the bed behavior and in the drying efficiency for different suspensions or pastes also suggests the importance of an investigation on the relationship of the adhesion characteristics of the particle–suspension or paste with the drying process, yielding efficient powder production. In other words, the drying of a paste with efficient production or coating of particles is a function of the characteristics of particle–suspension adhesion, which is related to the particle wettability by the paste.

Wettability is described as the ability of a drop to wet and spread upon the surface of a particle. It is what governs the adhesion and growing rate of particles in coating processes (Iveson et al., 2001; Link and Schlünder, 1997; Pont et al., 2001; Tenou and Poncelet, 2002).

The wettability of a solid by a liquid depends on the contact angle between the three phases: solid, liquid, and gas. The contact angle (θ) is a property of the solid–liquid–gas system and depends on the characteristics of the solid particle and paste or suspension (liquid), such as hygroscopicity, roughness, and surface tension.

Adhesion is correlated with wettability and occurs due to physical, chemical, and/or mechanical bonds in the solid–liquid interface. Under some conditions (plain, homogeneous, isotropic, and nondeformed surface) the adhesion energy can be quantified by adhesion work (τ_{ad}) defined as

$$\tau_{ad} = \sigma_{lv}(1 + \cos\theta) \tag{6.3}$$

where σ_{lv} is the liquid–vapor surface tension.

High values of adhesion work indicate good wetting and low values indicate bad wetting for each solid–fluid set (Adamson, 1990; Iveson et al., 2001; Neumann and Kwok, 1999).

When θ is 0°, the work of adhesion of solid to liquid reaches a maximum and is equal to the work of liquid cohesion, which can spread indefinitely on the surface because the system is energetically indifferent to whether the liquid is in contact with itself or with the solid. In contrast, if θ is 180° then τ_{ad} is 0 and there is no loss of Gibbs energy to separate the solid from liquid. The liquid does not wet the solid and does not spread over it.

Typical spouted bed equipment with inert particles used for the drying of paste, like the one shown in Figure 6.3, can also be utilized for particle coating. Inert particles of different sizes, shapes, densities, and flowabilities, as well as suspensions

TABLE 6.1
Physical Properties of Inert Particles

Material	Glass beads	ABS®	PP	PS
Shape	Round	Round	Irregular	Wedge
d_p (mm)	2.13	2.90	2.91	4.58
ϕ	0.81 ± 0.05	0.77 ± 0.06	0.74 ± 0.06	0.90 ± 0.07
Φ (°)	31.03 ± 1.15	40.27 ± 2.22	37.67 ± 2.12	36.67 ± 1.22
ρ_{true} (kg/m³)	2491.5 ± 0.4	1022.1 ± 0.8	905.3 ± 0.6	1060 ± 0.6
ρ_{ap} (kg/m³)	2491.7	1013.6	905.3	1049.1
ε_p (%)	0	0.83	0	1.03
FI (mm)	18	14	18	10
$\sigma_p \times 10^3$ (N/m)	85	34	30	32

with different formulations were used by Marques (2007) to analyze the influence of particle–suspension adhesion during the atomization of the suspension over a bed of inert particles. The inert particles used were characterized by shape, average diameter, sphericity, repose angle, bulk, real and apparent densities, porosity, flowability, and surface energy (Table 6.1). Three different formulations of aqueous polymeric suspensions were used, which showed differences in their properties of surface tension, solids concentration, density, and apparent viscosity obtained at a deformation rate of 1000/s (Table 6.2). The particle contact angle values for the three suspensions analyzed are presented in Table 6.3. The operating conditions were set after a fluid dynamic analysis and some preliminary experiments, and they were also based on the results of previous works (Donida et al., 2005; Table 6.4).

Through the analysis of the process behavior it is possible to sort the particles into two groups: glass particles and ABS, which were coated by the suspensions because of their characteristics of wettability (contact angles <70°), and particles of

TABLE 6.2
Physical Characteristics of Suspensions

Characteristics	Suspension		
	1	2	3
C_s (kg/kg)	0.1270 ± 0.006	0.1050 ± 0.001	0.1024 ± 0.003
ρ_{susp} (kg/m³)	1042.8 ± 5.70	1054.5 ± 5.8	1015.1 ± 3.0
$\sigma \times 10^3$ (N/m)	60.09 ± 1.20	46.27 ± 0.61	67.50 ± 0.65
n	0.8754 ± 0.021	0.8226 ± 0.024	0.8330 ± 0.059
m (Ns^n/m²)	0.3235 ± 0.004	0.1220 ± 0.010	0.1109 ± 0.035
μ_{ap} (Ns/m²)	0.1368	0.0358	0.0349

TABLE 6.3
Particle–Suspension Contact Angles

Material	Contact Angle		
	Suspension 1	Suspension 2	Suspension 3
Glass	37.82 ± 1.42	34.94 ± 2.44	37.86 ± 2.34
ABS	70.35 ± 1.35	54.29 ± 3.48	70.0 ± 1.45
PS	76.68 ± 2.38	57.50 ± 3.31	79.5 ± 1.57
PP	78.98 ± 2.33	70.98 ± 2.41	80.9 ± 1.21

PS and PP, which promoted drying of the suspension (contact angle >76°) that is due to their superficial characteristics.

These results are in agreement with those of Adeodato et al. (2004), who investigated the adhesion of an aqueous polymeric suspension to several inert particles (glass, ABS, placebo, PP, PS, and LDPE) in a spouted bed with the suspension atomized at the top. They observed that the suspension did not adhere to particles with large angles of contact with the suspension (≥80°), such as PS, PP, and LDPE; therefore, the suspension was elutriated. The particles that were coated by the suspension (glass, ABS, and placebo) had contact angles <80°. Figure 6.4 shows the results.

For PP and PS there is no formation of a coating layer because of the weak adhesion between the particle and suspension. The film formed on the PP particles is not uniform and can be easily removed by interparticle attrition. The film formed on the PS surface is more homogeneous, which must be related to the strong influence of the solid surface energy on the process efficiency. The PS particles have higher surface energy that favors particle–liquid interaction and the above behavior (Marques et al., 2006).

It can be concluded that the efficiency of the inert particle (and therefore of the drying operation) is directly related to the surface characteristics of the particle–paste pair and the choice of the inert particle must be based on these characteristics. (The 80° limit of the contact angle is a good indication of the good efficiency of inert particles for a drying operation.) Studies about the influence of solid surface energy

TABLE 6.4
Spouted Bed Operating Conditions

Bed height (m)	0.155
Spout velocity (m/s)	1.15 × Ujm
Spouting air temperature (°C)	60
Suspension flow rate (ml/min)	6.0
Atomization pressure (psig)	20
Process time (min)	30

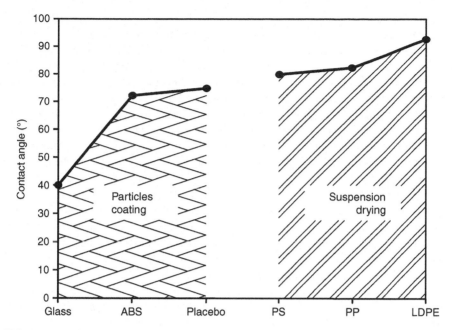

FIGURE 6.4 Influence of contact angle on the particle–suspension adhesion.

combined with the friability of the film on the efficiency of powder production are challenges for further investigations.

Once the inert particles have been chosen, their properties (size, shape, rest angle, density, and porosity) and the conditions of operation will directly influence process performance. Optimization of the process may be carried out for each case.

6.3 DRYING OF TROPICAL FRUIT PULPS

A significant amount of fruit production is wasted every year throughout the world (mainly in developing countries) because of the lack of adequate and low-cost processing (mainly drying) that would permit its industrialization and marketing. In Brazil, the loss is up to 40% for some cultivations (it should typically not exceed 10%).

The drying of fruit pulps in a spouted bed is a lower cost alternative to spray drying. This method is appropriate for small-scale processing, can process different kinds of pulps (even mixed pulps) in the same equipment, and has the same level of product quality as spray drying.

Knowledge of the chemical composition of fruit pulps, mainly the reducing sugars and lipids, is essential to obtain an efficient drying process. Lima et al. (2000) identified the influence of reducing sugars, lipids, and fibers on the drying efficiency of the following tropical fruit pulps: umbu (*Spondias tuberosa*), cajá (*Spondia lutea*), cajá-manga (*Spondias dulcis*), pinha (*Armona squamosa*), seriguela (*Spondia pupurea*), mango (*Mangifera indica*), West Indian cherry (*Malpighia punicifolia*), and avocado (*Persea americana*). Martinez et al. (1996) observed the influence of reducing sugars and lipids on the performance of milk drying.

Medeiros et al. (2002) attempted to determine the influence of paste composition on spouted bed hydrodynamics and on drying performance. The effect of varying the chemical composition of the pulp was directly related to variations in physical properties and surface properties, which causes the spouted bed drier behavior to change.

Taking natural mango pulp as the basis, modified pulps with controlled concentrations of reducing sugars, lipids, fibers, starch, and pectin were obtained. Starch, fibers, and pectin were included in the analysis of the process because of the important influence of these components on the physical properties of liquid foods and the presence of starch and pectin in almost all tropical fruit pulps (with high concentrations of starch in some of the pulps).

The effects that the concentration of these components in the pulps had on the fluid dynamics of the spouted bed and on the drying process were investigated using a factorial design for the experiments. The operating conditions were fixed in all experiments. The efficiency of powder production, the stable spout pressure drop, and the minimum spouting flow rate were the responses analyzed. Details of the experimental procedure and materials utilized to control the pulps compositions can be found in Medeiros et al. (2002). HDPE particles (group D in Geldart's diagram) with a diameter of 3×10^{-3} m were used as the inert material (density = $950 \, \text{kg/m}^3$, sphericity = 0.76, static bed porosity = 0.29, apparent density = $679 \, \text{kg/m}^3$). The concentrations levels utilized in the factorial design of the experiments are provided in Table 6.5. The operating conditions were the following: air inlet temperature of $70 \pm 1°C$, 2.5 ± 0.1 kg load of inert particles, 50 ± 0.5 g pulp feeding, 40-min drying time, an 1.25 ± 0.05 ratio of air-flow rate to minimum spouting flow rate.

Examples of drying curves are presented in Figure 6.5 (absolute humidity at the outlet of the bed as a function of time) and Figure 6.6 (evaporation rate as a function of time). Very fast evaporation occurs in the initial minutes of the drying procedure, as can be observed. Later the absolute humidity of the air at the outlet of the bed reaches the value of absolute humidity entering the bed and the evaporation rate is brought down to zero. This behavior was seen in most of the experiments, and total evaporation occurred in the first 10 min of drying.

The evaporation rates were not influenced by pulp composition or physical properties. Some small differences in the values of the air humidity at the outlet and

TABLE 6.5
Concentration Levels of Components

	Concentration Levels ($C/C_{water} \times 100\%$)				
	C_1 Reducing Sugars	C_2 Lipids	C_3 Fibers	C_4 Starch	C_5 Pectin
−1	7.67	0.78	0.50	0.52	0.67
0	13.81	3.81	1.25	2.59	1.25
+1	19.95	6.84	2.00	4.66	1.82

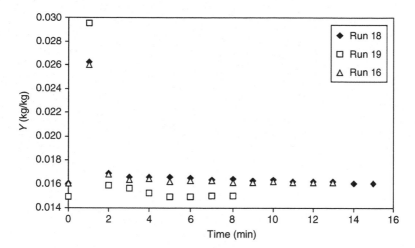

FIGURE 6.5 Absolute air humidity as a function of time for runs 16, 18, and 19. Run 16: $C_1 = 19.95$, $C_2 = 6.84$, $C_3 = 2.00$, $C_4 = 4.66$, $C_5 = 1.82$. Run 18: $C_1 = 13.81$, $C_2 = 3.81$, $C_3 = 1.25$, $C_4 = 2.59$, $C_5 = 1.25$. Run 19: $C_1 = 13.81$, $C_2 = 3.81$, $C_3 = 1.25$, $C_4 = 2.59$, $C_5 = 1.25$. (Adapted from Medeiros, M.F.D. et al., *Dry. Technol.*, 20, 855–881, 2002. With permission of Taylor & Francis.)

evaporation rates were noted that were attributable to noncontrolled changes in psychrometric conditions (Medeiros et al., 2002).

By integrating the curves for the evaporation rates as a function of time we can obtain the mass of water evaporated during the drying, considering for each experiment the time interval in which water evaporation is occurring. Figure 6.7

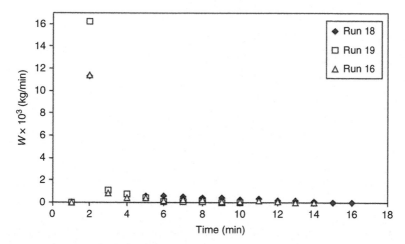

FIGURE 6.6 Evaporation rates as a function of time for runs 16, 18, and 19. Run 16: $C_1 = 19.95$, $C_2 = 6.84$, $C_3 = 2.00$, $C_4 = 4.66$, $C_5 = 1.82$. Run 18: $C_1 = 13.81$, $C_2 = 3.81$, $C_3 = 1.25$, $C_4 = 2.59$, $C_5 = 1.25$. Run 19: $C_1 = 13.81$, $C_2 = 3.81$, $C_3 = 1.25$, $C_4 = 2.59$, $C_5 = 1.25$. (Adapted from Medeiros, M.F.D. et al., *Dry. Technol.*, 20, 855–881, 2002. With permission of Taylor & Francis.)

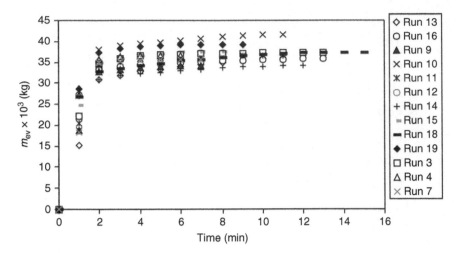

FIGURE 6.7 Cumulative curves for evaporated water mass as a function of time. (Adapted from Medeiros, M.F.D. et al., *Dry. Technol.*, 20, 855–881, 2002. With permission of Taylor & Francis.)

exemplifies cumulative curves for water evaporated as a function of time in several experiments of fruit pulp drying at different compositions.

Independent of the pulp composition, the masses of water evaporated in the first 2 min varied from 30 g (experiment 13) to 38 g (experiments 10 and 19), which corresponds to fractions of evaporation between 84.5 and 95.8% (Figure 6.7). The evaporation rate after that period is very low, showing the high capacity of evaporation of the spouted bed with inerts.

Concerning the fluid dynamics, the feeding of pulps followed by intense modification of the fluid dynamic behavior of the bed was characterized by variations in pressure drop, air velocity, fountain height, and expansion in the annulus. Depending on the intensity of these variations, the bed can undergo serious instability problems and spout collapse. In some experiments, the problems with instability, although attenuated by evaporation of water, persisted even when the evaporation rates were zero. Note that composition influences the bed fluid dynamics not only during feeding but also instability problems caused by the presence of pulp can persist even when the bed is practically dry. In this case, the bed is modified by the presence of fine solids.

The degree of saturation and stability of the bed during drying of pastes in spouted beds, when the inert particles are coated with liquid films, is referred to in the most important works on the fluid dynamic changes of the bed (Passos and Mujumdar, 2000; Schneider and Bridgwater, 1993; Spitzner Neto and Freire, 1997). Thus, the models and equations proposed in these works to estimate cohesion forces, correlated with the degree of bed saturation and with other fluid dynamic variables of the bed at minimum spouting conditions, are not appropriate to simulate the bed dynamic behavior when the film adhering to the inert dries very fast (as in the drying of fruit pulps).

Modifications of the physical properties of the pulps (density, apparent viscosity, and surface tension) due to water evaporation and material heating make it unviable

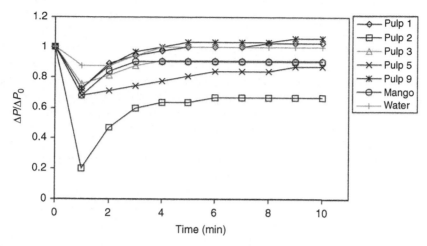

FIGURE 6.8 Influence of pulp feeding with a higher concentration of pectin (pulp 1), reducing sugars (pulp 2), lipids (pulp 3), fiber (pulp 5), starch (pulp 9), natural mango, and water on the bed pressure drop.

to discuss the effects of these properties (evaluated for the pulps at their original moisture content) on the bed fluid dynamic and drier performance.

Figure 6.8 illustrates the changes caused in bed pressure drop by the feeding of different pulps, water, and natural mango pulp relative to the dry bed. All of the results show a sudden decrease in pressure drop at the moment when the pulps were added to the bed. The decrease can be explained by agglomeration, which along with the viscosity of the paste affects particle circulation as well as the passage of air through the bed, increasing the air fraction that crosses the spout region and consequently the air velocity in this region. Because there is a lower resistance to the air passage, the pressure drop decreases. The same effect was observed by Spitzner Neto and Freire (1997) with continuous feeding of paste.

After the pulps dry on the inert particle surface, the pressure drop is not restored to its initial value but stabilizes at a lower value. The solid films covering the particle surface give rise to solid bridges, which increase the adhesion forces between the particles. Thus, the influence of the dried pulp on the bed fluid dynamics is evident.

Simultaneous changes in the bed by the addition of pastes (pulps) can usually be found, which are characterized by an increase in superficial air velocity, a decrease in pressure drop, and a more expanded fountain and annulus. These phenomena were observed in every experiment, except for experiment 2 (presence of reducing sugars), in which chance of collapse was characterized by a lower and less dense fountain than the one observed with the bed of inert particles (Medeiros et al., 2002). The adhesion characteristics prevailed and caused different behavior in the case of the presence of reducing sugars.

A statistical analysis of the results demonstrated that the composition of all of the components had significant effects on the efficiency of powder production, except for the concentration of fibers. The concentration of fibers did not have significant effects on any of the response variables analyzed in the fluid dynamic and drying experiments.

Sugar causes a decrease in efficiency, and its effect is the most important. Lipids, starch, and pectin favor powder production. Among the variables that favorably influence efficiency, the concentration of starch is the most important. The combined effects of reducing sugars–lipids and starch–pectin on the efficiency of powder production are significant. The results on the influences of lipids and reducing sugars are in agreement with those found in Martinez et al. (1996).

Medeiros et al. (2002) utilized statistical analysis of the dynamic parameters to show that the composition variables do not have significant effects on the minimum spouting flow rate, but the effects of sugar and starch concentrations on the stable spout pressure drop are significant. Beyond the individual effects of reducing sugar and starch on the stable spout pressure drop, the interactions of reducing sugars– lipids, lipids–starch, and lipids–starch–pectin also had significant effects on Δp_s. Thus, interactions between the composition variables can cancel or decrease the main effects of the isolated concentrations.

This analysis pointed out the possibility of using a mix of fruit pulps to optimize the fluid dynamic behavior of the spouted bed, resulting in higher drying efficiency due to the synergy of the components of different pulps.

6.4   DRYING OF GRAINS

Seeds and grains, which have high nutritional value, are important products in the food industry. Grains, in general, are harvested in the field and have high moisture contents. For storage and conditioning, it is essential to decrease the moisture content of the grain by an efficient drying process. The choice and design of the dryer, the operating conditions, and the influence of these parameters on the final product quality are very important characteristics to take into account in an analysis of grain drying (Inprasit and Noomhorm, 2001; Sangkram and Noomhorm, 2002).

The main grain drying applications for the SBD are beans and paddy. Investigations show good results in terms of final moisture content. Good quality products with homogeneous moisture content and material temperature were obtained by various studies (Lima and Rocha, 1998; Madhiyanon et al., 2001a, 2002; Robins and Fryer, 2003; Wiriyaumpaiwong and Devahastin, 2005; Wiriyaumpaiwong et al., 2003).

However, because of the limitation of its scale of production, the spouted bed has received limited attention for application on a large scale. One of the few industrial-scale spouted beds was built and tested by Madhiyanon et al. (2001b) for the continuous drying of paddy. The dryer capacity was around 3500 kg/h of paddy with initial moisture contents in the range of 14 to 25%. They evaluated the consumption of thermal energy, which was in the range of 3.1 to 3.8 MJ/kg of evaporated water, comparable to that of other commercial dryers. In addition to the difficulty of scale-up, most grains become fragile when submitted to a moving, fluidized, or SBD, which is due to exposure to severe attrition conditions. Thus, the use of an intermittent drying process could be profitable and even necessary to maintain product quality (especially when dealing with grain seeds).

Intermittence in the operation is when the air supply to the bed or the active fluid dynamic regime is stopped from time to time (e.g., decreasing the air-flow rate from stable spouting to a value that establishes a fixed bed regime). Therefore, the bed of grains is not continuously under the active hydrodynamic regime; consequently, less grain damage and energy savings are achieved. Nevertheless, it is necessary to analyze the influence of intermittence on the final grain water content.

Oliveira (1999) analyzed the behavior and efficiency of bean drying in the intermittent spouted or fixed bed and in the completely intermittent spouted bed or rest regimes and compared the results to those obtained in a conventional spouted bed without intermittence. *Phaseolus vulgaris* L. beans (commonly called *carioca* beans) were utilized, which is the most common variety of beans, responsible for about 95% of the world production. They are a very important source of calories in the human diet, mainly in the underdeveloped countries of tropical and subtropical regions.

A simple spouted bed drying system and bed dimensions used by Oliveira (1999) are demonstrated in Figure 6.9a and b, respectively. Two types of intermittent processes were conducted: spouted or fixed bed (process I) and spouted bed or rest (process II). In process I the bed (after attaining thermal equilibrium) is loaded with grain and the air-flow rate adjusted to a stable spout condition. The air-flow rate is then decreased from time to time until the spout collapses and the bed of beans remains fixed. The air-flow rate is maintained constant under the fixed bed condition for a period of time defined as the intermittence time (t_i). After that time, the bed is again spouted by increasing the air-flow rate to the previous value of a stable spout condition for the same period t_i. The same procedure is repeated during a 240-min drying process.

Process II is similar to process I, but at the t_i the air flow to the bed is totally interrupted. After the t_i of interruption of air to the bed, the air supply is restored under the stable spout condition. This procedure is repeated during 240 min of drying time.

The values of the initial moisture content of the grains (X_0) loaded onto the bed correspond to freshly picked beans at different times of the year (20 and 30% db). The drying air temperatures (T_{ar}) were 60 and 80°C, and the t_i values were 20 and 40 min.

The lowest value of the final moisture content (X_f) of the beans obtained for process I is 9.02% db, which corresponds to operating conditions in which T_{ar} is 80°C, X_0 is 20% db, and t_i is 20 min. For process II the lowest final moisture content was 11.58% db for operating conditions of T_{ar} at 80°C, X_0 at 20% db, and t_i at 40 min. For both processes, the lowest moisture content was obtained at the highest temperature, as expected. Although higher final moisture content values were obtained for the spouted bed or rest intermittent process, both values of the lowest X_f (9.02 and 11.58% db) are in the range specified for good conditions of carioca bean storage $(9.0\% \le X_f \le 13.0\%$ db; Cavalcanti Mata, 1997). Figures 6.10 and 6.11 are examples of drying curves and grain temperature curves, respectively (all experiments showed the same qualitative behavior).

The T_{ar} had a significant influence on the drying curves, especially for drying times above 40 min. These results were expected, because temperature has a significant influence on the falling rate period in drying processes.

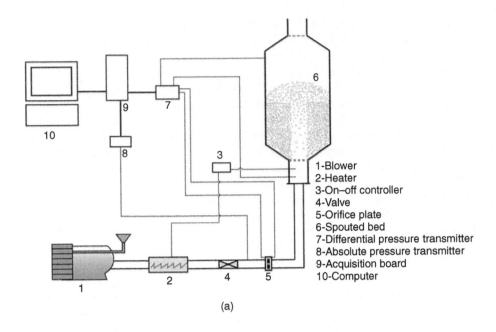

1-Blower
2-Heater
3-On–off controller
4-Valve
5-Orifice plate
6-Spouted bed
7-Differential pressure transmitter
8-Absolute pressure transmitter
9-Acquisition board
10-Computer

(a)

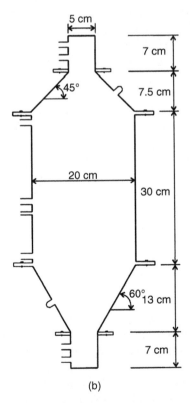

(b)

FIGURE 6.9 Spouted bed (a) drying system and (b) dimensions.

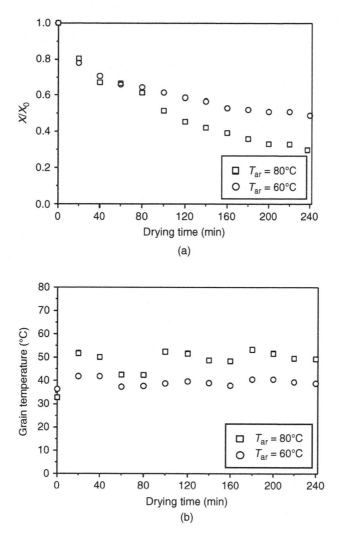

FIGURE 6.10 Influence of air temperature on the drying kinetics for drying process I: spouted or fixed bed. Conditions: $t_i = 40$ min, $X_0 = 30\%$ db.

It is important to point out that some experiments have drying curves showing rewetting points attributable to vapor condensation inside the bed during the intermittence period. These points occurred particularly for process II, where the air supply is totally interrupted in the intermittence period. These rewetting points are shown more explicitly mainly in the drying runs that operated under conditions of a high initial moisture content of the grains of around 30% db.

Figures 6.12 to 6.15 show comparisons of the drying of beans in fixed, spouted, and intermittent spouted beds, respectively. Lower grain temperatures are obtained when intermittent process II is applied, which is a satisfactory result, because lower grain temperatures are required to maintain their unchanged physical and chemical

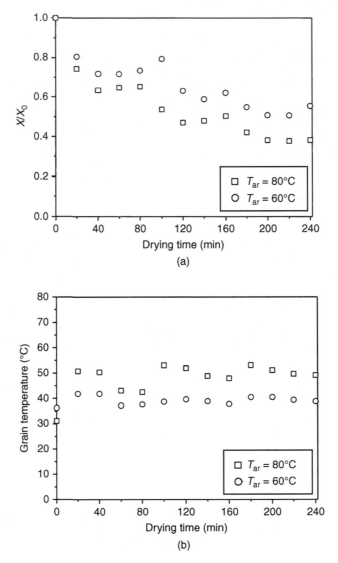

FIGURE 6.11 Influence of air temperature on grain temperature for drying process II: spouted bed or rest. Conditions: $t_i = 40\,\text{min}$, $X_0 = 30\%$ db.

properties. The drying kinetics is almost the same for the fixed and spouted beds, but different behaviors are observed for intermittent processes I and II.

Points of rewetting appear for intermittent process II (spouted bed or rest). Consequently, the bean moisture content is higher at the same drying time for this type of intermittence. For intermittent process I (spouted or fixed bed) the rewetting points are not as evident. However, a higher moisture content is still observed for the same drying time compared to the drying conducted in the spouted bed without intermittence or the fixed bed.

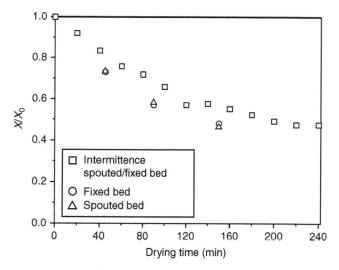

FIGURE 6.12 Comparison of drying curves for *carioca* beans in fixed and spouted beds with and without intermittence ($T_{ar} = 80°C$, $X_0 = 20\%$ db, $t_i = 40$ min).

The final moisture contents at 240 min of drying are close in all cases under the same operating conditions, especially when comparing spouted bed and spouted or fixed bed (process I).

In addition to the advantage of intermittent drying in terms of product quality, it is important to evaluate whether intermittent drying promotes more energy savings than spouted bed drying. The energy evaluation in a drying process can be made

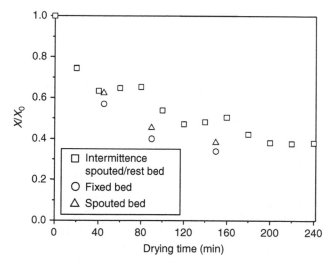

FIGURE 6.13 Comparison of drying curves for *carioca* beans in fixed and spouted beds with and without intermittence ($T_{ar} = 80°C$, $X_0 = 30\%$ db, $t_i = 40$ min).

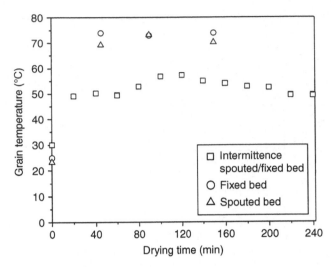

FIGURE 6.14 Comparison of grain temperature of *carioca* beans in fixed and spouted beds with and without intermittence ($T_{ar} = 80°C$, $X_0 = 20\%$ db, $t_i = 40$ min).

based on energy coefficients presented in the literature (Dewettinck et al., 1999; Lima and Rocha, 1998; Pakowski and Mujumdar, 1995).

Drying efficiency (DE) is defined as the energy flux necessary to heat the grains and to evaporate the water in relation to the total energy flux supplied to the operation [Eq. (6.4)]. The drying coefficient (DC) is defined as the amount of water evaporated by the amount of energy supplied to the operation [Eq. (6.8)]. The fluid dynamic

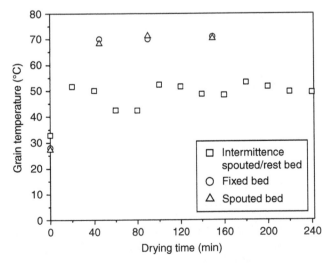

FIGURE 6.15 Comparison of grain temperature of *carioca* beans in fixed and spouted beds with and without intermittence ($T_{ar} = 80°C$, $X_0 = 30\%$ db, $t_i = 40$ min).

drying coefficient (FC) is the ratio between the energy spent to vaporize the water and the mechanical energy required to pump the air [Eq. (6.9)]. It was used to compare the fluid dynamic aspects of drying processes I and II.

$$DE = \frac{E_{h,g} + E_{v,w}}{ET} \times 100 \tag{6.4}$$

The necessary energy flux to heat the grain from its initial temperature to the final temperature ($E_{h,g}$) can be calculated as proposed by Strumillo and Kudra (1986):

$$E_{h,g} = m_{g,dry}(c_{gf} \times T_{gf} - c_{g0} \times T_{g0})/t_{h,g} \tag{6.5}$$

The necessary energy flux for evaporation of the moisture ($E_{v,w}$) is

$$E_{v,w} = \Delta H_{v,w} \times (m_{g,w}/t_d) \tag{6.6}$$

The total energy flux supplied to the operation (ET) is considered as the sum of the thermal and mechanical energy fluxes.

The thermal energy flux (E_{ter}) supplied to the operation was calculated as

$$E_{ter} = \dot{m}_{ar} \times (c_{ar} \times T_{ar} - c_{amb} \times T_{amb}) \tag{6.7}$$

$$DC = \frac{m_{g,w}/t_d}{ET} \tag{6.8}$$

$$FC = \frac{E_{v,w}}{EM} \times 100 \tag{6.9}$$

A comparison of the energy analysis for intermittent processes I and II and for spouted bed drying of beans without intermittence are presented in Tables 6.6

TABLE 6.6

Drying Efficiency and Energy Coefficients: Spouted or Fixed Bed and Spouted Bed or Rest Regimes

Run	Drying Process I: Spouted or Fixed Bed			Drying Process II: Spouted Bed or Rest		
	DE (%)	DC (kg/kJ) $\times$ 10^5	FC (%)	DE (%)	DC (kg/kJ) $\times$ 10^5	FC (%)
1	14.81	1.33	3.59	21.06	1.98	5.37
2	11.83	1.32	3.57	17.18	1.72	4.67
3	5.19	1.51	3.94	7.39	2.10	5.47
4	7.04	1.06	2.76	7.76	1.63	4.25
5	10.60	1.08	2.94	15.69	1.27	3.49
6	9.21	1.06	2.88	11.15	1.43	3.93
7	10.35	0.95	2.47	23.47	1.34	3.47
8	12.66	0.94	2.46	16.02	1.19	3.09

TABLE 6.7

Comparison of Drying Processes in Spouted Bed without and with Intermittence

			Process: Spouted Bed				
T_{ar} (°C)	X_0 (% db)	Grain Mass (g)	$\dot{m}_{ar}$ (kg/min)	X_f (% db)	DE (%)	DC (kg/kJ) × 10⁵	FC (%)
83	20	3000	2.27	9.13	7.56	1.6	1.04
83	30	3000	2.33	9.25	9.29	2.5	1.66
		Intermittent Drying Process I: Spouted or Fixed Bed					
80	20	3200	2.44	9.02	11.83	1.32	3.57
80	30	3200	2.44	9.17	14.81	1.33	3.59
		Intermittent Drying Process II: Spouted Bed or Rest					
80	20	3200	2.44	11.58	17.18	1.72	4.67
80	30	3200	2.44	11.85	21.06	1.98	5.37

and 6.7, respectively. Higher DE values were obtained at T_{ar} of 80°C, X_0 of 20 or 30% db, and t_i of 40 min for processes I and II. A high intermittence time causes a decrease in the total energy supplied to the grains whereas the high air temperatures and initial grain moisture content cause an increase in the mass of moisture evaporated, thus increasing the energy efficiency of drying. It can also be verified that a high DE value was obtained at T_{ar} of 60°C, X_0 of 20 or 30% db, and t_i of 20 min in the case of process II. This result indicates that the low total energy input (low air temperature and interruption of the air to the dryer in the ranges analyzed) does not jeopardize the evaporation of moisture. Comparing processes I and II, the energy efficiency of drying was higher for process II under all operating conditions.

The highest DC and FC coefficients were obtained for conditions consisting of T_{ar} at 60°C, X_0 at 30% db, and t_i at 40 min. Higher values for the coefficients were obtained with t_i at 40 min. The long intermittence time causes a decrease in the total energy input and resulted in a decrease in mechanical energy, thus favoring the increase in DC and FC coefficients.

The general rule applying to drying efficiency is spouted bed or rest > spouted or fixed bed > spouted bed, which results in the choice of the intermittent process spouted bed or rest to obtain the lowest energy consumption or high energy efficiencies.

The numbers obtained for energy efficiency and coefficients are close to the values found for industrial continuous convective dryers (Raghavan et al., 2004; Vieira et al., 2005).

6.5 LATEST ADVANCES IN SPOUTED BED DRYING

The following sections present a summary of recent developments in spouted bed technology. Quality, novel equipment designs, computational simulation, and control are the main focus of the recent research applying this drying technique.

6.5.1 PRODUCT QUALITY IN SPOUTED BED DRYING OF PASTES OR SUSPENSIONS

The quality of food products must follow specifications for esthetics and nutritional properties. Because of the thermal sensitivity of biological materials, such as foodstuffs, the operating conditions of drying are limited. Beyond an analysis of the drying process, knowledge of the changes in and degradation of the product as a consequence of drying is among the main objectives of recent works on drying technology. In addition, difficulties in the drying process due to modification of the material inside the dryer (i.e., shrinkage) were recently studied (Almeida et al., 2006).

The quality of foodstuffs is characterized by flavor, texture, and color. If color is unacceptable to the consumer, the product is rejected, regardless of its taste and texture, and modifications of protein and variations in physical properties may be quickly recognized by color changes. Quantification of important components that are present in a determined product (e.g., vitamins, carotenoids, etc.) complements its qualification. Standard methodologies for their determination are available in the specialized literature.

Ascorbic acid (vitamin C) is an important nutrient in fruits and vegetables. It is more sensitive to heat, oxygen, and light than other compounds, such as vitamins A and E and some phenolic compounds. Therefore, vitamin C content is usually utilized as the parameter that defines the best drying method or condition for retaining the nutritional quality of fruits and vegetables. The loss of vitamin C during drying depends on the type of process, the physical properties of the product, and the time–temperature conditions utilized in each type of dryer. In addition, quality requirements according to the final use of the product (such as particle size distribution, rehydration capacity, final moisture content, storing possibilities, etc.) are also very important.

The drying of fruit pulps can result in browning reactions and pigment destruction (Abers and Wrolstad, 1979). Recent works on product quality in the spouted bed drying of pastes and suspensions can be found in current research.

Cunha et al. (2006) evaluated the influence of operating conditions on the efficiency of powder production in the drying of mango pulp. The effects of air temperature, feed rate, and annulus and spout gas flow rate were evaluated using a full factorial design for the experiments. In addition, the powder morphology as well as vitamin C and carotenoid contents of some samples were analyzed after drying. A two-dimensional spout fluidized bed (0.05 × 0.32 m) with a square inlet air orifice (0.05 × 0.05 m) was utilized. The remaining area was dedicated to annulus aeration and fluidization.

Although temperature has the greatest influence on the maintenance of quality in fruit drying (high temperature jeopardizes the maintenance of nutritional properties) for the range analyzed by Cunha et al. (2006), higher drying temperatures favored the retention of carotenoids, in addition to diminishing product browning and moisture content. The ranges utilized for mango pulp drying were temperatures of 55 to 75°C, feed rates of 0.5 to 2.5 g/min, and spout and annulus air velocities (or flow rates) of 6–6.6 m/s (or 15–16.4 l/s) and 1.81–1.9 m/s (or 22.6–24.1 l/s), respectively. Tests were conducted with intermittent feeding, because the mango pulp could not be fed continuously. The elevated concentration of sugars caused the bed dynamics to collapse. Temperature was the variable that had the greatest influence on the process, and it was directly proportional to the product powder efficiency. Higher feed mass and air-flow rates at the annulus also contributed to an improvement in product quality and process efficiency.

Benali and Amazouz (2006) also investigated product quality aspects for the drying of vegetable starch solutions in a continuous conical jet spouted bed with inert particles of Teflon. The Teflon load varied from 8 to 12 kg and the starch solution feed rate was in the range of 60 to 120 kg/h. Color and level of starch damage were utilized to define the quality of the dry product. Method 76-31 of the American Association for Cereal Chemists was applied to measure starch damage. For the material tested, the damage index should be <2.5% and the lightness should be ≥93. The results showed very good performance for drying in the continuous conical jet spouted bed, resulting in a product with lightness and damage indexes in the ranges established as industrial targets for vegetable starch (95.0–96.1 and 1.86–1.88, respectively, for bottom atomization and 94.1–94.4 and 1.98–2.22, respectively, for top atomization of the starch solution). In addition, a comparison of the proposed dryer with a commercial flash dryer (90–92 and 3.5–5.1, respectively) indicated its competitiveness in terms of product quality.

6.5.2 NOVEL AND COMBINED EQUIPMENT

The objectives of the design of new equipment or the utilization of combined drying techniques are to improve drying efficiency and to enable the use of the spouted bed on a real scale.

Some of the most recent research dealing with new equipment and techniques is for a mechanically spouted bed (MSB; Pallai et al., 2001), triangular and hexagonal spouted beds (TSB and HSB; Hung-Nguyen et al., 2001), a combined microwave spouted bed (MWSB; Nindo et al., 2003), and a sorption drying technique in a jet spouted bed (Wachiraphansakul and Devahastin, 2007).

The new construction of the previous MSB (Pallai et al., 1995) includes an internal tube of varying length, providing intensive drying and well-controlled heat and mass transfers (Pallai et al., 2001). In the modified equipment, alterations of screw length and bed height can be made to control the particle size distribution and to decrease the bed pressure drop when drying pastes or suspensions. Increasing the screw length and maintaining an unchanged bed height results in an improvement of wearing and grinding effects; decreasing the spouted bed height, maintaining the screw length, and including the internal tube above the bed surface results in a decrease in the bed pressure drop; and increasing the screw length together with a decrease in spouted bed height result in both improvement of the wearing effect and a decrease in the bed pressure drop. One of the most important advantages of the modified dryer construction is the possibility to decrease the spouted bed height and thereby also the pressure drop without a reduction in the wearing time and without affecting the drying efficiency. Details of the original MSB and the modified recent construction can be found in Pallai et al. (2001).

The TSB has a different design (Hung-Nguyen et al., 2001), whose main aim is scaling up. Satisfactory laboratory results on drying grains were determined for the building of a pilot plant with a capacity of 500kg/batch. The pilot-scale TSB dryer, which was designed and constructed by Hung-Nguyen et al. (2001), basically consists of three plane walls connected to a 60° slanted base. A moveable draft plate is installed in one corner in order to control the separation distance and thus determine the flow rate of grains entering the spout region and to separate the spout and downward flowing grains. For the pilot TSB, a separation distance kept below 150mm worked the best.

An HSB dryer formed of a combination of six TSB units, which can work in continuous or batch operation, was proposed by Hung-Nguyen et al. (2001). Evaluating process performance and product quality when drying paddy with high moisture content, the authors expected that paddy (moisture content = 24 to 27%) can reach 18 to 20% moisture in continuous drying in an HSB. The results point to a potential application of the HSB dryer as a first stage in grain treatment in a two-stage drying method. The second stage of drying may apply a low air temperature so as not to jeopardize the grain quality (temperatures above 80°C are not recommended for grains having moisture contents lower than 20%).

Combined drying methods are important alternatives to take advantage of and implement different drying technologies. One application for the combination of drying in spouted beds with microwave application (MWSB) is the drying of asparagus slices (Pallai et al., 1995). MWSB drying at a power level of 4 W/g was faster than tray drying, increased retention of the beneficial bioactive compounds in the dried asparagus, and had the highest rehydration ratio when compared to tray, spouted bed, Refractance Window®, and freeze drying. For drying processes with hot air, the highest retention of total antioxidant activity was verified for an MWSB at 60°C and 2 W/g. The results show the promising potential of the combined MWSB drying technique as an efficient and fast process for obtaining good quality dried food products.

An alternative to deal with sticky and pulpy products, reducing agglomeration and improving drying efficiency, is the sorption drying technique. The use of an adsorbent facilitates mass transfer from the moisture material and consequently agglomeration is reduced; in addition, lower temperatures can be employed in sorption drying. In the spouted bed drying of a sticky material, the addition of spoutable adsorbent particles enhances the stable spout dynamic regime. Sorption drying was applied to dry okara (soy milk residue) in a jet spouted bed (Wachiraphansakul and Devahastin, 2007).

Soy milk residue (okara) is widely utilized as animal feed and in the East as human food because of its large quantity of proteins and fibers. The residue is a highly cohesive mass with a complex chemical composition. In addition, okara protein is considered to be of higher quality than that of other soy products. Its moisture content is around 75 to 80%, which makes it highly perishable and sticky (Taruna and Jindal, 2002; Wachiraphansakul and Devahastin, 2005), thus being a candidate for the sorption drying technique. High temperature drying of okara at 90 to 130°C was conducted in a jet spouted bed of silica gel (sorbent) to produce okara for animal feed, and low temperature drying (55 to 65°C) was performed to obtain dried okara to be used as an ingredient in human foods (Wachiraphansakul and Devahastin, 2007).

The combined effects of the operating conditions influenced the drying rate, and the effect of temperature was stronger for high superficial air velocities and low silica heights. The particle circulation in the bed increased under these conditions. In comparing the drying with and without adsorbent, higher drying rates were obtained with the inclusion of silica, mainly due to higher temperatures (90°C). At low temperatures the material moisture diffusivity is lower and the moisture may not be available for absorption; at high temperatures the effect of convection prevails and competes with the effects of sorption.

Drying rates increase with higher absorbent mass/wet material mass ratios. Nevertheless, there is a limit above which some segregation problems may occur. For okara drying, Wachiraphansakul and Devahastin (2007) observed segregation when the ratio was 0.3, mainly for lower drying air temperature and velocity and higher initial bed heights. Thus, process optimization when using the sorption drying technology implies not only the analysis of operating conditions but also the quantity of adsorbent utilized.

Concerning the specific energy consumption, the inclusion of adsorbent tends to reduce index values.

Summarizing the results on application of this drying technology to okara drying, the following should be pointed out:

1. Adding sorbent particles to the spouted bed enhanced the drying process in terms of the drying kinetics and quality of the dried okara.
2. The convection–sorption drying of okara in the jet spouted bed resulted in better quality in terms of oxidation level, rehydration ability, and protein solubility.
3. Addition of sorbent particles did not modify the results for urease activity (significance level of 95%).
4. The operating conditions must be optimized together with the percentage of adsorbent added to the bed to enhance the results for the drying rates, quality indexes, and energy consumption.

6.5.3 Modeling, Simulation, and Control

Mathematical modeling, simulation methods, and control are reported in recent work on SBD technology. Two types of approaches are encountered for modeling the SBD processes: the first approach includes models and simulations that utilize commercial codes, mainly applying the computational fluid dynamics (CFD) technique, and the second approach involves models that do not apply commercial codes.

Because it is not the objective of this chapter to detail mathematical modeling, simulation, and control of the SBD, a summary of the recent advances in this area is provided.

An example of the second type of mathematical modeling can be found in Costa et al. (2001). A mathematical model and a computer program (Fortran language) are presented to simulate the drying of suspensions in conical spout–fluid beds of inert particles. Mass and energy balances for three phases (air, inert particles, and suspension), air flow and particle circulation models, and an equation for the drying kinetics comprise the model. The effects of cohesive forces were also included in the model. Previously developed models were adopted for fluid flow (Silva et al., 1997) and solids (Oliveira, 1997) circulation. The basic assumptions of this model are as follows. The column is thermally isolated and the suspension is injected at the top of the column. The inlet air-flow rates are high enough to maintain the stable spouting regime and promote inert particle attrition. Drying occurs only in the annulus and fountain regions; as a consequence, the air humidity and the suspension moisture content are constant in the spout region. In this spout region, the dry suspension layer that covers the inert particles is brittle enough to break off because of particle attrition and collision; and the powder produced in the spout region is in thermal equilibrium with the spout air. As soon as this powder is produced, it is carried by air to a filter or cyclone attached to the dryer exit. At the dryer exit, air and powder are in thermal equilibrium. Temperature gradients inside the inert particles are neglected (validated by the heat transfer model developed previously). In the annulus and fountain regions, the water vapor concentration at the suspension layer surface is given by the suspension sorption equilibrium curve.

Data from the literature on evaporation of water and drying of blood were applied in the simulations to validate the model. Interparticle forces were neglected in these data. The results showed that the simulations described the experimental values well.

CFD modeling can be applied to describe fluid dynamics and heat and mass transfers in SBD. An example of this technique is encountered in the work of Szafran and Kmiec (2004) for the drying of grains in a spouted bed with a draft tube. Drying kinetics is described by classical and diffusional models for constant and falling rate periods. Conditions assumed for the simulations are as follows: a continuous phase treated as an ideal gas, an incompressible solid phase, an incompressible flow, adiabatic walls and nonslip wall conditions for all phases, an inlet condition considered as a uniform velocity profile, a pressure boundary condition considered at the outlet, and axisymmetric boundary conditions applied along the axis of symmetry. The heat and mass transfer models were added to FLUENT 6.1 as a compiled executable program.

CFD modeling provides detailed information about transient flow behavior, distributions of solid-phase volume fractions, temperatures in the fluid and solid phases, Nusselt number, zones of high heat transfer rates in the column, mass fractions in the fluid and solid phases, and zones in the column where different periods of drying occur. To evaluate the CFD technique, simulations were performed using inlet conditions obtained in a spouted bed column 0.7 m high, having a diameter of 0.17 m (cylindrical part) and an inlet orifice of 0.03 m. The cone base angle was 50° and a draft tube was fixed inside the column (internal diameter = 0.027 m, height = 0.116 m, distance from bottom = 0.068 m). Even when dealing with highly wetted particles, the results showed zones where internal mass transfer resistances were predominant as well as zones with no mass transfer. CFD simulations predicted the mass-transfer rate well but underpredicted the heat transfer rate, with the values being close to those in the experimental data. To overcome the problem of a long computational time with the CFD technique, the so-called point-by-point procedure can be used, mainly for long drying time batch processes (Szafran and Kmiec, 2005). This procedure cannot predict short-time transient dependences (e.g., the preliminary period of drying).

Detailed information about flow behavior in two-dimensional and cylindrical spouted beds obtained from CFD modeling can be found in Duarte et al. (2004). The simulations used data obtained from the literature for both bed geometries that were analyzed: cylindrical utilizing glass beads and two-dimensional utilizing soybean seeds. The simulated results for the velocity and voidage profiles showed good agreement with the experimental ones.

For drying process control when using SBD to dry pastes, an adaptive control algorithm, generalized predictive control (GPC), has shown good performance as an advanced control strategy.

Corrêa et al. (2002) obtained a robust and stable controller by applying a self-tuning control using the GPC algorithm for spouted bed drying of egg paste. The manipulated variables were the paste flow rate and the electric power of the heat exchanger to maintain the air humidity and temperature set points. The ratio of egg powder moisture to the difference between inlet and outlet air humidity was efficient as a tool to estimate product moisture, and the GPC control strategy maintained the specified product quality (moisture content of the egg powder of up to 4%).

NOMENCLATURE

VARIABLES

c	specific heat (J/kg °C)
C_s	solids concentration (kg/kg)
DC	drying coefficient (kg/J)
DE	drying efficiency (%)
d	diameter (m)
db	dry basis
E	energy flux (W)
FC	fluid dynamic coefficient (%)
FI	flowability index (m)
H	height (m)
ΔH	latent heat of vaporization (J/kg)
m	mass (kg)
$\dot{m}$	mass flow rate (kg/s)
k	power law rheology parameter (Ns^n/m^2)
n	power law rheology parameter
Δp	pressure drop (N/m^2)
Q	flow rate (m^3/s)
T	temperature (°C)
t	time (s)
W	evaporation rate (kg/s)
X	moisture content (kg/kg)
Y	absolute humidity (kg/kg)

GREEK SYMBOLS

ε	porosity (%)
ϕ	sphericity
Φ	angle of repose (°)
η	efficiency (%)
μ	viscosity (Ns/m^2)
θ	contact angle (°)
ρ	density (kg/m^3)

| σ | surface tension (N/m) |
| τ | work (N/m) |

SUBSCRIPTS

ad	adhesion
amb	ambient
ap	apparent
ar	air
bulk	bulk
d	drying
ev	evaporated
f	final
g	grain
g,dry	dry grain
g,w	wet grain
h,g	grain heating
i	intermittence
lv	liquid–vapor
M	mechanical
max	maximum
ms	minimum spout
o	outlet
p	particle
paste	paste
po	powder
s	spout
T	total
true	true
v,w	water evaporation
0	initial

REFERENCES

Abers, J.E. and Wrolstad, R.E., Causative factors of color deterioration in strawberry preserves during processing and storage, *J. Food Sci.*, 44, 75–78, 1979.

Adamson, A.W., *Physical Chemistry of Surfaces*, John Wiley & Sons, New York, 1990.

Adeodato, M.G., Donida, M.W., and Rocha, S.C.S., Adhesion of an aqueous polymeric suspension to inert particles in a spouted bed, *Dry. Technol.*, 22, 1069–1086, 2004.

Almeida, M.M., Silva, O.S., and Alsina, O.L.S., Fluid-dynamic study of deformable materials in spouted bed dryer, *Dry. Technol.*, 24, 499–508, 2006.

Alsina, O.L.S., Lima, L.M.R., and Morais, V.L.M., Fluid-dynamic variables study of spouted bed drying of acerola pulp, in *Proceedings of the XXIV Brazilian Congress of Particulate Systems* (in Portuguese), Vol. 1, Barrozo, M.A.S. and Damasceno, J.J.R., Eds., Federal University of Uberlândia, Uberlândia, Brazil, 1996, pp. 289–294.

Benali, M. and Amazouz, M., Drying of vegetable starch solutions on inert particles: Quality and energy aspects, *J. Food Eng.*, 74, 484–489, 2006.

Cavalcanti Mata, M.E.R.M., Effect of high temperature–short time stationary bed drying on the conditioning of bean seeds (*Phaseolus vulgaris* L.), "carioca" variety: Experimental analyzes modeling and simulation (in Portuguese), Ph.D. thesis, State University of Campinas, Campinas, Brazil, 1997.

Corrêa, N.A., Corrêa, R.G., and Freire, J.T., Self-tuning control of egg drying in spouted bed using GPC algorithm, *Dry. Technol.*, 20, 813–828, 2002.

Costa, E.F. Jr., Cardoso, M., and Passos, M.L., Simulation of drying suspensions in spout–fluid beds of inert particles, *Dry. Technol.*, 19, 1975–2001, 2001.

Cunha, R.L., de la Cruz, A.G., and Menegalli, F.C., Effects of operating conditions on the quality of mango pulp dried in a spout fluidized bed, *Dry. Technol.*, 24, 423–432, 2006.

Dewettinck, K., Messens, W., Deroo, L., and Huyghebaert, A., Agglomeration tendency during top-spray coating with gelatin and starch hydrolysate, *Lebensm.-Wiss. Technol.*, 32, 102–106, 1999.

Donida, M.W., Rocha, S.C.S., and Bartholomeu, F., Influence of polymeric coating suspension characteristics on the particle coating in a spouted bed, *Dry. Technol.*, 23, 1–13, 2005.

Duarte, C.R., Murata, V.V., and Barrozo, M.A.S., Study of the spouted bed fluid dynamics using CFD, in *Drying 2004—Proceedings of the 14th International Drying Symposium (IDS 2004)*, Silva, M.A. and Rocha, S.C.S., Eds., Ourograf, Campinas, Brazil, 2004, pp. 581–588.

Geldart, D., *Gas Fluidization Technology*, John Wiley & Sons, New York, 1986.

Hung-Nguyen, L., Driscoll, R.H., and Srzednicki, G., Drying of high moisture content paddy in a pilot scale triangular spouted bed dryer, *Dry. Technol.*, 19, 375–387, 2001.

Inprasit, C. and Noomhorm, A., Effect of drying air temperature and grain temperature of different types of dryer and operation on rice quality, *Dry. Technol.*, 19, 389–404, 2001.

Iveson, S.M., Litster, J.D., Hapgood, K., and Ennis, B.J., Nucleation, growth and breakage phenomena in agitated wet granulation process: A review, *Powder Technol.*, 117, 3–39, 2001.

Kachan, G.C. and Chiapetta, E., Dehydration of tomato paste in a spouted bed dryer (in Portuguese), in *Proceedings of the VIII Brazilian Congress of Chemical Engineering*, Vol. 2, Pires, J.L.C., Ed., University of São Paulo, São Paulo, Brazil, 1988, pp. 510–523.

Kalwar, M.I., Kudra, T., Raghavan, G.S.V., and Mujumdar, A.S., Drying of grains in a drafted two dimensional spouted bed, *J. Food Process Eng.*, 13, 321–332, 1991.

Kalwar, M.I. and Raghavan, G.S.V., Spouting of two-dimensional beds with draft plates, *Can. J. Chem. Eng.*, 70, 887–894, 1992.

Kalwar, M.I. and Raghavan, G.S.V., Batch drying of shelled corn in two-dimensional spouted beds with draft plates, *Dry. Technol.*, 11, 339–354, 1993.

Lima, A.C.C. and Rocha, S.C.S., Bean drying in fixed, spout and spout–fluid beds: Comparison and empirical modeling, *Dry. Technol.*, 16, 1881–1902, 1998.

Lima, M.F.M. and Alsina, O.L.S., Drying of umbu pulp in spouted bed: Characteristic curves, in *Drying '92*, Mujumdar, A.S., Ed., Hemisphere Publishing, New York, 1992, pp. 1508–1515.

Lima, M.F.M., da Mata, A.L.M., Lima, L.M.F., and Moreno, M.T.S., Dehydration of beetroot (*Beta vulgaris* L.) in spouted bed, in *Proceedings of the XII Brazilian Congress of Chemical Engineering* (in Portuguese), Brazilian Association of Chemical Engineers, Ed., Federal University of Rio Grande do Sul, Porto Alegre, Brazil, CD, 1998.

Lima, M.F.M., Rocha, S.C.S., Alsina, O.L.S., Jerônimo, C.E.M., and da Mata, A.L.M., Influence of the material chemical composition on the spouted bed drying of fruit pulps performance, in *Proceedings of the XIII Brazilian Congress of Chemical Engineering* (in Portuguese), Maciel Filho, R. and Rocha, S.C.S., Eds., Sonopress-Rimo Ind. Com. Fonográfica Ltda., São Paulo, Brazil, CD, 2000.

Link, K.C. and Schlünder, E.U., Fluidized bed spray granulation and film coating. A new method for the investigation of the coating process on a single sphere, *Dry. Technol.*, 15, 1827–1843, 1997.

Madhiyanon, T., Soponronnarit, S., and Tia, W., A two-region model for batch drying of grains in a two-dimensional spouted bed, *Dry. Technol.*, 19, 389–404, 2001a.

Madhiyanon, T., Soponronnarit, S., and Tia, W., Industrial-scale prototype of continuous spouted bed dryer, *Dry. Technol.*, 19, 207–216, 2001b.

Madhiyanon, T., Soponronnarit, S., and Tia, W., A mathematical model for continuous drying of grains in a spouted bed dryer, *Dry. Technol.*, 20, 587–614, 2002.

Markowski, A.S. and Kaminski, W., Hydrodynamic characteristics of jet spouted beds, *Can. J. Chem. Eng.*, 61, 377–381, 1983.

Marques, A.M.M., Influence of particle–suspension adhesion on the spouted bed coating with bottom atomization (in Portuguese), M.S. thesis, State University of Campinas, Campinas, Brazil, 2007.

Marques, A.M.M., Donida, M.W., and Rocha, S.C.S., Effect of solid surface energy on particles coating and suspension drying, in *Drying 2006—Proceedings of the 15th International Drying Symposium*, Farkas, I., Ed., Szent István University Publisher, Gödöllő, Hungary, 2006, pp. 177–183.

Martinez, O.L.A., Brennam, J.G., and Nirajam, K., Spouted bed dryer for liquid foods, *Food Control*, 4, 41–45, 1993.

Martinez, O.L.A., Brennam, J.G., and Nirajam, K., Study on drying of food in a spouted bed with inert particles, in *Proceedings of the 1st Ibero American Food Congress* (in Spanish), Ortega-Rodríguez, E., Ed., Brazilian Society of Food Science and Technology, Campinas, Brazil, 1995, pp. 73–81.

Mathur, K.B. and Epstein, N., *Spouted Beds*, Academic Press, New York, 1974.

Medeiros, M.F.D., Rocha, S.C.S., Alsina, O.L.S., Medeiros, U.K.L., and da Mata, A.M.L., Drying of pulps of tropical fruits in spouted bed: Effect of composition on dryer performance, *Dry. Technol.*, 20, 855–881, 2002.

Mujumdar, A.S., Spouted beds: Principles and recent developments, in *Proceedings of the XVII Brazilian Congress of Particulate Systems*, Vol. 1, Sartori, D.J.M. and Silveira, A.M., Eds., Federal University of São Carlos, São Carlos, Brazil, 1989, pp. 3–13.

Neumann, A.W. and Kwok, D.Y., Contact angle measurement and contact angle interpretation, *Adv. Colloid Interface Sci.*, 81, 167–249, 1999.

Nindo, C.I., Sun, T., Wang, S.W., Tanga, J., and Powers, J.R., Evaluation of drying technologies for retention of physical quality and antioxidants in asparagus (*Asparagus officinalis* L.). *Lebensm.-Wiss. Technol.*, 36, 507–516, 2003.

Oliveira, C.A., Study of bean (*Phaseolus vulgaris* L.) drying in spouted bed with intermittent operating conditions (in Portuguese), M.S. thesis, State University of Campinas, Campinas, Brazil, 1999.

Oliveira, I.M. and Passos, M.L., Simulation of drying suspension in a conical spouted bed, *Dry. Technol.*, 15, 593–604, 1997.

Oliveira, J., Theoretical and experimental studies of solids and fluid flow in conical spouted beds (in Portuguese), M.S. Thesis., Federal University of Minas Gerais, Belo Horizonte, Brazil, 1997.

Pakowski, Z. and Mujumdar, A.S., Basic process calculations in drying, in *Handbook of Industrial Drying*, Mujumdar, A.S., Ed., Marcel Dekker, New York, 1995, pp. 71–112.

Pallai, E., Szentmarjay, T., and Mujumdar, A.S., Spouted bed drying, in *Handbook of Industrial Drying*, Mujumdar, A.S., Ed., Marcel Dekker, New York, 1995, pp. 453–488.

Pallai, E., Szentmarjay, T., and Szijjártó, E., Effect of partial processes of drying on inert particles on product quality, *Dry. Technol.*, 19, 2019–2032, 2001.

Passos, M.L., Massarani, G., Freire, J.T., and Mujumdar, A.S., Drying of pastes in spouted beds of inert particles: Drying criteria and modeling, *Dry. Technol.*, 15, 605–624, 1997.

Passos, M.L. and Mujumdar, A.S., Effect of cohesive forces on fluidized and spouted beds of wet particles, *Powder Technol.*, 110, 222–238, 2000.

Patel, K., Bridgwater, J., Baker, C.G.J., and Schneider, T., Spouting behaviour of wet solids, in *Drying '86*, Mujumdar, A.S., Ed., Hemisphere Publishing, New York, 1986, pp. 183–189.

Pont, V., Saleh, K., Steinmetz, D., and Hemati, M., Influence of the physicochemical properties on the growth of solids particles by granulation in fluidized bed, *Powder Technol.*, 120, 97–104, 2001.

Raghavan, V.G.S., Rennie, T.J., Sunjka, P.S., Orsat, V., Phaphuangwittayakul, W., and Terdtoon, P., Energy aspects of novel techniques for drying biological materials, in *Drying 2004—Proceedings of the 14th International Drying Symposium (IDS 2004)*, Silva, M.A. and Rocha, S.C.S., Eds., Ourograf, Campinas, Brazil, 2004, pp. 1021–1028.

Ré, M.I. and Freire, J.T., Drying of pastelike materials in spouted beds, in *Proceedings of the 6th International Drying Symposium (IDS88)*, Roques, M.A. and Mujumdar, A.S., Eds., Hemisphere Publishing, New York, 1988, pp. 426–431.

Reyes, A.E., Diaz, G., and Blasco, R., Experimental studies of slurries on inert particles in spouted bed and fluidized bed dryers, in *Drying '96*, Mujumdar, A.S., Ed., Hemisphere Publishing, New York, 1996, pp. 605–612.

Reyes, A.E., Diaz, G., and Blasco, R., Experimental studies of slurries drying on inert particles in fluid–particles contact equipment, in *Proceedings of the Inter-American Drying Conference (IADC)*, Silva, M.A. and Rocha, S.C.S., Eds., Gráfica Paes, Campinas, Brazil, 1997, pp. 150–157.

Reyes, A.E. and Massarani, G., Evaluation of pastes drying in conical spouted beds, in *Proceedings of the XXI Brazilian Congress on Particulate Systems* (in Spanish), Vol. 2, Passos, M.L., Ed., Federal University of Minas Gerais, Belo Horizonte, Brazil, 1993, pp. 479–483.

Robins, P.T. and Fryer, P.J., The spouted bed roasting of barley: Development of a predictive model for moisture and temperature, *J. Food Eng.*, 59, 199–208, 2003.

Sangkram, U. and Noomhorm, A., The effect of drying and storage of soybean on the quality of bean, oil and lecithin production, *Dry. Technol.*, 20, 2041–2054, 2002.

Schneider, T. and Bridgwater, J., The stability of wet spouted beds, *Dry. Technol.*, 11, 277–301, 1993.

Silva, C.M., Rodrigues, M.E. Jr., and Passos, M.L., Simulation of the fluid flow in conical spout–fluid beds (in Portuguese), *Sci. Eng. J.*, 6, 55–62, 1997.

Spitzner Neto, P.I. and Freire, J.T., Study of pastes drying in spouted beds: Influence of the paste presence on the process, in *Proceedings of the XXV Brazilian Congress of Particulate Systems*, Vol. 1, Ferreira, M.C. and Freire, J.T., Eds., Federal University of São Carlos, São Carlos, Brazil, 1997, pp. 185–190.

Strumillo, C. and Kudra, T., *Drying: Principles, Applications and Design*, Gordon and Breach Science Publishers, Montreux, France, 1986.

Szafran, R.G. and Kmiec, A., CFD modeling of heat and mass transfer in a spouted bed dryer, *Ind. Eng. Chem. Res.*, 43, 1113–1124, 2004.

Szafran, R.G. and Kmiec, A., Point-by-point solution procedure for the computational fluid dynamics modeling of long-time batch drying, *Ind. Eng. Chem. Res.*, 44, 7892–7898, 2005.

Taruna, I. and Jindal, V.K., Drying of soy pulp (okara) in a bed of inert particles, *Dry. Technol.*, 20, 1035–1051, 2002.

Tenou, E. and Poncelet, D., Batch and continuous fluid bed coating—Review and state of the art, *J. Food Eng.*, 53, 325–340, 2002.

Tia, S., Tangsatitkulchai, C., and Dumronglaohapun, P., Continuous drying of slurry in a jet spouted bed, *Dry. Technol.*, 13, 1825–1840, 1995.

Vieira, M.G.A., Estrella, L., and Rocha, S.C.S., Energy efficiency and drying kinetics of recycled paper in convective drying, in *Proceedings of the 3rd Inter-American Drying Conference (IADC 2005)*, Orsat, V., Raghavan, G.S.V., and Kudra, T., Eds., Montreal, Canada, CD, 2005.

Wachiraphansakul, S. and Devahastin, S., Drying kinetics and quality of soy residue (okara) dried in a jet spouted bed dryer, *Dry. Technol.*, 23, 1229–1242, 2005.

Wachiraphansakul, S. and Devahastin, S., Drying kinetics and quality of okara dried in a jet spouted bed of sorbent particles, *Food Sci. Technol.*, 2, 207–219, 2007.

Wiriyaumpaiwong, S., Soponronnarit, S., and Prachayawarakorn, S., Soybean drying by two-dimensional spouted bed, *Dry. Technol.*, 21, 1735–1757, 2003.

7 Application and Development of Osmotic Dehydration Technology in Food Processing

John Shi and Sophia Jun Xue

CONTENTS

7.1   INTRODUCTION

Dehydration is an important processing procedure to preserve raw food material and products in the food industry. The basic objective in dehydration of foods is the removal of water from the raw material to extend the shelf life or reduce the load for the following operations. Preservation methods like drying, canning, and freezing have been applied to prolong the shelf life of foods, but these methods produce food products that are low in quality compared to their original fresh state. A tough and woody texture, slow or incomplete rehydration, and loss of juiciness typical of fresh food, as well as unfavorable changes in color and flavor are the most common quality defects of dehydrated food (Karel, 1975). It is generally well known that the highest quality dry food products are produced by freeze-drying. However, this technique is one of the most expensive unit operations. Freezing of plant materials such as fruits and vegetables is an energy intensive process. The large amount of energy spent in this process includes the energy involved in freezing, packaging, transportation, and storage of the large amount of water present in the fresh material (Rao, 1977). One way to deal with this problem is to partially concentrate or dehydrate the product before freezing so that the overall energy requirement of the process is reduced (Huxsoll, 1982). During the few last decades many studies have been conducted to remove water from fresh food materials for longer shelf life and better handling properties; to decrease transport and storage costs by volume and weight reduction; and to improve the quality of the food products, which has been the subject of intensive research in the food engineering field.

Researchers have looked for new ways to process foods to improve the quality of preserved food products. One of these new methods is osmotic dehydration. Osmotic dehydration has shown the potential to obtain better food products by removing water at low temperature using economical means (Bolin et al., 1983) and by the quality improvement of the final products (Fito et al., 1998; Flink, 1975). Although osmosis is a well-known process, in recent years there has been increasing interest in it. Osmotic dehydration consists of immersing a water-containing cellular material in an aqueous solution of osmotic agents and allowing part of the natural water to transfer from the product to the osmotic solution. Solutions of osmotic agents have high osmotic pressure and low water activity. The most commonly used osmotic agents are sucrose and sodium chloride (NaCl). Other osmotic agents such as lactose, maltodextrin, ethanol, glucose, glycerine, and corn syrups have also been used (Giangiacomo et al., 1987; Hawkes and Flink, 1978). From an organoleptic point of view, sugar solutions have been used for the dehydration of fruits and salts have been used for vegetables and meat. Aqueous solutions of mixed osmotic agents have also been utilized to dehydrate fruits, vegetables, and meat, such as sucrose with lactose, maltodextrin, and sodium chloride for apple and papaya ring dehydration (Argaiz et al., 1994); NaCl with sucrose for diced potato dehydration (Lenart and Flink, 1984); and solutions of NaCl and ethanol in water for dehydration of diced carrots (Le Maguer and Biswal, 1984).

In the last two decades fresh fruits and vegetables have been increasing in popularity for consumption compared to canned and frozen fruits (Shewfelt, 1987). A growing trend is emerging, however, as many of the large food processing companies have decided to market living, respiring plant tissue. To satisfy the growing market demand for commodities in a "freshlike" state, minimal processing, such as low-level irradiation, packaging, and so forth, will be increasingly used. Initial contamination has been of primary importance in such processes, but good quality shelf-stable products have been obtained. The use of an osmotic treatment as a unit operation has already been widely applied in the food area.

Most previous studies have focused attention on rapid and effective removal of desired amounts of water from food materials such as fruits. A high osmotic rate would make the process more efficient and practical. Some methods have been employed to speed up water transfer, such as using a high concentration of osmotic solution, high solution temperature, prolonging treatment time, and so forth. However, intensification by increasing the temperature and concentration gradient is limited because the nutrient content of processed foods has become an important consumer concern. The benefits of reduced heat processing are being realized in a wide range of processed fruit products. There is increasing demand for processed fruits with quality very similar to fresh fruits.

7.2   PRINCIPLES OF OSMOTIC DEHYDRATION

The principle of osmosis as a natural phenomenon of water removal from biological material has been known for a long time. In recent years there has been an increase in interest in this process. Osmotic treatment is actually a combination of dehydration and impregnation processes, which can minimize negative modifications of fresh food components. This process may provide the possibility of modifying the functional properties of food material, improving the overall quality of the final products, creating attractive new products, and potential energy saving.

After immersing a water-rich fresh food material in a hypertonic solution (sugar, salt, etc.), the driving force for water removal is the concentration gradient between the solution and the intracellular fluid. If the membrane is perfectly semipermeable, the solute is unable to transfer through the membrane into the cells. However, it is difficult to obtain a perfect semipermeable membrane in food material because of their complex internal structure and possible damage during processing. Thus, osmotic dehydration is actually a multicomponent transfer process of two simultaneous, countercurrent solution flows and one gas flow. The solution flow out of food material is water mixed with solutes such as organic acids, reducing sugars, minerals, and flavor compounds that affect the organoleptic and nutritional characteristics of the final products. Soluble solids present in the osmotic solution are taken up by the food material. Gas flow is also out of intercellular space (Figure 7.1).

During osmotic dehydration, the cell walls act as semipermeable membranes, which allow small molecules such as water to pass through, but not larger molecules such as sugar. Water continues to pass through the cell wall membrane until the concentration of water molecules is the same on both sides. Differences in chemical potentials of water and solutes in the solid–solution system result in fluxes of several components of the material and solution. The osmotic pressure gradient is imposed by a concentrated external solution (sugar, salt, etc.). Through the control of main variables in the process, mass transfer behavior will appear in different patterns: osmotic dehydration (dewatering) or impregnation soaking (swelling). By using a highly concentrated osmotic solution (usually 50 to 80 g solute/100 g solution), water flows intensively out of the product into the osmotic solution. The transport of the product's natural solutes is always accompanied by water transfer. In contrast, other natural solutes transfer from the osmotic solution into the food product. Water removal in this situation is much greater than osmotic solute uptake. Under these

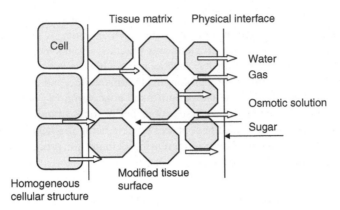

FIGURE 7.1 Schematic cellular material tissue representation and mass transfer pattern.

circumstances, the process is under an "osmotic dehydration situation" as shown in Figure 7.1. During the osmotic process, the product's solution removal and osmotic solute uptake are simultaneous and countercurrent flows and isothermal without any phase change, but they are accompanied by structural changes such as shrinkage or swelling.

If the food product is immersed in a low concentration osmotic solution, the osmotic solution uptake is greater than water removal. Thus, the process is in an "impregnation soaking situation" (or swelling and rehydration). Under this particular situation, the gradient of moisture concentration drives the osmotic solution into the product while the gradient of solutes transfers them into the osmotic solution. Therefore, some pigments, flavors, and nutrients are usually transferred out of the product during soaking and rehydration processes. After the absorbing solution, the product becomes swollen and rehydrated. There is a mixture of various transport mechanisms, and the contributions of the different mechanisms to the total transport vary as a function of the position within the solid and the solution, changing as drying progress (Bruin and Luyben, 1992). Research on osmotic dehydration has come a long way since Ponting and collaborators pioneered it (Ponting, 1973; Ponting et al., 1966). Numerous studies have been performed on a variety of food materials. This application has been utilized on fruits, vegetables, and fish as an effective preservation method. The past few decades have brought significant developments in the process of osmotic treatments and their importance in the food processing industry. Researchers have covered many different aspects to further the understanding and applications of this preprocessing technique.

7.3 INDUSTRIAL APPLICATION OF OSMOTIC TREATMENT IN FRUIT AND VEGETABLE PROCESSING

Recently, there has been increasing interest in the food industry about the possibility of modifying the functional properties of food materials and improving the overall quality of the products. The osmotic dehydration method shortens the drying time and reduces the ascorbic acid loss during drying (Uzuegbu and Ukeka, 1987). Fresh fruits are increasing in popularity compared to canned and frozen fruits (Shewfelt, 1987). In order to satisfy the growing market demands for commodities in a freshlike state, for minimal processing, and for "natural food," osmotic dehydration as a unit operation in fruit processing has been widely applied in the fruit industry.

The process of osmosis has often been proposed as a first step followed by any kind of drying operation such as air drying, vacuum drying, or freeze-drying. The plant material is immersed in an aqueous solution of compounds such as glycerol, ethanol, sugar, or/and salt to partially dehydrate the food in order to reduce the load of water at further drying steps and to improve the quality of the final product. Osmotic dehydration generally will not give a product with sufficiently low moisture content to be considered self-stable. Thus, this process is considered as a pretreatment or as an intermediate step prior to conventional stabilizing processes such as air, vacuum, or freeze-drying or freezing, chilling, pasteurization, or other processes for fruit, vegetable, fish, and meat products.

Ponting et al. (1966) suggested a process for a 50% weight reduction of apples by osmotic dehydration after which the apples were either frozen, air dried, or vacuum dried. Sucrose and invert sugar were used in solution. The apple products had superior quality and did not need sulfur dioxide treatment. Ponting (1973) also pointed out that partially dehydrating fruit pieces or slices by osmosis followed by vacuum drying produces products with better flavor than freeze-dried ones. This combination process should cost less. Farkas and Lazar (1969) studied the feasibility of osmotic dehydration of apples prior to freezing. Good quality apple products were obtained by a dehydrofrozen process. Hope and Vitale (1972) described the application of osmotic dehydration for concentration of bananas, plaintain, and ripe mangoes using a 67% (w/w) sucrose solution and green mangoes using a 25% (w/w) sodium chloride solution. Hawkes and Flink (1978) used osmotic dehydration preconcentration prior to air or vacuum drying, which retained the structural integrity of the tissue. Garcia and McFelley (1978) found that pretreatment in a 65°Bx sucrose solution did not affect fruit quality. Dixon et al. (1976) applied a combination of osmotic dehydration and vacuum drying to produce dry apple slices.

Flink (1975) and Chirife et al. (1973) found that the dried product flavor quality depended primarily on the initial solids content. If the solids concentration was increased, the retention of volatiles was greatly improved. An osmotic treatment of fruits in a 60 to 65°Bx sugar solution was used to increase the initial solids content prior to freeze-drying. Hawkes and Flink (1978) investigated osmotic dehydration as an initial step in freeze dehydration and emphasized that the organoleptic quality of freeze-dried fruit products can be improved by increasing the solids content. This also results in a reduction of the water content of the sample to be freeze-dried, which reduces processing time. They also studied the possibility of adding sucrose, lactose, salt, and so forth into the solution. Flink (1980) pointed out that osmotic dehydration can be used as a step prior to freeze-drying to yield fruit products with improved stability. Thus, less freeze-drying time was required. Le Maguer and Biswal (1984) suggested a dehydrocooling process, which would simultaneously cool and remove water from vegetables before they are frozen by traditional methods.

The coolant–dehydrant used in this process is a solution of sodium chloride and ethanol in water. A methodology has been developed for measuring the movement of solutes and water in the vegetable pieces to better understand the nature of dehydrocooling. This process is appropriate for continuous operation using existing designs in such closely related areas as extraction and other separation process. Specific applications of dehydrocooling include processing of carrots, peas, sweet corn, snap beans, and so forth. Experimental results indicate that approximately 22% of the water can be removed from carrots using a 15% NaCl–water solution at room temperature, thus reducing the energy requirements for freezing by 22% (Le Maguer and Biswal, 1984).

Andreotti et al. (1985) proposed a combination of osmotic dehydration with appertization to improve canned fruit preserves. Torreggiani et al. (1988) assessed the feasibility of a process for obtaining high quality fruit in a sugar solution on a plant production scale. Karel (1975) noted that intermediate moisture foods have received attention since the development of new products based on technological principles such as lowering water activity, retarding microbial growth by adding antimicrobial agents, and so forth. Intermediate moisture foods can be produced with satisfactory quality using an osmotic dehydration treatment. Maltini and Torreggiani (1986) examined the possibility of obtaining shelf-stable products with no need for further treatments and less preservatives by means of an osmotic treatment in a sucrose solution. The fruit products tasted like fresh ones. They were shelf stable with water activity between 0.94 and 0.97 and water content in the range of 65 to 75% after vacuum packaging and pasteurization and remained stable for many months at ambient temperature. This process causes minimal changes in the sensory and physical–chemical characteristics of fruits. Compared to traditional intermediate moisture foods that usually have water activity between 0.65 and 0.90 and water content from 20 to 50%, good quality and shelf-stable fruit products were obtained by the osmotic dehydration process.

Thus, the simplicity of the osmotic dehydration process without the use of expensive equipment and requirements for less or no energy makes it suitable for large-scale fruit preservation, particularly where sugar is abundantly available. The potential industrial applications of osmotic dehydration are given in the Figure 7.2. Various industrial applications of osmotic dehydration on fruit processing and preservation are provided in Figure 7.3.

Osmotic treatments often provide better quality intermediate moisture fruits than air drying, especially in fruits and vegetables that sensitive to heat (Torreggiani, 1995). An example is osmotically dehydrated fruit pieces that have been successfully used to avoid the presence of synergized whey in yogurt (Giangiacomo et al., 1994). The dehydration levels that are reached depend on specific needs, but products between 80 and 50% moisture can be obtained through this treatment (Torreggiani et al., 1999). The solutes in the solution diffuse into the material simultaneously, contributing to the depression of water activity. Because no phase change is involved and osmotic dehydration is usually carried out at temperatures below 50°C, the products suffer less damage than those treated with conventional direct dehydration methods (Raoult-Wack, 1994). Osmotically dehydrated products are better quality after further drying or freezing.

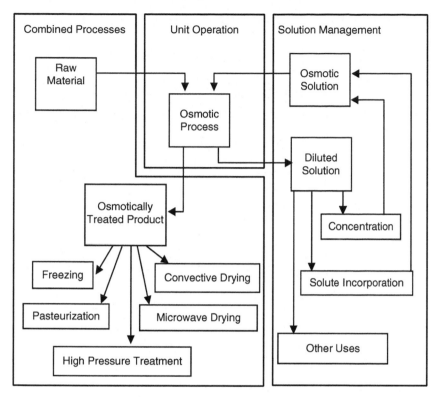

FIGURE 7.2   Osmotic treatment in food processing. (Modified from Spiess, W.E.L., *Osmotic Treatments for the Food Industry. Proceedings of the 1st Osmotic Treatment Seminar,* FEUP Ediçoes, Porto, Portugal, 1999, pp. 9–17. With permission.)

Some new technologies have also been introduced into the osmotic dehydration process such as the combined method of heat treatment and acoustic oscillation, the combined method of alternating heating and cooling of fruits (Rogachev and Kislenko, 1972), and vacuum osmotic dehydration (Fito, 1994; Fito et al., 1996). A pilot-scale belt-type osmotic contactor was designed and used in a continuous osmotic dehydration process to attain maximum water removal from fresh fruits and vegetables with minimum contact time. The carrot was chosen as a representative vegetable and the osmotic solution was a ternary sucrose–NaCl–water mixture with a total solids concentration of about 50 to 55% (w/w) (Azuara et al., 1998). A continuous cross-current coupled with a countercurrent osmotic dehydration process is very effective and efficient. A substantial reduction of the processing time is the most significant improvement in osmotic dehydration. The optimum condition was 44% sucrose/7% NaCl for a 16-min contact time with 26% water loss, 24% mass loss, 2.1% sugar gain, and 0.7% salt gain. The application of a continuous osmotic operation gave very satisfactory design parameters for the dehydration of carrot cubes from the initial water concentration to the desired final concentration in the range studied. A continuous osmotic contactor may encourage the study of a greater variety of vegetables or other foods. Some countercurrent equipment is available for contacting

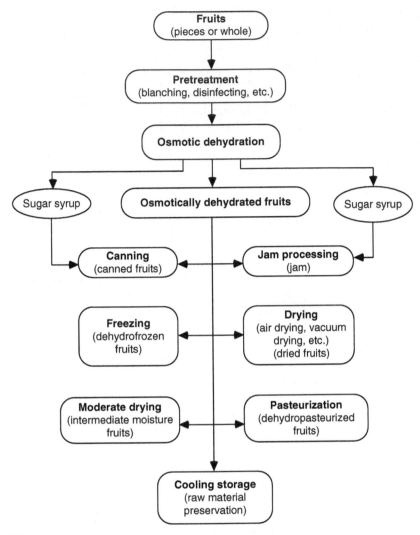

FIGURE 7.3 Various industrial applications of osmotic dehydration on fruit processing and preservation.

operations. This equipment has the additional advantage that it can be part of an on-line production system.

7.4 MECHANICS OF MASS TRANSFER IN OSMOTIC DEHYDRATION PROCESS

Research on the mass transfer mechanism of osmotic treatment of food material is being carried out on different aspects of the osmotic process to study the mass transfer kinetics. Mass transport in a food–osmotic solution system is affected by numerous complicated and often interactive physical and chemical processes. Operation parameters (temperature, time, pressure, concentration, and composition of solution), material properties (porosity, geometry, and microstructure), and multicomponent mass transfer characteristics (fluxes and their interaction) are recognized as the main process variables affecting the mass transfer property.

Models have been developed in the last 15 years for describing the osmotic process and for evaluating the transport of interacting solutes in porous materials (Fito, 1994; Fito and Chiralt, 2000; Fito et al., 1996, 1998; Marcotte and Le Maguer, 1991; Marcotte et al., 1991; Shi and Le Maguer, 2002a, b, 2003). Most of these models are based on the assumption that the mass transfer is the rate limiting step, and its rate can be approximately predicted by appropriate mathematical solutions of the simplified unsteady-state second Fickian equation as a diffusion process. Osmotic treatment may be defined as a solid–liquid contacting operation in which a solid food material is immersed in a concentrated aqueous solution. Differences in the chemical potential of water and various solutes in the system result in fluxes of the components between the material and solution. There is an internal exchange and interaction of components with their network. Moreover, the transfer mechanism, diffusion rates, and equilibrium moisture content for each of these components are generally very different, which affects the kinetics and final equilibrium of the system.

There are two major mass transfer phenomena involved in osmotic dehydration: the movement of solute into the material and the flow of water out of the tissue. The kinetics of mass transfer is usually described through water loss and solids or solutes gain and weight reduction. Water loss and solids gain can be measured through the rate of water and solute flow, respectively, over time or through the amount of water lost or solute gained after a certain period of time based on the amount of initial material. Solids gain can also be studied through solute penetration techniques that determine the amount of solute uptake by the material with time. Studies on model agar cubes were also performed to describe the "dewatering" and "impregnation" processes (Raoult-Wack, Guilbert, et al., 1991; Raoult-Wack, Jourdain, et al., 1991; Raoult-Wack et al., 1992). Some models described not only water loss and solids gain but also reduction of water activity and shrinkage of the samples (Hough et al., 1993). Water loss and solids gain were modeled using different approaches.

Several investigations have dealt with different aspects of the osmotic process to study the mass transfer kinetics. The concentration and composition of the osmotic solution, temperature, time, pressure, natural material properties, and shrinkage are recognized as the main process variables in developing mass transfer modeling. If the concentration of the osmotic solution remains constant and the resistance at the surface is negligible, Fick's second law can be applied (Magee et al., 1983). Other studies (Conway et al., 1983) also modeled these phenomena with Crank's equation (Crank, 1975). This approach however is limited to processes in which the rate of mass transfer is simply diffusion, so the flow is unidirectional and there are negligible interactions between the components in the tissue and the solute during diffusion (Fito, 1994; Fito and Chiralt, 2000; Fito et al., 1996, 1998; Le Maguer, 1997; Marcotte and Le Maguer, 1991; Marcotte et al., 1991). Osmotic dehydration requires accurate prediction of water removal from foods for industrial applications. Simultaneous flow of solutes and water and interactions between components require a theoretical approach based on the application of linear irreversible thermodynamic processes.

The estimation of the moisture diffusion from drying data of food materials was accomplished by different models proposed from the slopes of the drying curves by computer optimization, which reported a wide variation of values because of the complexity of the food material and the different methods of estimation. Most of these studies focused on the formulation of appropriate mass balance (partial differential) equations governing solute transport, the derivation of appropriate analytical or numerical solutions of those equations, and the relation of the transport equations with observed solute distributions versus time.

In the kinetic models, the driving force is based on the concentration gradients of solutes and water in the system. This approach ignores processes of mass transfer that may have an important impact on water and solute exchanges between the food material and the external solution. Osmotic treatment is a much more complicated process than one controlled by mass transfer under a concentration gradient. The processes that are independent of the concentration gradients can play an important role in the impregnation with solutes and the resulting functional modification of the food material.

Modification of the composition of the food material produces structural changes (shrinkage, porosity reduction, and cell collapse) and modifies the mass transfer behavior in the cellular tissue. The complexity of the nonequilibrium state inside the food solid (consisting of gases, extracellular liquid, intercellular liquid, and solid matrix) also affects the mechanisms and kinetics of the transfer phenomena. Depending on the physical characteristics of the cellular tissue and the operating conditions (temperature, pressure), osmosis (providing that membrane integrity exists), diffusion, flux interactions, and shrinkage can all be acting at the same time and contributing to the complexity of the process to different extents. Models involving the concepts of irreversible thermodynamics in order to integrate the contribution of each component and a description of natural tissue action were studied recently (Fito and Chiralt, 2000; Shi and Le Maguer, 2002a, b, 2003; Yao and Le Maguer, 1996).

7.4.1 Mass Transfer Phenomena

When a piece of tissue is placed in contact with an osmotic solution, the solute begins to diffuse into the extracellular space and water begins to diffuse in the opposite direction. If there was no transmembrane flow, equivolumetric diffusion would take place. As long as the mass transfer coefficient at the solution side is large enough, the concentration at the subsurface of the tissue can be considered to become almost immediately equal to the concentration of the solution. After the initial fast changing period, a slow changing quasi-steady state is eventually reached. The water transfer property in the food material immersed in osmotic solution may be described by several different transport mechanisms, depending on the nature of the material, type of moisture bonding, moisture content, temperature, pressure in the capillary pores, and so forth (Fish, 1958). In general, liquid diffusion occurs in nonporous solids whereas capillary movement occurs in porous solids. Transport of water in liquid solution takes place only by molecular diffusion, a relatively simple phenomenon. In capillary–porous biological materials, mass transfer occurs in gas-filled cavities, capillaries, cell walls, and extracellular and intracellular spaces. Biological material contains a variety of individual soluble components. When cellular biological materials are immersed in a high concentration of osmotic solution, osmotic treatment is actually a multicomponent transfer process in which simultaneous, countercurrent solution flows with a combination of dehydration, leaching, and impregnation processes occurring in the biological tissue matrix. The mass transfer process of each component in the solid–liquid system is affected by operation parameters and by the presence of other components.

When a cellular solid material is immersed in hypertonic solution (sucrose solution), the cells in the first layer of the material contact the hypertonic solution and begin to lose water because of the concentration gradient between the cells and hypertonic solution; then, they begin to shrink (Figure 7.1). After the cells in the first layer lose water, a "chemical potential difference of water" between the first layer of cells and second layer of cells is established. Subsequently, the second layer cells begin to pump water to the first layer cells and then shrink. The phenomena of mass transfer and tissue shrinkage are spread from the surface to the center of the material as a function of operation time. Finally, the cells in the material center lose water and the mass transfer process tends to equilibrate after a long period of solid–liquid contact. Mass transfer and tissue shrinkage occur simultaneously. Thus, for a certain operating time, mass transfer and tissue shrinkage are only related to a specific part of the whole material (Fito and Chiralt, 2000; Shi and Le Maguer, 2002a, b, 2003; Yao and Le Maguer, 1996).

7.4.2 MODELING APPROACH

The study of multicomponent mass transfer has been important in many areas of chemical engineering for a long time. A wide variety of multicomponent flux equations have been proposed either on the basis of a physical model or from an irreversible thermodynamic model. Interest in multicomponent mass transfer phenomena in osmotic treatment is relatively recent in food engineering. Multicomponent mixtures display a number of unusual interaction phenomena such as osmosis, diffusion barriers, and reverse diffusion that can occur in mass transfer processes, but not present in a binary system. After food material is immersed in osmotic solution, the water is transported by several mechanisms simultaneously: molecular diffusion, liquid diffusion, vapor diffusion (through gas flow), hydrodynamic flow, capillary transport, surface diffusion, and most frequently a combination of these mechanisms. The transfer processes of food material can be considered as follows:

1. Water and solutes are transported by diffusion in the osmosis process because of concentration gradients.
2. Water and solutes are transported by capillary flow because of differences in the total system pressure caused by external pressure, shrinkage, and capillarity.
3. Hydrodynamic flow occurs in pores.
4. Water vapor diffusion occurs within partly filled pores because of the capillary–condensation mechanism.
5. Water diffusion occurs at pore surfaces because of gradients at the surfaces.

A number of modeling approaches may be classified according to what fluxes are assumed to occur internally and how the model deals with the process of a simultaneous, countercurrent internal transport mechanism of water removal and solute uptake. Mass transfer kinetics depend on both the process parameters and the structural properties of products. In biological materials such as fruits, vegetables, fish, and meat, the water content, cell maturity, tissue structure, porosity, and geometry of pieces are related to water removal, product solute loss, and osmotic solute uptake. Mass transfer occurs far from equilibrium and is accompanied by significant shrinkage and deformation and probably by flow interaction (Fito and Chiralt, 2000; Lenart and Dabrowska, 1999; Shi and Le Maguer, 2002a, b). A two-film model based on generalized multicomponent mass transfer theories is a good approach to analyze the coupled mass transfer at an isothermal solid–liquid interface, based on the knowledge of concentration gradients and the relation with cell structure in an isothermal solid–liquid contacting system (Shi and Le Maguer, 2002a, b).

7.4.2.1 Pure Water Flow in Cells

As shown in Figure 7.1, when cells in the first layer are contacted with hypertonic solution, the first layer cells lose water because of the concentration gradient. However, at this moment, the second layer cells are still turgid. The cell wall and membrane exert pressure (turgid pressure) on the cellular solution and cause water transport to neighboring cells through the plasmalemma. According to Nobel (1983), the chemical potential difference of water inside the cellular volume (at a full turgid condition) and extracellular space can be considered as

$$\Delta\mu_w = RT \left(\ln a_w^c + \ln a_w^L\right) + V_w(P^c - P_0) \tag{7.1}$$

where $\Delta\mu_w$ is the chemical potential difference of water inside the cellular volume and extracellular volume, R is the gas constant, T is the absolute temperature, a_w^c is the water activity of the cellular solution inside the cells, a_w^L is the water activity of the cellular solution in the extracellular space, V_w is the water partial molar volume, P^c is the hydrostatic pressure (turgid pressure) inside the cells above atmospheric pressure, and P_0 is the atmospheric pressure in extracellular space. The pure water flux across the cell membrane is expressed by

$$J_1^0 = \psi\left[\frac{RT}{M_w}(X_{su} - X_{ss}^c) + \frac{V_w}{M_w}\left(P^c - P_0 \frac{1}{2}\right)\right] \tag{7.2}$$

where ψ is a parameter related to water permeability across the cell membrane, X_{ss}^c is the soluble solid concentration in the cell, and X_{su} is the sugar concentration in extracellular space.

7.4.2.2 Mass Fluxes Relative to Interface

In the description of the interface mass transfer process, the mass transfer flux J_i of component i relative to interface I is considered as (Cussler, 1984)

$$J_i^I = \rho_i(v_i - v^I) \tag{7.3}$$

where ρ_i is the mass concentration of component i, v_i is the velocity of component i with respect to a stationary coordinate reference frame, and v^I is the interface velocity.

The velocity of the interface with reference to the center of material is considered to be

$$v^I = \frac{1}{A_0}\frac{dV}{dt} \tag{7.4}$$

where V is the total volume of solid material immersed in the liquid and A_0 is the surface area of the interface.

7.4.2.3   Structure Shrinking Condition

If v_3 is the velocity of component 3 at the surface of the cellular material with respect to a fixed frame of reference (center of the solid), the mass fluxes of component 1 (water) and 2 (osmotic solute) can be written as the following equations (King, 1970):

$$v_3 = -J_1^3 V_1 - J_2^3 V_2 \tag{7.5}$$

In such conditions, mass fluxes J_i^3 and J_i^v are relative to the fixed coordinate or v of component 3 (solid frame). The following expression is for the mass fluxes of water and osmotic solute transfer in one dimension:

$$J_1^v = J_1^3 - \rho_1(J_1^3 V_1 + J_2^3 V_2) \tag{7.6}$$

$$J_2^v = J_2^3 - \rho_2(J_1^3 V_1 + J_2^3 V_2) \tag{7.7}$$

$$J_i^3 = \rho_i\,(v_i - v_3) = J_i^v - \left(\frac{\rho_i}{\rho_3}\right)J_3^v \tag{7.8}$$

$$J_3^v = -J_1^v\left(\frac{V_1}{V_3}\right) - J_2^v\left(\frac{V_2}{V_3}\right) \tag{7.9}$$

The mass fluxes of water and osmotic solute transfer related to the solid frame velocity of component 3 in one dimension are as follows:

$$J_1^3 = J_1^v\left(1 + \frac{\rho_1 V_1}{\rho_3 V_3}\right) + J_2^v\left(\frac{\rho_1 V_2}{\rho_3 V_3}\right) \tag{7.10}$$

$$J_2^3 = J_1^v\left(\frac{\rho_2 V_1}{\rho_3 V_3}\right) + J_2^v\left(1 + \frac{\rho_2 V_2}{\rho_3 V_3}\right) \tag{7.11}$$

Two-film theory assumes that equilibrium exists at the interface, with resistance to mass transfer in both the solid product and the liquid phases. Experiments and analyses showed that interface mass transfer may play an important role in solid–liquid contacting operations, thus exchanging mass across their common interface (Shi and Le Maguer, 2003). Despite the great efforts made, one of the problems in modeling the osmotic process is still the lack of a full understanding of the fundamentals of mass transfer in biological cellular structures.

7.5 MATERIAL PROPERTIES OF CELLULAR MATERIAL IN OSMOTIC TREATMENTS

Biological structures and behavior are difficult to describe in quantitative terms. Thornley (1976) stated that the central problem of biology for the quantitative description of a structure is that the rate of change of a structure is determined by the environment and by the structure itself. In osmotic dehydration of plant tissues, these complex structures pose a challenge to food scientists and engineers in the optimization process and design of equipment. The structure and properties of the tissues affect the mass transfer phenomena occurring in plant tissues during osmotic dehydration. There is a wide variation in the physical nature of cellular materials. The eventual cellular dehydration will strongly depend on the fruit's initial biological microstructural characteristics, especially the intercellular space present in the fruit tissue (Islam and Flink, 1982; Lenart and Flink, 1984). The complex cell wall structure of plant materials acts as a permeable membrane. In contrast, membranes are a major resistance to mass transfer (Le Maguer and Yao, 1995). Upon removal of some initial water, shrinkage is most likely to occur during the osmotic treatment process. Shrinkage of biological materials during osmotic treatment takes place simultaneously with moisture loss and solute gain, and thus may affect the mass transfer rate because of gradient effects. Structural changes occur when fresh fruit material is immersed in osmotic solution, as shown in Figure 7.4. A cell unit consists of a cell, extracellular space, and cell wall. The phenomena of volume change and mass transfer (water loss and solids gain) can affect the different physical properties of the cellular material.

The data sets used for the analysis of materials that have more or less the same experimental conditions (e.g., same temperature, solute concentration) are listed in Table 7.1. The distribution of the data is determined using a histogram analysis. From the initial flux, rate of water loss, or solids gain, J at $t = 0$ is described as fast, average, or slow in a simple model proposed by Azuara et al. (1996).

The values of the kinetic constant for water loss (S_1), kinetic constant for osmotic solute gain (S_2), water flux (J_w, the rate of mass transfer of the substance per unit area), and flux of the solids (J_s) of seven kinds of fresh fruit materials are presented in Tables 7.2 and 7.3. Apple "a" had the highest S_1 value and the peach exhibited the

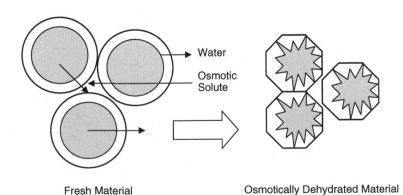

Water

Osmotic
Solute

Fresh Material   Osmotically Dehydrated Material

FIGURE 7.4 Mass transfer and structure changes of cells.

TABLE 7.1
List of Materials and Their Osmotic Treatments and Conditions

Material	Variety	Osmotic Solution	Temp. (°C)	Dimensions	Ref.
Peach	Baby Gold	50% (w/w) corn syrup/ sucrose, 70°Bx	25	2 cm thick slices	Giangiacomo et al. (1987)
Cherry	Bianca di Verona	50% (w/w) corn syrup/ sucrose, 70°Bx	25	1.8-cm diameter	Giangiacomo et al. (1987)
Apricot (a)	Reale	Sucrose, 70°Bx	25	1 cm thick halves	Giangiacomo et al. (1987)
Apricot (b)	Reale	Fructose + glucose + sucrose, 65.2°Bx	25	1 cm thick halves	Giangiacomo et al. (1987)
Apple (a)	Golden	Sucrose, 70%	30	2 × 1 cm thick disks	Azuara et al. (1996)
Apple (b)	Macintosh	Sucrose, 60%	23	3–4 mm thick slices	Hawkes and Flink (1978)
Apple (x)	Granny Smith	Sucrose, 60%	30	2 cm × 2 mm thick disks	Experimental material

highest S_2 value. In terms of fluxes, peaches showed the highest value for both water and solids initial fluxes. The values for apricots and apples stayed at the lower end of the range. From the rates and fluxes of water and solids obtained in Tables 7.2 and 7.3, several correction parameters can be calculated including the mass transfer coefficient (K_x^*) and the cell osmotic behavior constant (ϕ_c), as shown in Table 7.4.

The classification of the mass transfer behavior of seven different materials based on the osmotic constant of the cell ϕ_c is provided in Figure 7.5. The classification was based on the values of the osmotic behavior constant of the cell (ϕ_c). The values of the ϕ_c ratio describe the bulk flow (water loss) transport over the diffusion (solids gain) for every material at the initial time. If the ratio ϕ_c is >1, there is more dehydration than impregnation of solids. The higher this value is, the faster the rate of dehydration. If the ratio ϕ_c is <1, there is more impregnation than dehydration because diffusion

TABLE 7.2
Results of Water Flux Analysis

Material	$M_{w-\infty}$ (%)	S_1 (l/s)	Volume (m³)	Area (m²)	Density (kg/m³)	M/A (kg/m²)	J_w [kg/(m² s)]
Peach	0.47	3.17E-04	6.78E-05	1.04E-02	1012	6.57	9.77E-04
Cherry	0.40	1.57E-04	2.77E-06	2.86E-04	1067	10.34	6.48E-04
Apricot a	0.44	1.00E-04	2.44E-06	1.47E-03	1048	1.75	7.76E-05
Apricot b	0.51	1.00E-04	2.44E-06	1.47E-03	1048	1.75	8.82E-05
Apple a	0.49	1.31E-03	1.57E-06	9.42E-04	787	1.31	8.41E-04
Apple b	0.56	2.45E-04	5.00E-06	4.52E-03	800	0.89	1.21E-04
Apple x	0.58	6.96E-04	6.28E-07	7.54E-04	1001	0.83	3.35E-04

Source: From Le Maguer, M. et al., *Food Sci. Technol. Int.*, 187–192, 2003. With permission.

TABLE 7.3

Results of Solids Flux Analysis

Material	$M_{ss-\infty}$ (%)	S_2 (l/s)	Volume (m³)	Area (m²)	Density (kg/m³)	M/A (kg/m²)	J_s [kg/(m² s)]
Peach	0.049	2.35E-03	6.78E-05	1.04E-02	1012	6.57	7.55E-04
Cherry	0.063	1.62E-04	2.77E-06	2.86E-04	1067	10.34	1.06E-04
Apricot a	0.250	6.05E-04	2.44E-06	1.47E-03	1048	1.75	2.64E-04
Apricot b	0.228	5.90E-04	2.44E-06	1.47E-03	1048	1.75	2.35E-04
Apple a	0.221	3.95E-04	1.57E-06	9.42E-04	787	1.31	1.14E-04
Apple b	0.263	6.55E-04	5.00E-06	4.52E-03	800	0.89	1.53E-04
Apple x	0.524	3.70E-04	6.28E-07	7.54E-04	1001	0.83	1.62E-04

Source: From Le Maguer, M. et al., *Food Sci. Technol. Int.*, 187–192, 2003. With permission.

dominates over the bulk flow. Materials with a ϕ_c of >1 are in class I (fast). Those with a ϕ_c of <1 are in class III (slow). Those with ϕ_c values around 1 were classified as having an average rate of dehydration and are in class II. As with the behavior of J_w, J_s, and S_2, the peach still exhibits a fast rate of mass transfer. However, the cherry has a faster rate of dehydration than the peach. Apple "a" followed in class II with the two apricots. They have an average rate of dehydration. Apples "x" and "b" are at the end of the graph in class III and are classified as slow (Le Maguer et al., 2003).

This classification allows the determination of the processing conditions that should be employed on a material, depending on the quality of the end product. For example, if the purpose of osmotic dehydration is to reduce the moisture content, the appropriate osmotic concentration must be chosen so that it facilitates dehydration more than impregnation. If the dehydration capacity of the material is slow, then

TABLE 7.4

Mole Fraction of J_w and J_s at the Interface with Corrected Mass Transfer Coefficient (K_x^*)

Material	J_w [kmol/(m² s)]	J_s [kmol/(m² s)]	C_i	R	Φ	K_x^* [kmol/(m² s)]	ϕ	ϕ_c (Corrected)
Peach	5.43E-05	2.21E-06	0.0612	0.116	0.946	4.5E-04	1.06	5.41
Cherry	3.60E-05	3.10E-07	0.0652	0.107	0.950	3.3E-04	0.73	5.98
Apricot a	4.31E-06	7.72E-07	0.0691	0.014	0.993	2.5E-04	0.07	1.21
Apricot b	4.90E-06	6.88E-07	0.0672	0.026	0.987	1.6E-04	0.09	0.98
Apple a	4.67E-05	3.34E-07	0.0431	0.598	0.784	7.8E-05	0.95	1.35
Apple b	6.74E-06	4.46E-07	0.0558	0.137	0.937	4.6E-05	0.13	0.41
Apple x	1.86E-05	4.73E-07	0.0424	0.448	0.826	4.0E-05	0.37	0.51

Source: From Le Maguer, M. et al., *Food Sci. Technol. Int.*, 187–192, 2003. With permission.

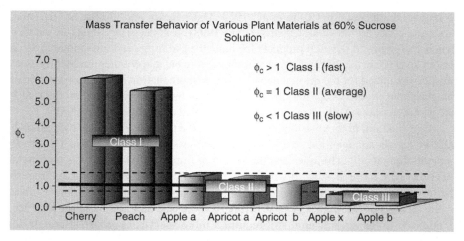

FIGURE 7.5   Classification of materials based on ϕ_c values. (Modified from Le Maguer, M. et al., *Food Sci. Technol. Int.*, 187–192, 2003. With permission.)

using very high concentrations may not be necessary. The same considerations should also be applied for impregnation purposes (such as in candying).

7.6   FUTURE DEVELOPMENT

Research in applications of osmotic dehydration to food processing in technology and in component transfer mechanisms is being carried out in several countries. The technology of osmotic dehydration is useful to provide consumers with safe products of excellent quality. The new osmotically dehydrated products and industrial applications require appropriate manufacturing equipment and procedures at the industrial level (Fito and Chiralt, 2000). When analyzing the results of water transfer obtained from normal pressure treatment and vacuum treatment in osmotic dehydration, water transfer was much faster with a vacuum operation (Fito, 1994; Fito et al., 1996, 1998; Dalla Rosa et al., 1982; Lerici et al., 1985). Fito and Chiralt (1995, 2000) described this mass transfer mechanism under vacuum treatment during osmotic dehydration of fruits as a "hydrodynamic penetration mechanism." Analyses of this mechanism in osmotic dehydration of fruits provided a better understanding of mass transfer phenomena in porous fruit tissues that are immersed in a sugar solution. Osmotic dehydration studies were also carried out using ultrasound to increase the mass transfer rates (Rastogi and Niranjan, 1998). The use of high-pressure pretreatment enhanced mass transfer rates in pineapples. This was attributed to the breaking up of the cell walls, thus speeding up water loss. Cell wall break-up was observed with differential interference contrast microscopy (Rastogi et al., 1999). A high intensity electrical field pulse was also applied to increase mass transfer rates in carrots (Spiess, 1999). Knowledge of the characteristics of the materials and the mass transfer relationships needs to be further understood.

Integration of advances in the combined processes (raw materials and osmosis dehydrated products), unit operation, and solution management should be through complete system analysis. This will make technical and economic evaluations closer to industrial applications (Bolin et al., 1983; Dalla Rossa and Giroux, 2001; Valle, Aranguiz et al., 1998; Valle, Cuadros et al., 1998). It is necessary to set up pilot or bench prototypes to develop products for industry and to upgrade the modeling to a design level (Qi et al., 1999). Design of multicomponent and countercurrent processes must be optimized for the desired dehydration or impregnation rates. Incorporation of the glass-transition concept into the stabilization of osmotically treated products also needs more studies (Valle, Aranguiz et al., 1998; Valle, Cuadros et al., 1998).

An evaluation of the distribution of phases in the tissue will provide clear knowledge of the phenomena that control most of the mass transfer processes. The development of practical techniques and the advent of improved imaging tools will make this procedure more feasible. It will be important to link the microscopy findings to the processing parameters of the operation and other compositional and mechanical characteristics of the tissues. The structure of the tissue and its role in the kinetics, in addition to the equilibrium, still require a more specific approach (Barat et al., 1999; Islam and Flink, 1982).

The osmotic solution becomes diluted because of loss of sugar as well as uptake of water from fresh food material. In practical applications, one problem for large-scale production using an osmotic concentration process is a large quantity of diluted syrup left at the end of the process. The syrup needs to be reconcentrated and reused in order to make the procedure economically feasible. Sugar solutions tend to darken and caramelize when heated for concentration. It is preferable simply to concentrate the sugar solution by adding more sugar without heating. Sugar solutions also cannot be kept more than 1 week at room temperature because they are likely to ferment. It is possible to integrate the osmotic dehydration process with other processes, such as canning and bottling of fruits, alcoholic production, and other fermentation processes as possible means of economically utilizing the spent sugar solution, in which the spent sugar solution can be satisfactorily used. Thus, the economic feasibility of this process probably depends on the availability of cheap sugar and reuse of spent sugar solution.

However, osmotic dehydration requires accurate prediction of water removal from foods for industrial applications. Osmotic dehydration of food materials is a less understood area of food dehydration. Simultaneous flow of solutes and water and interaction between components requires a theoretical approach based on the application of linear irreversible thermodynamic processes. Despite the great achievements made in the fundamental research and industrial applications around the world, simultaneous flows of solutes and water and interaction between components require a further theoretical approach to allow this technique to be applied on a large scale in the food industry.

NOMENCLATURE

VARIABLES

a	water activity
A	surface area (m^2)
I	interface
J	mass flux (kg m^2/s)
K	mass transfer coefficient (kg/m^2/s)
M	total mass (kg)

P pressure (kg/m^2)
v molecular average velocity (m/s)
V water partial specific volume (m^3)
R gas constant
S kinetic constant
t time (s)
T temperature (°C)
X transfer direction (m)

GREEK SYMBOLS

μ chemical potential
ρ density (kg/m^3)
ϕ osmotic behavior constant
ψ cell wall elastic modulus constant (N/m^2)

SUBSCRIPTS

i, j index denoting component number
I interface
0   initial
ss   soluble solid
su   sugar
w   water
∞   infinity
1, 2, 3   components 1, 2, and 3

SUPERSCRIPTS

c   cell
L   equilibrium in liquid phase
0   initial
v fixed volume frame of reference

REFERENCES

Andreotti, R., Tomasicchio, M., de Giorgi, A., and Palmas, D., Conservazione di pesche parzialmente disidratate per osmosi diretta, *Ind. Conserve.*, 60, 96–103, 1985.

Argaiz, A., Lopez-Malo, A., Paou, E., and Welti, J., Osmotic dehydration of papaya with corn syrup solids, *Drying Technol.*, 12, 1709–1725, 1994.

Azuara, E., Beristain, C.I., and Garcia, H.S., Development of a mathematical model to predict kinetics of osmotic dehydration, *J. Food Sci. Technol.*, 29, 239–242, 1996.

Azuara, E., Beristain, C.I., and Gutierrez, G.F., A method for continuous evaluation of osmotic dehydration, *Lebensm.-Wiss. Technol.*, 31, 317–321, 1998.

Barat, J.M., Albors, A., Chiralt, A., and Fito, P., Equilibration of apple tissue in osmotic dehydration: Microstructural changes, *Dry. Technol.*, 17, 1375–1386, 1999.

Bolin, H.R., Huxsoll, C.C., Jackson, R., and Ng, K.C., Effect of osmotic agents and concentration on fruit quality, *J. Food Sci.*, 48, 202–212, 1983.

Bruin, S. and Luyben, K.C., Drying shrinkage and stresses, *Proc Eighth International Drying Symposium IDS '92*, Quebec, 155–216, 1992.

Chirife, J., Karel, M., and Flink, J., Studies on mechanisms of retention of volatile in freeze-dried food models: The system PVP-*n*-propanol, *J. Food Sci.*, 38, 671–678, 1973.

Conway, J., Castaigne, F., Picard, G., and Vovan, X., Mass transfer considerations in the osmotic dehydration of apples, *Can. Inst. Food Sci. Technol. J.*, 16, 25–22, 1983.

Crank, J., *The Mathematics of Diffusion*, Clarendon Press, Oxford, U.K., 1975.

Cussler, E.L., *Diffusion: Mass Transfer in Fluid System*, Cambridge University Press, Cambridge, U.K., 1984.

Dalla Rossa, M. and Giroux, F., Osmotic treatments and problems related to the solution management, *J. Food Eng.*, 49, 223–236, 2001.

Dalla Rosa, M., Pinnavaia, G., and Lerici, C.R., La disidratazione della frutta mediante osmosi diretta. *Ind. Conserve.*, 57, 3–12, 1982.

Dixon, G.M., Jen, J.J., and Paynter, V.A., Tasty apple slices result from combined osmotic dehydration and vacuum drying process, *Food Prod. Dev.*, 10, 60–67, 1976.

Farkas, D.-F. and Lazar, M.-E., Osmotic dehydration of apple pieces: Effect of temperature and syrup concentration on rates, *Food Technol.*, 23, 90–92, 1969.

Fish, B.P., Diffusion and thermodynamics of water in potato starch, paper presented at the Conference on Fundamental Aspects of the Dehydration of Foodstuffs, Aberdeen, Scotland, 1958.

Fito, P., Modelling of vacuum osmotic dehydration of food, *J. Food Eng.*, 22, 313–328, 1994.

Fito, P., Andres, A., Chiralt, A., and Pastor, R., Coupling of hydrodynamic mechanism and deformation–relaxation phenomena during vacuum treatments in solid porous food–liquid system, *J. Food Eng.*, 21, 229–240, 1996.

Fito, P. and Chiralt, A., An update on vacuum osmotic dehydration, in *Food Preservation by Moisture Control: Fundamentals and Applications, ISOPOW Practicum II*, Barbosa-Cánovas, G.V. and Welti-Chanes, J., Eds., Technomic Publishing Company, Lancaster, PA, 1995, p. 351.

Fito, P. and Chiralt, A., Vacuum impregnation of plant tissues, in *Design of Minimal Processing Technologies for Fruits and Vegetables*, Alzamora, S.M., Tapia, M.S., and Lopez-Malo, A., Eds., Aspen Publishers, Gaitherburg, MD, 2000.

Fito, P., Chiralt, A., Barat, J., Salvatori, D., and Andres, A., Some advances in osmotic dehydration of fruit, *Food Sci. Int.*, 4, 329–338, 1998.

Flink, J.M., Process conditions for improved flavor quality of freeze dried foods, *J. Agric. Food Chem.*, 23, 1019–1026, 1975.

Flink, J.M., Dehydration of carrot slices: Influence of osmotic concentration on drying behaviour and product quality, in *Food Process Engineering*, Vol. 1, Linko, P., Malkki, Y., Olkku, J., and Larinkari, J., Eds., Applied Science, London, 1980, p. 412.

Garcia, R., and McFelley, J.C., Determining the cellular concentration and osmotic potential of plant tissues, *Am. Biol. Bull.*, 40, 119–128, 1978.

Giangiacomo, R., Torreggiani, D., and Abbo, E., Osmotic dehydration of fruit. 1. Sugars exchange between fruit and extracting syrups, *J. Food Process. Preserv.*, 11, 183–195, 1987.

Giangiacomo, R., Torreggiani, D., Erba, M.L., and Messina, G., Use of osmodehydrofrozen fruit cubes in yoghurt, *Int. J. Food Sci.*, 3, 345–350, 1994.

Hawkes, J. and Flink, J.M., Osmotic concentration of fruit slices prior to freeze-dehydration, *J. Food Process. Preserv.*, 2, 265–284, 1978.

Hope, G.W. and Vitale, D.G., Osmotic dehydration, a cheap and simple method of preserving mangoes, bananas and plantains (IDRC-004c), Food Research Institute, Canada Department of Agriculture, Ottawa, 1972.

Hough, G., Chirife, J., and Marini, C., A simple method for osmotic dehydration of apples, *Lebensm.-Wiss. Technol.*, 26, 151–155, 1993.

Huxsoll, C.C., Reducing the refrigeration load by partial concentration of food prior to freezing. *Food Technol.*, 5, 98–102, 1982.

Huxsoll, C.C., Bolin, H.L., and King, A.D., Jr., Physicochemical changes and treatments for lightly processed fruits and vegetables, in *Quality Factors of Fruit and Vegetables, American Chemical Society Series No. 405*, Jen, J.J., Ed., American Chemical Society, Washington, DC, 1989, pp. 203–215.

Islam, M.N. and Flink J.N., Dehydration of potato II. Osmotic concentration and its effect on air drying behaviour, *J. Food Technol.*, 17, 387–403, 1982.

Karel, M., Osmotic drying, in *Principles of Food Science—Part II*, Fennema, O.R., Ed., Marcel Dekker, New York, 1975.

King, C.J., *Freeze Drying of Foods*, CRC Press, Cleveland, OH, 1970.

Le Maguer, M., Mass transfer modeling in structured foods, in *Food Engineering*, Fito, P., Ortega-Rodríguez, E., and Barbosa-Cánovas, G.V., Eds., Chapman & Hall, New York, 1997.

Le Maguer, M. and Biswal, R.N., Multicomponent diffusion and vapour–liquid equilibria of dilute organic components in aqueous sugar solutions, *Am. Inst. Chem. Eng. J.*, 18, 513–519, 1984.

Le Maguer, M., Shi, J., and Fernandez, C., Characterization of mass transfer behavior of plant tissues during osmotic dehydration, *Food Sci. Technol. Int.*, 9, 187–192, 2003.

Le Maguer, M. and Yao, Z.M. Mass transfer during osmotic dehydration at the cellular level, in *Food Preservation by Moisture Control: Fundamentals and Applications, ISOPOW Practicum II*, Barbosa-Cánovas, G.V. and Welti-Chanes, J., Eds., Technomic Publishing Co. Inc., Lancaster, 216–222, 1995.

Lenart, A. and Dabrowska, R., Kinetics of osmotic dehydration of apples with pectin coatings, *Dry. Technol.*, 17, 1359–1373, 1999.

Lenart, A. and Flink, J.M., Osmotic concentration of potato. I. Criteria for the end-point of the osmosis process, *J. Food Technol.*, 19, 45–63, 1984.

Lerici, C.R., Pinnavaia, G., Dalla Rosa, M., and Bartolucci, L., Osmotic dehydration of fruit: Influence of osmotic agents on drying behavior and product quality, *J. Food Sci.*, 50, 1217–1219, 1985.

Magee, T.R.A., Hassaballah, A.A., and Murphy, W.R., Internal mass transfer during osmotic dehydration of apple slices in sugar solutions, *Ir. J. Food Sci. Technol. (Dublin)*, 7, 147–155, 1983.

Maltini, E. and Torreggiani, D., A new application of osmosis: The production of shelf-stable fruits by osmosis, paper presented at the Progress in Food Engineering 246th Event for European Food Chemistry and Engineering, Milan, Italy, 1986.

Marcotte, M. and Le Maguer, M., Repartition of water in plant tissues subjected to osmotic processes, *J. Food Proc. Eng.*, 13, 297–320, 1991.

Marcotte, M., Toupin, C., and Le Maguer, M., Mass transfer in cellular tissues. Part I: The mathematical model, *J. Food Eng.*, 13, 199–220, 1991.

Nobel, P.S., *Biophysical Plant Physiology and Ecology*, W.H. Freeman, San Francisco, 1983.

Ponting, J.D., Osmotic dehydration of fruits—Recent modifications and applications, *Process. Biochem.*, 8, 18–23, 1973.

Ponting, J.D., Watters, G.G., Forrey, G.G., Jackson, R.R., and Stanley, R., Osmotic dehydration of fruits, *Food Technol.*, 20, 125–131, 1966.

Qi, H., Le Maguer, M., and Sharma, S.K., Design and selection of processing conditions of a pilot scale contactor for continuous osmotic dehydration of carrots, *J. Food Process. Eng.*, 21, 75–88, 1999.

Rao, M.A., Energy consumption for refrigerated, canned and frozen peas, *J. Food Process. Eng.*, 1, 149–155, 1977.

Raoult-Wack, A.L., Recent advances in the osmotic dehydration of foods, *Trends Food Sci. Technol.*, 5, 255–259, 1994.

Raoult-Wack, A.L., Guilbert, S., Le Maguer, M., and Rios, G., Simultaneous water and solute transport in shrinking media—Part 1. Application to dewatering and impregnation soaking processes analysis, *Dry. Technol.*, 9, 589–612, 1991.

Raoult-Wack, A.L., Jourdain, P., Guilbert, S., and Rios, G., Technique de fluidisation inverse appliquee aux procedes de dehydration impregnation par immersion (DII), *Coll. Recentes Progres Genie Proced.*, 5(14), 329–334, 1991.

Raoult-Wack, A.L., Lenart, A., and Guilbert, S., Recent advances in dewatering through immersion in concentrated solutions ("osmotic dehydration"), in *Drying of Solids*, Mujumdar, A.S., Ed., VSP International Science Publishers, Zeist, The Netherlands, 1992.

Rastogi, N.K., Eshtiaghi, M.N., and Knorr, D., Accelerated mass transfer during osmotic dehydration of high intensity electrical field pulse pretreated carrots, *J. Food Sci.*, 64, 1020–1023, 1999.

Rastogi, N.K. and Niranjan, K., Enhanced mass transfer during osmotic dehydration of high pressure treated pineapple, *J. Food Sci.*, 63, 508–511, 1998.

Rogachev, V.I. and Kislenko, I.I., Combined methods of decreasing water activity in fruits under preservation, *Acta Aliment.*, 2, 245–252, 1972.

Shewfelt, R.L., Quality of minimally processed fruits and vegetables, *J. Food Qual.*, 10, 143–149, 1987.

Shi, J. and Le Maguer, M., Mass transfer in cellular material at solid–liquid contacting interface, *Lebensm.-Wiss. Technol.*, 35, 367–372, 2002a.

Shi, J. and Le Maguer, M., Analogical cellular structure changes in solid–liquid contacting operations, *Lebensm.-Wiss. Technol.*, 35, 444–451, 2002b.

Shi, J. and Le Maguer, M., Mass transfer flux at solid–liquid contacting interface, *Food Sci. Technol. Int.*, 9, 193–197, 2003.

Spiess, W.E.L., Improvement of overall food quality by application of osmotic treatments in conventional and new processes: Concerted action FAIR-CT96-1118, in *Osmotic Treatments for the Food Industry. Proceedings of the 1st Osmotic Treatment Seminar*, Sereno, A.M., Ed., FEUP Ediçoes, Porto, Portugal, 1999, pp. 9–17.

Thornley, J.H.M., *Mathematical Models in Plant Physiology. A Quantitative Approach to Problems in Plant and Crop Physiology*, Academic Press, London, 1976.

Torreggiani, D., Technological aspects of osmotic dehydration in food, in *Food Preservation by Moisture Control: Fundamentals and Applications, ISOPOW PRACTICUM II*, Barbosa-Cánovas, G.V. and & Welti-Chanes, J., Eds., Technomic Publishing Co. Inc., Lancaster, 281–304, 1995.

Torreggiani, D., Forni, E., Maestrelli, A., and Quadri, F., Influence of osmotic dehydration on texture and pectic composition of kiwifruit slices, *Dry. Technol.*, 17, 1387–1397, 1999.

Torreggiani, D., Maltini, E., and Forni, E., Osmotic pre-treatments: A new way to directly formulate fruit and vegetable ingredients, in *Osmotic Treatments for the Food Industry. Proceedings of the 1st Osmotic Treatment Seminar*, Sereno, A.M., Ed., FEUP Ediçoes, Porto, Portugal, 1999, pp. 19–28.

Torreggiani, D., Forni, E., and Rizzolo, A., Osmotic dehydration of fruit. Part II. Influence of the osmotic time on the stability of processed cherries, *J. Food Process. Preserv.*, 12, 27–31, 1988.

Uzuegbu, J.O. and Ukeka, C., Osmotic dehydration as a method of preserving fruits to minimize ascorbic acid loss, *J. Food Agric.*, 1, 187–193, 1987.

Valle, J.M. del, Aranguiz, V., and Leon, H., Effects of blanching and calcium infiltration on PPO activity, texture, microstructure and kinetics of osmotic dehydration of apple tissue, *Food Res. Int.*, 31, 557–569, 1998.

Valle, J.M. del, Cuadros, T.R.M., and Aguilera, J.M., Glass transitions and shrinkage during drying and storage of osmosed apple pieces, *Food Res. Int.*, 31, 191–204, 1998.

Yao, Z. and Le Maguer, M., Mathematical modelling and simulations of mass transfer in osmotic dehydration processes. Part I: Conceptual and mathematical models, *J. Food Eng.*, 29, 349–360, 1996.

8 Fundamentals and Tendencies in Freeze-Drying of Foods

J.I. Lombraña

CONTENTS

8.1 INTRODUCTION

The ancient Peruvian Incas of the Andes knew the basic process of freeze-drying food. Freeze-drying or lyophilization is the sublimation or removal of water content from frozen food. The dehydration occurs under a vacuum, and the wet zone is frozen during the whole process so that nearly perfect preservation results. The Incas stored their potatoes and other food crops on the mountain heights above Machu Picchu (Carter, 1964). The cold mountain temperature froze the food and the water inside slowly vaporized under the low air pressure of the high altitudes; basically, the foods were freeze-dried.

During freeze-drying, the solvent (generally water) is removed from the frozen product under a vacuum. Sublimated vapor is evacuated by the action of mechanical pumps or steam ejectors.

As a general rule, lyophilization produces the best quality dehydrated foods. This is attributable to the rigidity of the frozen state during water removal because it is essentially a sublimation process. The result is the apparition of a porous matrix that keeps the initial structure and enables rapid complete rehydration, reestablishing the original food. In the application to foods, freeze-drying presents the additional advantage of avoiding losses of volatiles and flavors. Moreover, the low processing temperature and the rapid transition of the whole product from an initial hydrated structure to the nearly complete dehydrated matrix minimizes the deterioration reactions that occur during conventional drying, such as nonenzymatic browning, enzymatic reactions, and protein denaturation. A small amount of water, called bound water, inevitably remains in foods after freeze-drying. The process temperature should be kept below the glass-transition temperature of the food, corresponding to bound water, to avoid undesirable effects (Mellor, 1978). The high running costs of the freeze-drying process is compensated by the lack of postoperation treatments like freezing or refrigeration during storage.

Two characteristics are necessary for the application of lyophilization as a drying method: the food value and the quality of the dehydrated food. The presence of these two factors leads to a wide application of freeze-drying to biological and biomedical products such as blood plasma, hormone solutions, serum, surgical transplants (arteries, bones, skin), and living cells (bacteria, yeasts).

In addition, freeze-drying is especially interesting for processing foods that present drying difficulties (coffee, soups, extracts, and some fruits) where the quality of the freeze-dried products (i.e., aroma, rehydration, bioactivity, etc.) is considerable higher than those processed by other drying techniques.

Freeze-drying requires very low pressure (vacuum) to obtain acceptable drying rates. If the product has water as a solvent, freeze-drying can be performed at 4.58 mmHg pressure and 0°C. Nevertheless, for foods with bound water the process requires temperatures below 0°C for its complete solidification. Most foods can be freeze-dried at a pressure below 2 mmHg and a temperature below −10°C.

Lyophilization from a processing point of view can be considered as the sequence of three operations: freezing to low temperatures below the melting point, drying by sublimation at a pressure lower than that corresponding to triple point, and desorption of bound water. Finally, storage should be carried out under controlled conditions (inert atmosphere and humidity) in opaque containers.

If these operations are performed under controlled optimized conditions, freeze-dried foods can thus be kept for long periods of time while maintaining their physical, chemical, and organoleptic properties, similar to those of fresh products and for indefinite periods.

8.2 FUNDAMENTALS

The most remarkable characteristic of freeze-drying is the high quality of the dehydrated foods, which is incomparable to any other drying process. Freeze-dried foods are characterized by their rapid rehydration and their texture and flavor that is similar to fresh foods. These quality properties are derived from (1) the low temperature involved in the process avoids chemical alterations and (2) moisture is mostly removed as vapor from a frozen solid leaving behind a dehydrated porous matrix, which avoids undesirable reactions and allows rapid rewetting.

Lyophilization is essentially a sublimation process. The basic elements for a freeze-drying installation are the vacuum chamber, an evacuating system, and a heating device to supply the necessary heat for sublimation and vaporization (Gutcho, 1977). Figure 8.1 shows the basic combination of these elements that usually are complemented with monitoring and operation control devices.

Several kinds of equipment are used for vapor evacuation from steam ejectors to numerous kinds of mechanical pumps. The gas volumes that are required to move at the low pressures of freeze-drying are enormous. For example, a drying rate of only 1 kg/h gives rise to vapor volumes of about 3000 m^3/h, which is well over the ability of gas-ballast mechanical pumps. Steam ejectors could be a solution when the plant has large amounts of economical steam and cold water available. These steam ejectors were often found in the first big freeze-drying installations (Cotson and Smith, 1963). Nevertheless, when using ejectors the vacuum level attained is only moderate. The basic typical equipment for vacuum systems consists of a mechanical pump after a condenser system to retain vapor, which means additional consumption of energy for refrigeration. It is thus necessary to obtain sufficient pressures to provoke the

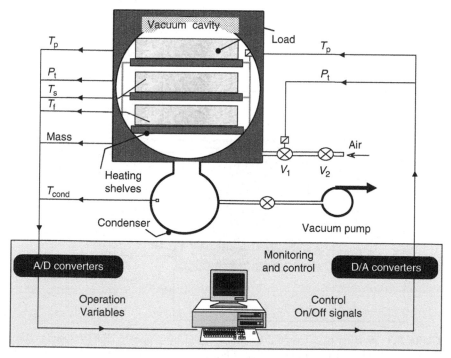

FIGURE 8.1 Scheme of a freeze-drying installation showing the basic elements.

212   Advances in Food Dehydration

necessary driving force between the product and the condenser coil surface, consequently increasing the product temperature to the maximum allowed (near melting
temperature) and lowering the condenser temperature to the minimum to obtain the
most favorable drying rate. Therefore, the condenser temperature should be kept
well below the sublimation temperature of the product (–45°C) and even lower
(up to –90°C) in some research prototypes. This requires the use of special refrigeration systems with two or more refrigerants working in combined refrigerating cycles
(Gutcho, 1977).

Freeze-drying is a mass transfer controlled operation during nearly all of the
primary drying. Nevertheless, in some occasions it could be a problem to supply a
certain amount of heat, as it is when applying mainly radiant heat. Heat transfer
resistance of a dry layer as well as a vacuum are primary factors conditioning the
heat supply up to the sublimation front.

The first freeze-dryers tried to reduce the dry layer and resistance by applying
heat directly by conduction from the product surface closely in contact with the
heating source. The product was disposed in thin layers between two heating plates.
The equipment proposed by Smithies (1959) was equipped with this kind of heating
device that consisted of an ingenious system of spiked metal plates pushed into meat.
This system presents great advantages in reducing the resistance to heat transfer,
intensifying the heat application by conduction. At the same time, by drilling holes
down the spikes, the resistance of water flow also decreased. The idea of a food layer
between heating plates applied to both surfaces of a product is illustrated in Figure 8.2.
This device reportedly gave a substantial reduction in drying time. However, the
hygienic and mechanical disadvantages were an insurmountable barrier in modern
freeze-dryers in spite of the good results obtained in meat and fish processing.

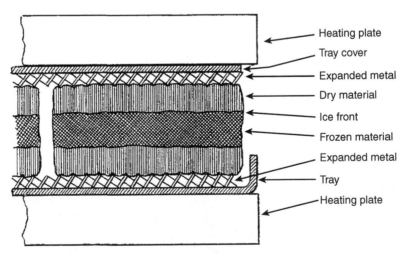

FIGURE 8.2 Section of a freeze-dried food heated by a double contact system. (From
Cotson, S. and Smith, D.B., *Freeze Drying of Foodstuffs*, Columbia Press, Manchester, U.K.,
1963. With permission.)

The heat transfer resistance could be reduced to a great extent by using radiofrequency heating (microwave energy). This technique, which will be treated in detail later, was already investigated in the early 1950s (Harper and Tapel, 1957) and it is currently adopted in certain drying facilities because of its enormous advantages. However, the difficulties of the temperature control of the frozen nucleolus are still a serious drawback for the application of microwaves in the food industry.

Radiant heat is present to a greater or smaller extent in lyophilization facilities currently used for different industrial applications, not only in foods but also in biomedicals and pharmaceuticals. The amount of heat supplied by radiation is given by the Stefan–Boltzman law. In accordance with it, the energy removed is proportional to the difference between the fourth power of the absolute temperature of the heat source (plate) and the product. This heating mechanism presents certain advantages with respect to others because it is independent of the media properties (pressure and position) because it supplies heat through the free surface of the product and makes the subsequent mass flux easy. Radiant energy in the wavelength range from 10^4 to 10^5 Å has better penetration, resulting in shorter drying cycles (Levinson and Oppenheimer, 1948). Nevertheless, radiation heating is strongly affected by the color and texture of the absorbing surface, often a characteristic of raw foods, which could give rise to nonuniform heating problems.

8.3 MATHEMATICAL DESCRIPTION OF FREEZE-DRYING

Products undergoing lyophilization can be very different in size and type. This is the case of solutions in vials (Boss et al., 2004), such as extracts and infusions. Other cases are granulated products and even raw foods. Thus, a general approach to allow the description and application of freeze-drying physics to different common situations encountered during food processing will be made in this section, and disposition and thermophysical properties will be taken into account. In most situations, the application of theoretical models is done on a product amount that represents the load in the drying cabinet. In this case, a general situation representing a finite amount of any geometry will be treated. Because of easiness of the treatment, the amount of product will be related to the sphere geometry through the sphericity parameter. Thus, many geometrical aspects of the advancing sublimation front can be followed with a sole dimension parameter analogous to the radius in a same volume spherical particle of radius r_{eq}. The sphericity of a body is defined as the ratio of the surface of a sphere having the same volume to the surface of that body. As an example, sphericity ψ of a cube of side L is

$$\psi = \frac{L}{2r_{eq}} = \frac{E}{r_{eq}} \tag{8.1}$$

Generalizing, any body can be defined by electric field E ($= \psi r_{eq}$), a geometrical parameter characteristic of the solid in a spherical analogy, so that its volume V having an outer surface S would be

$$V = \frac{S}{3}E = \frac{4}{3}\pi r_{eq}^3 \psi \tag{8.2}$$

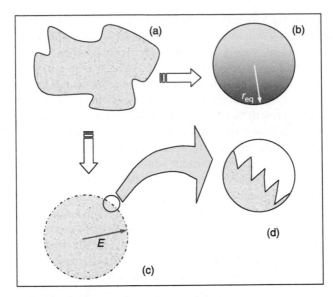

FIGURE 8.3   Analogy parameters of the cross section of an irregular particle and sphere: (a) irregular body, (b) equivalent sphere volume, (c) analogous spherical body of the same volume and geometry, and (d) expanded detail of surface.

Figure 8.3 shows a cross section of a particle of irregular geometry and a sphere of radius r_{eq} with the same volume. At the beginning of freeze-drying, the whole volume is frozen. As freeze-drying progresses, a dry layer of thickness δ appears, so

$$\delta = E_f^0 - E_f \tag{8.3}$$

Consequently, the velocity of sublimation front advance v and that of the reference sphere with the same volume can be related as

$$v = \frac{\partial E_f}{\partial t} = \psi \frac{\partial r_{eq,f}}{\partial t} \tag{8.4}$$

The dry layer thickness advances up to the complete disappearance of the frozen nucleus. Moisture removal is performed by sublimation, which in most of the cases represents more than half of the drying duration and more than 90% moisture. Once the frozen nucleus has been sublimated, secondary drying arises in which the low residual moisture that remains on the porous matrix will be removed by evaporation.

A very important characteristic of freeze-drying is the apparition of an outer dry zone and an inner frozen one. Many authors (Liapis and Bruttini, 1994; Millman, 1985; Song et al., 2005) consider two well-defined zones separated by a moving interface of infinitesimal thickness. This physical model is quite realistic in the freeze-drying of solutions in vials. In this case, the dried product would present a structure originated from the perfect dispersion of solute within the solvent. Thus, in such a circumstance it is possible to consider rigorously a direct transition of solid to vapor. Nevertheless, many raw food materials, such as meat, fish, fruits, and vegetables, have a characteristic structure. Consequently, their dry zone would not be homogeneously distributed (anisotropy), depending on the corresponding texture of the product: cells, fibers, and other typical formations. In these cases, water is not completely free but bound. Thus, it seems reasonable that in those circumstances water could not pass directly to vapor because they would be retained in the corresponding structure (Kochs et al., 1993). Moreover, ice dendritic formations appear at certain freezing rates (Kochs et al., 1991; Kurz and Fisher, 1989). In summary, it can be concluded that a sublimation interface of infinitesimal thickness could be a reasonable assumption, although in many cases important discrepancies would arise with respect to its application in reality. This leads to the consideration of a more realistic "diffuse" sublimation front, in which moisture decreases gradually outward.

Primary drying is characterized by the presence of frozen water, and a majority of it is removed from the product matrix by sublimation. Secondary drying appears once there is no frozen water left in the porous matrix and the dry zone is extended over the whole product. Here, the mechanism for moisture removal is by desorption of bound water and the subsequent transport as vapor through channels present in the dry porous matrix.

During drying, the energy required for sublimation and desorption of bound water must be supplied by an adequate heating source. In many cases it consists of a radiant plate at high temperature placed in the top side of the product load or from the bottom side by conduction. If the global heating flux supplied by the heating mechanisms over the surface of the product is constant and no restrictions exist for the vapor flow, the decrease of the frozen nucleus will progress, keeping a geometrical similarity to the original form and being placed in the geometrical center of the product. Nevertheless, uniformly distributed heating is very difficult to find in most cases and there are often great differences in the heating supply according to the freeze-dryer design, position, and arrangement of the load within the vacuum cabinet. The decrease of the ice nucleus would thus depend on the heating flux in each spatial direction, which is especially important in the freeze-drying of products contained in vials (Schelenz et al., 1995).

Good running of the freeze-drying operation involves keeping the temperatures of the dry and frozen zones below certain prefixed levels, assuring good quality in the dehydrated product. Thus, depending on the type of product, it is necessary to define its proper constraint temperatures for the dry and frozen product, which would be the scorch temperature and the melting point, respectively. Keeping the product temperature below these levels during the operation is the most important rule to follow for achieving a correct freeze-drying process. If not, vaporization of the melted ice would take place, which in turn would produce a collapsed structure and shrinkage of the affected zones, with negative effects on the quality properties of the dehydrated product (i.e., physicochemical alterations and dehydration difficulties). The melting point of many foods is around $-10°C$ because it depends on the amount of bound water, which is similar for a great variety of foods (Lombraña and Díaz, 1987). Conversely, the scorch temperature can be quite different depending on the product. In addition, the temperature of the dry zone must be kept below the scorching temperature during both primary and secondary drying. The value for the scorch point oscillates habitually between 10 and 50°C, depending on the criteria used to determine such temperatures. There are several studies on the quality of freeze-dried production that were performed on measurable parameters such as color, vitamin, protein, and nutritional compounds contents. Once the constraint temperatures are determined, a good strategy would consist of achieving the freeze-drying operation in a minimum operating cycle. Efficient running of the operation requires the application of the maximum heat supply at each moment, taking into account the constraints of the product and limitations of the installation. Keeping the product under the thermal limitations requires careful regulation of the heating supply or pressure (Lombraña et al., 1997). In some cases where heat transfer limits the process, the scorch temperature will be the restriction temperature limiting the process; in other cases where mass transfer is in control, the melting point will be the temperature to watch more closely. Both situations may be controlled by reducing the heating power, but in mass transfer limited cases it should be accompanied by a reduction of pressure within the possibilities of the vacuum equipment, positively influencing the mass transfer and subsequently decreasing the frozen core temperature.

Numerous researchers have used mathematical modeling to describe freeze-drying on the basis of applying heat and mass transfer laws and balances (Liapis and Bruttini, 1995). Most of the freeze-drying models are hardly conditioned by geometry aspects. This is the case of lyophilization in vials, a situation frequently treated in the literature because of its importance for some food and pharmaceutical industrial processes (Kochs, 1991; Nastaj, 1991; Song, 2005). Because of the great variety of situations and products in which freeze-drying is applied, a general approach to describe most freeze-drying situations will be explained here. The most common suppositions in previous works (Liapis and Litchfield, 1979; Sadikoglu and Liapis, 1997) follow:

1. Sublimation of frozen moisture happens at the interface between frozen and dry zones.
2. Vapor pressure is in equilibrium at the sublimation temperature of the product.
3. To a lesser degree than vapor, moisture can migrate in the liquid state either during primary or secondary drying.
4. Vaporized water diffuses through the porous matrix of the dry layer, which is filled with low-pressure stagnant air. The net flux of air is assumed to be negligible.
5. Shrinkage of the material is assumed to be negligible.
6. The bound water content in the dry porous matrix is in equilibrium with the local temperature.
7. Properties such as the density, heat capacity, and thermal conductivity vary with the moisture content following a linear relationship.

Taking into account the assumptions just indicated, the energy balance during primary drying can be expressed according to the following:

$$N_{wm} C_{pl} \nabla T + N_{wg} C_{pg} \nabla T + \frac{\partial M_m}{\partial t}(f_v \rho_d \Delta H_s) + \rho C_p \frac{\partial T}{\partial t} + T \frac{\partial(\rho C_p)}{\partial t} = k_e \nabla^2 T \qquad (8.5)$$

In this equation, C_{pl}, C_{pg}, and C_p are the specific heat capacities of water in liquid phase, of vapour and of food, respectively. The equation distinguishes between the flux of water in the solid matrix (N_{wm}) and that of water vapour (N_{wg}). The third term express the energy to sublimate the ice through the variation of the ice content in the solid matrix (M_m). Parameter f_v indicates the degree of moisture removed by sublimation, in a certain location, respect to the total (N_w) according to:

$$f_v = \frac{N_{wg}}{N_w} \tag{8.6}$$

The continuity equations for moisture removal can be expressed by the following equation:

$$\frac{\partial M}{\partial t} = D_{el}\nabla^2 M \tag{8.7}$$

Water fluxes in gas and liquid forms must be considered in Eq. (8.5):

$$N_{wg} = -D_g\nabla p_w \tag{8.8}$$

$$N_{wm} = -D_{ml}\nabla M_m \, \rho_d \tag{8.9}$$

$$N_w = N_{wg} + N_{wm} = -D_{el}\nabla M_m \rho_d \tag{8.10}$$

According to the last equation, global moisture is referred to the water content either as gas or associated with the solid matrix:

$$M = \frac{\varepsilon p_w M_w}{RT\rho_d} + M_m \tag{8.11}$$

The moisture retained in the solid matrix or M_m content varies sharply from the dry zone to the frozen zone. In the former, the M_m is very low and composed only of the bound water (M_b) whereas in the frozen zone it is the initial moisture (M^0). The residual moisture (MR) at each location of the solid can be defined as

$$MR = \frac{M - M_b}{M^0 - M_b} \tag{8.12}$$

which is equal to 1 in the frozen zone.

A third zone of variable thickness could be considered, depending on the type of product, in which the MR would vary gradually from 0 to 1. The thickness of this hybrid zone, in which moisture may be present in the three states, would be practically nonexistent for a uniformly retreating ice front (URIF model) or extended all along the piece (Lombraña and Villarán, 1993).

To solve the mathematical model formulated by the former equations, the following initial ($t = 0$) conditions will be considered in the whole product vapor, pressure, moisture, and temperature: $p_w = p_w^0$, $T = T^0$, and $M = M^0$.

The boundary conditions (at the external surface of the product and the geometrical center) necessary to solve the characteristic equations of the energy balance and the continuity equation are the following:

$$(\nabla T)_{center} = 0 \tag{8.13}$$

$$(\nabla p_w)_{center} = 0 \tag{8.14}$$

$$k_e \nabla T|_{surf} = q|_{surf} \tag{8.15}$$

$$N_w|_{surf} = N_{wg}|_{surf} \tag{8.16}$$

$$D_e \rho_d \nabla M|_{surf} = N_w|_{surf} \tag{8.17}$$

$$D_e \rho_d \nabla M|_{surf} = N_w|_{surf} = K_m(p_w|_{surf} - p_w|_{cond}) \tag{8.18}$$

$$M|_{surf} = \frac{\varepsilon p_w|_{surf} M_w}{RT \rho_d} + M_b \tag{8.19}$$

The transition zone from the frozen to dried layers where the sublimation of ice takes place is assumed to be in equilibrium with the temperature at the corresponding location. Consequently, the vapor pressure at the sublimation interface will be obtained by means of the corresponding thermodynamic solid–vapor equilibrium function for water (see Table 8.1) depending on the sublimation temperature.

$$p_w = f_w(T) \quad \text{when } M_m > M_b \tag{8.20}$$

TABLE 8.1
Values and Estimations of Thermophysical and Transport Properties of Skim Milk in Freeze-Drying

C_1 (m)	3.865×10^{-4}	k_d (W/m K)	$2.596 \times 10^{-4} P + 3.9806 \times 10^{-2}$
C_2	0.921	k_w (W/m K)	2.56
C_{pg} (J/kg K)	1616.6	M_b^0	0.645 (at 233.15 K)
C_{pd} (J/kg K)	2590	M_b	$\dfrac{\exp\{2.6[1.36 - 0.036(T - T^0)]\}}{100}$
C_{pw} (J/kg K)	1930		
D_g (m²/s)	$\kappa_1 + \kappa_2 p_w \dfrac{\nabla P}{\nabla p_w}$	M_w	18.00
		M_{in}	29.80
	$D_w = \dfrac{2.16}{P}\left(\dfrac{T}{273.15}\right)^{1.8}$	R (J/kmol K)	8314
	$D_w^0 = D_w P$	T_m (K)	263.15
		T_{scor} (K)	313.15
	$K_w = C_1\left(\dfrac{RT}{M_w}\right)^{0.5}$	$f_w(T)$	$133.3224\,\{\exp(-2445.56/T + 8.23121$
			$\log_{10} T - 0.01670T + 1.20514 \times$
			$10^{-5} T^2 - 6.757169)\}$
	$K_{in} = C_1\left(\dfrac{RT}{M_{in}}\right)^{0.5}$	ε	0.785
	$K_{mx} = y_{in} K_w + y_w K_{in}$	ρ_d (kg/m³)	$328\left(\dfrac{1 + M_b}{1 + M_b^0}\right)$
	$\kappa_1 = \dfrac{C_2 D_w^0 K_w}{C_2 D_w^0 + K_{mx} P}$	ρ_d (kg/m³)	1030
	$\kappa_2 = \dfrac{C_2 D_w^0 K_w}{C_2 D_w^0 + K_{mx} P} + \dfrac{C_{ol}}{\mu_{mx}}$	ΔH_s (J/kg)	2840×10^3
		ΔH_s (J/kg)	2687×10^3

The equation for $f_w(T)$ can be found in Table 8.1. Although the sublimation zone could be variable in thickness, here it is treated as an infinitesimal sublimation front and positioned through parameter E_f. This facilitates the velocity analysis of the retreating ice front as shown later, when searching a spherical equivalency in which the outer surface is that of the solid.

Thermophysical properties such as ρ and C_p and properties of transport (i.e., k_e and effective diffusivity for primary and secondary drying, D_{e1} and D_{e2}) change strongly with moisture. Thus, for the estimation of such properties one can use the following equations:

$$k_e = k_d + MR(k_w - k_d) \tag{8.21}$$

$$\rho = \rho_d + MR(\rho_w - \rho_d) \tag{8.22}$$

$$C_p = C_{pd} + MR(C_{pw} - C_{pd}) \tag{8.23}$$

where MR is defined in Eq. (8.12) and M_b is the bound water of the foodstuff, which is a function of the temperature of the porous matrix. During primary drying, the free water content is removed and bound water remains in the solid matrix. This changes with temperature and, once all free water has been removed, heating is employed for increasing the solid porous matrix temperature from that of sublimation to the temperature corresponding to the outer surface of the product. As the sublimation front retreats, the temperature of a certain location increases with the subsequent decrease in bound water content.

With respect to the diffusion coefficients, either D_{e1} or D_{e2} are influenced by moisture through the diffusivity components in the porous matrix (D_{m1} and D_{m2}) according to Eqs. (8.8) to (8.10). The values of D_{m1} and D_{m2} would be taken as constant during the primary and secondary drying periods, respectively.

From a macroscopic point of view, it is interesting to know where the theoretical position of the sublimation front is. This position can be estimated using a spherical analogy through the sphericity and parameter E, according to Eq. (8.1). Such a position would be coincident with the real one in the case of freeze-drying solutions or completely isotropic materials. Nevertheless, in the case of raw foods with textured structures, as is the case of fruits and meats, it is not realistic to consider a clear sublimation front. In the case of fuzzy sublimation, the theoretical position could be considered as an averaged representative position. Thus, the theoretical dry zone volume in a freeze-drying particle is expressed as a function of time by

$$V_\delta = V_f^0 - V_f = \frac{S^0 - S_f}{3}(E^0 - E_f) = \frac{S^0 - S_f}{3}\delta \tag{8.24}$$

being

$$V_f = V_f^0\, MR \tag{8.25}$$

$$\frac{r_{eqf}}{S_f} = \frac{r_{eq}}{S} = \psi \tag{8.26}$$

where E_f^0 and E_f are the initial and final characteristic dimensions of the clear-front frozen nucleus, respectively, and δ is the thickness of the dried zone. Equations (8.3) and (8.4) would thus be formulated for the outer dried layer of thickness δ and for the velocity of the retreating ice front, respectively.

The velocity of the retreating ice front of the piece could be expressed in terms of the mass or drying rate:

$$v_m = N_w|_{surf} S = v\frac{S_f}{3}\rho_d M^0 \tag{8.27}$$

where S_f is the real surface of the frozen nucleus, taking into account Eq. (8.4). Moreover, if the average moisture of the solid (M_{av}) is known, then

$$(M^0 - M_{av})V\rho_d = \int_0^t v_m\,dt \tag{8.28}$$

8.3.1 SECONDARY DRYING

The essential characteristic during secondary drying is that the frozen zone has been removed completely. The transport equations for heat and mass formulated previously continue to be valid in this case except that the residual moisture corresponds only to the bound water. Consequently, the energy balance is given by Eq. (8.5) with the following remarks:

1. The enthalpy of sublimation must be substituted by that of evaporation.
2. Physical properties can be considered constant during the entire period ($\rho_d = \rho_d$ and $C_p = C_{pd}$).
3. Transport properties take the value corresponding to the second drying period:

$$k_e = k_d \tag{8.29}$$

$$D_{e2} = D_g\frac{M_w}{RT}\frac{\nabla p_w}{\nabla M} + D_{m2}\rho_d\frac{\nabla M_m}{\nabla M} \tag{8.30}$$

There is an important difference between effective diffusivities D_{e1} and D_{e2}. In the secondary drying period, diffusivity is basically coincident with the diffusivity of water vapor in the gas phase (D_g) whereas during primary drying, the total flux of water relative to the porous matrix (N_{wm}) could be important in the zone near sublimation where the gradient ∇M_m is high.

The initial conditions at the beginning of secondary drying (temperature, pressure, and moisture content) would correspond to the profiles of such variables at the end of primary drying. It is important to know these profiles to initiate secondary drying.

The moisture of the porous matrix decreases from the values at the end of primary drying to that of equilibrium, depending on the temperature. In contrast, the concentration of water in the solid matrix is coincident to the bound water content all along the process:

$$M_m = M_b = f(T) \tag{8.31}$$

The previous equation system describing freeze-drying either in primary or secondary drying can be solved using different methods of finite differences. In the case of microwave heating, an internal heat generation term (see dielectric heating section) must be added to the right side of the energy balance [Eq. (8.5)].

8.4 PHYSICAL PROPERTIES AND FREEZE-DRYING CONTROL

When controlling the freeze-drying process, physical properties (density, specific heat, and thermal conductivity) and transport properties (diffusivity and mass transfer coefficient) must be taken into account. Both kinds of properties undergo strong changes during the process in accordance with the MR [Eqs. (8.12) and (8.21) to (8.23)].

Table 8.1 provides the more significant properties during freeze-drying of skim milk as an example. This product has been studied by several groups (Boss et al., 2004; Sheehan and Liapis, 1998; Song et al., 2005). Approximate equations for estimation of the properties and data from review articles can be used for other foods (Maroulis and Saravacos, 2003; Saravacos and Kostaropoulos, 2002).

The pressure in the chamber and the heating plate temperature should be considered as the control variables for running a lyophilization process. In the case of microwaves, the heating plate temperature is substituted by the electric power input.

Excessively long drying cycles is one of the main drawbacks that significantly affect running costs and subsequently the economical viability of the freeze-drying operation. Thus, an important objective in running a lyophilization process is to minimize the drying time and reduce costs, increasing its application possibilities. Depending on the food and the application, variables such as the product geometry and size modification by pretreatments must be defined together with an adequate operating strategy whereby it is possible to modify the control variables (pressure and heating temperature or power).

The aim of every control strategy is to run the operation searching optimization. This requires adequately varying the control variables, achieving freeze-drying of the product in a minimum amount of time but without altering quality. Such parameters should be prefixed according to the food type and within the characteristic limitations of the installation: the minimum pressure attained by the vacuum equipment and the heating power or temperature for a certain drying facility.

Generally, optimization criteria are applied to primary drying because it presents more alternatives in the control of process variables. Most of the water removal is also achieved in this period, around 90% in most cases.

A rigorous optimization of freeze-drying requires a continuous variation of the pressure and heating intensity while searching for an instantaneous maximum drying rate during the entire process. There are references of optimization strategies in the literature for freeze-drying of soluble coffee (Boss et al., 2004) using optimization routines like that of Schittkowski (1985), which are based on the solution of nonlinear programming problems. Boss et al. (2004) searched for an optimization on the basis of finding the best combination of plate temperature, pressure, and thickness of a product (not variable but constant during primary drying). Nevertheless, a higher reduction in the process time can be obtained with strategies based on varying the heating and pressure along the process (Lombraña, 1997).

The minimum processing time requires the maximum possible heat supply but avoiding the imposed constraints. These are defined through the melting point (T_m) and scorch point (T_{scor}), which refer to the wet–frozen and dried zones, respectively, which must be taken into account to assure a minimum quality level. Depending on which is the active constraint (T_m or T_{scor}), two well-defined periods can be distinguished: the mass transfer controlled period and the heat transfer controlled period.

In the *mass transfer controlled period*, apart from the transitory initial period in which the product temperature increases to the prefixed value corresponding to the operational policy imposed, the drying rate is adjusted during primary drying to avoid exceeding T_m. Thus, the heat supply must be controlled to equilibrate heat demand corresponding to the maximum mass flux for the cabinet pressure and condenser temperature of the installation. During this period, the cabinet pressure is at a rather high level with a slight tendency to decrease. In such circumstances, the pressure value is the result of the high drying rate and the pumping ability of the installation. Operation takes place as described earlier until all of the frozen moisture is removed at the end of primary drying.

Figure 8.4 shows how during the primary period plate temperature is decreased sharply in a short interval to avoid the temperature of the frozen product overtaking the T_m in the freeze-drying of skim milk in vials (Sheehan and Liapis, 1998). This is a consequence of the mass transfer conditions imposed in the cabinet. Better evacuation conditions or equivalent lower pressure lead to the possibility of increasing the

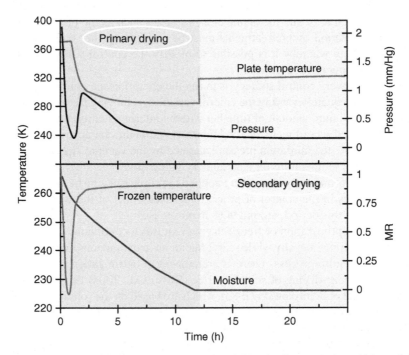

FIGURE 8.4 Typical profiles of the process variables in freeze-drying. (Adapted from Sheehan, P. and Liapis, A.I., *Biotechnol. Bioeng.*, 60, 712–728, 1998. With permission.)

plate temperature with the subsequent increase of the drying rate. Once primary drying is finished, constraint T_m will no longer be active and the heating temperature can be increased with the only limitation of the scorch point of the dried product.

The *heat transfer controlled period* corresponds basically to secondary drying. Once the moisture of the wet–frozen zone is removed, there is no restriction referring to T_m and the plate temperature can be increased immediately to higher values with the sole restriction imposed by T_{scor}. The plate temperature will be kept at a constant value that is high enough to accelerate drying but without overtaking the T_{scor} constraint up to the complete removal of bound water. The plate temperature in this period is usually very similar to the scorch temperature and varies depending on the heat transfer characteristic between the heating plate and product. As seen in Figure 8.4, during this period the plate temperature is kept constant at an intermediate value between the maximum and the minimum corresponding to primary drying while the pressure decreases slightly, reaching a minimum at the end of the process.

In summary, constraint T_m is usually active during the primary period and limits the drying rate within the characteristics of the installation. During secondary drying, it is T_{scor} that is the active constraint and will be directly responsible for the drying rate during the last part of freeze-drying.

8.5 MICROWAVE FREEZE-DRYING

Microwave heating is treated separately from the other methods because it represents a significant change in the method of running the freeze-drying operation. Moreover, despite its great possibilities that were known a long time ago (Cotson and Smith, 1963; Rey, 1975), it is currently a technology neither completely applied nor developed in many aspects. In a microwave field, the amount of heat applied to the product depends on the dielectric properties of the product and its variation as a function of temperature.

Another important favorable aspect for this kind of heating is that an inverse temperature gradient from the sublimation temperature to the free surface appears. This situation is not a problem when the product is heated through the dry layer, which happens in most cases, except when the vapor exit is limited as in freeze-drying of solutions in vials. In these cases rigorous temperature control at the frozen core is required to avoid melting. This inverse temperature gradient, if compared to a conventional thermal gradient, favors mass transfer because it progresses in the same direction as the moisture gradient.

The selective character of the product for absorbing dielectric energy is also favorable to the drying process. The absorbing capacity of dielectric energy is strongly affected by the moisture content. Thus, the loss factor (ϵ''), one of the most important properties that define the microwave absorption of materials, is extremely low in dried foods compared to ice or water. When employing microwave heating, a peripheral dried zone will stay at a lower temperature while the wet frozen zone will absorb the energy more intensively at the dried–frozen interface. Nevertheless, there are problems related to the great increase of the loss factor with temperature and the phase change from ice to liquid. Therefore, a tendency for the frozen zone to melt if vapor removal is not adequate (i.e., evacuated at the same speed) could arise.

Correct and controlled microwave heating will avoid thawing of the inner frozen zone and make possible a regular advancing of the dried–frozen interface following the geometry of the retreating nucleus up to complete removal of the frozen zone at the end of primary drying.

Good control of the frozen nucleus temperature is a very important aspect for success in running freeze-drying under microwave heating. The most recent research on microwave drying and its industrial applications has incorporated adequate techniques for correct monitoring of the temperature of the frozen nucleus and the product, within an electric field, based on optic fiber and infrared thermometry. Using such techniques, the product temperature can be followed precisely at the dried and frozen zones during freeze-drying. The collected information can be used to adequately regulate the process with a computerized control system (Rodríguez et al., 2005).

In the application of microwave heating it is necessary to explain some fundamental aspects of this heating method. Microwave heating employs high frequency alternate current from 300 MHz to 3 GHz. As a consequence of the polarization movement at such frequencies, in which polar molecules of dielectric compounds (i.e., water) attempt to align themselves with the rapidly alternating field, great frictional heat is developed that constitutes a heat source by itself that is not dependent on the thermal gradient. In this case, the product placed in the dielectric active zone would generate heat, which is distributed internally by conduction. The capacity of absorbing heat per unit product volume (P_{mw}) is obtained by the Lambert's law equation:

$$P_{mw} = 2\pi f \, \epsilon_0 \epsilon'' E^2 \qquad (8.32)$$

where f is the wave frequency (Hz), ϵ is the permittivity, ϵ_0 is the vacuum permittivity, and E is the electric field (V/m) of the microwave at a certain location. Electric field E is expressed with respect to that of a reference E_0 by

$$E = E_0 \exp(-x_L/d_p) \qquad (8.33)$$

where x_L is the distance with respect to the electric reference field E_0.

The penetration depth (d_p) and the loss tangent (tan δ) are the most significant properties responsible for microwave heating. Moreover, the loss factor (ϵ'') is obtained from the dielectric constant (ϵ') by

$$\epsilon'' = \tan \delta \cdot \epsilon' \qquad (8.34)$$

The penetration depth is defined as the depth below the surface of the material at which the electrical field strength is $1/e$ ($e = 2.3010$) that of free space (Ohlsson, 1989; Von Hippel, 1954):

$$d_p = \frac{\lambda_o}{2\pi} \sqrt{\frac{2}{\epsilon'[(1+\tan^2\delta)^{0.5} - 1]}} \qquad (8.35)$$

Dielectric properties ϵ' and ϵ'' constitute the real and imaginary components of the known complex permittivity that gives an overall view of the capacity to transform the electric energy of a microwave generator.

TABLE 8.2

Values of Characteristic Dielectric Properties for Raw Potato as a Function of Temperature

Temp. (°C)	ϵ''		tan δ		d_p (cm)	
	915 MHz	2.54 GHz	915 MHz	2.54 GHz	915 MHz	2.54 GHz
0	16	21	0.23	0.3	5.5	1.6
20	19	15	0.29	0.23	4.5	2.2
40	24	13	0.41	0.23	3.4	2.2
60	31	14	0.58	0.27	2.5	2
80	35	17	0.8	0.34	2	1.7
10	48	18	1.06	0.41	1.6	1.5

Source: From Mudgett, R.E., *Food Technol.*, 20, 84–98, 1986. With permission.

Permittivity ϵ or tan δ are two forms that together express dielectric properties ϵ' and ϵ''. Property ϵ' indicates the material's ability to capture energy, and property ϵ'' is a measure of its ability to dissipate electrical energy into heat. Table 8.2 shows the influence of temperature on the principal dielectric properties at two frequencies that are often used.

Frequency has a big impact on penetration depth, which is higher at 913 MHz than at 2.45 MHz. Nevertheless, when the temperature increases the difference of d_p with

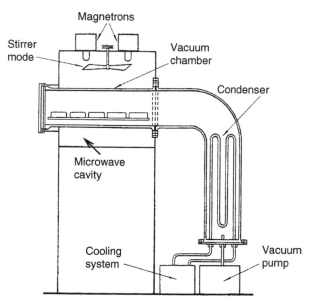

FIGURE 8.5   Scheme of a microwave freeze-dryer showing basic elements. (From Cotson, S. and Smith, D.B., *Freeze Drying of Foodstuffs*, Columbia Press, Manchester, U.K., 1963. With permission.)

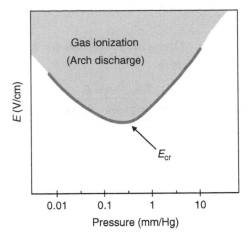

FIGURE 8.6 Combinations of electric field and pressure for zone arch discharge risk.

frequency becomes minimal. Although several frequencies have been used (Schmidt, 1958), the frequency of 2.45 GHz is almost the only one used in industrial and commercial prototypes. Figure 8.5 provides a scheme of a microwave freeze-dryer. A glass pipe is introduced into the vacuum chamber, which maintains the vessel at a pressure as low as 200 or 300 μm (26 or 39.5 Pa). A condensation system is included in the freeze-dryer.

The same equations used for mathematical freeze-drying can be used for the mathematical description of lyophilization with microwave heating. In addition, the heat generated per unit volume must be introduced through Eq. (8.32), taking into account the electric field for a certain coordinate given by Eq. (8.2). In it, x_δ would be the distance to the electric field of the outer adjoining volume.

A serious drawback can arise when applying a microwave under vacuum, that is, gaseous ionization that may derive the phenomena of arch discharge with unfavorable consequences of scorching at the surface of freeze-dried materials (Arsem and Ma, 1985). This can be avoided by reducing the power input or varying the total pressure. Figure 8.6 shows the boundary lines that, depending on the frequency, separate the upper or gas discharge zone from the lower one, where such a phenomenon can be avoided.

8.6 INDUSTRIAL FREEZE-DRYING

There are no data about the first freeze-drying installation. Meryman (1976) and Jennings (1999) refer to Sackell as the first person who used the term of freeze-drier in an experimental apparatus used to dehydrate samples for biomedical purposes. Nevertheless, there was no real freeze-drying installation until the 1930s, when biologist Greaves designed a freeze-drier to dehydrate blood serum. However, the big installations to create freeze-dried coffee, for example, appeared later with companies like Nestlé and Taster's Choice Coffee.

Later, several patents related to freeze-drying were attributable to the Canadian engineer James Mercer between 1966 and 1972 (Mercer and Rowell, 1972; Tagashi and Mercer, 1966). Such patents have been fundamental in the design of continuous freeze-drying installations for the largest food companies during the last 40 years.

More than 400 different types of freeze-dryers were designed for industrial purposes since the 1960s. There are references to a great variety of freeze-dried foods such as fruits, meats, and extracts. Other foods (e.g., lettuce and watermelon) are not recommended for freeze-drying because of their high water content and poor freeze-drying properties. Freeze-dryers can be classified into any of the following types.

8.6.1 Lab or Pilot Plant Freeze-Dryers

Lab or pilot plant freeze-dryers are employed in the food and pharmaceutical industries to research the possibilities for preserving different kinds of foods or medicines. In other cases, the target is tested on a pilot scale before full-scale production. Most of these freeze-dryers are compact and have facilities for refrigerating, heating, and vacuum in a sole cavity. Usually the condenser coils are incorporated in the cavity (Figure 8.7). The food is introduced inside the cabinet in carts whose shelves are connected to the corresponding flexible conduits to enable the passing of heating fluid. The production of a pilot freeze-dryer does not usually exceed 20 kg of frozen loaded product, and it can operate at a pressure of around 200 μm (26 Pa) or lower in the research prototypes.

8.6.2 Batch Freeze-Dryers

These are the most extended and often found prototypes for industrial purposes. There are two fundamental versions: trays and multicabinets. These cover product

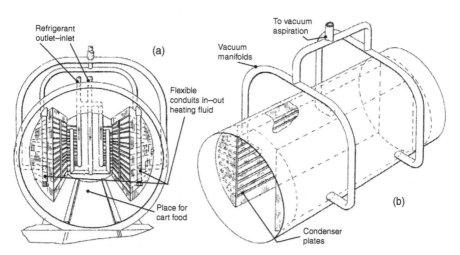

FIGURE 8.7 Scheme of a pilot plant freeze-dryer. (From Gutcho, M.H., *Freeze Drying Processes for the Food Industry*, Noyes Data, Park Ridge, NJ, 1977. With permission.)

abilities up to 2000 kg in a case where the tray surface is near 150 m^2. The heating device is usually incorporated into the trays, and its surface is a fundamental production parameter. Consequently, depending on the heating surface, the following prototypes can be considered: low production (from 7 to 27 m^2), medium production (up to 70 m^2), and high production (up to 140 m^2). This type of equipment can reach drying capacities of around 300 kg/h, but the duration of the drying cycles will depend on the maximum product temperature.

The design of a batch freeze-dryer has characteristics similar to the pilot one but with bigger dimensions according to the production capacity. It consists of the following essential elements:

- Vacuum cabinet (vessel) with internal radiant plates and condenser coils
- Aluminum product trays and rack for easy loading and unloading
- External services for panel heating, vacuum generation, thawing of the condenser, and refrigerant separation
- Control system
- Refrigerant system for the condenser, usually ammonia based

The machines having low and medium capacities condensate all the sublimated water without stopping the operation. However, those of greater production could incorporate some deicing devices. The function of such a special condenser is thawing and releasing the sublimated vapor during operation.

A batch plant is usually programmed to minimize drying time and to attain the highest possible production. In some cases, a single cabinet cannot reach certain production levels. Such a difficulty could be augmented in plants processing many special or complex products. Thus, when major production and flexibility for treating a great variety of foods are required, other batch prototypes are used. A good solution for these cases are multibatch freeze-dryers that consist of a number of batch cabinets programmed with staggered, overlapping drying cycles. Each cabinet can be loaded with product of similar or different characteristics and served by a centralized system for tray heating, refrigeration of condensers, and vacuum pumping. The process can be individually controlled for each cabinet from an external control panel. Numerous industrial freeze-dryers have been used for a long time in this way as multicabinet batch plants (Goldblith et al., 1975).

8.6.3 CONTINUOUS OR TUNNEL FREEZE-DRYERS

Continuous installations can be used to process any product, depending on its nature and where the economics of scale and operation require large production rates. This type of freeze-dryer commonly uses radiant heating combined with automatic loading, transport, and unloading of the product tray that passes through a tunnel cabinet. The frozen product is continuously loaded onto portable trays (trolleys), which are then placed into an automated control lock. The air lock is rapidly evacuated and the trays are immediately moved into the vacuum vessel and transported between the radiant heating panels so that sublimation occurs at the optimum rate for the product. The heating panels are controlled from the outside. Thus, the heat history of the product

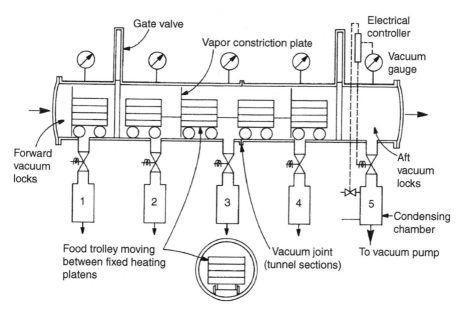

FIGURE 8.8   Continuous processing in a tunnel freeze-drier. (From Liapis, A.I. and Bruttini, R., in *Handbook of Industrial Drying*, Marcel Dekker, New York, 1995. With permission.)

can be adjusted to its specific needs during processing. A discharge air lock in the other end releases the dry product from the vacuum tunnel and equalizes the pressure vessel to the atmospheric one (Figure 8.8).

There are installations with two transport lines that can reach over an 800 m² tray area. No contact of the product load with the equipment elements takes place inside the vessel. Therefore, the freeze-dryer installation can operate uninterrupted for different types of foods without opening up the machine for cleaning. Consequently, adequate systems for tray filling and emptying must be provided in the installation. Manufacturing techniques based on the use of continuous equipment have been developed for instant coffee (Liapis and Bruttini, 1995).

8.6.4   SPECIAL FREEZE-DRYERS

Vacuum spray dryers are mentioned in this section because these facilities are especially used for processing beverages and liquid extracts such as coffee, tea infusions, and milk. Liquids are fed and sprayed at the top of a cylindrical vacuum tower at about 0.5 mmHg pressure. Drops of sprayed liquid will freeze into 150-μm diameter particles during falling because they lose about 15% of their original moisture by evaporation. Freeze-drying occurs in these frozen particles once collected in the vacuum cabinet at the bottom of the tower (Figure 8.9).

The installation operates in continuous mode, and it is especially suitable for high production of freeze-dried liquids. The most remarkable drawback in certain cases is that considerable losses of flavors and volatiles occur that are related to the flash evaporation during the first part of the process. This problem can present considerable

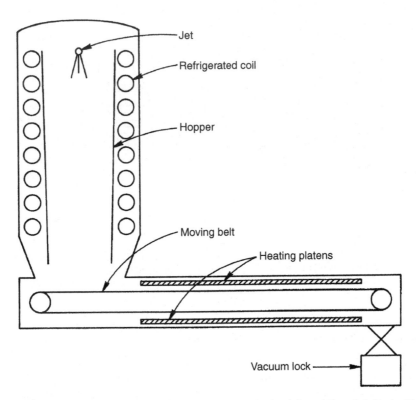

FIGURE 8.9 Spray freeze-drying installation. (From Liapis, A.I. and Bruttini, R., in *Handbook of Industrial Drying*, Marcel Dekker, New York, 1995. With permission.)

consequences in the lyophilization of coffee and other liquids characterized by their aromatic content (Liapis and Bruttini, 1995).

8.7 FREEZE-DRYING COSTS

The study of a freeze-drier requires the analysis of the following process steps:

Freezing: This provides the necessary initial condition for low temperature drying.

Vacuum: After freezing, the product is placed under a vacuum. This enables the frozen solvent in the product to sublimate without passing to a liquid phase.

Heat: Heat is applied to freeze the food to accelerate sublimation.

Condensation: A low-temperature condenser removes the vaporized solvent from the vacuum chamber by converting it back to a solid. This completes the separation process.

From these steps, either heating or vacuum condensation is the most significant operation in a freeze-dryer. According to their characteristics, it will depend on essentially the length, energy requirements, and subsequent running costs.

In estimating the efficiency and costs of a freeze-drying plant, the following circumstances must be thoroughly analyzed because of their great significance: operating reliability, loss of products, easy and good quality process control, and efficient vapor retention to facilitate vacuum operation. Depending on the freeze-dryer installation, such aspects could have more or less significance. Independent of the prototype, the condensation temperature is an essential factor that directly affects the two most important operations (heat transfer and vapor removal) involved in lyophilization.

The condensation temperature requires regulation to attain equilibrium between the heat and mass fluxes according to the possibilities of the heating supply and vacuum ability of the facility. If the latter is large enough to obtain high vacuum levels (or very low absolute pressure), the required condensation temperature does not need to be extremely low except in the case of high drying rates and heating supplies. As a rule, the higher the heat absorbed by the load is, the lower the temperature.

In any process an economic study requires knowledge of the interrelation of operating variables and costs. A good analysis of costs is necessary to select the most adequate variables for optimization purposes while searching for a compromise between efficiency and quality.

Fixed costs are mostly related to capital depreciation, loan charges, maintenance, insurance, and taxes. These could exceed running costs twofold (Mellor, 1978; Saravacos and Kostaropoulos, 2002). Because fixed costs are independent of production and type of foods, they are not usually taken into account in design analysis.

Variable costs in a food process are composed of storage, packaging, raw material, energy, and investment. Among these costs the last three (running costs) can change significantly, depending on the type of food and characteristic of installation. Moreover, their significance will be higher with the value of the food. In the freeze-drying of high value foods, the raw materials part does not usually exceed 66% of the variable costs whereas in low value foods this same cost fraction hardly reaches 20% (Ratti, 2001). In low value foods, the running costs can reach an important level (~70 or 80%) within the variable costs. Thus, the economic viability of freeze-drying requires an in-depth analysis of all factors to compete against other drying technologies, also taking into account the advantage of the excellent quality of freeze-dried products.

The following fundamentals factors should be considered to enhance the efficiency of freeze-drying in order to extend its applicability not only to high value foods but also to other less valuable foods:

1. Investments of the installation should not be excessive. Very expensive installations leading to important depreciation costs above 30% would not be recommendable.
2. The running costs are related to the operating characteristics of installation. All those installations, operating in continuous, with an important degree of automation, would be recommendable so that the costs related to operations of loading, cleaning and emptying will be minimal.
3. Use of efficient operational strategies that lead to reasonable lengths of drying cycles and, consequently, with subsequent decrement of energy costs for refrigeration of the condenser, heating, and vacuum pumping.

In conclusion, the quality of dehydrated foods obtained by freeze-drying is a very important factor to take into account in conjunction with the running costs. Thus, a moderate or even high cost could be justified when considering the advantages of the quality of freeze-dried products with respect to other low quality products obtained by low cost drying techniques.

NOMENCLATURE

VARIABLES

C_{ol}	constant dependent only upon structure of porous matrix for D'Arcy flow permeability
C_1	constant dependent upon structure of porous medium (m) for relative Knudsen permeability
C_2	constant dependent upon structure of porous medium (m) for free gas bulk diffusivity
C_p	specific heat capacity of food (J/kg K)
D_e	effective diffusivity of moisture during freeze-drying (m²/s)
D_g	diffusivity of water vapor in gas phase (m²/s)
D_m	diffusivity of liquid water through solid matrix (m²/s)
d_p	penetration depth (m)
D_w	free gas diffusivity of water in air (m²/s)
D_w^0	$D_w^0 = D_w P$ (N/s)
E	electric field (V/m) and characteristic dimension in spherical analogy (m)
E_0	reference electric field (V/m)
f_v	local fraction of vapor flux
$f_w(T)$	liquid–vapor equilibrium function (N/m²)
k	thermal conductivity of product (W/m K)
k_e	effective thermal conductivity of product (W/m K)
K_m	external mass transfer for water vapor (kg/N s)
K_{in}	Knudsen diffusivity for air (W/m K)
K_{mx}	Knudsen diffusivity for binary mixture (W/m K)
K_w	Knudsen diffusivity for water vapor (W/m K)
L	side of cube (m)
M	moisture content of product (kg water/kg dry basis)
M_b	bound water of product (kg water/kg dry basis)
M_{in}	molecular weight of air (kg/kmol)
M_w	molecular weight of water (kg/kmol)
MR	moisture ratio or remaining frozen moisture, Eq. (8.12)
N_w	total flux of water (kg/m² s)
p_w	partial pressure of water vapor (N/m²)
P	absolute pressure (N/m²)
P_{mw}	heat dissipated by microwave heating (J/m³)
q	heat flux (W/m²)
r_{eq}	equivalent radius of sphere (m)
R	ideal gas constant (J/kmol K)

S	external surface of particle (J/kmol K)
t	time (s)
T	temperature (K)
T_g	glass-transition temperature (K)
T_m	melting temperature (K)
T_{scor}	scorch temperature (K)
V	volume of particle (m^3)
x_L	distance from reference electric field (m)

GREEKS

δ	thickness of dried layer (m) and loss angle
ΔH_s	sublimation heat (J/kg)
ΔH_v	vaporization heat (J/kg)
ε	porosity of dried product
ϵ	permittivity
ϵ'	dielectric constant
ϵ''	loss factor
κ_1	bulk diffusivity constant (m^2/s)
κ_2	self diffusivity constant (m^2/s)
λ_o	wavelength in free space (m)
μ_{mx}	viscosity of vapor phase in porous of dried layer (kg/ms)
ρ	density of food (kg/m^3)
$\tan \delta$	loss tangent
υ	velocity of retreating ice front (m/s)
υ_m	velocity of retreating ice front in terms of mass (kg/s)
ψ	sphericity

SUPERSCRIPT

0	initial value

SUBSCRIPTS

av	average value
center	center of freeze-drying piece
cond	surface of condenser coils
surf	outer surface of freeze-drying piece
f	frozen nucleus
g	relative to water in gas phase
d	relative to dried layer
l	relative to liquid water
m	relative to porous matrix
w	relative to frozen water for ρ, C_p, and k
δ	relative to dry layer
1	primary drying
2	secondary drying

REFERENCES

Arsem, H.B. and Ma, Y.H., Aerosol formation during the microwave freeze dehydration of beef, *Biotechnol. Progr.*, 1, 104–110, 1985.

Boss, E.A., Filho, R.M., and Vasco de Toledo, E.C., Freeze drying process: Real time model and optimization, *Chem. Eng. Process.*, 43, 1475–1485, 2004.

Carter, G.F., *A Cultural Geography*, Holt, Rinehart and Winston, New York, 1964.

Cotson, S. and Smith, D.B., Eds., *Freeze Drying of Foodstuffs*, Columbia Press, Manchester, U.K., 1963.

Goldblith, S.A., Rey, L., and Rothmayr, W.W., *Freeze Drying and Advanced Technology*, Academic Press, London, 1975.

Gutcho, M.H., *Freeze Drying Processes for the Food Industry*, Noyes Data, Park Ridge, NJ, 1977.

Harper, J.C. and Tapel, A.L., Freeze-drying of foods products, *Adv. Food Res.*, 7, 171, 1957.

Jennings, T.A., *Lyophilization: Introduction and Basic Principles*, Interpharm Press, Denver, CO, 1999.

Kochs, M., Körber, C., Heschel, I., and Nunner, B., The influence of the freezing process on vapour transport during sublimation in vacuum freeze drying of macroscopic samples, *Int. J. Heat Mass Transfer*, 36, 1727–1738, 1993.

Kochs, M., Körber, C., Nunner, B., and Heschel, I., The influence of the freezing process on vapor transport during sublimation in vacuum freeze drying, *Int. J. Heat Mass Transfer*, 34, 2395–2408, 1991.

Kurz, W. and Fisher, D., *Fundamentals of Solidification*, 3rd ed., Trans Tech Publications, Aedermannsdorf, Switzerland, 1989.

Levinson, S.O. and Oppenheimer, F., Drying of frozen material by heat radiation, U.S. Patent 2,445,120, 1948.

Liapis, A.I. and Bruttini, R., Theory for the primary and secondary drying stages of the freeze drying of pharmaceutical crystalline and amorphous solutes: Comparison between experimental data and theory, *Sep. Technol.*, 4, 144–155, 1994.

Liapis, A.I. and Bruttini, R., Freeze drying, in *Handbook of Industrial Drying*, 2nd ed., Mujumdar, A.S., Ed., Marcel Dekker, New York, 1995, pp. 309–343.

Liapis, A.I. and Litchfield, R.J., Optimal control of a freeze dyer. I: Theoretical development and quasi steady state analysis, *Chem. Eng. Sci.*, 34, 975–981, 1979.

Lombraña, J.I., De Elvira, C., and Villarán, M.C., Analysis of operating strategies in the production of special foods in vials by freeze drying, *Int. J. Food Sci. Technol.*, 32, 107–115, 1997.

Lombraña, J.I. and Díaz, J.M., Solute redistribution during the freezing of aqueous solutions under instability conditions, *Cryo-Letters*, 8, 244–259, 1987.

Lombraña, J.I. and Villarán, M.C., Kinetic modeling of sublimation and vaporization in low-temperature dehydration processes, *J. Chem. Eng. Jpn.*, 26, 389–394, 1993.

Maroulis, Z.B. and Saravacos, G.D., *Food Process Design. Food Science and Technology Series*, Vol. 126, CRC Press, New York, 2003.

Mellor, J.D., *Fundamentals of Freeze Drying*, Academic Press, London, 1978.

Mercer, J.L. and Rowell, L.A., Continuous freeze drying system, U.S. Patent 3,648,379, 1972.

Meryman, H.T., Historical recollections of freeze drying, *Dev. Biol. Stand.*, 36, 29–32, 1976.

Millman, M.J., Laipis, A.I., and Marchelo, J.M., An analysis of the lyophilization process using a sorption–sublimation model and various operational policies, *AIChE J.*, 31, 1594–1604, 1985.

Mudgett, R.E., Microwave properties and heating characteristics of foods, *Food Technol.*, 40(6), 84–98, 1986.

Nastaj, J., A mathematical modelling of heat transfer in freeze drying, in *Drying '91*, Mujumdar, A.S. and Filková, I., Eds., Elsevier Science Publishers, Amsterdam, 1991, pp. 405–413.

Ohlsson, T., Dielectric properties and microwave processing, in *Food Properties and Computer-Aided Engineering of Food Processing Systems*, Singh, R.P. and Medina, A., Eds., Kluwer Academic, Amsterdam, 1989. pp. 7–92.

Ratti, C., Hot air and freeze-drying of high-value foods: A review, *J. Food Eng.*, 49, 4, 311–319, 2001.

Rey, L., *Advances in Freeze-Drying: Lyophilisation; Recherches et Applications Nouvelles*, Hermann, Paris, 1975.

Rodríguez, R., Lombrana, J.I., and Kamel, M., Kinetic and quality study of mushroom drying under microwave and vacuum, *Dry. Technol.*, 23, 2197–2213, 2005.

Sadikoglu, H. and Liapis, A.I., Mathematical modeling of the primary and secondary drying stages of bulk solution freeze drying in trays: Parameter estimation and model discrimination by comparison of theoretical results with experimental data, *Dry. Technol.*, 15, 791–810, 1997.

Saravacos, G.D. and Kostaropoulos, A.E., *Handbook of Food Processing Equipment*, Springer, New York, 2002.

Schelenz, G., Engel, J., and Rupprecht, H., Sublimation during lyophilization detected by temperature profile and x-ray technique, *Int. J. Pharm.*, 113, 133–140, 1995.

Schittkowski, K., NLPQL: A FORTRAN-subroutine for solving constrained non-linear programming problems, *Ann. Operat. Res.*, 5, 485–500, 1985.

Schmidt, W., Microwave generators coupled to a loaded cavity for dielectric heating of foodstuffs, *Electron. Appl.*, 19, 147–164, 1958.

Sheehan, P. and Liapis, A.I., Modeling of the primary and secondary drying stages of the freeze drying products in vials: Numerical results obtained from the solution of a dynamic and spatially multidimensional lyophilization model for different operational strategies, *Biotechnol. Bioeng.*, 60, 712–728, 1998.

Smithies, W.R., Design of freeze drying equipment for the dehydration of foodstuffs, *Food Technol.*, 13, 610–614, 1959.

Song, C.S., Nam, J.H., Kim, C.J., and Ro, S.T., Temperature distribution in a vial during free-drying of skim milk, *J. Food Eng.*, 67, 467–475, 2005.

Tagashi, H.J. and Mercer, J.L., Freeze dried product and method, U.S. Patent 3,293,766, 1966.

Von Hippel, A.R., *Dielectric and Waves*, MIT Press, Cambridge, MA, 1954.

9 Rehydration and Reconstitution of Foods

Alejandro Marabi and I. Sam Saguy

CONTENTS

9.1   INTRODUCTION

In the last decade, there has been a continuous rise in the demand for convenience foods, including dehydrated products, that is mainly attributable to modern lifestyles. This trend has been accompanied by a decrease in the ability, desire, or time to prepare food and an increase in financial means, leading consumers to choose foods that are readily available and convenient and require only minimal or no preparation before consumption (Tillotson, 2003). One of the methods utilized to produce convenience foods is dehydration of fresh products. The resulting dried foods present important advantages, such as low transportation costs, extended shelf-life stability, and ease of use. The latter represents an important consumer expectation, and it is often translated as the need to add only hot or cold water to the dry ingredients and mix or simmer for a short period of time in order to achieve rehydration and reconstitution of the product. High quality and flavor, nutritional value, and resemblance to the fresh product are typical important traits that need to be taken into account in order to meet consumer expectations and quality standards. Safety, nutritional, and sensory aspects of foods are often related to the rehydration process as well as to the severity of the drying process used. To achieve maximum quality and the desired hedonic and food characteristics, optimum processing conditions should be targeted, namely, those that result in minimal nutritional losses and optimal sensory traits (Rahman, 2005).

Numerous dried products are either consumed or further utilized for various industrial applications. Today, dried vegetables are used extensively in instant dry soups, ready-to-eat meals, snack foods, and seasoning blends, to name only a few. Vegetables that are commonly dehydrated and used for rehydration include carrots, potatoes, peppers, mushrooms, corn, onions, beets, parsley, horseradish, garlic, green beans, and celery. Fruits, legumes, cereal grains, fish, pasta, meat, and extruded breakfast cereals, among others, are also used in dried form or after rehydration, usually in water or milk.

During the rehydration process, the dry material, which is submerged in water or some other aqueous medium, undergoes several simultaneous physicochemical changes (e.g., in moisture and solids content, porosity, volume, temperature, gelatinization, and texture). The rehydration involves various processes running in parallel, including imbibition of liquid into the dried material, transport of the liquid through the porous network and diffusion through the solid matrix, swelling of certain domains in the solid matrix, and leaching of soluble solids into the external liquid. As a result of all of these processes, rehydration of dry foods is a very complex phenomenon that involves several different, simultaneously occurring physical mechanisms.

An important distinction needs to be made between two terms that are often used in the literature; although similar, their meaning should be clearly stated. The first term is *rehydration*, which relates to the amount of water a dry food is able to absorb in a given period of time. This term is usually associated with the kinetics and technological aspects of the process. The second term is *reconstitution*, which denotes the state of the product after the uptake of liquid and is related to the sensory aspects of the process; that is, how much the reconstituted sample resembles the original fresh product before it was dried. Note that in most cases (breakfast cereals being a typical exception, an al dente product may be another such exception) the aim is to shorten the rehydration and reconstitution time and minimize the effort required by the consumer to achieve a fast, convenient, and tasty meal.

Past studies of the rehydration process were carried out in conjunction with, and by analogy to, studies of drying kinetics. However, in the last decade, it has been frequently demonstrated that the rehydration process is not the reverse of dehydration. Consequently, the need to better understand different aspects of the rehydration of dried foods has been reemphasized. This trend is clearly evidenced by a significant increase in the number of publications focusing on rehydration (Figure 9.1).

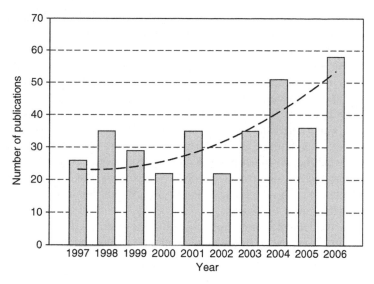

FIGURE 9.1   Number of studies published in books, journals, and conference proceedings during the last 10 years according to the FASTA database in June 2007 (keyword: rehydration).

The objective of this chapter is to highlight the most relevant issues related to the rehydration and reconstitution of dried foods. In addition, detailed information is provided on measurement techniques, microstructural properties of dried foods and changes during rehydration, effects of various rehydration conditions on the kinetics and sensory aspects of the process, and mathematical modeling of the rehydration kinetics. A special section also deals with recent advances in the field and specific recommendations for future research. However, the wetting, dissolution, and rehydration of powders, as well as other related topics, are not covered here because they require special attention and are beyond the scope of this chapter. Hence, our treatment focuses on the rehydration and reconstitution of nondissolving solid foods in liquid media.

9.2   REHYDRATION OF FOODS

The rehydration of dried foods is a fundamental unit operation in the food industry. It is also performed on the consumer scale, for which a fast and easy process is often of the utmost importance. The quality of rehydrated and reconstituted products is affected by the drying conditions and rehydration processes utilized, ultimately influencing consumer acceptance. During the drying process, physicochemical changes, including textural and structural modifications, migration of solutes, and loss of volatiles and nutrients, occur in an irreversible manner and have an impact on the quality of the final product. Therefore, the drying process needs to be understood and controlled in order to create a dried product with optimal nutritional, sensorial, and rehydration characteristics. Other chapters in this book provide detailed descriptions of the different drying methods utilized in the food industry and their effects on the final product, and they are therefore not detailed here. Note, however, that in order to obtain satisfactory rehydration and reconstitution properties, the drying and rehydration conditions need to be mutually optimized. The rate and extent of rehydration may be used as an indicator of the quality of the dried product: foods that are dried under optimum conditions suffer less damage and rehydrate more rapidly and completely than poorly dried foods.

Food rehydration is a topic of extensive interest, particularly in the case of legumes, fruits and vegetables, cereal grains, dried fruits, extruded cereal products, pasta, breakfast cereals, fish, and so forth. This interest is reflected in the steady increase in the number of studies published in the last few years (Figure 9.1) that seek to better understand and improve the process. Nevertheless, the number of articles is quite small relative to those relating to the drying of food products or to other unit operations, such as deep-fat frying.

The most fundamental aspect of rehydrating a dried food is related to water absorption, and it is expressed by the ratio commonly termed the rehydration capacity (RC; Lewicki, 1998b):

$$RC = \frac{X_t}{X_0} \tag{9.1}$$

where X_t is the weight after a given rehydration time and X_0 is the initial weight of the dry sample. Typical values for the rehydration capacity range between ~2 and 7.

The other criterion utilized to describe reconstitution (RE) is defined as

$$RE = \frac{X_t}{X_i} \tag{9.2}$$

where X_i is the initial weight of the sample before drying.

Typical values are often lower than 1, mainly because of the changes occurring during the drying process. For samples dried under optimal conditions, a value very close to 1 can be achieved in a relatively short time if the rehydration conditions are also optimized.

As might be expected (and discussed further on), during rehydration, solids can leach from the food product to the medium. To overcome this problem, the concept of nondissolvable solids (NDS), defined as the amount of solids after all dissolvable solids have leached out, was proposed (Marabi, Dilak et al., 2004). The amount of NDS can be determined when adequate time has passed, allowing the establishment of quasiequilibrium: at this stage, the solids remaining in the sample are practically nondissolvable. For realistic situations, the NDS should be determined after the shortest time possible to ensure sample integrity and avoid its disintegration (Marabi, Dilak et al., 2004). Thus, the rehydration ratio (RR) can be defined as

$$RR = \frac{W_t - W_0}{W_\infty - W_0} \tag{9.3}$$

where W_t, W_0, and W_∞ are the water contents at time t, initial time t_0, and infinite time t_∞, respectively.

Typical RR values for short rehydration times (e.g., 3 min) range from 0.2 to 0.6 for air-dried (AD) and freeze-dried (FD) carrots, respectively. Values of ~0.3 were observed for FD peas, whereas FD corn, potatoes, and beets yielded RR values close to 0.8. The equilibrium RR value is usually independent of the drying method and asymptotically reaches a value close to 1.

The principal factors affecting the rehydration kinetics and reconstitution may be divided into two main groups: those related to the dry particulate solid, and those related to the external rehydration medium. In the first group, the particle geometry, water content, porosity, tortuosity, density, and physical state of the sample can play a paramount role, affecting the speed and quality of the reconstitution. In the second group, the medium temperature, mixing regime, density, viscosity, composition, and presence of insoluble matter can play a determining role.

The most common approach to describing the mechanism of liquid uptake during the rehydration of dry foods is diffusion. By making several assumptions and solving Fick's laws, the value of the effective diffusivity (D_{eff}) can be derived from experimental data. However, in one study the derived values included, in addition to transport due to diffusion, contributions of all other mechanisms (e.g., bulk flow, capillary flow) that play a significant role during the process and take place simultaneously (see Waananen et al., 1993, for the possible mechanisms occurring during drying). One of these mechanisms may prevail in each specific case, depending mainly on the microstructure of the product and the physicochemical properties of the liquid (Saguy et al., 2005b).

Typical D_{eff} values for moisture in foods range from 10^{-8} to 10^{-12} m²/s, generally being closer to 10^{-10} m²/s (Doulia et al., 2000; Maroulis et al., 2001), and they are influenced by temperature (Bello et al., 2004; Cunningham et al., 2007; Garcia-Pascual et al., 2005, 2006; Kaymak-Ertekin, 2002; Sanjuan, Carcel et al., 2001; Sanjuan, Clemente et al., 2001), water content (Maroulis et al., 2001; Vagenas and Karathanos, 1991), pressure (Karathanos et al., 1991), physical properties (Marousis et al., 1991; Waananen and Okos, 1996), and the dried food's structure (Ruiz-Cabrera et al., 1997). In turn, the latter is significantly affected by the pretreatments and drying processes utilized in its production. It is generally accepted that the main resistance to liquid ingress is usually encountered in the dried food's matrix. However, increased viscosity of the liquid medium and the presence of nondissolved and/or dispersed particles, as well as fat, may also have detrimental effects on rehydration rates.

Rehydration is a complex phenomenon that requires careful consideration. Previous data have not been consistent and include various methods to describe both rehydration and reconstitution. Careful consideration is needed when data from various sources are compared.

9.3 REHYDRATION MEASUREMENTS AND INDICES UTILIZED

As mentioned earlier, in the last few years the study of the rehydration process of dried foods has attracted a great deal of attention. When comparing published works on rehydration, it is common to find inconsistencies in the measuring procedures and calculations of the rehydration indices for dried foods. The most significant sources of variability in the experimental procedures when measuring the rehydration of dried foods reside in the ratio between the dry sample and the liquid, ranging from

1:5 to 1:50; the temperature of the rehydrating media, ranging from 5 to 100°C; the time of the experiment, usually between a few minutes and 24 h; and the hydrodynamic conditions, which vary from stagnant liquid to high-mixing regimes. Moreover, different studies examine either one or several particles, which are returned, or not, to the rehydration liquid after weighing, according to the method utilized. The samples are sometimes blotted on paper, drained in a sieve, or centrifuged to remove excess liquid on their surface (Lewicki, 1998b). These differences make comparisons between different studies difficult. Consequently, use of the calculated rehydration indices in a modeling attempt might lead to erroneous results.

Compilations of data for drying methods and weight ratios, temperature, and time of immersion have shown that the rehydration process can be expressed by a variety of different indices (e.g., gH_2O/g dried product, gH_2O/g dry solids, g rehydrated product/g dried product, gH_2O/g rehydrated product, gH_2O/g fresh product, etc.; Oliveira and Ilincanu, 1999). Another common reason for discrepancies in the reported values is the leaching of soluble solids, which takes place while the liquid is being absorbed into the dried food sample. A number of studies (Ilincanu et al., 1995; Lewicki, 1998a; Lucas et al., 2007; Machado et al., 1997, 1998; Marabi, Dilak et al., 2004; Oliveira and Ilincanu, 1999; Vidal et al., 1992) have demonstrated the magnitude of this phenomenon. Thus, it is clear that it should be considered when reporting values based on the weights of the sample before, during, or after rehydration.

These examples clearly demonstrate the need to standardize the way in which the data are collected and presented. It is therefore recommended that information be provided on a dry-weight basis, taking into consideration the possibility of solids leaching out from the original dried product. This is further discussed in Section 9.10 [see also Eq. (9.3)]. Alternatively, the solids leached during the rehydration process have not been considered, and the consequent error in the mass of the rehydrated sample, which is not entirely due to water uptake, should be addressed.

9.4 MATHEMATICAL MODELING

As pointed out previously, the rehydration and reconstitution of dried foods is a complex process. Experimental data for the rehydration of a large number of foods under different conditions exist in the literature and can be utilized to model the process. The use of different mathematical models facilitates the understanding of some of the process characteristics, provides insight into the governing mechanisms taking place, and could ultimately lead to improved processing conditions and end-user products. Liquid uptake is typically modeled by applying an empirical and/or quasimechanistic approach based on a physical model, which can be exemplified in the application of the Weibull distribution function and Fick's second law of diffusion, respectively. The latter is mathematically formulated from the process's basic physical principles. It requires an in-depth understanding of the process, together with a careful evaluation of the physical properties of the food matrix. The efficiency of these models is typically limited to approximations because of the complex process interrelationships that are very difficult to quantify. In contrast, the development of empirical models requires considerably less effort. The basic idea of empirical-process characterization is to consider the process as a "black box," varying specific input setup parameters, measuring output quantities, and deriving the adequate correlations. In this way, statistically designed experiments offer a valid basis for developing an empirical model, which establishes a formal relation between the input and output quantities. However, empirical models are limited by their boundary conditions and are nontransferable (Saguy et al., 2005b).

In recent years, the field of food science has been experiencing an encouraging transition from an empirically based approach to physically based models that basically consider food as a porous medium (see Sections 9.12 and 9.13). Nevertheless, most rehydration studies reported in the literature still include physically based and empirical models to describe the mechanisms of liquid uptake, solids leaching, and the kinetics of the processes. Fick's laws of diffusion and their derived equations account for the vast majority of the models utilized in food science, as can be observed from recent publications (e.g., Ade-Omowaye et al., 2002; Bello et al., 2004; Bilbao-Sainz et al., 2005; Cunningham et al., 2007; Debnath et al., 2004; Falade and Abbo, 2007; Garcia-Pascual et al., 2005, 2006; Kaymak-Ertekin, 2002; Nayak et al., 2006; Sanjuan, Carcel et al., 2001; Sanjuan et al., 1999). Diffusivity values for various foods undergoing several processes can be found in different compilations (Doulia et al., 2000; Saravacos and Maroulis, 2001). However, recent data indicate that the actual mass-transport process may not be restricted to diffusion and could include several other mechanisms. Other theories, based mainly on the Washburn equation, are being used to model liquid movement into dried foods (Aguilera et al., 2004; Moreira and Barrufet, 1998; Saguy et al., 2005a; Weerts et al., 2003a,b), which are described in more detail in Section 9.12. The close similarities between dried foods and porous media, and the fact that other scientific fields (e.g., soil and petroleum sciences, chemical engineering) utilize well-established physically based models for describing transport processes, are leading food scientists to undertake a multidisciplinary approach to further develop their understanding of the transport phenomena occurring in different processes.

Application of the diffusion model and a few other empirical models used for studying the rehydration process is reviewed in the following subsections.

9.4.1 DIFFUSION MODEL

The diffusion model is a combination of physical and empirical approaches, founded on Fick's first and second laws:

$$J_x = -D_{eff} \frac{dW}{dx} \tag{9.4}$$

$$\frac{\partial W}{\partial t} = D_{eff} \frac{\partial^2 W}{\partial x^2} \tag{9.5}$$

where J_x is the flux ($kg\,H_2O/m^2\,s$), W is the moisture content ($kg\,H_2O/m^3$), x is the spatial coordinate (m), t is the time (s), and D_{eff} is the effective diffusion coefficient of water (m^2/s).

The D_{eff} is derived from Fick's second law and is an apparent value that comprises all of the factors involved in the process. This approach is quite common and allows the straightforward derivation of D_{eff}. However, it is well documented that mechanisms of mass transfer other than molecular or Fickian diffusion may also occur. One of the main drawbacks of utilizing Fick's laws is that most assumptions made for solving the partial differential equation [Eqs. (9.4) and (9.5)] are at best only partially valid. As a consequence, valuable insight is lost or cannot be retrieved from the experimental data. Some of the common assumptions and simplifications often made for solving Fick's second law include the following:

- The process is controlled by diffusion and no other transport mechanisms are active.
- Fick's law holds for the diffusion of unreacted water.
- The diffusion coefficients are independent of moisture concentration.
- The matrix is uniform and isotropic.
- The initial moisture content is uniform in the sample and the surface attains saturation moisture instantly.
- The boundary-layer resistance is much smaller than the internal resistance.
- Mass-diffusivity and reaction-rate constants are independent of concentration.
- Shrinkage and swelling during the process are negligible.
- The sample is considered an approximation of a sphere, cylinder, or slab and is representative of the actual geometry.
- Heat transfer equations are ignored.

Despite the fact that these assumptions are only partially valid, at best, the Fickian approach is frequently utilized to model a wide range of transport phenomena in foods, including drying (Doymaz, 2007; Falade and Abbo, 2007; Jambrak et al., 2007; Kaymak-Ertekin, 2002), rehydration (Cunningham et al., 2007; Falade and Abbo, 2007; Garcia-Pascual et al., 2006; Meda and Ratti, 2005; Nayak et al., 2006; Sanjuan, Clemente et al., 2001), and leaching of soluble solids (Debnath et al., 2004), among others. These phenomena include both the transport of liquids in solids and the movement of solutes in the liquid phase. Modeling of the rehydration process by means of Fick's laws of diffusion provides some insight into the rehydration process and allows a comparison of how different process parameters affect the overall transport phenomena. The utilization of Fick's laws will probably continue in the future, because it is a solidly based method. However, the pitfalls should be carefully considered when utilizing this approach in order to obtain a more accurate description of the underlying processes.

9.4.2 EMPIRICAL AND SEMIEMPIRICAL MODELS

Several other empirical and semiempirical mathematical models are utilized when studying the rehydration process. In particular, there are six models that are often used, the equations for which are found in Table 9.1:

1. *Exponential model* [Eq. (9.6); Misra and Brooker, 1980]
2. *Peleg's model* [Eq. (9.7); Abu-Ghannam and McKenna, 1997; Bilbao-Sainz et al., 2005; Cunningham et al., 2007; Garcia-Pascual et al., 2006;

TABLE 9.1

Empirical Models Frequently Used in Curve Fitting of Rehydration Data

Model	Formula	Eq.
Exponential model	$\dfrac{M_t - M_e}{M_0 - M_e} = \exp(-P_1 t^{P_2})$	(9.6)
Peleg's model	$M_t = M_0 + \dfrac{t}{(P_3 + P_4 \times t)}$	(9.7)
First-order kinetics	$\dfrac{M_t - M_e}{M_0 - M_e} = \exp(-P_5 t)$	(9.8)
Becker's model	$M_t - M_0 = \Delta M_0 + \dfrac{2}{\sqrt{\pi}}(M_s - M_0)\left(\dfrac{S}{V}\right)\sqrt{D_{\text{eff}}}\sqrt{t}$	(9.9)
Weibull distribution function	$\dfrac{M_t}{M_e} = 1 - \exp\left[-\left(\dfrac{t}{\alpha}\right)^\beta\right]$	(9.10)
Normalized Weibull distribution function	$\dfrac{W_t - W_0}{W_e - W_0} = 1 - \exp\left[-\left(\dfrac{t \times D_{\text{eff}} \times R_g}{L^2}\right)^\beta\right]$	(9.11)

Note: See Nomenclature for terms.

Source: Reprinted from Saguy, I.S. et al., *Trends Food Sci. Technol.* 16, 495–506, 2005b. Copyright 2005 Elsevier. With permission.

Giraldo et al., 2006; Hung et al., 1993; Machado et al., 1998; Maskan, 2002; Peleg, 1988; Planinic et al., 2005; Ruiz-Diaz et al., 2003; Sacchetti et al., 2003; Sopade et al., 1992; Turhan et al., 2002]

3. *First-order kinetics* [Eq. (9.8); Chhinnan, 1984; Gowen et al., 2007; Krokida and Marinos-Kouris, 2003; Krokida and Philippopoulos, 2005; Machado et al., 1998; Maskan, 2001; Pappas et al., 1999]

4. *Becker's model* [Eq. (9.9); Becker, 1960; Fan et al., 1963; Lu et al., 1994]

5. *Weibull distribution function* [Eq. (9.10); Cunha, Oliveira, and Ilincanu, 1998; Cunha, Oliveira, and Oliveira, 1998; Cunningham et al., 2007; Garcia-Pascual et al., 2006; Machado et al., 1997, 1998, 1999; Ruiz-Diaz et al., 2003; Sacchetti et al., 2003]

6. *Normalized Weibull distribution function* [Eq. (9.11); Marabi, Dilak et al., 2004; Marabi, Jacobson et al., 2004; Marabi and Saguy, 2004, 2005; Marabi et al., 2003]

Among the empirical models, the Weibull distribution function is frequently used and has recently been improved to describe the rehydration of dried foods (see Section 9.5).

All of these empirical models provide an excellent basis for curve fitting and allow process representation as a function of the physical properties and conditions of both the rehydration medium and the rehydrating food particles. However, they offer rather limited insight into the fundamental principles involved, hindering an understanding of the transport mechanisms that are actually taking place.

9.4.3 CAPILLARY FLOW IN POROUS MEDIA

In an attempt to study the rehydration process further without having to rely on the common postulation that it is governed by Fickian diffusion, numerous investigations have been conducted on the utilization of mathematical models based on capillary flow in porous media to describe the data obtained experimentally. These models are derived from the well-known Washburn equation (also referred to as the Lucas–Washburn or Washburn–Rideal equation). Rehydration studies of dried plant tissues reveal a very complex phenomenon involving different transport mechanisms, including molecular diffusion, convection, hydraulic flow, and capillary flow (Saravacos and Maroulis, 2001). Several research groups have also suggested that the ingress of liquid into a dried food sample occurs by various mechanisms occurring in tandem (Chiralt and Fito, 2003; Hallstrom, 1990; Marabi and Saguy, 2005; Marabi et al., 2003; Martinez-Navarrete and Chiralt, 1999; Oliveira and Ilincanu, 1999). The contribution of mechanisms involving mass flux attributable to a temperature gradient (Soret effect) is considered insignificant and is therefore often disregarded (Datta, 2007a). Several studies utilizing the Washburn equation to represent the movement of liquids into porous food matrices are reviewed here. Some of the most common models that are based on the Washburn equation and that have been applied in food science are listed in Table 9.2 along with their equations.

A model of capillary rise in porous media was used to explain the linear relationship between oil uptake and contact angle for deep-fat fried foods (Pinthus and Saguy, 1994). A similar approach was used later (Moreira and Barrufet, 1998) to analyze the mechanism of oil absorption of tortilla chips during cooling after frying. The assumption was that a "microscopically uniform" porous medium is formed,

TABLE 9.2
Washburn-Based Models Frequently Utilized in Rehydration of Foods

Model	Formula	Eq.
Horizontal capillary (short time)	$h = \sqrt{\dfrac{r\gamma\cos(\delta)}{2\tau^2\mu}}t$	(9.12)
Equilibrium liquid rise (long time)	$y(t) = \dfrac{k_1}{k_2}\left[1 - \exp\left(-\dfrac{k_2 y(t)}{k_1}\right)\exp\left(-\dfrac{k_2^2 t}{k_1}\right)\right]$ $k_1 = \dfrac{r\gamma\cos(\delta)}{4\mu} \quad k_2 = \dfrac{r^2 g\rho}{8\mu}$	(9.13)
Capillary imbibition (initial slope of uptake vs. $\sqrt{t}$)	$C = \phi\rho\sqrt{\dfrac{r\gamma\cos(\delta)}{2\mu}}$	(9.14)
Porous media (mixed-form unsaturated capillary flow)	$\dfrac{\partial\psi}{\partial t} = \nabla(K\nabla h) - \nabla[K\sin(\varphi)]$	(9.15)

Note: See Nomenclature for terms.
Source: Reprinted from Saguy, I.S. et al., *Trends Food Sci. Technol.*, 16, 495–506, 2005b. Copyright 2005 Elsevier. With permission.

that is, a simplification was made by assuming that an average pore-size distribution existed. More recently, the rates of sunflower oil penetration into beds of chocolate crumbs were measured from the oil front observed in a horizontal position. In contrast to the two previous studies, the time dependency of the process was studied by means of the Washburn model, which agreed with the measured data. It was concluded that the penetration rates were solely governed by the Laplace capillary pressure (Carbonell et al., 2004). Another study showed that fat migration in chocolate conforms to the general shape of the capillary-rise curve predicted by the Washburn equation (Aguilera et al., 2004). For the imbibition of water into dried porous foods, it was also shown that the process followed the general Washburn equation (Lee et al., 2005; Saguy et al., 2005a). However, discrepancies related to the utilization of a single "effective" cylindrical capillary and a constant contact angle were also reported (Saguy et al., 2005a). An additional factor that may be responsible for the inaccuracies encountered when comparing experimental data with the Washburn equation may be related to the tortuosity of the pores within the food sample. The pore network is often regarded as a bundle of cylindrical and straight capillaries with a determined effective radius. Thus, the Washburn equation may be utilized in its original form (Aguilera et al., 2004) or otherwise corrected with a tortuosity factor [e.g., Eq. (9.12) in Table 9.2; Carbonell et al., 2004].

A few other studies arrived at similar conclusions with regard to the mechanisms of water movement during various common processes in the food industry. For instance, it was proposed that during vacuum osmotic drying (VOD) the water transfer results from a combination of traditional Fickian diffusion and vacuum capillary flow, especially during the first few hours. Shi and Fito-Maupoey (1994) proposed that the capillary flow function was closely related to the porosity of the apricot fruit being tested. Hills et al. (1996) used a radial nuclear magnetic resonance (NMR) microimaging technique to study rehydration of pasta, and they concluded that the rehydration was a non-Fickian process. Transport mechanisms other than diffusion were reported during OD of apples. The proposed alternatives included capillary penetration or another fast transport mechanism occurring near the interface of the samples (Salvatori et al., 1999).

More recently, and as a consequence of the these findings, a further step was undertaken in which various groups started applying the theory of capillary imbibition for modeling the rehydration of foods. A capillary-flow approach was utilized (Weerts et al., 2003a,b) to model the temperature and anisotropy effects during the rehydration of tea leaves according to Eq. (9.15) in Table 9.2. The predicted values agreed well with the experimental data derived from NMR measurements, leading to the conclusion that the physically based constitutive relationships of water activity and hydraulic conductivity could be utilized to overcome the simplification of modeling water transport as a process governed by Fick's laws. The approach could be extended to include gravity and osmotic pressure effects and also coupled with heat and solute transport in porous media for modeling heat, water, and chemical transport in general hydration and drying operations of porous food materials.

Singh et al. (2004) noted in their study that one of the salient features of fluid transport through biological systems is the complex flow path presented by the biopolymeric matrix, thus expanding the approach. Similarly, it was recently proposed that use of an "effective" single cylindrical capillary radius is the simplest possible model to describe capillary penetration into a porous medium and that this model may be insufficient in many cases (Saguy et al., 2005a), especially when a significant distribution of pore sizes exists (Marmur and Cohen, 1997), as previously shown for dried foods (Karathanos et al., 1996). More specifically, and depending upon the type of food and its processing history, it may contain pores ranging in radius from 0.1 to 300 μm, but pores in the 10 to 300 μm range are the most common (Bell and Labuza, 2000).

Note that in the field of chemical engineering the capillary model has also been widely studied and is continuously being improved. One such work by Marmur and Cohen (1997) utilized the kinetics of liquid penetration to characterize various porous media. They showed that it is possible to independently evaluate the effects of the pore radius (r) and cosine contact angle [$\cos(\delta)$], thus offering a possible solution for the cited discrepancies. This was demonstrated for the case of a single vertical cylindrical capillary and for the case of a parallel assembly of vertical cylindrical capillaries (i.e., when a significant distribution of pore sizes exists). The developed model might be extended to evaluate the capillary mechanism taking place during the rehydration of food samples, leading to more accurate and representative results. Studies of vertical liquid penetration were recently presented based on Eq. (9.14) (Saguy et al., 2005a) and Eq. (9.13) (Lee et al., 2005), which are provided in Table 9.2. However, in neither case was the contact angle value appropriately resolved.

Although numerous models have been proposed and frequently applied, additional improvements are required before modeling of the process based on the real mechanisms that are occurring becomes possible. Nevertheless, recent studies toward more fundamental physically based approaches have taken on a central role. Hence, significant progress is anticipated.

9.5   UTILIZING WEIBULL DISTRIBUTION IN MODELING REHYDRATION PROCESS

The Weibull distribution function (Rosin and Rammler, 1933; Weibull, 1939) is a model that is quite useful to fit experimental data describing the rehydration of dried food products. Basically, it describes the process as a sequence of probabilistic events. This empirical approach followed a previous model suggested by Rosin and Rammler (1933) and used for the description of particle-size distributions. Typically, the Weibull distribution function is described by two parameters: the scale parameter (α), which is related to the reciprocal of the process rate constant, and the shape parameter (β). The scale parameter defines the rate and represents the time needed to accomplish approximately 63% of the process. When β is 1, the Weibull distribution function reduces to first-order kinetics. Many attempts have been made to utilize the shape parameter of the Weibull distribution function as an indicator of the mechanism (e.g., diffusion, external resistance, or matrix relaxation) of liquid uptake during rehydration because different values of β lead to different curves and as such can describe the mechanisms taking place (Cunha, Oliveira, and Ilincanu, 1998; Marabi et al., 2003). Utilization of the Weibull distribution showed excellent fit in describing the rehydration of a variety of dried foods, and it adequately described rehydration processes controlled by different mechanisms (Ilincanu et al., 1995; Machado et al., 1997, 1998, 1999). For a sphere geometry, the diffusion process was clearly identified when the β value is 0.6, whereas processes controlled by external resistance or relaxation phenomena could not be differentiated from the values of β that were close to 1 (Cunha, Oliveira, and Ilincanu, 1998).

The Weibull two-parameter distribution function is described as (Cunha, Oliveira, and Ilincanu, 1998)

$$\frac{M_t - M_0}{M_\infty - M_0} = 1 - \exp\left[-\left(\frac{t}{\alpha}\right)^\beta\right] \tag{9.16}$$

where M_t, M_0, and M_∞ are the moisture content at time t, time 0, and infinite time, respectively.

The left-hand side of Eq. (9.16) is also equivalent to the defined rehydration ratio. Note that the use of NDS is recommended.

Some studies that have applied the Weibull distribution (Ilincanu et al., 1995; Oliveira and Ilincanu, 1999) did not consider factors such as the sample thickness, which is a paramount variable in any practical rehydration processes. Other factors that could affect the derived Weibull parameters also need to be considered, including the effect of water uptake at equilibrium, the experimental error of the acquired data, the geometry of the sample, and so forth. To account for these effects, a normalized Weibull distribution was proposed by Marabi et al. (2003). The Weibull distribution [Eq. (9.16)] was modified and the rate parameter was normalized with a characteristic dimension for the thickness:

$$\frac{M_t - M_0}{M_\infty - M_0} = 1 - \exp\left[-\left(\frac{t}{\alpha'}\right)^\beta\right] \tag{9.17}$$

and

$$\alpha' = \frac{L^2}{D_{\text{calc}}} \tag{9.18}$$

where α' is the normalized scale parameter (s), L is the half-slab thickness or radius for spherical and cylindrical samples (m), and D_{calc} is the calculated diffusion coefficient of water (m²/s).

The D_{eff} could then be derived from

$$D_{\text{eff}} = R_g^{-1} \times D_{\text{calc}} \tag{9.19}$$

where R_g is the geometry factor (dimensionless).

This model was utilized to show that for rehydrating AD carrots the derived β value closely agreed with the value representing diffusion (~0.8 for a planar geometry). In contrast, the values of β obtained for FD carrots did not match any of the modeled values corresponding with these mechanisms. Thus, it was proposed that liquid uptake may also occur via capillary flow, which is due to the very high porosity of the samples (Marabi et al., 2003).

An extensive summary of the different derived values for the parameters of the Weibull model are presented in work by Saguy and Marabi (2005). They found that the shape parameter describing the rehydration process of different dried foods ranges from ~0.4 to 0.8, depending mainly on the drying method utilized and the type of food (vegetable, fruit, cereal, etc.) analyzed. The rate parameter also ranged from a few seconds to several hours. The excellent fit of the normalized Weibull distribution function to experimental data was observed in many different cases (e.g., Marabi et al., 2003; Marabi, Jacobson et al., 2004; Marabi and Saguy, 2004, 2005).

In recent years, an increasing amount of work has made use of the Weibull distribution function to model the rehydration kinetics of foods (Cunningham et al., 2007; Garcia-Pascual et al., 2005, 2006). This is mainly because of its ease of application, the flexibility of the model, and the derived excellent fit. However, special caution is needed because this is a fitting model and it cannot replace the physically based model required for a better understanding of the mechanisms involved.

9.6 PRETREATMENTS AND THEIR EFFECTS ON FOOD STRUCTURE

The physical properties and microstructure of dried food products play a key role in determining the kinetics of the rehydration process and should be carefully considered during the predrying and drying steps. Because these might be the only processes used to produce the dry foods, they determine their final quality by affecting the microstructure of the material and consequently, the overall transport phenomena taking place. In general, two contrasting needs can be described in relation to the rehydration process: (1) fast water or medium uptake and reconstitution (typically encountered in instant products such as soups with vegetables and noodles) to a state as similar as possible to that of the predrying food conditions, or (2) slow water or medium uptake and the maintenance of attributes such as crispiness for longer periods, as in breakfast cereals and the like. Some examples of treatments applied before drying that could potentially affect the final structure of the dried product include blanching, osmotic dehydration (OD), utilization of sulfites, high-intensity electrical field pulses (HELP/PEF), high pressure (HP), vacuum impregnation (VI), and application of ultrasound and γ-irradiation.

Blanching is probably the most commonly used physical pretreatment before drying. Its main goals are inactivation of the enzymes that can cause deterioration in the quality of the dried product (e.g., enzymatic browning), partial softening of the tissues, a decrease in drying time, and removal of intercellular air (Lewicki, 1998a). The main effect on product texture is related to the activity of the pectin methyl esterase enzyme. Blanching may cause modifications in the final structure of the dried food, affecting its texture and rehydration characteristics. For instance, blanching of potato slices yields a significantly more compact, less porous structure with lower effective water diffusivity (Mate et al., 1998). Higher apparent diffusivity values were found during rehydration of dried peppers blanched in water and treated with a sulfite solution compared to nontreated samples. Peppers placed in an osmotic salt solution also had a lower apparent diffusivity compared to the nontreated samples. However, in this case, the derived values were not significantly different, indicating the need for further study (Kaymak-Ertekin, 2002). By contrast, the blanching temperature significantly affected the equilibrium moisture content in broccoli florets. Conventional blanching resulted in the highest equilibrium moisture contents whereas stepwise blanching treatments at 55, 60, and 65°C yielded the lowest (Sanjuan, Clemente et al., 2001). Water blanching was also compared with freezing of green beans and carrots prior to drying: this resulted in lower water uptake after 20 min of rehydration in boiling water. In contrast, HP treatment before drying resulted in lower water uptake (Eshtiaghi et al., 1994), suggesting that microstructure plays an important role.

Another study that looked at a combination of several pretreatments reported the optimization of blanching time and preservative (sodium chloride, potassium metabisulfite, sodium benzoate) and its pretreatment concentration levels (0.5, 1.0, and 1.5%) for solar-dehydrated cauliflower (Kadam et al., 2006). Blanching time was found to significantly affect the rehydration ratio, whereas it was nonsignificant in terms of the evaluated sensory attributes (color, odor, taste, and overall acceptability) of the rehydrated cauliflower. The authors reported that dehydrated cauliflower obtained from a combined treatment consisting of 3 min of blanching and dipping in 1.0% potassium metabisulfate was optimal with respect to all aspects of the physicochemical properties as well as the sensory attributes.

The effect of microwave (MW) heating on the physical properties and microstructure of apples before air dehydration showed that this pretreatment results in very hard, rather collapsed samples with a high bulk density. However, the rehydration capacity increased by 25 to 50% for these samples compared to apples dehydrated in air only (Funebo et al., 2000). These data show that, although the apple structure collapsed, it maintained its structural memory and was able to better absorb water compared to its AD counterparts.

The use of HELP/PEF as a predrying treatment has attracted a great deal of attention because it is generally recognized to increase the permeability of the plant's cell walls with a concomitant positive influence on mass transfer in further processes. Indeed, as expected, a combination of HELP and OD resulted in an increased rehydration capacity over prolonged periods compared to shorter OD times (Taiwo et al., 2002b). The enhanced rehydration capacity of HELP plus OD treated apple slices was attributed to their less compact structure, due to the absorbed sugar. When the samples were treated with HELP alone, the lower rehydration capacity was partly attributed to greater shrinkage: this, in turn, was due to faster water loss during air drying as a result of increased membrane permeabilization (Angersbach and Knorr, 1997; Taiwo et al., 2002a).

Another example of the effect of OD was demonstrated with carrot cubes. The parameters were different concentrations of sodium chloride solution, the temperatures, and different process durations; water loss and solute gain were monitored. However, this study also looked at other attributes that play an important consumer role. Hence, the process was optimized by response surface methodology not only for maximum water loss but also for the rehydration ratio and overall acceptability, minimum solute gain, and shrinkage of the rehydrated product. The optimum conditions for the various process parameters were an 11% salt concentration, a 30°C osmotic-solution temperature, and a 120-min process duration of (Singh et al., 2006).

In a subsequent study, the effects of ternary solutions of water, sucrose, and sodium chloride at different solution concentrations, temperatures, and process durations were investigated. In this case, the optimum conditions of various process parameters were found to be 50°Bx and 10% (w/v) aqueous sodium chloride, a 46°C solution temperature, and a 180-min process duration (Singh et al., 2007). This example highlights the need to look at the whole process of drying and rehydration using a systematic approach.

Other effects of various pretreatments on the structural changes in different fruits were studied. In particular, the glass-transition temperature (T_g) of the dried samples was related to different structural features. These investigations included the effect of VI and MW application on structural changes that occurred during air drying of apples (Contreras et al., 2005) and the influence of osmotic pretreatment and MW application on properties of AD strawberries (Contreras et al., 2007). For apples, the T_g was affected by the type of drying and generally showed higher values for VI samples. In non-VI samples, an increase in T_g was observed when MWs were applied and when the air temperature increased. However, no clear correlations between the T_g values and the mechanical parameters of the samples were established (Contreras et al., 2005). In the case of AD strawberries, pulsed VOD (PVOD) pretreatment and the changes in pectin solubility that occurred during drying affected both the average molecular weight of the solutes present in the fruit's aqueous phase and the cell-bonding forces supporting the cellular structure. These two factors had a great impact on the T_g of the samples and on the mechanical response of the dried and rehydrated products. All samples with 10% water content obtained from the different treatments were in a rubbery state at the usual temperatures of storage and commercialization, which inhibited sample fracture during handling but increased sample deterioration rate. MW-treated samples, which showed higher T_g values, therefore provided maximum stability. These samples were also the most firm and rigid when dried, although their structure became seriously affected during the drying process, which gave rise to a very soft texture after rehydration. PVOD-treated samples showed the lowest T_g values and higher susceptibility was therefore observed, but their structure was much better preserved during drying and they showed the best mechanical behavior after rehydration (Contreras et al., 2007).

The effects of calcium chloride and osmotic treatment on drying kinetics and rehydration properties of tomatoes were investigated (Lewicki et al., 2002). Dried tomatoes from all treatments showed a poor capacity for rehydration, which was attributed to the restrictive effect of the Ca^{2+} bridges formed in the tomato tissue on polymer hydration and swelling. The increase in mass (weight) during rehydration was hindered by the pretreatment with calcium chloride. Raw tomatoes dried by convection increased in mass more than fourfold during rehydration, whereas tomatoes pretreated with calcium chloride increased only 2.7-fold. A combination of calcium treatment and osmosis resulted in an even lower increase in mass, not exceeding a factor of 2. Tomatoes dried by convection absorbed only 30% of the water that had been present in the raw material. Pretreatment with calcium chloride reduced this value to 20%, and a combination of calcium treatment with osmosis resulted in 15% water absorption.

The effects of osmotic pretreatments (10% NaCl for 1 h or 50% sucrose for 3 h) on the rehydration kinetics of dried onions over a range of temperatures (25 to 65°C) were reported by Debnath et al. (2004). The effective diffusion coefficients for water and solute were determined by considering the rehydration process to be governed by Fickian diffusion. The osmotic pretreatment resulted in a decrease in the diffusion coefficient of water as well as an increase in the diffusion coefficient of solute during rehydration. The decrease in the diffusion coefficient of water was related to an increase in the proportion of ruptured and shrunken cells caused by the osmotic treatment, which in turn resulted in a reduced ability of the dried onion tissue to absorb water. The increase in the diffusion coefficient for solids during rehydration was higher in the case of osmotic-pretreated material, because some of the solids absorbed during OD were not retained in the cell matrix and dissolved faster than the constitutive onion dry matter. The microstructures of untreated, osmotic-pretreated, and AD samples were also compared by means of scanning electron microscopy (SEM). The osmotic pretreatment resulted in compactness in the microstructures, especially in the 50% sucrose-treated sample, compared to the control, leading to a decreased rate of water absorption during rehydration.

More recently, the effect of ultrasound and blanching pretreatments on weight and moisture loss or gain upon drying and rehydration of Brussels sprouts and cauliflower were reported by Jambrak et al. (2007). The drying time after ultrasound treatment was shortened for all samples, relative to the untreated ones. The rehydration properties were improved for the combination of FD and ultrasound-treated samples.

These studies clearly highlight the impact of several pretreatments on both drying and rehydration. Although no general rules are available, the need for a system approach in which the whole process is considered is obvious. In this approach, the consumer has already been integrated. Thus, combining the drying or rehydration processes and the resulting sensory attributes to meet consumer needs and expectations is not only recommended but it is also probably the only way to truly optimize the process.

9.7 DRYING METHODS AND THEIR EFFECTS ON PRODUCT'S PHYSICAL PROPERTIES

Selection of the most adequate drying method for a given food product is usually based on several different considerations. A few of main factors include the economic aspects, raw material characteristics, availability of drying facilities and their proximity to the raw materials, and final product quality. The final moisture content of the dried product can also dictate the method selection. The moisture content of the final product usually needs to be close to or below 5% to ensure adequate shelf-life stability for long periods. It is well established that the process conditions selected to dry fresh foods have a significant influence on the physical properties of the final product (Krokida and Maroulis, 1997). The drying method employed determines most of the particle characteristics that will affect the rehydration kinetics, as well as nutritional value, sensory perception, and consumer acceptance. The most common drying method used is the removal of water by means of forced-air drying convection. A significant number of research groups have studied the parameters related to this method and affecting product quality, such as cell shrinkage and tissue damage, case hardening, porosity, density, color changes, and so forth (Karathanos et al.,

1996; Lewicki, 2006; Lewicki and Pawlak, 2005; Mayor and Sereno, 2004; Nieto et al., 1998; Ramos et al., 2004; Zogzas et al., 1994). Other investigations have compared the properties of the dried products obtained by different drying methods or their combination, including freeze-drying, air drying, and vacuum-puffed drying (VPD), MW heating, and fluidized-bed drying (Gowen, Abu-Ghannam, Frias, Barat et al., 2006; Gowen, Abu-Ghannam, Frias et al., 2006; Krokida and Maroulis, 1997; Lin et al., 1998; Litvin et al., 1998; Maskan, 2000, 2001; Ravindra and Chattopadhyay, 2000; Ruiz-Diaz et al., 2003; Torringa et al., 2001).

The main advantage of air drying is its reduced cost and the relatively simple equipment required (i.e., a cabinet or continuous conveyor-belt dryers). The disadvantages of this method are related to the irreversible changes that take place, for example, loss of nutrients due to their thermal degradation and the shrunken, low-porosity product obtained, often resulting in very slow rehydration kinetics. More expensive drying methods, such as freeze-drying, are needed in the case of dried particles intended for use in premium products with high added value. The main advantage of these methods is that collapse is almost completely avoided because the initial structure of the solid matrix is retained during sublimation of the water. The FD product is characterized by its low bulk density, high porosity, very fast rehydration kinetics, and improved consumer acceptance. Other drying methods, such as fluidized beds and drum dryers, are also used for fruits and vegetables. Some of these, as well as novel food dryers, are discussed elsewhere in this book.

Physical characterization of the microstructure of the dried material is usually carried out with techniques such as pycnometry, mercury porosimetry, SEM, transmission electron microscopy (TEM), and confocal laser scanning microscopy. These techniques provide both quantitative and qualitative information on the density; open, closed, and total porosity; pore-size distribution; damage produced on the cellular tissue during drying; and so forth. The very powerful x-ray microcomputerized tomography (x-ray MCT) technique has recently become commercially available. This technique provides very useful information, both quantitative and qualitative, on the internal structure of the dried material and some of its remarkable features. It has been reviewed and compared to traditional imaging methods (Trater et al., 2005). Several recent publications have utilized this technique to successfully characterize the microstructure of different food products (Babin et al., 2006; Falcone et al., 2004, 2005; Trater et al., 2005; Haedelt et al., 2005; Lim and Barigou, 2004; Mousavi et al., 2007), providing quantitative information that can be used to model different transport phenomena. One of the main advantages of this technique is that it circumvents the need for sample preparation in contrast to SEM or TEM. However, the imaging time needed to obtain high-quality images and the processing time needed to obtain relevant quantitative data such as porosity, connectivity, tortuosity, and so forth, might be rather long and demanding in terms of computational resources. Images obtained by x-ray MCT for three different drying methods (AD, FD, and VPD) applied to carrots are presented in Figure 9.2. High-density areas (solid matrix) are white, and the pores are black.

Figure 9.2a is a slice through an AD carrot sample. The dark pores in the middle of the sample are internal and do not connect directly to the outer regions. The gray strips on the left side of the sample are low-density regions. The AD carrot sample is

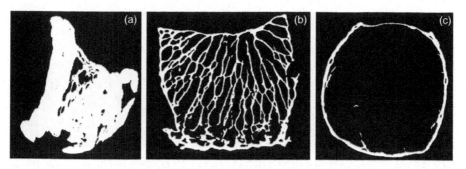

FIGURE 9.2 Single slices through three individual carrot samples: (a) AD, (b) FD, and (c) VPD. High-density areas (solid matrix) are white, and pores are black. (From Grader, A.S., Marabi, A., Wallach, R., and Saguy, I.S., unpublished results, 2007.)

shrunken as a consequence of the drying conditions utilized and becomes very dense in some locations. In contrast, the FD carrot (Figure 9.2b) retains its original web structure, revealing very highly interconnected porosity. The VPD sample (Figure 9.2c) shows a dense outer shell enclosing an inner region. In most of the VPD samples, there is also a connecting pore between the inner total volume and the external domain. However, in some particles, the internal volume is completely isolated. Moreover, there are detectable pores that are isolated inside the wall of the VPD sample, but their volume fraction is not significant in comparison to the inner volume. The collapse of the sample can also be appreciated, as reflected by irregular shapes compared to the FD sample, in which the initial shape is preserved.

Figure 9.3 provides three-dimensional renditions of portions of different carrot samples. Significantly different carrot microstructures can be clearly observed for the three different drying methods. The pore size, spatial distribution, and wall structure are affected by the treatments, which in turn have a great impact on the rehydration kinetics (see Section 9.9). Figure 9.4a shows the internal pores that are not connected to the external space around the AD sample. The closed pores in the sample are not evenly distributed throughout the volume, a feature that is not revealed by other conventional imaging methods and one that might have a significant effect

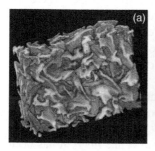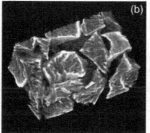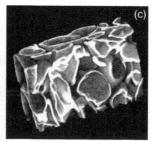

FIGURE 9.3 Three-dimensional renditions of carrot samples: (a) AD, (b) FD, and (c) VPD. (From Grader, A.S., Marabi, A., Wallach, R., and Saguy, I.S., unpublished results, 2007.)

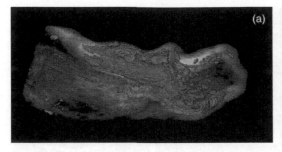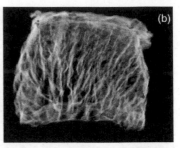

FIGURE 9.4 X-ray MCT images of dried samples of (a) AD carrot and (b) FD carrot. (From Grader, AS., Marabi, A., Wallach, R., and Saguy, I.S., unpublished results, 2007.)

on the rehydration properties. The same analysis was performed on the FD sample (Figure 9.4b). In this case the connected porosity makes up most of the volume of the sample, and the unconnected porosity value is minimal.

These data indicate that x-ray MCT is a very powerful tool for detailed observations of the microstructure of porous foods. Moreover, it is expected to become a pertinent method in the near future to obtain detailed information that can be further utilized in computational simulations of the imbibition of liquids into porous media. These data will open new avenues for ultimately quantifying and modeling internal changes during both drying and rehydration, thereby providing the necessary information for studying the mechanisms involved and enabling better control and quality improvements.

9.8 REHYDRATION CONDITIONS AND THEIR EFFECTS ON PROCESS KINETICS

As previously stated, the conditions under which the rehydration process takes place have a significant effect on the final characteristics of the reconstituted product and will ultimately determine its consumer acceptability. These conditions can be controlled and optimized in some given processes and for some food products. However, in other cases it is not possible to change or optimize the conditions for rehydration and reconstitution of the dried food, for example, in vending machines where the mixing or contact time between the particles and water is limited, in products intended for reconstitution at low temperatures, or when consumption occurs after a short period of reconstitution. Some process parameters that can be controlled include the temperature, composition of the liquid media (density, viscosity), presence of insoluble matter, and mixing regime.

The effect of the medium temperature on reconstitution [RE, Eq. (9.2)], water uptake [RR, Eq. (9.3)], and process kinetics has been extensively studied (Krokida and Marinos-Kouris, 2003; Lewicki et al., 1998; Lin et al., 1998; Machado et al., 1999; Meda and Ratti, 2005; Neubert et al., 1968; Sanjuan et al., 1999). It has been shown that higher temperatures lead to an increased rehydration rate but usually only marginally affect the rehydration capacity (i.e., the final amount of liquid absorbed by the dry food). The effect of temperature on the rehydration rate is due to both the decreased viscosity of the immersion medium and its impact on the product's structure. The overall effect of temperature is typically described by an Arrhenius-type relationship (Cunningham et al., 2007; Oliveira and Ilincanu, 1999).

The composition of the rehydration medium is another important factor affecting the process. Most studies utilize water as the rehydrating medium, whereas a few have utilized more complex rehydration media (del Valle et al., 1992; Horn and Sterling, 1982; Marabi, Jacobson et al., 2004; Marabi and Saguy, 2005; Neubert et al., 1968). A higher liquid viscosity is often a desired characteristic in such products as soups, providing improved mouth feel. This effect is often achieved with thickening agents, such as natural or modified starches or gums. Consequently, the liquid media may contain nondissolved particles, which in addition to increasing the viscosity, could potentially hinder the ingress of liquid into the solid dry matrix. It has been demonstrated that both increased viscosity and the presence of particles lead to slower rehydration kinetics of dry vegetables (Marabi and Saguy, 2005). One possible way of overcoming this effect is to utilize thickening agents with a delayed build-up of the viscosity, allowing for faster liquid uptake during the initial steps, which usually accounts for the biggest proportion of absorbed liquid.

The presence of ionic species in the dissolution media may also hinder the rehydration of dry particles. Lewicki (1998a) proposed that ions orient water molecules around them and decrease their availability to hydrate the dry food particles.

An additional factor influencing the rehydration process is the hydrodynamic conditions of the media. The rehydration process is usually performed either with no stirring or under constant mixing. It is generally recognized that the foremost resistance to water or liquid transfer into the dried matrix lies within the food material itself, and agitation is therefore not expected to play a major role, except with highly viscous immersion media (Oliveira and Ilincanu, 1999).

Some specific examples of these effects are presented here. For instance, mass transfer and physicochemical characteristics of FD strawberries rehydrated in sugar solutions showed that increased concentrations lead to a decrease in both water uptake and soluble solids losses (Mastrocola et al., 1997). A similar study on dehydrated apple cubes reconstituted in sugar solutions demonstrated that increasing the concentration of the rehydrating solution results in a decrease in water uptake and a loss of soluble solids in FD and AD apple cubes (Mastrocola et al., 1998). The rehydration capacity of apples in water and in yogurt yielded significantly higher uptake in the former medium (Prothon et al., 2001). Breakfast cereals immersed in water showed a faster rate of moisture uptake relative to that obtained in milk, which was attibutable to easier hydration in water. Furthermore, the equilibrium moisture content for immersion in water was almost twice that obtained with milk. This process was reported to markedly affected by temperature (Machado et al., 1997). Another study on breakfast cereals immersed in whole and skimmed milk at three different temperatures showed the key role played by milk fat. For short immersion times, uptake proceeded at a similar rate in both skimmed and whole milk, although the sorption mechanisms differed (Machado et al., 1999). Textural changes in ready-to-eat breakfast cereals during immersion in semiskimmed milk were also reported (Sacchetti et al., 2003).

The drying method has a significant impact on rehydration. To demonstrate this, we consider the effects of air drying and freeze-drying on the rehydration of carrots (Marabi, Jacobson et al., 2004). An almost eightfold higher bulk density of the AD sample compared to its FD counterpart was measured (1.28 and 0.16 g/cm³, respectively). This significant difference was attributable to the structural collapse and shrinkage that occur during air drying. The porosity of the FD samples was mainly composed of open pores and was about sixfold higher than its AD counterpart (0.90 and 0.15, respectively). It is interesting to note that the closed porosity for FD carrots was quite low (~0.02) because the sample maintained its structure during dehydration. In contrast, the closed porosity in the AD samples was significantly higher (~0.24), probably caused during shrinkage and structure collapse. The closed porosity could play an important role in slowing down the rehydration rate. The drying method markedly affected the rehydration ratio of the dried particles, resulting in much faster water uptake by the porous FD samples (Figure 9.5). The RR values after rehydration for 180 s were ~0.15 and 0.55 for AD and FD carrots, respectively. The much higher water uptake of the FD samples was expected and previously documented (Farkas and Singh, 1991; Karathanos et al., 1993; Lin et al., 1998; Marabi et al., 2003; Saca and Lozano, 1992). Apart from the clear effect of porosity, Figure 9.5 also reveals that the mixing speed (0 to 1500 r/min) had only a marginal effect on the kinetics of water uptake for both AD and FD samples. The RR values were not significantly affected by increasing the speed. Thus, we concluded that the resistance to liquid uptake for low-viscosity media lies within the food material itself, and the rate-limiting factor in that case is not related to the rehydration medium or mixing speed. However, given the significant difference in the porosity of the samples, capillary

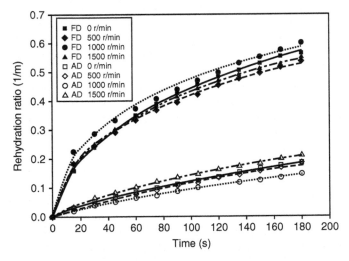

FIGURE 9.5 Effect of revolutions per minute on the rehydration kinetics of AD and FD carrot samples in water at 85°C. Lines depict values derived by the normalized Weibull model. (From Marabi, A., Jacobson, M. et al., *Eur. Food Res. Technol.*, 218, 339–344, 2004. With permission from Springer Science and Business Media.)

imbibition could play an important role at the early stages, leading to an almost instantaneous uptake of water (Benavente et al., 2002; Weerts et al., 2003a). Except for viscous immersion media and a few potentially unique cases, mixing is not expected to play a major role in rehydration kinetics.

As expected, the average effective diffusivity in the FD carrots was ~2 orders of magnitude higher than in their AD counterparts (1.2×10^{-9} and $5.2 \times 10^{-11}\,m^2/s$, respectively). This difference is probably attributable to the dissimilar porosity. The derived values were not affected by mixing, as expected when water uptake is an internally controlled phenomenon.

Another example provided by Marabi, Jacobson et al. (2004) is the effect of mixing on liquid uptake of FD and AD carrots in a medium containing 3% starch and having an apparent viscosity of 110 mPa s at 85°C (Figure 9.6). The data showed markedly different behavior in comparison with water (Figure 9.5). In the viscous medium, increased mixing speed significantly improved uptake in both AD and FD samples. The differences between the samples are still obvious; however, under conditions of no mixing, liquid uptake by the FD sample was significantly reduced and was quite similar to that of the AD sample at the higher mixing speed (1000 r/min).

Marabi and Saguy (2005) provided another example also highlighted here that considered the effect of nondissolved particles. When FD carrots were rehydrated in four different starch solutions (Figure 9.7), the fastest rehydration was obtained with water. Medium uptake in the presence of native corn or potato starches showed very similar kinetics, although it was significantly lower than that for water, followed by native rice starch; a pregelatinized, commercial starch (Ultra Sperse 2000, National Starch and Chemicals Co., Bridgewater, NJ) resulted in the slowest rehydration.

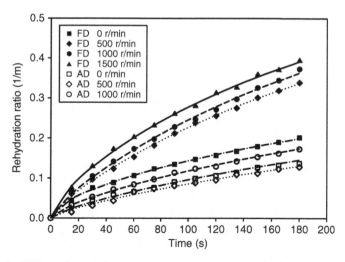

FIGURE 9.6 Effect of revolutions per minute on the rehydration kinetics of AD and FD carrot samples in a 3% starch solution (110 mPa s) at 85°C. Lines depict values derived by the normalized Weibull model. (From Marabi, A., Jacobson, M. et al., *Eur. Food Res. Technol.*, 218, 339–344, 2004. With permission from Springer Science and Business Media.)

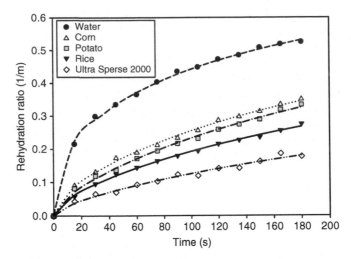

FIGURE 9.7 Rehydration of FD carrots in different starch solutions (30 g/kg, 30°C, 500 r/min) and water. Symbols indicate experimental data and lines indicate the normalized Weibull distribution model. (Reproduced from Marabi, A. and Saguy, I.S., *J. Sci. Food Agric.*, 85, 700–706, 2005. Copyright Society of Chemical Industry. With permission.)

Note that the presence of starch in the rehydrating media dampened the benefits of the FD sample by drastically reducing the rate of liquid uptake.

The effect of processing conditions is also expressed via changes in the D_{eff} (also termed *water diffusivity*). This important parameter is derived from Fick's second law and is an apparent value that comprises all of the factors involved in the process. The theoretical aspects are discussed separately later. However, at this point, we will cover the changes observed due to the rehydration process parameters.

For instance, a comparison of the effective diffusivities (Figure 9.8) revealed a significantly higher value for rehydration in water relative to all starches. Rehydration in potato and corn starch solutions yielded almost identical effective diffusivities, followed by rice starch and Ultra Sperse 2000. Note that the effective diffusivity in water was sixfold higher than in medium containing Ultra Sperse 2000. This finding could be significant when designing a rehydration process for ready-to-eat products. Moreover, the choice of the most appropriate thickening agent should take into account the effect of rehydration kinetics, not merely the physical and sensory properties obtained with a particular product. Correlating the viscosity of the starch solutions with the derived effective diffusivity did not yield a clear trend. The solution containing Ultra Sperse 2000 had the highest apparent viscosity (129 mPa s) and the slowest liquid uptake (Figure 9.7) or effective water diffusivity (Figure 9.8). However, although the viscosity of the solution containing corn starch (37 mPa s) was almost twofold those of solutions with either potato starch (20 mPa s) or rice starch (19 mPa s), it demonstrated faster liquid uptake (Figure 9.7). Thus, viscosity probably plays a lesser role in affecting rehydration kinetics. Nevertheless, it is also possible that because the viscosity of the solution containing Ultra Sperse 2000 was

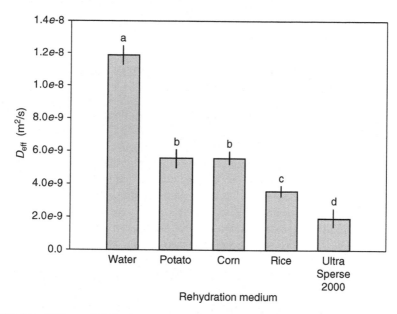

FIGURE 9.8 Effects of different starches (30 g/kg) on the D_{eff} (30°C, 500 r/min). Different letters indicate significant differences ($p < .05$). (Reproduced from Marabi, A. and Saguy, I.S., *J. Sci. Food Agric.*, 85, 700–706, 2005. Copyright Society of Chemical Industry. With permission.)

3.5- to 6.7-fold higher (i.e., 129 vs. 19 to 37 mPa s), it was the primary factor affecting the rehydration kinetics. In contrast, a trend could be observed with regard to the particle size of the starch. Rice starch, which has the smallest average particle diameter (11 µm), yielded the slowest liquid uptake. Potato starch showed two peaks (35 and 70 µm). The granule size of the starch as the limiting factor could explain the similar uptake kinetics for the solution containing corn starch (Figure 9.7), based on the quite similar average particle diameters (41 and 47 µm for potato and corn, respectively). Thus, apart from very viscous solutions, the particle size of the thickening agent could play a significant role in the rehydration process.

In this respect, we hypothesized that small particles can clog the penetration of the liquid into the capillaries of the FD samples whereas bigger particles are unable to penetrate the pores, and therefore can act as a filter aid. This theory is supported by the fact that most of the pores of FD carrots are smaller than 30 µm, with two peaks at 1.1 and 21 µm (Karathanos et al., 1996).

No correlation was found between the rehydration uptake kinetics and the 10th, 50th (median), and 90th percentiles of the gelatinized starch particle diameters. Thus, it seems that more than just one mechanism is responsible for the observed differences in the liquid uptake kinetics, leading to a complex interaction of factors.

The effect of mixing and medium viscosity on the values of the derived D_{eff} showed marked differences (Marabi, Jacobson et al., 2004). The D_{eff} values of both the AD and FD carrots rehydrated in water were higher than the corresponding values in starch for all mixing speeds tested. Furthermore, mixing had a significant effect on uptake for the viscous media, implying that an external resistance model prevails (see following).

9.9 PHYSICAL PROPERTIES OF DRIED FOODS AND THEIR EFFECT ON REHYDRATION

As shown earlier and in other publications (e.g., Aguilera, 2000, 2004; Aguilera et al., 2003; Chiralt and Fito, 2003; Fito and Chiralt, 2003; Marousis et al., 1991), the structure of foods and biological materials has a sizeable effect on the transport phenomena taking place in different processes. The physical properties of the dried foods that affect the rehydration process include the sample's size and geometry, chemical composition and physical state, moisture content, porosity, tortuosity, and density. These properties are affected by the predrying treatments, dehydration methods, and conditions under which they are produced.

One of the most important parameters is porosity, and its influence on rehydration has been widely studied (Farkas and Singh, 1991; Karathanos et al., 1993; Krokida and Maroulis, 1997, 2001; Machado et al., 1998; Saca and Lozano, 1992; Witrowa-Rajchert and Lewicki, 2006). Typically, the bulk volume of a food sample can be measured by different methods, including volume displacement by immersion in *n*-heptane, toluene, sand, or glass beads, or it can be derived from the outer geometric dimensions. The application of sophisticated image analysis can also be considered. The solid volume of the sample is often measured using a pycnometer, utilizing helium gas displacement. Helium is capable of penetrating all open pores up to the diameter of the gas molecule (~3 Å); therefore, all larger pores can be quantified. The solid density is then obtained by grinding and sieving dry samples and determining the true volume with a gas pycnometer. The pore-size distribution in a dried food is often quantified with a mercury porosimeter. Tortuosity is also a fundamental physical property of a product, but it is seldom reported and/or included in the modeling of transport processes through porous foods. It is the factor that compensates for the path through longer pores. Typical values for the tortuosity are between 1.5 and 5. In porous solids, the D_{eff} is an important parameter that is defined as (Gekas, 1992)

$$D_{eff} = \frac{\varepsilon D}{\tau} \tag{9.20}$$

where ε is the porosity, D is the true diffusion coefficient of water (m/s^2), and τ is the tortuosity.

Thus, a method that enables quantification of the tortuosity of dried foods is paramount for enhancing the accuracy of modeling water-transport phenomena and may also reveal other transport mechanisms involved during the rehydration process. As mentioned in Section 9.7, three-dimensional virtual samples obtained by x-ray MCT can serve as the basis for developing flow models that include the impact of tortuosity on mass transport.

Water transport is an important physical process during drying, storage, and rehydration, significantly affecting the quality and utilization of many food products. Of particular importance is the transport of water within porous foods. Principles of heat and mass transfer during drying are typically applied, mainly because external mass transfer is well understood and can be analyzed more readily (Saravacos and Maroulis, 2001). The water transport mechanisms will be covered separately. This section focuses on the porous structure of the dehydrated materials that plays a paramount role in the modeling of mass transfer applications in dehydrated foods (Vagenas and Karathanos, 1991). The porosity of a dried material is specific for each product and can be related to the extent of drying, shrinkage, and possible collapse. The latter could lead to a significant reduction in the number and size of the pores (Lewicki, 1998a), with a consequent unfavorable impact on the rehydration properties. The effect of drying on product porosity has been studied quite intensively (Farkas and Singh, 1991; Karathanos et al., 1993, 1996; Krokida and Maroulis, 1997, 1999). However, it is interesting to note that most studies report only bulk porosity.

To demonstrate the effect of porosity on the rehydration kinetics (Marabi and Saguy, 2004), carrots were dehydrated to a final moisture content of ~3% (wet basis) by using one or a combination of drying processes (air and freeze-drying). The combination of these two drying methods and various processing times furnished a wide range of porosities from ~0.90 to 0.14 for the fully FD or AD samples, respectively. The bulk and open porosities decreased with the amount of time for the AD process. These data are quite important because the rehydration kinetics are probably directly linked to the open porosity rather than to the sample's overall bulk porosity, if the presence of the closed porosity is significant. Note that the decrease in open pores may also be involved in determining the pertinent mechanisms (e.g., diffusion, imbibition, capillary flow) controlling the water uptake. Hence, although the closed and open porosities are important product characteristics, only open pores are important for water uptake. The rehydration data obtained for the samples with different initial porosities (Figure 9.9) depict faster rehydration for samples with

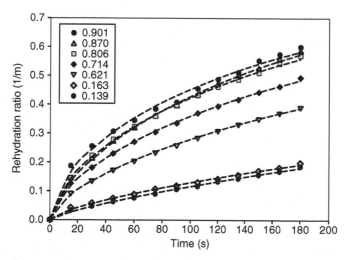

FIGURE 9.9 Rehydration of carrots with different initial bulk porosities (water, 85°C, 500 r/min). Symbols indicate experimental data and lines indicate the normalized Weibull distribution model. (Reproduced from Marabi, A. and Saguy, I.S., *J. Sci. Food Agric.*, 84, 1105–1110, 2004. Copyright Society of Chemical Industry. With permission.)

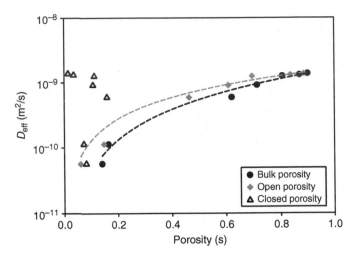

FIGURE 9.10 Effect of bulk, open, and closed porosity on the D_{eff} of the samples. (Reproduced from Marabi, A. and Saguy, I.S., *J. Sci. Food Agric.*, 84, 1105–1110, 2004. Copyright Society of Chemical Industry. With permission.)

higher bulk porosity. The effective diffusivity was derived utilizing the normalized Weibull distribution (Marabi et al., 2003). The values were then plotted as a function of bulk, open, and closed porosities. A clear and similar trend was only observed with the first two porosities (Figure 9.10). These data show that the effective moisture diffusivity increases with the porosity. In fact, the values obtained for the FD carrots were ~2 orders of magnitude higher than those for their AD counterparts, and they were not affected by the closed porosity. We therefore concluded that the main factor affecting water uptake during rehydration was the carrot's initial open porosity (Marabi and Saguy, 2004).

The mechanism of capillary imbibition plays an important role in the early stages of rehydration, leading to almost instantaneous water uptake (Benavente et al., 2002; Weerts et al., 2003a). This is probably the case when comparing the rehydration curves of FD versus AD samples. However, it is logical to assume that there is a critical porosity value above which the mechanism of water uptake is more similar to a porous medium than the one represented by Fickian diffusion.

The overall impact of the drying methods on the porosity, physical properties, and appearance was also studied and advantageous effects were found in terms of structure and rehydration kinetics. For instance, a combination of MW energy with convective air drying resulted in a more porous dehydrated chickpea. It improved the porosity of the final dehydrated product and samples experienced less shrinkage than those dried by convective hot air, leading to faster rehydration kinetics (Gowen, Abu-Ghannam, Frias, Barat et al., 2006). Another study utilized four different methods (hot-air drying, MW-assisted convective drying, freeze-drying, and vacuum-drying) to dehydrate cranberries (Beaudry et al., 2004). Hot-air drying produced dried cranberries with the best visual appearance, FD cranberries had the highest rehydration ratios, and the other methods presented similar RR values.

The greater rehydration capacity of the FD samples was because the internal structure of the fruit remained quite undisturbed. This, in turn, was attributable to the structural rigidity of the frozen product, which prevents the collapse of the solid matrix remaining after drying. Therefore, FD products tend to have a porous and nonshrunken structure with excellent rehydration capacity.

The drying kinetics and rehydration characteristics of MW–vacuum dried and convective hot-air dried mushrooms were also reported (Giri and Prasad, 2007). MW–vacuum drying resulted in a 70 to 90% decrease in drying time, and the dried products had better rehydration characteristics than their convective AD counterparts. The rehydration properties were improved by drying at lower system pressure and higher MW power. As the pressure level decreased the rehydration ratio increased, which was due to the increased drying rate and the creation of pores induced by the vacuum conditions. The higher rehydration ratio at higher MW power was attributed to the development of greater internal stresses during drying at higher power levels. The quick MW-energy absorption caused rapid evaporation of water, creating a flux of rapidly escaping vapor that helps in preventing shrinkage and case hardening, thus ultimately improving the rehydration characteristics (Giri and Prasad, 2007).

The use of γ-irradiation in combination with air drying and its effects on the rehydration properties of potatoes were also investigated (Wang and Chao, 2003; Wang and Du, 2005). Higher irradiation doses (between 0 and 10 kGy) resulted in lower rehydration ratios in AD potatoes. The authors concluded that irradiation pretreatment does not improve the rehydration quality of potatoes. Similarly, another study reported the combined effect of γ-irradiation and osmotic treatment on mass transfer during the rehydration of carrots (Nayak et al., 2006). An increase in γ-irradiation dose resulted in a decrease in water and an increase in solute diffusion during rehydration. This was attributed to the permeabilization of the cell membranes upon exposure to γ-irradiation. The pretreatment with water, as opposed to the osmosed samples (50°Bx), resulted in increased water gain and reduced solute loss during rehydration for irradiated samples. The equilibrium moisture and solids contents were dependent on the treatment conditions.

The drying and rehydrating characteristics of carrots dried in a solar cabinet, fluidized bed, and MW ovens were also reported (Prakash et al., 2004). The rehydration ratio of carrots dried by a fluidized bed oven at 70°C was the highest whereas that of carrots dried in the solar cabinet was the lowest. There was a decrease in the rehydration ratio with increasing storage time. This was attributed to the dried carrot's absorption of moisture from the environment during storage. Drying at higher temperature or MW power was improved the rehydration properties, because of less shrinkage of the dried carrots. Carrots were also dried in MW and halogen lamp–MW combination ovens and the rehydration properties of the samples were evaluated (Sumnu et al., 2005). These samples were reported to have higher rehydration capacities than those subjected to hot-air drying, because of the high internal pressure produced by MW heating that can cause the carrot slices to expand and puff up.

Methods of drying have also been reported to affect the behavior of apple cubes upon rehydration. Convection-dried apples imbibed water and swelled during rehydration, with a collapse in structure observed at the beginning of the wetting process.

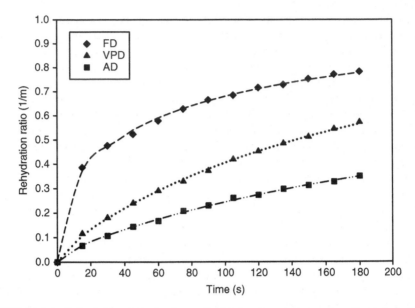

FIGURE 9.11 Rehydration kinetics (up to 3 min) for three test samples. Symbols indicate experimental data and lines indicate fitting of the normalized Weibull distribution. (From Marabi, A. and Saguy, I.S., unpublished data.)

An analysis of both mass and volume increases showed that at the beginning of rehydration the capillaries and pores were filled with water and thereafter swelling of the biopolymers occurred. FD samples increased in mass much faster than those dried by convection. Rehydration caused collapse of the structure, which was not rebuilt during further wetting (Lewicki and Wiczkowska, 2006).

The effects of three different drying methods on the samples' physical structure (Figure 9.3) and rehydration kinetics (Figure 9.11) were also compared. The different structures and solid porosities had a significant impact on the rehydration process for short times. These samples were the same as those studied by x-ray MCT and shown in Figures 9.2 and 9.3. The FD sample showed the fastest uptake of water, followed by the VPD and AD samples, in that order. However, with the information derived from the x-ray MCT analysis, it might be possible not only to correlate the physical properties of the samples with kinetics parameters derived from rehydration curves but also to develop more accurate models of flow in porous media, which are scarce in the field of food science. This topic is currently under investigation by several groups, and new breakthroughs are anticipated (see Section 9.12).

Xu et al. (2005) applied a two-stage convective-air and vacuum freeze-drying technique to bamboo shoots in order to optimize the cost of the drying process. They combined hot air flow with vacuum freeze-drying in two different ways: hot air-flow drying (FAD) followed by vacuum freeze-drying (VFD), and vacuum freeze-drying followed by hot air-flow drying. The combination of these two methods can overcome the disadvantages of AD products, that is, extremely hard texture, severe browning, low rehydration rate, low nutritive value, and so forth, while still retaining the benefits of vacuum freeze-drying, that is, preservation of characteristics such as color, aroma, taste, and shape of food and agricultural products. The quality of AFD products was worse than FD and FAD products, and FAD was found to be a suitable combined drying method for dehydration of bamboo shoots. Furthermore, its structure, rehydration, sensory, and nutritional characteristics were better than for AFD and AD and energy consumption was reduced.

These data show that the drying method and accompanying conditions play a paramount role in affecting the physical properties of the food products and consequently the rehydration process. Porosity is one of the most important parameters; yet, other properties such as the tortuosity, cell-size distribution, cell-wall thickness distribution, connectivity, voidage, and degree of anisotropy of the samples are also important. Significant breakthroughs are expected as x-ray MCT becomes more affordable. More data and information will provide the basis for insights and new understanding of the mechanisms taking place during both drying and rehydration.

9.10 LEACHING OF SOLIDS

During the rehydration process, substantial amounts of soluble solids that are initially present in the dried food may leach into the liquid medium while liquid is entering the solid matrix. Both the nutritional quality of the food product and its water uptake can be affected during this process. Complete solvation of the mass that has leached out is associated with relatively high amounts of water, and it can be expected that the smaller the amount of leaching is the higher the ability of the rehydrating material to absorb water (Lewicki, 1998a). Dry food products with highly porous structures will allow more water penetration that can also provoke the leaching of soluble solids in appreciable amounts. However, increased porosity does not always promote the leaching of soluble solids. This process is affected in part by the mass transfer taking place inside the solid matrix (e.g., inside the pores), with the soluble solids passing into solution, and the subsequent diffusion of the dissolved solids, transported against the current into the liquid, where they are absorbed into the bulk liquid.

Here we will review a few studies that demonstrate the leaching phenomenon. Leaching of soluble solids during the rehydration of dried onions at room and boiling temperatures was quite significant, in some cases reaching more than 70% of the initial dry matter. Predrying treatments were also led to significant differences in soluble solids loss during rehydration, and samples undergoing osmosis lost the most. However, the temperature of the rehydration medium did not influence the final retention of the dry matter (Lewicki et al., 1998). When AD and explosion-puffed (EP) bananas were rehydrated in boiling water for 20 min, more solids were leached from the AD samples, which showed a lower porosity, than from the EP product (Saca and Lozano, 1992). In another study, significantly lower dry-matter losses during rehydration was observed for dry peaches blanched for 5 min (Levi et al., 1988) compared with no blanching or longer blanching times (9 or 15 min, respectively). This effect was attributed to the integrity of the tissues that was probably affected by the heat treatment, leading to an increase in the leaching of water-soluble solids to the surrounding medium. In this case, the water-soluble solids were identified as sugars, acids, galacturonic acid monomers, short oligomers, and water-soluble pectin. As expected, increasing the concentration of glucose or corn syrup in the rehydrating solution led to decreased water uptake and loss of soluble solids in FD strawberry pieces (Mastrocola et al., 1997). In contrast, when the rehydration solution contains a relatively high concentration of solids, the direction of the solids' movement can be reversed, as was detected when rehydration was carried out in solutions containing more than 15% sugar. The rehydrating medium also has an impact on the amount of leached solids. For instance, more soluble solids were lost during the rehydration of breakfast cereals in water than in milk (Machado et al., 1999). The effect of temperature on the extent of solids leaching was also reported for cereals. The rate of solids leaching increased with temperature for both corn and peanut butter breakfast cereals (Machado et al., 1998).

Frosting on puffed breakfast flakes reduced water uptake compared to an unfrosted sample (Sacchetti et al., 2003). By contrast, the frosted product showed the highest loss of soluble solids during the soaking process in semiskimmed milk. Leaching of solutes from Rice Crispies into water at 40°C was also studied by NMR (Lucas et al., 2007). The leached solutes represented less than 0.6% of the immersion water (g dry matter/100 g immersion water). Nevertheless, in relation to the dry matter initially contained in the Rice Crispies, it represented 6.2 and 12.5% at 6 and 58 min of soaking, respectively. Clearly, such amounts cannot be disregarded because mass balances may be biased.

Despite the importance of the leaching process and its effect on the rehydration process, a significant number of studies do not take the leaching of solids to the surrounding solution into account in their calculations. As indicated, to overcome some of these drawbacks, the concept of NDS was defined and utilized in Eq. (9.3).

Predrying treatments (e.g., HELP, OD, HP, freezing, and blanching) influence dry-matter loss during rehydration (Rastogi et al., 2000; Taiwo et al., 2002a,b). Untreated samples lost more solids than HELP- and HP-treated ones. Other studies (Oliveira and Silva, 1992; Schwartzberg, 1975; Schwartzberg and Chao, 1982; Tomasula and Kozempel, 1989) included mathematical analyses of the leaching process, under the assumption that it could be described by Fick's second law. The dependence of the effective diffusivity on temperature was described by an Arrhenius-type equation for raw and previously frozen carrot cortex tissue (Oliveira and Silva, 1992). Recently, the Weibull probabilistic distribution was used to model both the water uptake and the soluble solids losses during rehydration (Ilincanu et al., 1995; Machado et al., 1997, 1998, 1999; Marabi et al., 2003). However, the nature of the leaching of soluble solids during short rehydration processes was not studied. Determination of the composition of the soluble-solids leaching from legumes and vegetables using high-performance liquid chromatography (HPLC) was reported (Rodriguez et al., 1999; Vidal et al., 1992, 1993).

More recently, the solids leaching from rehydrated AD and FD carrots was analyzed by HPLC and total organic carbon determination (Marabi, Dilak et al., 2004). Leaching of solids was significant at short rehydration times and amounted to up to ~35 and 40% of the initial sample weights for AD and FD samples, respectively (Figure 9.12). The leaching kinetics of AD and FD carrots during rehydration in 85°C water showed that sucrose was the main sugar that leached from both samples,

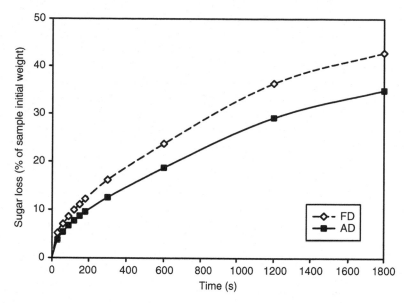

FIGURE 9.12   Total sugar (glucose + fructose + sucrose) loss as a function of rehydration time. (Reproduced from Marabi, A., Dilak, C. et al., *J. Food Sci.*, 69, FEP91–FEP96, 2004. With permission from Blackwell Publishing.)

whereas a minor difference between the AD and FD samples was observed in the case of fructose and glucose. The higher concentration of sucrose in the rehydrating medium was in accordance with its initial concentration in the dry samples.

   In conclusion, the leaching of solids may play a significant role during rehydration, especially for prolonged processes. Hence, it needs to be taken into consideration, especially when studying the kinetics of the process.

9.11   SENSORY ASPECTS

When considering dried foods that need to be reconstituted, it is essential to bear in mind that the utmost requirement is to meet consumers' expectations with respect to the sensory aspects of the final product. These may include color, texture, hardness, flavor, and even particle distribution in the final product. It is therefore essential to optimize the reconstitution conditions for each specific product, because predrying treatment, drying, and rehydration processes induce many changes in the structure and composition of dried foods (Marabi, Dilak et al., 2004; Marabi and Saguy, 2004). For rehydrated vegetables, the most pertinent properties are related to the texture and flavor of the particles. Typical tests of consumer affect are employed to quantify the product attributes and overall acceptance, and these tests can be adapted to assess the acceptability of the reconstituted foods.

   Organoleptic characteristics and textural softening of carrots were affected or progressed at different rates, depending on the physicochemical mechanisms that were used. Therefore, the impact of the dehydration process on the overall quality and acceptability should be considered in establishing the optimal process (Paulus and Saguy, 1980). The effect of low-temperature blanching on the texture and rehydration of dehydrated carrots was studied using an analytical method to measure the mechanical properties (Quintero-Ramos et al., 1992). The sensory evaluation of rehydrated carrots was reported, showing no significant difference in overall acceptability and texture for AD, FD, and vacuum–MW dried carrots rehydrated at 100°C for 10 min (Lin et al., 1998). In contrast, the data on aroma or flavor, appearance, and color revealed the importance of the different drying methods utilized. The rehydration kinetics of dried carrots was affected by several factors (e.g., drying method, rehydration medium composition, time, mixing speed, and porosity; Marabi, Dilak et al., 2004; Marabi, Jacobson et al., 2004; Marabi and Saguy, 2004, 2005; Marabi et al., 2003). For instance, FD carrots rehydrated faster than their AD counterparts while a high amount of carbohydrates leached out simultaneously. The combination of these factors and processes could lead to rehydrated products with diverse physical properties that, in turn, might translate to different and significant sensory characteristics. Hence, assessing such differences is an important step in merging the process with consumer preferences.

In one such study, a consumer test was designed to assess the overall product acceptability and relevant attributes that drive it as a function of the rehydration time for carrots dried by two different methods (AD and VPD). Water uptake was increased with soaking time and was higher for the VPD samples, even though no significant differences were found in the amount of water absorbed. Therefore, any differences observed in the sensory parameters must be attributable to the drying process used and its effects on the structure of the samples. Indeed, the evaluated trait descriptor values differed significantly between the AD and VPD carrots (Figure 9.13), except for flavor (150 s) and flavor intensity (120 s). In addition, immersion time showed different effects (Figure 9.13). AD samples were significantly affected by rehydration time for overall acceptability, texture, and hardness. By contrast, VPD carrots showed a significantly higher overall acceptability for all rehydration times (Figure 9.13a). The texture score of the AD carrots increased significantly with rehydration time (Figure 9.13b). As expected, greater differences in consumer acceptability were found for hardness, with the AD carrots being perceived as very hard. Rehydration time considerably softened the samples, and VPD carrots exhibited significantly lower hardness in comparison to their AD counterparts (Figure 9.13d). However, the hardness of the VPD samples remained unchanged. Flavor acceptance was only significantly higher for the VPD carrots at 60 and 120 s (Figure 9.13c). Flavor intensity was higher for the shortest immersion of the VPD carrots and decreased significantly with rehydration time only for these samples (Figure 9.13e). The decrease in flavor intensity for VPD samples might be a consequence of the high amount of solids leaching during the rehydration process (Marabi et al., 2006).

The data clearly indicated that the VPD sample was an acceptable product. Moreover, at the shortest rehydration time tested, only VPD samples enjoyed consumer acceptance. In contrast, AD samples needed to be rehydrated for longer than 150 s and thus may not be suitable for some types of products for which a shorter reconstitution time is expected. The hardness was also inversely correlated with the overall acceptability, texture, and flavor, indicating that this parameter

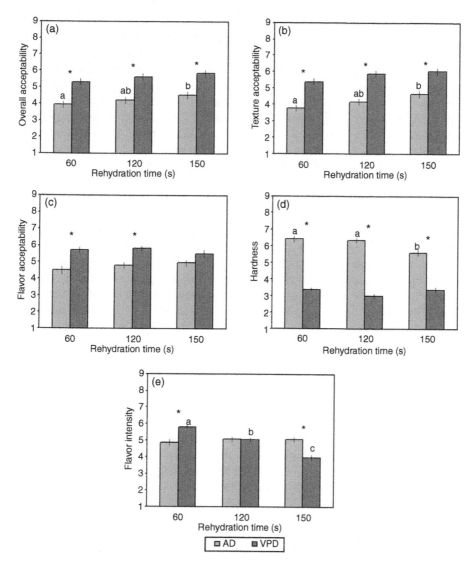

FIGURE 9.13   Average sensory scores for rehydrated AD and VPD carrots: (a) overall acceptability, (b) texture, (c) hardness, (d) flavor, and (e) flavor intensity. *Significant differences
between samples ($p < .05$). Different letters indicate significant differences within samples
($p < .05$). (Reprinted from Marabi, A. et al., *J. Food Eng.*, 72, 211–217, 2006. Copyright 2006
Elsevier. With permission.)

should be optimized to produce an appealing product. Optimizing the rehydration
time and dehydration process is therefore most important for AD products. The
longer rehydration time needed to achieve an acceptable hardness for the AD
carrots has significant commercial impact. The rehydration time is not very flexible
because of consumers' demand for a rapidly reconstituting product, so the need to
optimize the hydration process is paramount (Marabi et al., 2006).

Another study evaluated the effects of four drying methods on the quality of OD cranberries with respect to their rehydration characteristics and sensory aspects (Beaudry et al., 2004). In this case as well, the rehydration ratio of FD cranberries was significantly greater than that of the fruits dried by other methods. However, conventional hot-air drying seemed to confer a slight advantage over the other methods based on the overall appearance of the dried samples, probably due to their uniformity. Cranberries dried with MW produced a few darker fruits and a few fruits dried under vacuum were less aesthetic, probably because of excessive bleeding of the fruits during drying. In this case, the data obtained based on the taste of the samples showed no difference among the different drying methods.

The sensory characteristics of blanched carrots dried by three different methods were also evaluated (Prakash et al., 2004). Carrots subjected to fluidized-bed drying showed better color and rehydration properties, greater β-carotene retention, and higher overall sensory acceptability than those dried by MW oven or solar methods. Moreover, the sensory score decreased with increased storage time.

Sensory characteristics, which are designed to meet consumer expectations and acceptability, constitute the ultimate test for rehydrated products. Optimization is possible if all of the process aspects are considered and understood. A paradigm shift is needed so that a system approach is utilized. Such an approach should be designed to optimize each and every stage, namely, the raw material, drying, storage, and rehydration conditions that are utilized.

9.12 NEW APPROACHES AND RECENT ADVANCES

The rehydration of dried foods involves different physical mechanisms, among them water imbibition, internal diffusion (in the solid and in the pores), convection and diffusion at the surface and within large open pores, hydraulic flow, capillary flow, and relaxation of the solid matrix. The previous sections describe the most relevant factors affecting the rehydration of dried food particles. In addition, the various approaches frequently utilized to model this process are highlighted. Two main approaches are often used to describe liquid transport into the dried solid matrix: empirical or semiempirical models (e.g., Peleg's model, Weibull distribution function) and the application of Fick's laws of diffusion. Despite the significant and valuable insights these models have provided, several research groups have recently proposed that the rehydration of dry food particles cannot be explained and/or modeled solely by a Fickian mechanism. A growing body of data indicates that this simplification may not be fully warranted, or in some cases, might even result in misleading conclusions (Datta, 2007a,b; Saguy et al., 2005b; Weerts et al., 2003a,b). This realization is relatively new and important insights are coming to light from the merging of different scientific disciplines, including food and soil sciences, biophysics, environmental sciences, and petroleum and chemical engineering. The ultimate goal of these interdisciplinary efforts is to provide more advanced and accurate physical models describing the mechanisms taking place during the rehydration process.

The models recently developed and successfully applied to the flow of liquids into dried foods are based on theories that are usually termed the "capillary-flow approach" and/or "flow in porous media." The theories are already well developed in other fields, as are the methodologies needed to generate the required experimental data. However, there is still a considerable lack of data and, of more importance, standard methods allowing collection of the relevant physical properties of the food matrix and its interaction with the liquid. We anticipate that collaborative studies in the near future will provide the information needed to apply these theories.

The starting point for developing models that account for fluid flow in porous media is Darcy's equation and its Navier–Stokes analog (Datta, 2007a; Saguy et al., 2005b; Weerts et al., 2003b). In addition, the main roles of the food particles' physical properties (e.g., pore-size distribution, heterogeneity, tortuosity), the embedding liquid (e.g., density, viscosity, temperature), and the interface (e.g., contact angle) are also considered in order to quantify the changes occurring during drying and to facilitate the modeling of the rehydration process. These studies on fluid flow in porous media have presented adequate detail on the development of the models, and the reader is referred to them for more detailed information.

Some of the recent studies focusing on the various transport mechanisms taking place in porous media include the following:

1. Molecular diffusion of gases, including water vapor; Darcy flow of gases due to pressure; and Darcy flow of liquid due to gas and capillary pressures were studied (Datta, 2007a).
2. A variation of the Lucas–Washburn equation was also utilized to model the rehydration of dried foods (Saguy et al., 2005a; Lee et al., 2005).
3. The complexity of utilizing these mechanisms together with the input parameters that need to be derived from experimental data was investigated (Datta, 2007a,b; Saguy et al., 2005b). This category includes (a) density, porosity, thermal conductivity, and specific heat of the solid material; (b) molecular, capillary, and effective diffusivities; (c) moisture isotherms; (d) vapor and liquid permeability; and (e) water-potential curves. Data in (d) and (e) are either nonexistent for foods or very difficult to obtain. Consequently, future research should address these topics.

To utilize the theory of flow in porous media, a characteristic water curve (analogous to a food-sorption isotherm, also known as a moisture-retention curve) is needed. The characteristic curve describes the functional relationship between water content and matric potential under equilibrium conditions. This curve is an important property related to the distribution of pore space (sizes, interconnectedness), which is strongly affected by texture and structure, as well as by other related factors, including organic-matter content. The retention curve indicates the amount of water in the porous medium at a given matric potential. It is also a primary hydraulic property required for modeling water flow in porous media. Retention curves are highly nonlinear functions and are relatively difficult to obtain with any accuracy. They are measured experimentally. The traditional method of determining the water-retention curve involves establishing a series of equilibria between water in the porous-medium sample and a body of water at known suctions. Continuous retention-curve functions are needed to solve the flow equations. Numerous well-known empirical functions (e.g., Brooks and Corey, 1966; van Genuchten, 1980) have been developed, and similar functions and relationships will probably continue to be developed in the near future.

9.13 RESEARCH NEEDS

Future research needs to focus on the accomplishment of three main goals. The first could be described as a "system approach," whereby the drying and rehydration processes are combined. It is crucial to realize that rehydration is significantly affected by drying, and these therefore need to be studied by a unified approach. The second goal is to apply physically based models. Multidisciplinary research is needed in which theories and knowledge developed in other fields will be assimilated and integrated. The third goal is the development of new methods and the utilization of advanced equipment to quantify *in situ* changes in the microstructure of the food matrix and enable the collection of physical data.

More specifically, the following recommendations are proposed:

1. Additional studies are required to elucidate the various mechanisms occurring during the ingress of water or other liquids into a dry food. Capillary imbibition is very important in the early stages, leading to almost instantaneous uptake. Knowledge in this domain is evolving and will probably have a significant impact on our understanding of rehydration. The cooperation and transfer of know-how among numerous disciplines with more developed knowledge of transport phenomena and the physical mechanisms underlying such processes (e.g., soil sciences and physics, civil and petroleum engineering, geology, printing, etc.) and food science and engineering should lead to significant advances in our understanding of this important subject.

2. Application of porous-media theory to the modeling of food rehydration requires the use of a characteristic curve, which is time consuming and cumbersome. Bridging between sorption isotherms commonly utilized in food science and water-characteristic curves employed in soil physics could furnish a novel and integrated approach that would overcome some of the immense complexity that has hampered previous modeling attempts and open new avenues for studying and optimizing food rehydration. Development of the relevant data and, of more importance, standard methods enabling collection of the relevant parameters is required.

3. Computational fluid dynamics software must be developed specifically for foods. Such software would include specific problems such as evaporation, deformation of the solid matrix, and multiphase flow; development of methods to directly measure water-retention curves (drying and hydration) for food particles; validation of the Kelvin equation; examination of different methods for indirect determination of hydraulic conductivity; development of a database on food permeability; and the utilization of advanced imaging techniques to quantify the complex microstructure of dried foods, furnishing information on the porosity, pore-size distribution, and tortuosity (Datta, 2007b; Saguy et al., 2005b).

4. The final recommendation is the utilization of advanced techniques such as NMR and X-ray MCT to measure *in situ* and real-time changes within the food-matrix microstructure.

9.14 CONCLUSIONS

The overall objective of this chapter was to describe our present understanding of the rehydration process and to highlight the existing state of the art. Food rehydration was depicted as a complex process that involves numerous mechanisms. Collaborative multidisciplinary studies are required to provide the relevant information for applying new theories and modeling the phenomena. The vastness of the data summarized in this chapter calls for a new way of thinking and further development of advanced methods for measuring physical properties in order to provide a better understanding of the changes the product undergoes during both drying and rehydration. The field is entering a new era in which technology and scientific data can provide new insights and deeper knowledge. This will ultimately lead to much better food products that benefit from enhanced consumer appeal and acceptability.

NOMENCLATURE

ABBREVIATIONS AND VARIABLES

AD	air dried
AFD	hot air-flow drying followed by vacuum freeze-drying
C	liquid uptake coefficient (kg/m^2 s$^{0.5}$)
D	true diffusion coefficient of water (m^2/s)
D_{calc}	calculated diffusion coefficient of water (m^2/s)
D_{eff}	effective diffusion coefficient of water (m^2/s)
EP	explosion puffed
FAD	vacuum freeze-drying followed by hot air-flow drying
FD	freeze-dried
g	acceleration due to gravity (m/s^2)
h	pressure head (m)
HELP/PEF	high-intensity electrical field pulses
HP	high pressure
HPLC	high-performance liquid chromatography
J_x	flux (g H$_2$O/m^2 s)
K	hydraulic conductivity (m/s)
L	half-slab thickness or radius for spherical and cylindrical samples (m)
M_0	moisture content at time 0 (kg H$_2$O/kg dry substance)
M_e	moisture content at equilibrium (kg H$_2$O/kg dry substance)
M_t	moisture content at time t (kg H$_2$O/kg dry substance)
M_s	surface moisture content (kg H$_2$O/kg dry substance)
MW	microwave
NDS	nondissolvable solids
NMR	nuclear magnetic resonance

OD	osmotic dehydration
P_1, P_2	parameters in thin-layer rewetting equation (exponential model)
P_3	constant related to initial rate of sorption (kg dry substance/kg H_2O)
P_4	constant related to equilibrium moisture content (kg dry substance/kg H_2O)
P_5	rehydration rate (min^{-1})
PVOD	pulsed-vacuum osmotic dehydration
r	pore radius (m)
RC	rehydration capacity
RE	reconstitution
R_g	geometry factor (dimensionless)
RR	rehydration ratio
S	surface area (m^2)
SEM	scanning electron microscopy
t	time (s)
TEM	transmission electron microscopy
T_g	glass-transition temperature
V	volume (m^3)
VI	vacuum impregnation
VOD	vacuum osmotic drying
VPD	vacuum-puffed drying
W	water content (kg H_2O/m^3)
W_∞	g H_2O/g NDS at time t_∞ (s) (measured after ~18 to 24 h at ambient temperature)
W_0	g H_2O/g NDS at time t_0 (s)
W_t	g H_2O/g NDS at time t (s)
x	spatial coordinate (m)
X_0	initial weight of dry sample (g)
X_i	initial weight of sample before drying (g)
X_t	weight after a given rehydration time (g)
X-ray MCT	X-ray microcomputerized tomography
y	distance traveled by liquid meniscus at time t (m)

GREEKS

α	scale parameter (s)
α'	normalized scale parameter (s)
β	shape parameter (dimensionless)
γ	surface tension (N/m)
δ	contact angle (°)
ε	porosity
μ	viscosity (N s/m^2)
ρ	density (kg/m^3)
τ	tortuosity
φ	negative pressure head (water suction; m)

REFERENCES

Abu-Ghannam, N. and McKenna, B., The application of Peleg's equation to model water absorption during the soaking of red kidney beans (*Phaseolus vulgaris* L.), *J. Food Eng.*, 32, 391–401, 1997.

Ade-Omowaye, B.I.O., Rastogi, N.K., Angersbach, A., and Knorr, D., Osmotic dehydration behavior of red paprika (*Capsicum annuum* L.), *J. Food Sci.*, 67, 1790–1796, 2002.

Aguilera, J.M., Microstructure and food product engineering, *Food Technol.*, 54, 56–58, 2000.

Aguilera, J.M., Why food microstructure? *J. Food Eng.*, 67, 3–11, 2004.

Aguilera, J.M., Chiralt, A., and Fito, P., Food dehydration and product structure, *Trends Food Sci. Technol.*, 14, 432–437, 2003.

Aguilera, J.M., Michel, M., and Mayor, G., Fat migration in chocolate: Diffusion or capillary flow in a particulate solid? A hypothesis paper, *J. Food Sci.*, 69, R167–R174, 2004.

Angersbach, A. and Knorr, D., High intensity electric field pulses as pretreatment for affecting dehydration characteristics and rehydration properties of potato cubes, *Nahrung* 41, 194–200, 1997.

Babin, P., Della Valle, G., Chiron, H., Cloetens, P., Hoszowska, J., Pernot, P., Reguerre, A.L., Salvo, L., and Dendievel, R., Fast x-ray tomography analysis of bubble growth and foam setting during breadmaking, *J. Cereal Sci.*, 43, 393–397, 2006.

Beaudry, C., Raghavan, G.S.V., Ratti, C., and Rennie, T.J., Effect of four drying methods on the quality of osmotically dehydrated cranberries, *Dry. Technol.*, 22, 521–539, 2004.

Becker, H.A., On the absorption of liquid water by the wheat kernel, *Cereal Chem.*, 37, 309–323, 1960.

Bell, L.N. and Labuza, T.P., *Moisture Sorption—Practical Aspects of Isotherm Measurement and Use*, 2nd ed., American Association of Cereal Chemists, St. Paul, MN, 2000.

Bello, M., Tolaba, M.P., and Suarez, C., Factors affecting water uptake of rice grain during soaking, *Lebensm.-Wiss. Technol.*, 37, 811–816, 2004.

Benavente, D., Lock, P., Del Cura, M.A.G., and Ordonez, S., Predicting the capillary imbibition of porous rocks from microstructure, *Transp. Porous Media*, 49, 59–76, 2002.

Bilbao-Sainz, C., Andres, A., and Fito, P., Hydration kinetics of dried apple as affected by drying conditions, *J. Food Eng.*, 68, 369–376, 2005.

Brooks, R.H. and Corey, A.T., Properties of porous media affecting fluid flow, *J. Irrig. Drain. Am. Soc. Civil Eng.*, IR2, 61–68, 1966.

Carbonell, S., Hey, M.J., Mitchell, J.R., Roberts, C.J., Hipkiss, J., and Vercauteren, J., Capillary flow and rheology measurements on chocolate crumb/sunflower oil mixtures, *J. Food Sci.*, 69, E465–E470, 2004.

Chhinnan, M.S., Evaluation of selected mathematical models for describing thin-layer drying of in-shell pecans, *Trans. ASAE*, 27, 610–615, 1984.

Chiralt, A. and Fito, P., Transport mechanisms in osmotic dehydration: The role of the structure, *Food Sci. Technol. Int.*, 9, 179–186, 2003.

Contreras, C., Martin, M.E., Martinez-Navarrete, N., and Chiralt, A., Effect of vacuum impregnation and microwave application on structural changes which occurred during air-drying of apple, *Lebensm.-Wiss. Technol.*, 38, 471–477, 2005.

Contreras, C., Martin-Esparza, M.E., Martinez-Navarrete, N., and Chiralt, A., Influence of osmotic pre-treatment and microwave application on properties of air dried strawberry related to structural changes, *Eur. Food Res. Technol.*, 224, 499–504, 2007.

Cunha, L.M., Oliveira, F.A.R., and Ilincanu, L.A., Application of the probabilistic Weibull distribution to rehydration kinetics: Relationship between the model parameters and the underlying physical mechanisms, in *Proceedings of the 3rd Workshop of the Copernicus Project*, Oliveira, J.C. and Oliveira, F.A.R., Eds., Porto, 1998, pp. 9–13.

Cunha, L.M., Oliveira, F.A.R., and Oliveira, J.C., Optimal experimental design for estimating the kinetic parameters of processes described by the Weibull probability distribution function, *J. Food Eng.*, 37, 175–191, 1998.

Cunningham, S.E., McMinn, W.A.M., Magee, T.R.A., and Richardson, P.S., Modelling water absorption of pasta during soaking, *J. Food Eng.*, 82, 600–607, 2007.

Datta, A.K., Porous media approaches to studying simultaneous heat and mass transfer in food processes. I: Problem formulations, *J. Food Eng.*, 80, 90–95, 2007a.

Datta, A.K., Porous media approaches to studying simultaneous heat and mass transfer in food processes. II: Property data and representative results, *J. Food Eng.*, 80, 96–110, 2007b.

Debnath, S., Hemavathy, J., Bhat, K.K., and Rastogi, N.K., Rehydration characteristics of osmotic pretreated and dried onion, *Food Bioprod. Process.*, 82, 304–310, 2004.

del Valle, J.M., Stanley, D.W., and Bourne, M.C., Water absorption and swelling in dry bean seeds, *J. Food Process. Preserv.*, 16, 75–98, 1992.

Doulia, D., Tzia, K., and Gekas, V., A knowledge base for the apparent mass diffusion coefficient (D_{eff}) of foods, *Int. J. Food Prop.*, 3, 1–14, 2000.

Doymaz, I., Air-drying characteristics of tomatoes, *J. Food Eng.*, 78, 1291–1297, 2007.

Eshtiaghi, M.N., Stute, R., and Knorr, D., High-pressure and freezing pretreatment effects on drying, rehydration, texture and color of green beans, carrots and potatoes, *J. Food Sci.*, 59, 1168–1170, 1994.

Falade, K.O. and Abbo, E., Air-drying and rehydration characteristics of date palm (*Phoenix dactylifera* L.) fruits, *J. Food Eng.*, 79, 724–730, 2007.

Falcone, P.M., Baiano, A., Zanini, F., Mancini, L., Tromba, G., Dreossi, D., Montanari, F., Scuor, N., and d. Nobile, M.-A., Three-dimensional quantitative analysis of bread crumb by x-ray microtomography, *J. Food Sci.*, 70, E265–E272, 2005.

Falcone, P.M., Baiano, A., Zanini, F., Mancini, L., Tromba, G., Montanari, F., and d. Nobile, M.-A., A novel approach to the study of bread porous structure: Phase-contrast x-ray microtomography, *J. Food Sci.*, 69, FEP38–FEP43, 2004.

Fan, L.T., Chu, P., and Shellenberger, J.A., Diffusion of water in kernels of corn and sorghum, *Cereal Chem.*, 40, 303–313, 1963.

Farkas, B.E. and Singh, R.P., Physical properties of air-dried and freeze-dried chicken white meat, *J. Food Sci.*, 56, 611–615, 1991.

Fito, P. and Chiralt, A., Food matrix engineering: The use of the water–structure–functionality ensemble in dried food product development, *Food Sci. Technol. Int.*, 9, 151–156, 2003.

Funebo, T., Ahrne, L., Kidman, S., Langton, M., and Skjoldebrand, C., Microwave heat treatment of apple before air dehydration—Effects on physical properties and microstructure, *J. Food Eng.*, 46, 173–182, 2000.

Garcia-Pascual, P., Sanjuan, N., Bon, J., Carreres, J.E., and Mulet, A., Rehydration process of *Boletus edulis* mushroom: Characteristics and modelling, *J. Sci. Food Agric.*, 85, 1397–1404, 2005.

Garcia-Pascual, P., Sanjuan, N., Melis, R., and Mulet, A., *Morchella esculenta* (morel) rehydration process modelling, *J. Food Eng.*, 72, 346–353, 2006.

Gekas, V., *Transport Phenomena of Foods and Biological Materials*, CRC Press, Boca Raton, FL, 1992.

Giraldo, G., Vazquez, R., Martin-Esparza, M.E., and Chiralt, A., Rehydration kinetics and soluble solids lixiviation of candied mango fruit as affected by sucrose concentration, *J. Food Eng.*, 77, 825–834, 2006.

Giri, S.K. and Prasad, S., Drying kinetics and rehydration characteristics of microwave–vacuum and convective hot-air dried mushrooms, *J. Food Eng.*, 78, 512–521, 2007.

Gowen, A., Abu-Ghannam, N., Frias, J., and Oliveira, J., Modeling the water absorption process in chickpeas (*Cicer arietinum* L.)—The effect of blanching pre-treatment on water intake and texture kinetics, *J. Food Eng.*, 78, 810–819, 2007.

Gowen, A., Abu-Ghannam, N., Frias, J., and Oliveira, J., Optimisation of dehydration and rehydration properties of cooked chickpeas (*Cicer arietinum* L.) undergoing microwave–hot air combination drying, *Trends Food Sci. Technol.*, 17, 177–183, 2006.

Gowen, A.A., Abu-Ghannam, N., Frias, J.M., Barat, J.M., Andres, A.M., and Oliveira, J.C., Comparative study of quality changes occurring on dehydration and rehydration of cooked chickpeas (*Cicer arietinum* L.) subjected to combined microwave–convective and convective hot air dehydration, *J. Food Sci.*, 71, E282–E289, 2006.

Haedelt, J., Leo Pyle, D., Beckett, S.T., and Niranjan, K., Vacuum-induced bubble formation in liquid-tempered chocolate, *J. Food Sci.*, 70, E159–E164, 2005.

Hallstrom, B., Mass transport of water in foods—A consideration of engineering aspects, *J. Food Eng.*, 12, 45–52, 1990.

Hills, B.P., Babonneau, F., Quantin, V.M., Gaudet, F., and Belton, P.S., Radial NMR micro-imaging studies of the rehydration of extruded pasta, *J. Food Eng.*, 27, 71–86, 1996.

Horn, G.R. and Sterling, C., Studies on the rehydration of carrots, *J. Sci. Food Agric.*, 33, 1035–1041, 1982.

Hung, T.V., Liu, L.H., Black, R.G., and Trewhella, M.A., Water absorption in chickpea (*C. arietinum*) and field pea (*P. sativum*) cultivars using the Peleg model, *J. Food Sci.*, 58, 848–852, 1993.

Ilincanu, L.A., Oliveira, F.A.R., Drumond, M.C., Machado, M.F., and Gekas, V., Modeling moisture uptake and soluble solids losses during rehydration of dried apple pieces, in *Proceedings of the 1st Workshop of the Copernicus Project*, Oliveira, J.C., Ed., Escola Superior de Biotecnologia, Porto, Portugal, 1995, pp. 64–69.

Jambrak, A.R., Mason, T.J., Paniwnyk, L., and Lelas, V., Accelerated drying of button mushrooms, Brussels sprouts and cauliflower by applying power ultrasound and its rehydration properties, *J. Food Eng.*, 81, 88–97, 2007.

Kadam, D.M., Samuel, D.V.K., and Parsad, R., Optimisation of pre-treatments of solar dehydrated cauliflower, *J. Food Eng.*, 77, 659–664, 2006.

Karathanos, V.T., Anglea, S., and Karel, M., Collapse of structure during drying of celery, *Dry. Technol.*, 11, 1005–1023, 1993.

Karathanos, V.T., Kanellopoulos, N.K., and Belessiotis, V.G., Development of porous structure during air drying of agricultural plant products, *J. Food Eng.*, 29, 167–183, 1996.

Karathanos, V.T., Vagenas, G.K., and Saravacos, G.D., Water diffusivity in starches at high temperatures and pressures, *Biotechnol. Progr.*, 7, 178–184, 1991.

Kaymak-Ertekin, F., Drying and rehydrating kinetics of green and red peppers, *J. Food Sci.*, 67, 168–175, 2002.

Krokida, M.K. and Marinos-Kouris, D., Rehydration kinetics of dehydrated products, *J. Food Eng.*, 57, 1–7, 2003.

Krokida, M.K. and Maroulis, Z.B., Effect of drying method on shrinkage and porosity, *Dry. Technol.*, 15, 2441–2458, 1997.

Krokida, M.K. and Maroulis, Z.B., Effect of microwave drying on some quality properties of dehydrated products, *Dry. Technol.*, 17, 449–466, 1999.

Krokida, M.K. and Maroulis, Z.B., Structural properties of dehydrated products during rehydration, *Int. J. Food Sci. Technol.*, 36, 529–538, 2001.

Krokida, M.K. and Philippopoulos, C., Rehydration of dehydrated foods, *Dry. Technol.*, 23, 799–830, 2005.

Lee, K.T., Farid, M., and Nguang, S.K., The mathematical modelling of the rehydration characteristics of fruits, *J. Food Eng.*, 72, 16–23, 2005.

Levi, A., Ben Shalom, N., Plat, D., and Reid, D.S., Effect of blanching and drying on pectin constitutents and related characteristics of dehydrated peaches, *J. Food Sci.*, 53, 1187–1190, 1988.

Lewicki, P.P., Effect of pre-drying treatment, drying and rehydration on plant tissue properties: A review, *Int. J. Food Prop.*, 1, 1–22, 1998a.

Lewicki, P.P., Some remarks on rehydration of dried foods, *J. Food Eng.*, 36, 81–87, 1998b.

Lewicki, P.P., Design of hot air drying for better foods, *Trends Food Sci. Technol.*, 17, 153–163, 2006.

Lewicki, P.P., Le, H.V., and Pomaranska-Lazuka, W., Effect of pre-treatment on convective drying of tomatoes, *J. Food Eng.*, 54, 141–146, 2002.

Lewicki, P.P. and Pawlak, G., Effect of mode of drying on microstructure of potato, *Dry. Technol.*, 23, 847–869, 2005.

Lewicki, P.P. and Wiczkowska, J., Rehydration of apple dried by different methods, *Int. J. Food Prop.*, 9, 217–226, 2006.

Lewicki, P.P., Witrowa-Rajchert, D., Pomaranska-Lazuka, W., and Nowak, D., Rehydration properties of dried onion, *Int. J. Food Prop.*, 1, 275–290, 1998.

Lim, K.S. and Barigou, M., X-ray micro-computed tomography of cellular food products, *Food Res. Int.*, 37, 1001–1012, 2004.

Lin, T.M., Durance, T.D., and Scaman, C.H., Characterization of vacuum microwave, air and freeze dried carrot slices, *Food Res. Int.*, 31, 111–117, 1998.

Litvin, S., Mannheim, C.H., and Miltz, J., Dehydration of carrots by a combination of freeze drying, microwave heating and air or vacuum drying, *J. Food Eng.*, 36, 103–111, 1998.

Lu, R., Siebenmorgen, T.J., and Archer, T.R., Absorption of water in long-grain rough rice during soaking, *J. Food Process Eng.*, 17, 141–154, 1994.

Lucas, T., Le Ray, D., and Mariette, F., Kinetics of water absorption and solute leaching during soaking of breakfast cereals, *J. Food Eng.*, 80, 377–384, 2007.

Machado, M.F., Oliveira, F.A.R., and Cunha, L.M., Effect of milk fat and total solids concentration on the kinetics of moisture uptake by ready-to-eat breakfast cereal, *Int. J. Food Sci. Technol.*, 34, 47–57, 1999.

Machado, M.F., Oliveira, F.A.R., and Gekas, V., Modeling water uptake and soluble solids losses by puffed breakfast cereal immersed in water or milk, in *Proceedings of the Seventh International Congress on Engineering and Food—Part I*, Jowitt, R., Ed., Sheffield Academic Press, Sheffield, U.K., 1997, pp. A65–A68.

Machado, M.F., Oliveira, F.A.R., Gekas, V., and Singh, R.P., Kinetics of moisture uptake and soluble-solids loss by puffed breakfast cereals immersed in water, *Int. J. Food Sci. Technol.*, 33, 225–237, 1998.

Marabi, A., Dilak, C., Shah, J., and Saguy, I.S., Kinetics of solids leaching during rehydration of particulate dry vegetables, *J. Food Sci.*, 69, FEP91–FEP96, 2004.

Marabi, A., Jacobson, M., Livings, S., and Saguy, I.S., Effect of mixing and viscosity on rehydration of dry food particulates, *Eur. Food Res. Technol.*, 218, 339–344, 2004.

Marabi, A., Livings, S., Jacobson, M., and Saguy, I.S., Normalized Weibull distribution for modeling rehydration of food particulates, *Eur. Food Res. Technol.*, 217, 311–318, 2003.

Marabi, A. and Saguy, I.S., Effect of porosity on rehydration of dry food particulates, *J. Sci. Food Agric.*, 84, 1105–1110, 2004.

Marabi, A. and Saguy, I.S., Viscosity and starch particle size effects on rehydration of freeze-dried carrots, *J. Sci. Food Agric.*, 85, 700–706, 2005.

Marabi, A., Thieme, U., Jacobson, M., and Saguy, I.S., Influence of drying method and rehydration time on sensory evaluation of rehydrated carrot particulates, *J. Food Eng.*, 72, 211–217, 2006.

Marmur, A. and Cohen, R.D., Characterization of porous media by the kinetics of liquid penetration: The vertical capillaries model, *J. Colloid Interface Sci.*, 189, 299–304, 1997.

Maroulis, Z.B., Saravacos, G.D., Panagiotou, N.M., and Krokida, M.K., Moisture diffusivity data compilation for foodstuffs: Effect of material moisture content and temperature, *Int. J. Food Prop.*, 4, 225–237, 2001.

Marousis, S.N., Karathanos, V.T., and Saravacos, G.D., Effect of physical structure of starch materials on water diffusivity, *J. Food Process. Preserv.*, 15, 183–195, 1991.

Martinez-Navarrete, N. and Chiralt, A., Water diffusivity and mechanical changes during hazelnut hydration, *Food Res. Int.*, 32, 447–452, 1999.

Maskan, M., Microwave/air and microwave finish drying of banana, *J. Food Eng.*, 44, 71–78, 2000.

Maskan, M., Drying, shrinkage and rehydration characteristics of kiwifruits during hot air and microwave drying, *J. Food Eng.*, 48, 177–182, 2001.

Maskan, M., Effect of processing on hydration kinetics of three wheat products of the same variety, *J. Food Eng.*, 52, 337–341, 2002.

Mastrocola, D., Barbanti, D., Dalla Rosa, R.M., and Pittia, P., Physicochemical characteristics of dehydrated apple cubes reconstituted in sugar solutions, *J. Food Sci.*, 63, 495–498, 1998.

Mastrocola, D., Dalla Rosa, R.M., and Massini, R., Freeze-dried strawberries rehydrated in sugar solutions: Mass transfers and characteristics of final products, *Food Res. Int.*, 30, 359–364, 1997.

Mate, J.I., Quartaert, C., Meerdink, G., and van't Riet, K., Effect of blanching on structural quality of dried potato slices, *J. Agric. Food Chem.*, 46, 676–681, 1998.

Mayor, L. and Sereno, A.M., Modelling shrinkage during convective drying of food materials: A review, *J. Food Eng.*, 61, 373–386, 2004.

Meda, L. and Ratti, C., Rehydration of freeze-dried strawberries at varying temperatures, *J. Food Process Eng.*, 28, 233–246, 2005.

Misra, M.K. and Brooker, D.B., Thin-layer drying and rewetting equations for shelled yellow corn, *Trans. ASAE*, 23, 1254–1260, 1980.

Moreira, R.G. and Barrufet, M.A., A new approach to describe oil absorption in fried foods: A simulation study, *J. Food Eng.*, 35, 1–22, 1998.

Mousavi, R., Miri, T., Cox, P.W., and Fryer, P.J., A novel technique for ice crystal visualization in frozen solids using x-ray micro-computed tomography, *Int. J. Food Sci. Technol.*, 42, 714–727, 2007.

Nayak, C.A., Suguna, K., and Rastogi, N.K., Combined effect of gamma-irradiation and osmotic treatment on mass transfer during rehydration of carrots, *J. Food Eng.*, 74, 134–142, 2006.

Neubert, A.M., Wilson III, C.W., and Miller, W.H., Studies on celery rehydration, *Food Technol.*, 22, 1296–1301, 1968.

Nieto, A., Salvatori, D., Castro, M.A., and Alzamora, S.M., Air drying behaviour of apples as affected by blanching and glucose impregnation, *J. Food Eng.*, 36, 63–79, 1998.

Oliveira, F.A.R. and Ilincanu, L., Rehydration of dried plant tissues: Basic concepts and mathematical modeling, in *Processing Foods*, Oliveira, F.A.R. and Oliveira, J.C., Eds., CRC Press, Boca Raton, FL, 1999, pp. 201–227.

Oliveira, F.A.R. and Silva, C.L.M., Freezing influences diffusion of reducing sugars in carrot cortex, *J. Food Sci.*, 57, 932–934, 1992.

Pappas, C., Tsami, E., and Marinos, K.D., The effect of process conditions on the drying kinetics and rehydration characteristics of some MW–vacuum dehydrated fruits, *Dry. Technol.*, 17, 157–174, 1999.

Paulus, K. and Saguy, I.S., Effect of heat-treatment on the quality of cooked carrots, *J. Food Sci.*, 45, 239, 1980.

Peleg, M., An empirical-model for the description of moisture sorption curves, *J. Food Sci.*, 53, 1216–1219, 1988.

Pinthus, E.J. and Saguy, I.S., Initial interfacial tension and oil uptake by deep-fat fried foods, *J. Food Sci.*, 59, 804–807, 1994.

Planinic, M., Velic, D., Tomas, S., Bilic, M., and Bucic, A., Modelling of drying and rehydration of carrots using Peleg's model, *Eur. Food Res. Technol.*, 221, 446–451, 2005.

Prakash, S., Jha, S.K., and Datta, N., Performance evaluation of blanched carrots dried by three different driers, *J. Food Eng.*, 62, 305–313, 2004.

Prothon, F., Ahrne, L.M., Funebo, T., Kidman, S., Langton, M., and Sjoholm, I., Effects of combined osmotic and microwave dehydration of apple on texture, microstructure and rehydration characteristics, *Lebensm.-Wiss. Technol.*, 34, 95–101, 2001.

Quintero-Ramos, A., Bourne, M.C., and Anzaldua-Morales, A., Texture and rehydration of dehydrated carrots as affected by low temperature blanching, *J. Food Sci.*, 57, 1127–1128, 1992.

Rahman, M.S., Dried food properties: Challenges ahead, *Dry. Technol.*, 23, 695–715, 2005.

Ramos, I.N., Silva, C.L.M., Sereno, A.M., and Aguilera, J.M., Quantification of microstructural changes during first stage air drying of grape tissue, *J. Food Eng.*, 62, 159–164, 2004.

Rastogi, N.K., Angersbach, A., Niranjan, K., and Knorr, D., Rehydration kinetics of high-pressure pretreated and osmotically dehydrated pineapple, *J. Food Sci.*, 65, 838–841, 2000.

Ravindra, M.R. and Chattopadhyay, P.K., Optimisation of osmotic preconcentration and fluidised bed drying to produce dehydrated quick-cooking potato cubes, *J. Food Eng.*, 44, 5–11, 2000.

Rodriguez, S., Villanueva-Suarez, M.J., and Redondo, C.A., Effects of processing conditions on soluble sugars content of carrot, beetroot and turnip, *Food Chem.*, 66, 81–85, 1999.

Rosin, P. and Rammler, E., The laws governing the fineness of powdered coal, *J. Inst. Fuel.*, 7, 27–36, 1933.

Ruiz-Cabrera, M.A., Salgado-Cervantes, M.A., Waliszewski-Kubiak, K.N., and Garcia-Alvarado, M.A., The effect of path diffusion on the effective moisture diffusivity in carrot slabs, *Dry. Technol.*, 15, 169–181, 1997.

Ruiz-Diaz, G., Martinez-Monzo, J., and Chiralt, A., Modelling of dehydration–rehydration of orange slices in combined microwave/air drying, *Innov. Food Sci. Emerg. Technol.*, 4, 203–209, 2003.

Saca, S.A. and Lozano, J.E., Explosion puffing of bananas, *Int. J. Food Sci. Technol.*, 27, 419–426, 1992.

Sacchetti, G., Pittia, P., Biserni, M., Pinnavaia, G.G., and Rosa, M.D., Kinetic modelling of textural changes in ready-to-eat breakfast cereals during soaking in semi-skimmed milk, *Int. J. Food Sci. Technol.*, 38, 135–143, 2003.

Saguy, I.S. and Marabi, A., Rehydration of dried food particulates, in *Encyclopedia of Agricultural, Food, and Biological Engineering*, Heldman, D.R., Ed., Marcel Dekker, New York, 2005.

Saguy, I.S., Marabi, A., and Wallach, R., Liquid imbibition during rehydration of dry porous foods, *Innov. Food Sci. Emerg. Technol.*, 6, 37–43, 2005a.

Saguy, I.S., Marabi, A., and Wallach, R., New approach to model rehydration of dry food particulates utilizing principles of liquid transport in porous media, *Trends Food Sci. Technol.*, 16, 495–506, 2005b.

Salvatori, D., Andres, A., Chiralt, A., and Fito, P., Osmotic dehydration progression in apple tissue II: Generalized equations for concentration prediction, *J. Food Eng.*, 42, 133–138, 1999.

Sanjuan, N., Carcel, J.A., Clemente, G., and Mulet, A., Modelling of the rehydration process of broccoli florets, *Eur. Food Res. Technol.*, 212, 449–453, 2001.

Sanjuan, N., Clemente, G., Bon, J., and Mulet, A., The effect of blanching on the quality of dehydrated broccoli florets, *Eur. Food Res. Technol.*, 213, 474–479, 2001.

Sanjuan, N., Simal, S., Bon, J., and Mulet, A., Modelling of broccoli stems rehydration process, *J. Food Eng.*, 42, 27–31, 1999.

Saravacos, G.D. and Maroulis, Z.B., Transport of water in food materials, in *Transport Properties of Foods*, Marcel Dekker, New York, 2001, pp. 105–162.

Schwartzberg, H.G., Mathematical analysis of solubilization kinetics and diffusion in foods, *J. Food Sci.*, 40, 211–213, 1975.

Schwartzberg, H.G. and Chao, R.Y., Solute diffusivities in leaching processes, *Food Technol.*, 36, 73–86, 1982.

Shi, X.Q. and Fito-Maupoey, P., Mass transfer in vacuum osmotic dehydration of fruits: A mathematical model approach, *Lebensm.-Wiss. Technol.*, 27, 67–72, 1994.

Singh, B., Panesar, P.S., Gupta, A.K., and Kennedy, J.F., Optimisation of osmotic dehydration of carrot cubes in sucrose–salt solutions using response surface methodology, *Eur. Food Res. Technol.*, 225, 157–165, 2007.

Singh, B., Panesar, P.S., Nanda, V., Gupta, A.K., and Kennedy, J.F., 2006. Application of response surface methodology for the osmotic dehydration of carrots, *J. Food Process Eng.*, 29, 592–614.

Singh, P.P., Maier, D.E., Cushman, J.H., Haghighi, K., and Corvalan, C., Effect of viscoelastic relaxation on moisture transport in foods. Part I: Solution of general transport equation, *J. Math. Biol.*, 49, 1–19, 2004.

Sopade, P.A., Ajisegiri, E.S., and Badau, M.H., The use of Peleg's equation to model water-absorption in some cereal-grains during soaking, *J. Food Eng.*, 15, 269–283, 1992.

Sumnu, G., Turabi, E., and Oztop, M., Drying of carrots in microwave and halogen lamp-microwave combination ovens, *LWT Food Sci. Technol.*, 38, 549–553, 2005.

Taiwo, K.A., Angersbach, A., and Knorr, D., Influence of high intensity electric field pulses and osmotic dehydration on the rehydration characteristics of apple slices at different temperatures, *J. Food Eng.*, 52, 185–192, 2002a.

Taiwo, K.A., Angersbach, A., and Knorr, D., Rehydration studies on pretreated and osmotically dehydrated apple slices, *J. Food Sci.*, 67, 842–847, 2002b.

Tillotson, J.E., Convenience foods, in *Encyclopedia of Food Sciences and Nutrition*, Trugo, L. and Finglas, P.M., Eds., Academic Press, Oxford, U.K., 2003, pp. 1616–1622.

Tomasula, P. and Kozempel, M.F., Diffusion coefficients of glucose, potassium, and magnesium in Maine Russet Burbank and Maine Katahdin potatoes from 45 to 90 degrees, *J. Food Sci.*, 54, 985–989, 1989.

Torringa, E., Esveld, E., Scheewe, I., van den Berg, R., and Bartels, P., Osmotic dehydration as a pre-treatment before combined microwave–hot-air drying of mushrooms, *J. Food Eng.*, 49, 185–191, 2001.

Trater, A.M., Alavi, S., and Rizvi, S.S.H., Use of non-invasive x-ray microtomography for characterizing microstructure of extruded biopolymer foams, *Food Res. Int.*, 38, 709–719, 2005.

Turhan, M., Sayar, S., and Gunasekaran, S., Application of Peleg model to study water absorption in chickpea during soaking, *J. Food Eng.*, 53, 153–159, 2002.

Vagenas, G.K. and Karathanos, V.T., Prediction of moisture diffusivity in granular materials, with special applications to foods, *Biotechnol. Prog.*, 7, 419–426, 1991.

van Genuchten, M.Th., A closed-form equation for predicting the hydraulic conductivity of unsaturated soils, *Soil Sci. Soc. Am. J.*, 44, 892–898, 1980.

Vidal, V.C., Frias, J., and Valverde, S., Effect of processing on the soluble carbohydrate content of lentils, *J. Food Protect.*, 55, 301–304, 1992.

Vidal, V.C., Frias, J., and Valverde, S., Changes in the carbohydrate composition of legumes after soaking and cooking, *J. Am. Diet. Assoc.*, 93, 547–550, 1993.

Waananen, K.M., Litchfield, J.B., and Okos, M.R., Classification of drying models for porous solids, *Dry. Technol.*, 11, 1–40, 1993.

Waananen, K.M. and Okos, M.R., Effect of porosity on moisture diffusion during drying of pasta, *J. Food Eng.*, 28, 121–137, 1996.

Wang, J. and Chao, Y., Effect of gamma irradiation on quality of dried potato, *Radiat. Phys. Chem.*, 66, 293–297, 2003.

Wang, J. and Du, Y.S., The effect of gamma-ray irradiation on the drying characteristics and final quality of dried potato slices, *Int. J. Food Sci. Technol.*, 40, 75–82, 2005.

Weerts, A.H., Lian, G., and Martin, D., Modeling rehydration of porous biomaterials: Anisotropy effects, *J. Food Sci.*, 68, 937–942, 2003a.

Weerts, A.H., Lian, G., and Martin, D., Modeling the hydration of foodstuffs: Temperature effects, *AIChE J.*, 49, 1334–1339, 2003b.

Weibull, W., A statistical theory of the strength of materials, *R. Swed. Inst. Eng Res. Process.*, 151, 5–51, 1939.

Witrowa-Rajchert, D. and Lewicki, P.P., Rehydration properties of dried plant tissues, *Int. J. Food Sci. Technol.*, 41, 1040–1046, 2006.

Xu, Y., Zhang, M., Tu, D., Sun, J., Zhou, L., and Mujumdar, A.S., A two-stage convective air and vacuum freeze-drying technique for bamboo shoots, *Int. J. Food Sci. Technol.*, 40, 589–595, 2005.

Zogzas, N.P., Maroulis, Z.B., and Marinos-Kouris, D., Densities, shrinkage and porosity of some vegetables during air drying, *Dry. Technol.*, 12, 1653–1666, 1994..

10 Dehydration Processes for Nutraceuticals and Functional Foods

Susan D. St. George and Stefan Cenkowski

CONTENTS

10.1   INTRODUCTION

Awareness of the potential role of diet in the prevention and management of chronic and degenerative disease has led to a demand for high quality, safe, and effective healthy and beneficial food and food-based products. This growing demand has resulted in the formation and development of the nutraceutical and functional food industries. As with the development of any new industry, this creates the need for a greater knowledge base and associated methods and technology development.

Ongoing contributions to this knowledge and methods base include the identification of structure, activity, and quantification of major biologically active (bioactive) compounds. Until recently, food product quality was based primarily on physical and sensory characteristics; however, it can now extend to the preservation of key bioactive compounds. This knowledge can facilitate the assessment of the effects of processing on individual bioactive compounds in the final product. A variety of processing methods, including drying, are employed in the development of nutraceuticals and functional foods.

Because of the continual growth of the nutraceutical and functional food industries, there is a constant stream of studies reporting new research developments. The aim of this discussion is to provide a current overview of the use of drying in the development of safe and effective plant-based nutraceuticals and functional foods. This chapter is divided into four main sections: an overview of the nutraceutical and functional food industries, applicable drying technologies, drying methods and quality retention, and latest advances in the area.

10.2   NUTRACEUTICALS AND FUNCTIONAL FOODS

The traditional role of diet has been to provide energy and essential nutrients to sustain life and growth. However, with aging populations, longer life expectancies, and increasing health-care costs, the developed world is addressing the role of diet and lifestyle in the prevention and management of chronic and degenerative diseases (Oomah and Mazza, 2000). Preventable diseases such as certain cancers, heart disease, stroke, diabetes, diseases of the arteries, and osteoporosis have important dietary links (WHO, 2004). Part of a strategy to boost health and reduce the risk of disease involves incorporating foods and food products rich in healthful bioactive compounds into a healthy lifestyle, which includes exercise and a nutritious diet (Oomah and Mazza, 2000).

Several terms are used to describe the many natural products currently being developed for health benefits: functional foods, nutraceuticals, pharmafoods, designer foods, vitafoods, phytochemicals, phytofoods, medical foods, and "foodaceuticals" (foods combined with drugs to enhance body and mind; Oomah and Mazza, 2000; Small and Catling, 1999). Although the terms nutraceutical and functional food are the most commonly used worldwide, there is no definite consensus on their meaning. The definitions recognized by Health Canada (1998) will be employed to maintain consistency throughout the remainder of this discussion. The following sections will provide a description of the most commonly used terms in the nutraceutical and functional food industries, potential activity associated with certain bioactive compounds, product development, and compound stability.

10.2.1 NUTRACEUTICALS

A plant-based nutraceutical is a product isolated or purified from a food derived from plant material, and it is generally sold in medicinal forms (capsules, tablets, powders, and potions) that are not usually associated with foods (Health Canada, 1998). To be considered a nutraceutical, a product must also offer a physiological benefit or assist in the management or prevention of chronic and/or degenerative disease. Within Canada, nutraceuticals are included in the category of natural health products. Other natural health products include traditional herbal medicines; Chinese, Ayurvedic, and Native North American medicines; homeopathic preparations; and vitamin and mineral supplements.

10.2.2 FUNCTIONAL FOODS

A functional food appears similar to or is a conventional food that is consumed as part of a usual diet and has a physiological benefit or assists in the management or prevention of chronic and/or degenerative disease (Health Canada, 1998). A food can be made functional through the elimination of a compound having a negative physiological effect, increasing the concentration of beneficial compounds, adding a new compound observed to offer benefits, and partial replacement of a negative compound by a beneficial one without adversely affecting the nutritional value of the food (Gibson and Fuller, 1998).

10.2.3 PHYTOCHEMICALS

Plant chemicals (phytochemicals) are the bioactive compounds that contribute to the activity of plant-based nutraceuticals and functional foods (Oomah and Mazza, 2000).

Plants synthesize both primary and secondary metabolites (Webb, 2006; Wildman, 2001a). Primary metabolites include proteins, amino acids, chlorophyll, membrane lipids, nucleotides, and carbohydrates necessary for the existence of plants. Secondary metabolites have not been linked to plant processes (photosynthesis, respiration, etc.), and they were originally regarded as nonfunctional waste products. However, researchers now recognize that secondary metabolites, which fulfill important functions, may be associated only with certain plant species or taxonomically related groups. Functions include protecting the plant from herbivores, insects, fungi, bacteria, microbial infection, and ultraviolet light, as well as producing colorful pigments that attract insects and birds for pollination and seed dispersal. Secondary metabolites can be divided into three main groups: isoprenoid derivatives, phenolics, and sulfur and nitrogen containing compounds.

10.2.4 CLASSIFICATION AND ACTIVITY OF MAJOR BIOACTIVE COMPOUNDS

Combining primary and secondary plant metabolites enables a simple classification of bioactive compounds based on their chemical nature: isoprenoid derivatives, phenolic substances, fatty acids (FAs) and structural lipids, carbohydrates and derivatives, amino acid based substances, and minerals (Wildman, 2001b).

10.2.4.1 Isoprenoids

The isoprenoid derivative class of 25,000 different substances, one of the largest of the secondary metabolites, includes carotenoids, saponins, sterols, and simple terpenes (Wildman, 2001b). The carotenoids, including carotenes and xanthophylls (pigments that produce yellow, orange, and red colors), also play a significant role in photosynthesis and photoprotection. Saponins and sterols are gaining interest because of their potential cholesterol reducing ability (Wasowicz, 2003). Essential oils, a mixture of volatile monoterpenes and sesquiterpenes that impart the characteristic odor associated with specific plants, are often used in flavorings and perfumes (Webb, 2006).

10.2.4.2 Phenolics

Phenolics, a class of more than 8000 secondary metabolites, includes cinnamic and benzoic acid derivatives and simple phenols, coumarins, flavonoids and stilbenes, lignans and lignins, suberins and cutins, tannins, and tocopherols and tocotrienols (Shahidi and Naczk, 2004). Among the flavonoid groups are flavonols (quercitin, kaempferol, myricetin), flavones (rutin), flavanols (catechins, proanthrocyanidins), anthocyanidins (cyanidin), flavonones (hesperidin), and isoflavones (genistein, daidzein; Shahidi and Naczk, 2004; Webb, 2006). Tocopherol (α-T, β-T, γ-T, and δ-T) and tocotrienol (α-T3, β-T3, γ-T3, and δ-T3) isomers offer vitamin E activity (Shahidi and Naczk, 2004).

10.2.4.3 FAs and Structural Lipids

Plants contain structural lipids, which build cell membranes, and triacylglycerols (Kołakowska and Sikorski, 2003). Structural lipids consist of phospholipids (lecithin) and glycolipids. Lipids offer important functions in nutrient and antioxidant delivery (tocopherols, tocotrienols, and carotenoids). FA composition, which differs between plant products, determines the physical properties, stability, and nutritive value of lipids. Although many FAs contribute to health benefits, of particular interest are the essential polyunsaturated FAs (PUFAs), linoleic and α-linolenic acids (ALA), which cannot be synthesized by human or animals. These acids convert to eicosanoids that affect physiological reactions ranging from blood clotting to immune response. Oleic acid, a monounsaturated FA, is valued for its cholesterol lowering effect.

10.2.4.4 Carbohydrates and Derivatives

Among the carbohydrate and derivatives class, ascorbic acid (vitamin C) is one of the most utilized functional ingredients. Certain oligosaccharides may function as prebiotics that promote the growth of beneficial bacteria in the gastrointestinal tract. Another term for nonstarch plant polysaccharides is fiber, which can be separated into two groups: soluble and insoluble. Included within the insoluble fiber group are cellulose, hemicellulose, and the phenol lignan (Jalili et al., 2001).

10.2.4.5 Proteins and Amino Acids

The protein or amino acid based group includes intact protein (soy), polypeptides, amino acids, and nitrogenous and sulfur amino acid derivatives (capsaicinoids, isothiocyanates, allyl-s compounds; Wildman, 2001b). The amino acids arginine, ornithine, taurine, and aspartic acid have been investigated for their functional activity.

10.2.4.6 Minerals

Specific minerals (i.e., calcium, potassium, and trace minerals such as copper, selenium, manganese, and zinc) are recognized for their functionality and are being included as nutraceutical and functional food ingredients (Wildman, 2001b). The activity of trace minerals is attributable to antioxidation. However, as with many other natural compounds, toxicity may be an issue, depending on the dosage.

10.2.4.7 Activity of Bioactive Compounds

Organizing the compounds according to structure does not provide a clear indication of their respective functionality or activity. Major bioactive compounds, potential activity, and health benefits for selected foods are listed in Table 10.1. Much research is still required in human clinical testing to ascertain whether claims made for specific compounds or foods are valid. Comparing the bioactive compound classes that are shared by cranberries and blueberries, we would expect the same activity and health benefits. However, each of these fruits is composed of a unique set of compounds. A common thread that the four foods shown in Table 10.1 share is that they all possess antioxidant activity. Oxidative stress has been linked to many degenerative diseases including cancer, atherosclerosis, cardiovascular disease, arthritis, and diabetes. Antioxidants can restrict the effects of oxidation either directly (i.e., elimination of free radicals) or indirectly (i.e., prevention of radical formation; Shahidi and

TABLE 10.1
Major Bioactive Compounds, Activity, and Potential Health Benefits of Foods

Plant	Bioactive Compound Class	Activity	Condition Prevention or Management	Ref.
Cranberries	Phenolic acids, anthocyanins, flavonols, flavanols, procyanidins, proanthocyanidins	Antioxidant, antibacterial, antiinflammatory	Certain cancers, urinary tract infections	Vvedenskaya et al. (2004), Neto (2007)
Blueberries	Phenolic acids, anthocyanins, flavonols, flavanols, procyanidins, proanthocyanidins, tannins	Antioxidant, antiinflammatory, antiedemic, astringent, regeneration of rhodopsin	Platelet aggregation, vascular disorders, ulcers, diabetic retinopathy	Bruneton (1999), Fleming (2000)
Tomatoes	Carotenoids ascorbic acid, folate, polyphenols, vitamins A and E	Antioxidant, immuno-stimulatory	Certain cancers, artherosclerosis, coronary heart disease, boosting of immune system	Shi et al. (2002), Eskin and Tamir (2006)
Carrots	Carotenoids, insoluble fibers, vitamin A	Antioxidant, prooxidant, hypoglycemic	Cancers and cardiovascular disease (mixed results), Alzheimer disease	Eskin and Tamir (2006)

Naczk, 2004). There is some evidence, however, that in certain instances some compounds such as β-carotene can possess prooxidant activity (Eskin and Tamir, 2006).

10.2.5 PRODUCT DEVELOPMENT

A separate discussion of the development of nutraceuticals and functional foods is appropriate because they are available in very different forms. Nutraceuticals can be developed from raw plant materials or a portion of the product as extracted compounds or components. Processed raw materials can include or contain granules or powders of solids (e.g., garlic, cranberry, and ginseng root) and powdered or concentrated juice (e.g., cranberry and blueberry; Bruneton, 1999; Kim et al., 2002; Li et al., 2007). Purified and essential extracted compounds (e.g., peppermint oil and ALA) can be provided in a liquid or powder form (Baranauskiené et al., 2007; Sanguansri and Augustin, 2007). Products are generally standardized to contain a specific quantity of the active ingredient, for example, tablets standardized at 25 to 36% anthocyanin content (Fleming, 2000).

Functional foods have a similar appearance to conventional foods, but they contain bioactives inherent to the raw food or have added compounds or components from other sources. Fruit powders (blueberry, cranberry, concord grape, and raspberry) for use in extruded corn breakfast cereals were determined to be a potential source of anthocyanins if added in adequate proportions (Camire et al., 2007). Other functional foods include tea, fresh or dried fruits and vegetables, baked goods, bread, beverages, jellies, and granola bars, as long as they contain valuable bioactive compounds that provide health benefits beyond basic nutrition (Grabowski et al., 2007; Kim et al., 2002; Nindo et al., 2007).

The original raw product or a component thereof (e.g., juice or seeds) can be maintained intact by using it in the fresh or dried state. In some instances it is desirable to isolate specific bioactive compounds from the original product. Several processes have been investigated for the extraction and purification of bioactive compounds including supercritical fluid technology (Shi et al., 2007), pressurized low polarity water extraction (Cacace and Mazza, 2007), vacuum distillation (Bettini, 2007), and membrane separation (Kumar, 2007). Drying can play an integral role as a pretreatment and/or a finishing step for each of these extraction systems. Microencapsulation, the coating of food ingredients, enzymes, cells, and extracts in small capsules, is performed using various techniques including spray drying, fluidized and spouted bed coating, and freeze-drying (Desai and Park, 2005; Madene et al., 2006). It is being employed in the nutraceutical and functional food industries for retention of aroma and flavor; protection of core material from evaporation, degradation, and reactivity with surroundings; modification of the original material for ease of handling; controlled release; and flavor masking; and to allow for uniform dispersion in the final product.

10.2.6 Bioactive Compound Stability during Drying

Drying is a complex process that may involve simultaneous heat and mass transfer and the subsequent removal of moisture from the system (Cassini et al., 2006). The potential for enzyme-catalyzed reactions, nonenzymatic and Maillard reactions, protein denaturation, and nutrient loss (Roos, 2004) are associated with drying. Color degradation of fruits and vegetables during drying may be due to pigment loss or browning, which is caused by both enzymatic and nonenzymatic browning reactions (Krokida et al., 2001). Enzymes are deactivated at temperatures ranging from 60 to >100°C and activity drops as water activity (a_w) approaches 0.2 (Van Den Berg, 1986). Nonenzymatic browning is caused by caramelization of reducing sugars and ascorbic acid and by Maillard browning, a reaction between amines (amino acids or proteins) and carbonyls (sugars or flavors; Bell, 2001; Pokorný and Schmidt, 2003).

The main factor that can affect stability during the process of air drying (natural or forced) is oxidation, the rate of which can be influenced by oxygen, temperature, light, a_w, pH, and prooxidants (i.e., enzymes, minerals, and metals; DeMan, 1999). Oxidation is reduced to a minimum at an a_w value of 0.3, whereas at intermediate a_w values (0.5 to 0.8) the oxidation rate increases as a result of increased catalyst activity. Some compounds that are sensitive to oxidation and therefore to one or more of the oxidative catalysts include anthocyanins, carotenoids, vitamin C, PUFAs, and chlorophyll (Cui et al., 2004; Litwinienko and Kasprzycka-Guttman, 2000; Shi et al., 2002; Skrede and Wrolstad, 2002). The presence of antioxidants (vitamin C, tocols, and carotenoids) added to or present in the product can reduce the rate of oxidation (Cohen et al., 2000).

Researchers have determined that these reactions are time and temperature dependent and product degradation may be decreased with the combination of milder temperatures and reduced residence time in hot air driers (Nindo et al., 2007). Traditionally, culinary and medicinal herbs have been air dried because most products are heat labile and require low temperature drying (Kabganian et al., 2002; Kim et al., 2000; Rohloff et al., 2005). Many other technologies that reduce the potential of oxidation because of the removal of heat and/or air are also available. The ability to produce a uniform and effective product requires knowledge of its drying characteristics and a high level of control within the drying process.

10.3 APPLICABLE DRYING METHODS

Drying may be used as a processing stage to improve shelf life; for quality preservation (halt or reduce the growth of spoilage microorganisms and chemical reactions); for product enhancement; to simplify handling, storage, and transport; and as a pretreatment for subsequent processes (Cassini et al., 2006; Vega-Mercado et al., 2001). Many drying systems and technologies are available for drying plant products. Selection of the appropriate drying method depends on the original product and bioactive compound characteristics, availability of equipment and technology, required quality and characteristics of the final product, and economics. Because of the lack of information on drying characteristics and the complexity of many plant-based products and their compounds, studies on drying kinetics and quality have been performed on an individual basis. Drying methods that have been either used or investigated for the processing of plant-based food products that are of potential use in the nutraceutical and functional food industries are summarized in the following sections.

10.3.1 AIR, SUN, AND SOLAR DRYING

Air drying, a method by which the convective flow of ambient air removes moisture from a product, has been used for many crops including herbs such as peppermint (Rohloff et al., 2005), basil (Díaz-Maroto et al., 2004), and bay leaf (Díaz-Maroto et al., 2002). This method can be employed directly in the field (i.e., prewilting) and in vented structures. Exposing the product to sunlight employs radiant energy in the evaporation of water from products that are spread on a flat surface (e.g., concrete; Sokhansanj and Jayas, 2006). Sun drying is a traditional method in warm climates that is utilized for seaweed (Chan et al., 1997), rooibos tea (Joubert and de Villiers, 1997), figs, plums, coffee beans, cocoa beans, sweet peppers, pepper, and rice (Ibarz and Barbosa-Canovas, 2003).

Both of these natural methods offer a low cost alternative to mechanical systems, especially in areas where such equipment is not easily accessible. The disadvantages, however, include lack of control of the drying process; loss of product due to germination and nutritional changes; lack of uniformity in drying; contamination attributable to molds, bacteria, rodents, birds, and insects; potentially long drying times; and dependence on climate. Drying time can also vary depending on the product's physical characteristics (i.e., color, shape, structure) and initial and final moisture contents (Sokhansanj and Jayas, 2006). These methods may not be suitable for high value products that contain light- (sun drying) and oxygen-sensitive bioactive compounds.

Solar drying employs the sun's energy without direct exposure to UV light. Collected energy can be used to increase the air temperature by ~10°C (Sokhansanj and Jayas, 2006) or assist with air flow or drying systems such as a vacuum evaporator or osmotic drying system (Grabowski and Mujumdar, 1992). Farkas (2004) provided a comprehensive review of more than 20 solar drying configurations for products including grains, fruits, vegetables, herbs, and botanicals. Systems range from the basic solar heating of air upstream of the drying chamber to units with storage (water vessels, rock bed, foil), heat recovery systems, and/or auxiliary heating. A project establishing dryers in five Association of Southern Asian Nations countries (i.e., Indonesia, Malasia, Phillipines, Singapore, and Thailand) resulted in several economically viable solar and solar or heat pump configurations for the drying of teas, fruit, and other crops indigenous to the area (Halawa et al., 1997). Solar drying is the most applicable in hot and dry countries especially if the crop is to be harvested just prior to the dry season and where commercial drying is not available or feasible (Farkas, 2004).

10.3.2 Osmotic Drying and Pretreatments

Some berries and fruits, such as blueberries, cranberries, grapes, and tomatoes, have an external waxy layer that offers benefits such as protection of the fruit from environmental and external factors (Sunjka and Raghavan, 2004; Wildman, 2001a). The waxy layer also affects the flow of moisture from inside the fruit to its surface, slowing the drying rate and thus increasing drying time. The following methods can be used separately or in combination as an initial dewatering stage in combination with a finish drying system (Sunjka and Raghavan, 2004).

Osmotic drying involves the removal of water with an osmotic agent (hypertonic sugar or salt solution). This system consists of two countercurrent flows with water diffusion from the product to the surrounding solution and migration of solutes from the solution to the product (Sunjka and Raghavan, 2004). This is particularly useful for products that require application of sugar (e.g., cranberries) or salt (e.g., tomatoes) within the processing phase.

As shown in Chapter 1, three basic types of pretreatment methods have been investigated: chemical, mechanical, and thermal. Chemical methods involve dipping the product in alkaline or acidic solutions of oleate esters (Sunjka and Raghavan, 2004). Alkaline solutions develop fine cracks on product surfaces, whereas acidic solutions cause the wax platelets on the skin to dissociate. Mechanical methods include skin abrasion, puncturing, cutting, or peeling of the product (Sunjka and Raghavan, 2004). Thermal treatments include the exposure of the product to hot or cold fluids to alter the skin permeability (Grabowski et al., 2007; Sablani, 2006).

10.3.3 Convective Air Dryers

Cabinet and tunnel drying systems are based on convective heat and mass transfer facilitated by warm air flowing over the surface of a thin layer of product spread on mesh or solid trays (Sokhansanj and Jayas, 2006). These systems are typically used for sliced, chunked, or whole fruits and vegetables or other plant material such as seeds, flowers, roots, and leaves (Orsat and Raghavan, 2007; Vega-Mercado et al., 2001). The use of solid trays also allows the drying of fluids and semisolid products such as slurries, purees, and pomace. The main differences between cabinet and tunnel drying systems are configuration and capacity, with the latter being applicable to high volume products.

Belt and conveyor band dryers also employ convective drying mechanisms. However, the drying surface moves during the drying cycle and air flow is through rather than over the thin product layer (Sokhansanj and Jayas, 2006). Conveyor band dryers offer the flexibility of having separate regions with different air characteristics to accommodate product requirements throughout the drying cycle as well as continuous drying (Orsat and Raghavan, 2007). Product degradation is a major drawback associated with the systems discussed in this section because of potentially high drying temperatures, excessive air flow, and/or long drying times. Optimization of drying conditions according to product characteristics and behavior can assist in dealing with these issues.

10.3.4 FIXED BED AND FLUIDIZED SYSTEMS

A fixed deep bed system is one in which granular or solid products (i.e., grain, oilseeds, and legumes) remain stationary as the air flows through the layer (Vega-Mercado et al., 2001). Modifications, including fluidization and movement of the particles within the bed, has led to the development of fluidized, pulsed fluidized, vibrated fluidized, spouted bed, and rotary dryers. These methods offer an alternative for heat- and oxygen-sensitive products because of faster drying rates and concomitant reduced drying times and potential for lower drying temperatures (Kundu, 2004).

10.3.4.1 Fluidized Bed Drying

Fluidized bed dryers have a perforated floor through which air flows at a specific velocity to fluidize the product (Kundu, 2004). This dryer type is suitable for granular and solid products such as oilseeds (e.g., mustard, sunflower, soybean, groundnut, and rapeseeds). This method was also investigated for cranberries (Grabowski et al., 2002). Systems such as the pulsed fluidized bed and the vibrofluidized bed dryer produced better test results than the fluidized bed dryer for products that are difficult to fluidize, which is due to stickiness or polydispersity (Benali, 2004). The aerodynamic and mechanical functions of pulsed fluidized bed and vibrofluidized bed dryers, respectively, can assist in preventing agglomeration. The pulsed fluidized bed dryer, a novel and modified version of conventional fluidized bed systems, employs gas pulses to provide high-frequency retraction of the particle bed. These methods were investigated for the drying of particulate solids such as diced carrots, cranberries, blueberries, and onions (Benali, 2004; Grabowski et al., 2002, 2007). These systems can be run either continuously or in batch mode.

10.3.4.2 Spouted Bed Drying

A spouted bed dryer uses a high velocity jet for material agitation. Two distinct zones exist within a spouted bed dryer: a central jet with minimal product and high heat and mass transfer rates and a moving annular bed with lower heat and mass transfer rates. Originally developed for grain drying in the 1960s (Mathur and Epstein, 1974), the application has been investigated for soybeans (Wiriyaumpaiwong et al., 2004), medicinal plant extracts (Oliveira et al., 2006), and blueberries (Feng et al., 1999). With the addition of uniformly sized inert particles, liquids, slurries, pulps, and pastes that coat the inerts can also be dried (Orsat and Raghavan, 2007). As the product becomes dry the adhered material is removed by chipping, which is caused by collisions between inert particles. This system is currently used for small volumes in batch mode, because of difficulties involved with scale-up and continuous systems.

10.3.4.3   Rotary Drying

A rotary dryer consists of an angled cylindrical rotary drying chamber turning at slow speed. Drying air and wet solids are fed continuously at one end of the chamber and are discharged at the other end (Kundu, 2004). Allowing the air to flow in at the material outlet, resulting in a countercurrent flow, improves drying performance. The material is continuously churned, which is due to rotation, leading to uniform mixing with drying air. The retention time is a function of the speed and angle of inclination of the chamber. This system is traditionally used for drying chemicals. However, it has been used for some agricultural products such as dry beans, rice, nuts, seeds, cereal grains, and herbs (Benali, 2004). The rotational system leads to an increase in capital investment (Kundu, 2004).

10.3.4.4   Encapsulation with Fluidized Bed Technology

Special configurations of fluidized bed dryers, such as those shown in Figure 10.1, which were originally established for the production of pharmaceuticals, have been modified for the encapsulation of food ingredients, additives, and nutraceutical products (Desai and Park, 2005). The top spraying configuration (Figure 10.1a) is most widely used for the microencapsulation of droplets as small as 100 μm, because of its versatility, capacity, and relative simplicity. Similar sized particles have also been successfully coated using the bottom spray configuration (Figure 10.1b), also known as the Wûrster system. Figure 10.1c depicts another modification using a rotating drum that enhances gravitational forces and assists in fluidizing smaller particles. The fluidized bed drying coating process (Figure 10.2) involves fluidization of core particles, spraying of a coating solution, contact spreading and coalescence of the coating solution on the core, evaporation from the coating solution, and adherence of the coating material to the core (Madene et al., 2006). Coating materials can include carbohydrates (e.g., starch, maltodextrins, and corn syrup solids), cellulose (hemicellulose), gums (e.g., gum acacia, agar, and carrageenan), lipids (e.g., wax, paraffin, and oils), and proteins (e.g., gluten, casein, and gelatin; Desai and Park, 2005; Madene et al., 2006).

10.3.5   DRUM AND REFRACTANCE WINDOW® DRYERS

Drum drying consists of indirect moisture removal from a thin film of product on the surface of internally heated twin (or single drum) hollow metal cylinders that rotate

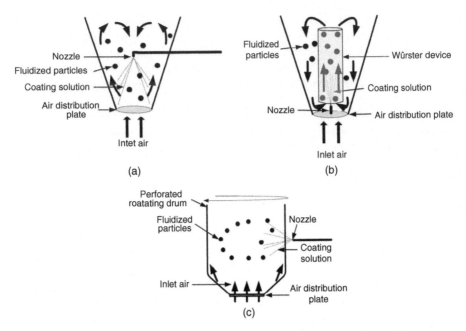

FIGURE 10.1 Fluidized bed coating systems with three configurations: (a) top spray, (b) bottom spray, and (c) tangential spray. (Adapted from Desai, K.G.H. and Park, H.J., *Dry. Technol.*, 23, 1361–1394, 2005. With permission.)

on a horizontal axis (Orsat and Raghavan, 2007; Vega-Mercado et al., 2001). Dried product is flaked off using a scraper. This system is applicable to viscous foods and purees that can withstand high temperatures for a short period. Drum dried powdered and flaked products are used in bakery goods, beverages, cereal, granola, and dairy foods (Vega-Mercado et al., 2001). This method was also investigated as a texturizing method for wheat, rice, and fababean mixed breakfast cereals or puffed baked snacks

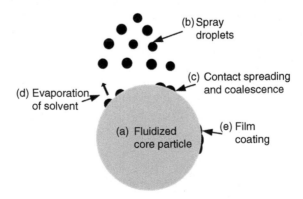

FIGURE 10.2 Schematic of fluidized bed coating process. (Adapted from Madene, A. et al., *Int. J. Food Sci. Technol.*, 41, 1–21, 2006. With permission.)

(Abdel-Aal et al., 1996) and for the processing of apple pomace (Constenla et al., 2002). The effectiveness of the drying system relies on the uniform thickness of the film applied to the drum surface, the speed of rotation, and the heating temperature. The main advantages of this system are the high drying rate and energy efficiency.

Refractance Window drying is a relatively new indirect drying technology that is used to evaporate moisture from foods (Nindo and Tang, 2007; Nindo et al., 2007; Vega-Mercado et al., 2001). This approach uses a plastic film to facilitate thermal energy transfer between a heating medium (i.e., water below the film) and a suspension (on the surface of the film). The thin plastic film, fabricated from Mylar, allows the transmission of infrared radiation in a wavelength that matches the absorption spectrum of water in the product. Refractance Window was developed to provide an alternative to long drying times and/or the use of high temperatures. Although the water is maintained at temperatures just below boiling at 95 to 98°C, product temperatures do not exceed 70°C. Shorter drying times are demonstrated by a reduction from hours (tray and freeze-drying) to 5 min (Refractance Window) for strawberry puree (Nindo and Tang, 2007). This system is applicable to the drying or preconcentration of liquid foods. It is also being pursued for its potential to process fruits, vegetables, and herbs into value-added powders, flakes, and concentrates.

10.3.6 SPRAY DRYING

The spray drying process consists of the conversion of a spray of pumpable liquid (i.e., juices, slurries, and purees) into a dry particulate (i.e., powder, granules, or agglomerate) by exposure to a hot (150 to 200°C) medium (Orsat and Raghavan, 2007; Vega-Mercado et al., 2001). Operating and dryer components that influence the final product include the feed rate, temperature of the inlet drying air, pressure of compressed air at the nozzle, air flow (i.e., cocurrent, countercurrent, or mixed flow), atomizer design, and air heating method. Spray dryers are the most widely used drying systems for the formation of powdered food additives and flavors in the dairy, beverage, and pharmaceutical industries (Benali, 2004; Desai and Park, 2005). This drying method has been investigated for many products including tomato pulp (Goula and Adamopoulos, 2005, 2006; Goula et al., 2004), flaxseed gum (Oomah and Mazza, 2001), medicinal plant extracts (Oliveira et al., 2006), and field pea protein (Tian et al., 1999).

Another application for spray drying includes the encapsulation of nutraceutical and functional foods and flavors. A summary of recent research on the use of spray drying for encapsulation is provided in Table 10.2. Spray drying microencapsulation (Figure 10.3) includes the following steps: preparation and homogenization of dispersion, atomization of the in-feed dispersion, and dehydration of the atomized particles. Shell materials are limited to gum acacia, maltodextrin, hydrophobically modified starch, and their mixtures because of solubility in water (Desai and Park, 2005). With different material preparations and operating conditions, other less water-soluble casings can be used: polysaccharides (alginate, carboxymethylcellulose, and guar gum) and proteins (whey, soy, and sodium caseinate). According to Madene et al. (2006), the advantages of spray drying encapsulation are low operating cost; high quality, stability, and rapid solubility of the capsules; and small size.

TABLE 10.2

Summary of Studies on Encapsulation through Use of Spray Drying

Compound	Encapsulant	Inlet/Outlet Temp. (°C)	Ref.
2-Acetyl-1-pyrroline	Gum acacia, maltodextrin	150/80	Apintanapong and Noomhorn (2003)
Shiitake flavors	Cyclodextrin, maltodextrin	160/84–93	Yoshii et al. (2005)
β-Carotene	Maltodextrin	170/95	Desobry et al. (1997)
Vitamin C	Chitosan	175/—	Desai and Park (2005)
Black carrot anthocyanins	Maltodextrin	160–200/107–131	Ersus and Yurdagel (2007)
Cinnamon oleoresin	Modified starch	160/120	Vaidya et al. (2006)
Cumin oleoresin	Gum arabic, maltodextrin, modified starch	160/120	Kanakdande et al. (2007)
Vitamin A	Modified starch	182/82	Xie et al. (2007)
Peppermint essential oil	Modified starch	200/120	Baranauskiené et al. (2007)

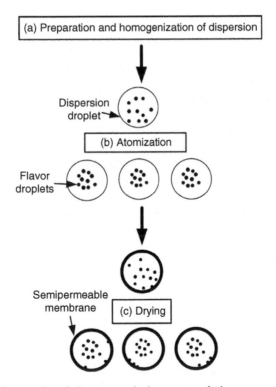

FIGURE 10.3 Schematic of the spray drying encapsulation process. (Adapted from Bhandari, B., in *Dehydration of Products of Biological Origin*, Science Publishers, Enfield, NH, 2004, pp. 513–533. With permission.)

The disadvantages include lack of uniformity, production of a very fine powder that requires further processing (i.e., fluidized bed coating), and lack of suitability for heat-sensitive material.

10.3.7   FREEZE-DRYING

Freeze-drying is a process by which a frozen product under vacuum conditions in an airtight container releases moisture in the form of vapor when exposed to thermal energy suitable for ice sublimation (Orsat and Raghavan, 2007). Originally developed in the 1940s for production of dry plasma and blood products, freeze-drying is viewed as the method that in most cases retains the structure, texture, appearance, and/or flavor associated with fresh products (Ratti, 2001; Vega-Mercado et al., 2001). This high level of quality preservation is due to minimum structural and chemical reactions associated with the absence of air and a low processing temperature. The major limitations of freeze-drying are capital and operating costs, long processing times, and low volume capability (Ratti, 2001).

With quality retention similar to fresh products, freeze-dried products are often used as a benchmark for other methods. Therefore, it is employed in many research studies on valuable products: seaweed (Chan et al., 1997), bay leaf (Díaz-Maroto et al., 2002), *Echinacea purpurea* (Kim et al., 2000), ginseng (Popovich et al., 2005), berries (Feng et al., 1999; Grabowski et al., 2002; Nsonzi and Ramaswamy, 1998), and soy products (Qing-guo et al., 2006; Tian et al., 1999). Freeze-drying has also been investigated for the microencapsulation of food ingredients and flavors (Desai and Park, 2005; Madene et al., 2006). The steps involved are mixing of core particles in a coating solution and freeze-drying of the mixture (Desai and Park, 2005). This method offers an alternative coating method for thermosensitive and water-unstable products. The costs involved are considerably higher than spray drying (Desobry et al., 1997).

10.3.8   VACUUM DRYING

Vacuum drying employs the lowering of the boiling point of water through the application of a vacuum, which allows evaporation of water at lower temperatures (Sokhansanj and Jayas, 2006). The product can be heated by steam, conduction, or radiation in the absence of oxygen (Orsat and Raghavan, 2007). Vacuum drying has been investigated for carrot slices and chives (Cui et al., 2004) and berries (Grabowski et al., 2002; Kongsoontornkijkul et al., 2006). Because of the costs involved, similar to freeze-drying, this operation is commonly used as a finish drying operation in conjunction with other drying technologies.

10.3.9   MICROWAVE DRYING

The mechanism involved with microwave drying consists of exposing a product to microwaves that penetrate directly into the material, resulting in heating from the inside out (Chua and Chou, 2005). The quick energy absorption by water molecules causes rapid evaporation of water, creating an outward flux of vapor. In addition to an increase in the drying rate, the outward flux helps to prevent the shrinkage of the tissue structure (Schiffmann, 1995). Microwave technology is applicable to many industrial processes such as baking, concentrating, cooking, curing, enzyme inactivation by blanching, pasteurizing, precooking, puffing, foaming, solvent removal, moisture leveling, and thawing–tempering (Vega-Mercado et al., 2001). Passardi et al. (2006) reported the advantages of one-step blanching and drying for Yerba maté. Chua and Chou (2005) compared intermittent and constant microwave treatment for carrots.

10.3.10 INFRARED DRYING

Far infrared radiation is a novel drying technology that employs radiation within a specific spectral range in the selective heating of food ingredients (Jun and Irudayaraj, 2003). Application of far infrared radiation has increased during the past few years because of the availability of commercial heaters with high emissivity; however, it is still the most efficient process for homogeneous products (i.e., soy protein and glucose). Determination of the optical properties of food products is necessary to increase energy efficiency, although there are few studies on it. Jun and Irudayaraj (2003) used filters to heat soy protein 6°C higher than glucose, whereas the opposite was true without filters. Far infrared radiation was also investigated for carrots (Chua and Chou, 2005) and ginseng (Kim et al., 2002).

10.3.11 COMBINATION SYSTEMS

Combination dryers and systems have been developed and investigated to capitalize on the benefits (i.e., low temperature, reduced exposure to oxygen, and short drying times) of certain systems while decreasing the impact of the disadvantages (i.e., oxidation, high costs, and long drying times) of other systems. Research performed on self-contained combination dryers that eliminate the exposure of the product to air included vacuum microwave drying of carrot slices and chives (Cui et al., 2004), echinacea (Kim et al., 2000), ginseng (Popovich et al., 2005), and soybeans (Qing-guo et al., 2006). Combining microwave energy with air-based drying systems, such as spouted bed and convective dryers, increases the drying rate, resulting in a shorter drying time as demonstrated for blueberries and cranberries (Beaudry et al., 2003; Feng et al., 1999).

Multimode systems may include a series of stages employing different drying technologies. Osmotic or pretreatments tend to be major components of multistage systems, because water removal is usually possible only to 50 to 55% moisture content. Therefore, a final drying method is required (Beaudry et al., 2003; Feng et al., 1999; Grabowski et al., 2007). Modification of vacuum–microwave combination systems included adding a pre- or postdrying stage such as convective or vacuum drying (Qing-guo et al., 2006; Cui et al., 2004). The addition of an extra drying method is to enhance drying at a specific stage throughout the drying process or to reduce costs associated with certain drying methods.

10.4 DRYING METHODS AND QUALITY RETENTION

The drying technologies in Section 10.3 are all at various stages of development with respect to applicability to the nutraceutical and functional food industries. Aspects such as the drying kinetics, dryer design and optimization, cost effectiveness, and energy efficiency are also being investigated in addition to product quality. Research linking drying technology and operation to retention or loss of key bioactive compounds or components is limited. The following sections focus on plant products with specific bioactive ingredients and the various drying methods that have been investigated. Because of the complexity of determining bioactive compound content, some researchers have measured a single compound (i.e., vitamin C representative of antioxidant activity) or other parameters such as color as a marker for anthocyanin and carotenoid content (Feng et al., 1999; Goula and Adamopoulos, 2006; Nsonzi and Ramaswamy, 1998).

10.4.1 BERRIES

We reviewed studies on the drying of cranberries and blueberries, which were based assessments on various parameters: moisture content, water activity, and/or solids gain (Grabowski et al., 2002, 2007; Sunjka and Raghavan, 2004); rehydration ratio (Feng et al., 1999; Grabowski et al., 2002, 2007; Nsonzi and Ramaswamy, 1998); texture (Beaudry et al., 2003; Nsonzi and Ramaswamy, 1998); bulk density (Feng et al., 1999); sensory attributes (i.e., taste and appearance; Beaudry et al., 2003; Grabowski et al., 2002, 2007); color (Feng et al., 1999; Nindo et al., 2007; Nsonzi and Ramaswamy, 1998); flavor volatility analysis (Feng et al., 1999); anthocyanin content (Grabowski et al., 2002, 2007); and ascorbic acid content (Nindo et al., 2007).

The most acceptable pretreatment among various chemical, thermal, and mechanical methods for cranberries was halving (Beaudry et al., 2003; Grabowski et al., 2002, 2007) or quartering (Sunjka and Raghavan, 2004) because it increased the rate of osmotic drying due to increased surface area and provided the best taste acceptability. Due to a wax layer that is 5 to 10 times thinner than cranberries, pretreatments were generally not required for blueberries (Grabowski et al., 2007). However, Feng et al. (1999) employed a chemical dipping solution of 2.5% ethyl oleate and 0.2% NaOH in conjunction with osmotic drying to prevent rupture of berries dried in a microwave spouted bed dryer.

The parameters used in osmotic drying (osmotic solution, temperature, time, and concentration) have a significant effect on quality and drying characteristics (Grabowski et al., 2007; Nsonzi and Ramaswamy, 1998; Sunjka and Raghavan, 2004). Optimal or applied osmotic treatments for the predrying of cranberries or blueberries are provided in Table 10.3. Although osmotic treatment did result in a shorter overall drying time, transfer of anthocyanins to the syrup was appreciable (Grabowski et al., 2002, 2007). Because of the concentration equilibrium, the use of a cranberry or blueberry processing syrup as the osmotic agent or recirculation of the osmotic syrup reduces loss of anthocyanins in the original product (Grabowski et al., 2002, 2007). In addition, the anthocyanin-rich osmotic syrup could be further processed into a marketable product (Grabowski et al., 2007).

Applied finish drying methods include convective air (Feng et al., 1999; Grabowski et al., 2002), freeze-drying (Feng et al., 1999; Grabowski et al., 2002), microwave spouted bed (Feng et al., 1999), vacuum (Grabowski et al., 2002), fluidized bed (Grabowski et al., 2002, 2007), pulsed fluidized bed (Grabowski et al.,

TABLE 10.3

Optimal Osmotic Dehydration Parameters for Cranberries and Blueberries

Sugar Agent	Concentration	Temp. (°C)	Time (h)	Ref.
		Cranberries		
High fructose corn syrup	2:1	23	24	Sunjka and Raghavan (2004)
High fructose corn syrup with 76°Bx	1:1	23	24	Beaudry et al. (2003)
Standard osmotic syrup with 67.5°Brix	1:5	50	5	Grabowski et al. (2002)
Sucrose syrup with 70°Bx	—	50	—	Grabowski et al. (2007)
		Blueberries		
Sucrose syrup	—	50	24	Feng et al. (1999)
Sucrose syrup with 70°Bx	—	50	6.8	Grabowski et al. (2007)
Sucrose syrup with 55°Bx	—	50	4.5	Nsonzi and Ramaswamy (1998)

2002, 2007), vibrofluidized bed (Grabowski et al., 2002, 2007), spouted bed (Feng et al., 1999), and microwave or convection drying (Beaudry et al., 2003). The anthocyanin content (70 to 90 mg/100 g) in blueberries was not significantly affected by the final drying method or conditions; however, the taste and rehydration ratio was superior for blueberries dried by a vibrofluidized bed at 85 versus 90°C (Grabowski et al., 2007). Temperature was not a factor for the quality of cranberries dried by a vibrofluidized bed, so 90°C was selected as the optimal temperature. The anthocyanin content in cranberries also did not differ significantly for a variety of drying methods including convective air, fluidized systems, and freeze-drying (Grabowski et al., 2002). Although freeze-drying offered superior color, rehydration coefficients, and taste acceptability, the vibrofluidized bed was determined as the most suitable method based on the energy efficiency and overall quality.

Based on sensory attributes, physical quality, and energy efficiency, Feng et al. (1999) selected microwave spouted bed technology at 70°C for chemically dipped or osmotically dried blueberries. Beaudry et al. (2003) recommended microwave or convection conditions of 0.75 W/g and a cycling period of 30 s on and 60 s off at an air temperature of 62°C for osmotically dried cranberries. Feng et al. (1999) also reported significant changes in the quantity of volatile compounds (gains and losses) in blueberries with microwave drying (13 and 2), convective air drying (10 and 5), and freeze-drying (2 and 5). The gains and losses related to microwave and convective air drying were attributed mainly to the chemical reactions associated with heat treatment, whereas the losses in freeze-dried samples were attributed to long drying times. Instability of volatiles was also reported by Nindo et al. (2007) in blueberry juice concentrated by Refractance Window; however, the evaporated volatiles could be recovered.

Reductions of 32 and 48% in vitamin C concentration occurred with drying for blueberry juice temperatures of 55.5 and 59.0°C, respectively (Nindo et al., 2007).

However, this loss was much less than the 70.1% reported for an industrial falling type evaporator providing a product temperature of 68°C. Low-pressure superheated steam drying was also beneficial in reducing vitamin C losses in the drying of Indian gooseberries (Kongsoontornkijkul et al., 2006). This method resulted in a 14% reduction in vitamin C compared to 20 and 23% associated with vacuum and convective air drying, respectively, at 75°C. Although low-pressure superheated steam drying resulted in greater losses during tea preparation because of a resulting higher porosity of berry flakes, a combined drying and tea preparation at a 70°C water temperature resulted in the best overall retention of vitamin C.

10.4.2 Tomatoes

Studies on the drying of whole (Azoubel and Murr, 2003), sliced (Unadi et al., 2002), halved or quartered (Goula and Adamopoulos, 2006; Lewicki et al., 2002), and pulped tomatoes (Goula and Adamopoulos, 2005, 2006; Goula ct al., 2004) were focused on the final product characteristics and/or retention of ascorbic acid under various drying conditions. The focus of a majority of studies on lycopene stability in tomatoes was on thermal treatment in liquids rather than moisture removal. However, Kaur et al. (2006) did investigate the effects of the drying temperature on color and lycopene degradation on tomato peel isolated from pomace.

A pretreatment of calcium chloride resulted in a 20% decrease in drying time and a 20% increase in water removal during subsequent osmotic drying of skinned quartered tomatoes (Lewicki et al., 2002). However, calcium bridges formed in the tomato tissue restricted polymer hydration and swelling, resulting in poor rehydration properties. Although a rehydration rate was not reported, Azoubel and Murr (2003) recommended a 10% sodium chloride osmotic solution for 2 h prior to drying of perforated cherry tomatoes. Sliced tomatoes without any pretreatment were dried within a tunnel dehydrator; higher throughput, lower energy consumption, and acceptable color were achieved for a counterflow process at 60°C versus batch two-step and parallel flow systems (Unadi et al., 2002).

Upon determination that standard spray drying of tomato pulp resulted in a 60 to 71% solid mass accumulation on the internal surfaces of the system (Goula et al., 2004), a modification employing a connection of the dryer inlet air intake to an air dehumidifier was added (Goula and Adamopoulis, 2005). This modification assisted with not only product recovery but also improved product properties such as moisture content, bulk density, and solubility. In a pulp concentration process using an evaporative rotary dryer, Goula and Adamopoulos (2006) reported a drying rate increase with a temperature difference between 75 and 90°C in conjunction with ascorbic acid losses of 45 and 60%, respectively. In the same study, spray drying of pulp resulted in a higher rate of ascorbic acid loss attributable to high exposure of individual droplets to oxidation. Pretreated tomato halves dried convectively at 80 and 110°C had the highest incidence of ascorbic acid loss at 90 and 95%, respectively. Similarly, losses of 21 to 47% lycopene concentration were reported for drying temperatures from 50 to 100°C, respectively, in tomato peel dried in a cabinet dryer (Kaur et al., 2006). Changes in lycopene concentration were directly correlated with color degradation, both a function of the drying temperature.

10.4.2 CARROTS

Carrots represent one of many products that are rich in carotenoids. Pretreatment and drying methods investigated for carrot pieces and slices include osmotic (Pan et al., 2003), convective air (Cui et al., 2004; Pan et al., 2003; Regier et al., 2005), oven (Pan et al., 2003), infrared radiation (Chua and Chou, 2005), microwave combination (Chua and Chou, 2005; Cui et al., 2004; Regier et al., 2005), heat pump (Chua and Chou, 2005), freeze-drying (Cui et al., 2004; Regier et al., 2005), and vacuum drying (Cui et al., 2004).

The methods were assessed through evaluation of total carotenoid, β-carotene, and/or lycopene concentrations (Cui et al., 2004; Pan et al., 2003; Regier et al., 2005) and color (Chua and Chou, 2005). Because of the solid texture of raw carrots, osmotic drying is viewed as an opportunity to decrease final drying times (Pan et al., 2003). The increase in water removal or solute gain, however, is proportional with carotene loss. The carotene concentration as a function of osmotic drying time is described by the following equation:

$$\ln C = -Kt + \ln C_0 \qquad (10.1)$$

where C is the carotene content; K is the nutrition loss rate constant, which is linearly related to the sucrose concentration with no presence of salt; t is the dehydration time; and C_0 is the initial carotene concentration. The solutions of 40% sucrose and 10% salts or 50% sucrose resulted in the highest carotene losses with K values of 0.0735 or 0.1777/h, respectively, compared to the 20% sucrose, 40% sucrose, and 15% salt solutions.

Regardless of degradation caused by osmotic drying, Pan et al. (2003) determined that vibrofluidized bed drying at 100°C with tempering for a minimum of 6h provided the shortest drying time compared to oven and belt drying for 3-mm carrot slices. Regier et al. (2005) reported complete retention of lycopene in blanched carrots dried with three different methods: convective air or inert gas (50 to 90°C), vacuum or microwave (400 W continuous at 5kPa), and freeze-drying (–60 or 30°C at 6 Pa). However, β-carotene losses up to 20% were noted for temperatures above 70°C for both convective air and inert gas drying, resulting in decreased total carotenoid concentrations. Long drying times of 8 to10h are associated with drying temperatures of <70°C, whereas vacuum–microwave samples dried within 2h. Major benefits of convective and vacuum–microwave drying were 10 and 75% increases, respectively, in lycopene concentration due to enhanced extractability.

Cui et al. (2004) determined that freeze-drying and microwave combination systems (i.e., vacuum–microwave, vacuum–microwave + convective, and vacuum–microwave + vacuum) resulted in 94.7 to 97.8% retention of total carotenoids compared to 70.8 to 85.5% for convective air drying (60 to 65°C) of 5-mm carrot slices. The 85.5% carotenoid retention is representative of blanched carrots, implying that enzymes in the unblanched carrots may have been responsible for further carotenoid degradation. Blanching was deemed unnecessary for the other methods.

10.4.4 BOTANICALS

10.4.4.1 Ginseng

Although North American and Korean ginseng contain more than 200 different compounds, the major bioactives are the ginsenosides (Kim et al., 2002; Martynenko et al., 2006; Popovich et al., 2005). Retention of total and individual ginsenosides was verified for freeze-drying (Popovich et al., 2005), convective air (Davidson et al., 2004; Kim et al., 2002; Popovich et al., 2005), vacuum–microwave (Popovich et al., 2005), and infrared drying (Kim et al., 2002). Freeze-drying and vacuum–microwave drying systems provided better retention and extraction efficiency of total ginsenosides than convective drying at 70°C (Popovich et al., 2005). The recommended air drying temperature for ginseng is 38°C, because higher temperatures are associated with degradation of ginsenosides and color and with higher energy costs (British Columbia Ministry of Agriculture, Food, and Fisheries, 1998). Peeling prior to drying (Martynenko et al., 2006) and use of a 38, 50, and 38°C staged drying system (Davidson et al., 2004) reduced convective air drying times, which could offer benefits in product quality. The addition of infrared technology did not provide greater retention of ginsenosides than convective drying at 50 and 60°C (Kim et al., 2002).

10.4.4.2 Echinacea

The roots of echinacea are widely accepted for their potential role in boosting of the immune system (Kabganian et al., 2002; Kim et al., 2000). Alkamides 1 and 2, echinacosides, and cynarin, some of the many bioactives identified in echinacea, are used as marker compounds for the evaluation of postharvest processing effects (Kabganian et al., 2002). The times achieved for moisture removal from 57 to 10% moisture content for roots dried in a convective oven at 23, 30, 40, 50, 60, and 70°C were 103.4, 55.7, 13.1, 11.2, 6.5, and 5.3 h, respectively. Echinacoside losses of 28 to 45% at temperatures of 30 to 65°C, respectively, were significant, whereas the effect on the other marker compounds was inconclusive. Kim et al. (2000) determined that freeze-drying and vacuum–microwave drying provided better retention of total alkamides (1, 2, 3/4, 5/6a/6, 7, and 8/9) than convective drying at 70°C. However, replacing the full vacuum in the vacuum–microwave dryer with a partial vacuum resulted in lower retention than convective drying. The decreased retention with convective and partial pressure vacuum–microwave drying may have been attributable to the high susceptibility of alkamides (1, 2, 3/4, 5/6a/6, and 7), to degradation at higher temperatures and oxygen levels.

10.4.4.3 Herbs

The retention of essential oil volatiles was reviewed for basil (Díaz-Maroto et al., 2004), bay leaf (Díaz-Maroto et al., 2002), and peppermint (Baranauskiené et al., 2007; Rohloff et al., 2005). The highest retention of total volatiles in basil occurred for air drying for 15 days compared to oven drying at 45°C for 15 h or freeze-drying for 24 h (Díaz-Maroto et al., 2004). The acceptability of air drying versus freeze-drying extended to bay leaf (Díaz-Maroto et al., 2002). Freeze-dried chopped leaves resulted in substantial volatile losses compared to the equally effective methods of air drying at 25°C for 3 weeks and convective drying at 45°C for 14 h. An air temperature of 30°C as opposed to 50 and 70°C was favored in the retention of peppermint volatiles (Rohloff et al., 2005). Because convective air drying at 30°C had results similar to 5-day prewilting with final drying at 30°C, the latter was preferred because of additional energy savings.

Each plant is unique in its chemical and physical structure, which explains the variety of parameters and drying methods discussed throughout this review. Table 10.4 provides a summary of the quality parameters and drying methods investigated for seaweed and Brazilian medicinal plants or their extracts: passion flower (*Passiflora alata*), *pata de vaca* (*Bauhinia forficata*), *espinheira santa* (*Maytenus ilicifolia*), and echinacea. An inlet air temperature of 150°C was used for both spouted bed and spray drying of the extracts of passion flower, pata de vaca, and espinheira santa; however, spouted bed drying resulted in negligible losses compared to 3.2 to 13.6% for spray drying (Oliveira et al., 2006).

As with herb volatiles, low temperature convective drying (50°C for 6.5 h) provided better retention of total alkamides in echinacea leaves compared to freeze-drying and vacuum–microwave drying (Kim et al., 2000). Seaweed showed different results with freeze-drying providing the highest retention of total amino acids, PUFAs, and vitamin C (Chan et al., 1997). However, based on the final product use and economic factors, sun drying, an alternative to freeze-drying, provided a similar nutritional value with some losses in ash, minerals, and vitamin C (possibly due to leaching effects and long exposure time to air). Seaweed samples oven dried at 60°C for 15 h had the lowest overall quality. Cui et al. (2004) reported a 61.7% loss of chlorophyll in chive leaves dried convectively at 60 to 65°C for 6 h, as opposed to systems that

TABLE 10.4
Summary of Studies on Herb Drying

Plant	Drying Method	Quality Parameter	Ref.
Seaweed	Sun, oven, and freeze-drying	Amino acids, FAs, minerals, and vitamin C	Chan et al. (1997)
Passion flower	Spray and spouted bed	Total flavonoids	Oliveira et al. (2006)
Pata de vaca	Spray and spouted bed	Total flavonoids	Oliveira et al. (2006)
Espinheira santa	Spray and spouted bed	Total tannins	Oliveira et al. (2006)
Echinacea	Freeze-drying, vaccum–microwave, and convective drying	Alkamides	Kim et al. (2000)
Chives	Freeze-drying, vacuum–microwave, vacuum–microwave + convection, vacuum–microwave + vacuum, and convection drying	Chlorophyll	Cui et al. (2004)

limit the potential of oxidation (i.e., freeze-drying, vacuum–microwave drying, vacuum–microwave + convective, and vacuum–microwave + vacuum). The latter methods offered more than 94% chlorophyll retention, allowing dryer selection to be based on operating costs and equipment availability.

10.4.5 Oilseeds and Legumes

10.4.5.1 Oilseeds

Flaxseed is recognized for its role as a dietary source of soluble fiber and ALA (Manthey et al., 2002; Oomah and Mazza, 2001). The total extractable lipids and stability were determined for spaghetti that contained ground flaxseed (Manthey et al., 2002). Losses that occurred in lipid content during the extrusion process may be linked to the binding of lipids to the gluten in wheat. Drying temperatures of 40 and 70°C had no affect on the lipid content or FA distribution, resulting in retention of ALA.

10.4.5.2 Legumes

Edamame is a soybean harvested at 80% maturity, and it is dried and eaten as a healthful snack (Qing-guo et al., 2006). The drying methods ranked from highest to lowest vitamin C and chlorophyll retention were freeze-drying, vacuum–microwave drying, convection plus vacuum–microwave drying, and convective air drying. Freeze-drying provided levels of vitamin C and chlorophyll retention of 82.1 and 95.8%, respectively, whereas the same levels for convective drying at 70°C were 35.5 and 24.5%. The soft texture represented by a low hardness value of the freeze-dried products, however, would not appeal to consumers who prefer a chewy texture. Considering other important factors such as the drying rate, texture, rehydration capacity, and costs, convection plus vacuum–microwave drying provided the best overall quality. This treatment consisted of preliminary drying for 20 min at 70°C followed by vacuum–microwave drying, resulting in retention of 66.2% of the vitamin C and 49.7% of the chlorophyll. Chickpeas contain certain antinutritional factors such as trypsin inhibitors, which lower their digestibility (Márquez et al., 1998). A heat treatment of 140°C for 6 h facilitated destruction of the majority of trypsin inhibitors in chickpea flour without causing protein denaturation, which did occur at longer durations of 12 and 24 h.

Field peas, a source of protein, contain high levels of the essential amino acid lysine (Tian et al., 1999). The stages involved with pilot-scale preparation of field pea isolate with salt solution included preparing flour from dehulled field peas, development of a slurry, decantation, clarification, ultrafiltration, diafiltration, and drying either by spray drying or freeze-drying. Freeze-dried and spray dried products did not differ in their solubility, but freeze-drying caused nonuniform particle size and an undesirable dark color.

10.5 LATEST ADVANCES

A variety of methods pertaining to the drying of plant products for use as ingredients in nutraceuticals and functional foods were discussed. The applicable novel technologies were Refractance Window (Nindo et al., 2007), far infrared (Chua and Chou, 2005; Kim et al., 2002), heat pump (Chua and Chou, 2005), and low-pressure superheated steam drying (Kongsoontornkijkul et al., 2006). These technologies were investigated for their potential to reduce thermal and oxidative degradation within products or for energy efficiency (heat pump drying and low-pressure super-heated steam). However, further research is required for each of these methods. Although products such as carrots, squash, asparagus, berries, mangoes, avocadoes, and blueberry and cranberry juice have been examined with Refractance Window drying, more optimization testing is required (Nindo and Tang, 2007). Jun and Irudayaraj (2003) determined with filters that individual compounds can be treated with far infrared radiation according to their absorption ability, but more testing is required for its applicability to complex food systems. Kongsoontornkijkul et al. (2006) determined that, although low-pressure superheated steam allowed better retention of ascorbic acid during drying, structural changes occurred that affected the stability during tea preparation of Indian gooseberries. This effect needs to be considered for application to other products. Chua and Chou (2005) ascertained that heat pump drying can be used as part of a mixed mode drying system; however, more investigation is required for optimization and applicability to other products.

In addition to the development and testing of novel drying systems, there is a trend to optimize or modify conventional techniques. Several combination systems that utilize convective drying in conjunction with vacuums and microwaves were determined to be good alternatives to freeze-drying (Cui et al., 2004; Kim et al., 2000; Qing-guo et al., 2006). Martynenko (2006) reported on a computer vision system to assess the color quality of ginseng for the prediction of the drying process. A correlation of image analysis with appearance and physical attributes (i.e., area, color, texture, moisture, and quality) allowed the approximate prediction of moisture content. Nanotechnology was considered for applications in the optimization of encapsulation and delivery systems that carry, protect, and deliver functional food ingredients (Weiss et al., 2006).

Optimization of drying system parameters can also be performed by determining the drying kinetics, compound and product degradation, energy efficiency, and product stability during and after drying. Chua and Chou (2005) evaluated the drying rate and energy efficiency for various drying systems, in addition to determining color degradation with drying. Nindo and Tang (2007) determined the product temperature, color, antioxidant activity, microbial destruction, and energy consumption in their evaluation of Refractance Window drying in blueberry and cranberry juice concentrations.

NOMENCLATURE

ALA	α-linolenic acid (ALA)
C	carotene content (mg/100 g dry material)
C_0	initial carotene concentration (mg/100 g dry material)
FA	fatty acid
K	nutrition loss content (/h)
PUFA	polyunsaturated fatty acid
t	dehydration time (h)

REFERENCES

Abdel-Aal, E.-S.M., Sosulski, F.W., Shehata, A.A.Y., and Youssef, M.M., Nutritional, functional and sensory properties of wheat rice and fababean blends texturized by drum drying, *Int. J. Food Sci. Technol.*, 31, 257–266, 1996.

Apintanapong, M. and Noomhorm, A., The use of spray drying to microencapsulate 2-acetyl-1-pyrroline, a major flavour component of aromatic rice, *Int. J. Food Sci. Technol.*, 38, 95–102, 2003.

Azoubel, P.M. and Murr, F.E.X., Effect of pretreatment on the drying kinetics of cherry tomato (*Lycopersicon esculentum* var. *cerasiforme*), in *Transport Phenomena in Food Processing*, Welti-Chanes, J., Vélez-Ruiz, J.F., and Barbosa-Cánovas, G.V., Eds., CRC Press, Boca Raton, FL, 2003, pp. 137–151.

Baranauskiené, R., Bylaité, E., Žukauskaité, J., and Venskutonis, R.P., Flavor retention of peppermint (*Mentha piperita* L.) essential oil spray-dried in modified starches during encapsulation and storage, *J. Agric. Food Chem.*, 55, 3027–3036, 2007.

Beaudry, C., Raghavan, G.S.V., and Rennie, T.J., Microwave finish drying of osmotically dehydrated cranberries, *Dry. Technol.*, 21, 1797–1810, 2003.

Bell, L.N., Stability testing of nutraceuticals and functional foods, in *Handbook of Nutraceuticals and Functional Foods*, Wildman, R.E.C., Ed., CRC Press, Boca Raton, FL, 2001, pp. 501–516.

Benali, M., Thermal drying of foods: Loss of nutritive content and spoilage issues, in *Dehydration of Products of Biological Origin*, Mujumdar, A.S., Ed., Science Publishers, Enfield, NH, 2004, pp. 137–152.

Bettini, M.d-F.M., Purification of orange peel oil and oil phase by vacuum distillation, in *Functional Food Ingredients and Nutraceuticals: Processing Technologies*, Shi, J., Ed., CRC Press, Boca Raton, FL, 2007, pp. 157–172.

Bhandari, B., Spray-drying: An encapsulation technique for food flavors, in *Dehydration of Products of Biological Origin*, Mujumdar, A.S., Ed., Science Publishers, Enfield, NH, 2004, pp. 513–533.

British Columbia Ministry of Agriculture, Food, and Fisheries, Farm mechanization factsheet (No. 280.380-1), Resource Management Branch, Ministry of Agriculture, Abbotsford, Canada, 1998, pp. 1–9.

Bruneton, J., *Pharmacognosy: Phytochemistry Medicinal Plants*, 2nd ed., Lavoisier Publishing, Paris, 1999.

Cacace, J.E. and Mazza, G., Pressurized low polarity water extraction of biologically active compounds from plant products, in *Functional Food Ingredients and Nutraceuticals: Processing Technologies*, Shi, J., Ed., CRC Press, Boca Raton, FL, 2007, pp. 135–155.

Camire, M.E., Dougherty, M.P., and Briggs, J.L., Functionality of fruit powders in extruded corn breakfast cereals, *Food Chem.*, 101, 765–770, 2007.

Cassini A.S., Marczak, L.D.F., and Noreña, C.P.Z., Drying characteristics of textured soy protein, *Int. J. Food Sci. Technol.*, 41, 1047–1053, 2006.

Chan, J.C.-C., Cheung, P.C.-K., and Ang, P.O., Jr., Comparative studies on the effect of three drying methods on the nutritional composition of seaweed *Sargassum hemiphyllum* (Turn.) C. Ag, *J. Agric. Food Chem.*, 45, 3056–3059, 1997.

Chua K.J. and Chou, S.K., A comparative study between intermittent microwave and infrared drying of bioproducts, *Int. J. Food Sci. Technol.*, 40, 23–39, 2005.

Cohen, M., Robinson, R.S., and Bhagavan, H.N., Antioxidants in dietary lipids, in *Fatty Acids in Foods and Their Health Implications*, 2nd ed., Chow, C.K., Ed., Marcel Dekker, New York, 2000, pp. 439–450.

Costenla, D., Ponce, A.G., and Lozano, J.E., Effect of pomace drying on apple pectin, *Lebensm.-Wiss. Technol.*, 35, 216–221, 2002.

Cui, Z.-W., Xu, S.-Y., and Sun, D.-W., Effect of microwave–vacuum drying on the carotenoids retention of carrot slices and chlorophyll retention of Chinese chive leaves, *Dry. Technol.*, 22, 563–575, 2004.

Davidson, V.J., Li, X., and Brown, R.B., Forced-air drying of ginseng root: 1. Effects of air temperature on quality, *J. Food Eng.*, 63, 361–367, 2004.

DeMan, J., *Principles of Food Chemistry*, 3rd ed., Aspen Publishers, Gaithersburg, MD, 1999.

Desai, K.G.H. and Park, H.J., Recent developments in microencapsulation of food ingredients, *Dry. Technol.*, 23, 1361–1394, 2005.

Desobry, S.A., Netto, F.M., and Labuza, T.P., Comparison of spray drying, drum drying, and freeze-drying for β-carotene encapsulation and preservation, *J. Food Sci.*, 62, 1158–1162, 1997.

Díaz-Maroto, M.C., Palomo, E.S., Castro, I., Viñas, M.A.G., and Pérez-Coello, M.S., Changes produced in the aroma compounds and structural integrity of basil (*Ocimum basilicum* L.) during drying, *J. Sci. Food Agric.*, 84, 2070–2076, 2004.

Díaz-Maroto, M.C., Pérez-Coello, M.S., and Cabezudo, M.D., Effect of drying method on the volatiles in bay leaf (*Laurus nobilis* L.), *J. Agric. Food Chem.*, 50, 4520–4524, 2002.

Ersus, S. and Yurdagel, U., Microencapsulation of anthocyanin pigments of black carrot (*Daucuscarota* L.) by spray drier, *J. Food Eng.*, 80, 805–812, 2007.

Eskin, M.N.A. and Tamir, S., *Dictionary of Nutraceuticals and Functional Foods*, CRC Press, Boca Raton, FL, 2006.

Farkas, I., Solar-drying of materials of biological origin, in *Dehydration of Products of Biological Prigin*, Mujumdar, A.S., Ed., Science Publishers, Enfield, NH, 2004, pp. 317–368.

Feng, H., Tang, J., Mattinson, D.S., and Fellman, J.K., Microwave and spouted bed drying of frozen blueberries: The effect of drying and pretreatment methods on physical properties and retention of flavor volatiles, *J. Food Process. Preserv.*, 23, 463–479, 1999.

Fleming, T., Ed., *Physicians Desk Reference: PDR for Herbal Medicines*, Medical Economics Company, Montvale, NJ, 2000.

Gibson, G.R. and Fuller, R., The role of probiotics and prebiotics in the functional food concept, in *Functional Foods. The Consumer, the Products and the Evidence*, Sadler, M.J. and Saltmarsh, M., Eds., Cambridge University Press, Cambridge, U.K., 1998, pp. 3–4.

Goula, A.M. and Adamopoulos, K.G., Spray drying of tomato pulp in dehumidified air: II. The effect on powder properties, *J. Food Eng.*, 66, 35–42, 2005.

Goula, A.M. and Adamopoulos, K.G., Retention of ascorbic acid during drying of tomato halves and tomato pulp, *Dry. Technol.*, 24, 57–64, 2006.

Goula, A.M., Adamopoulos, K.G., and Kazakis, N.A., Influence of spray drying conditions on tomato powder properties, *Dry. Technol.*, 22, 1129–1151, 2004.

Grabowski, S., Marcotte, M., Poirier, M., and Kudra, T., Drying characteristics of osmotically pretreated cranberries—Energy and quality aspects, *Dry. Technol.*, 20, 1989–2004, 2002.

Grabowski, S., Marcotte, M., Quan, D., Taherian, A.R., Zareifard, M.R., Poirier, M., and Kudra, T., Kinetics and quality aspects of Canadian blueberries and cranberries dried by osmo-convective method, *Dry. Technol.*, 25, 367–374, 2007.

Grabowski, S. and Mujumdar, A.S., Solar-assisted osmotic dehydration, in *Drying '92, Proceedings of the 8th International Drying Symposium (IDS '92)*, Vol. B, Mujumdar, A.S., Ed., Elsevier, Amsterdam, 1992, pp. 1689–1696.

Halawa, E.E.H., Brojonegoro, A., and Davies, R., An overview of the ASEAN–Canada project on solar energy in drying processes [Technical note], *Dry. Technol.*, 15, 1585–1592, 1997.

Health Canada, Nutraceuticals/functional foods and health claims on foods, policy paper, Therapeutic Products Programme and the Food Directorate from the Health Protection Branch, http://www.hc-sc.gc.ca/fn-an/alt_formats/hpfb-dgpsa/pdf/label-etiquet/nutra-funct_foods-nutra-fonct_aliment_e.pdf (accessed March 4, 2007).

Ibarz, A. and Barbosa-Canovas, G.V., *Unit Operations in Food Engineering*, CRC Press, Boca Raton, FL, 2003.

Jalili, T., Wildman, R.E.C., and Medeiros, D.M., Dietary fiber and coronary heart disease, in *Handbook of Nutraceuticals and Functional Foods*, Wildman, R.E.C., Ed., CRC Press, Boca Raton, FL, 2001, pp. 281–293.

Joubert E. and de Villiers, O.T., Effect of fermentation and drying conditions on the quality of rooibos tea, *Int. J. Food Sci. Technol.*, 32, 127–134, 1997.

Jun, S. and Irudayaraj, J., Selective far infrared heating system—Design and evaluation I, *Drying Technol.*, 21, 51–67, 2003.

Kabganian, R., Carrier, D.J., and Sokhansanj, S., Physical characteristics and drying rate of echinacea root, *Drying Technol.*, 20, 637–649, 2002.

Kanakdande, D., Bhosale, R., and Singhal, R.S., Stability of cumin oleoresin microencapsulated in different combination of gum arabic, maltodextrin and modified starch, *Carbohydr. Polym.*, 67, 536–541, 2007.

Kaur, D., Sogi, D.S., and Wani, A.A., Degradation kinetics of lycopene and visual color in tomato peel isolated from pomace, *Int. J. Food Prop.*, 9, 781–789, 2006.

Kim, H.-O., Durance, T.D., Scaman, C.H., and Kitts, D.D., Retention of alkamides in dried *Echinacea purpurea*, *J. Agric. Food Chem.*, 48, 4187–4192, 2000.

Kim, M.H., Kim, S.M., Kim, C.S., Park, S.J., Lee, C.H., and Rhee, J.Y., Quality of Korean ginseng dried by a prototype continuous flow dryer using far infrared radiation and hot air, *Can. Biosys. Eng.*, 44, 47–54, 2002.

Kołakowska, A. and Sikorski, Z.E., The role of lipids in food quality, in *Chemical and Functional Properties of Food Lipids*, Sikorski, Z.E. and Kołakowska, A., Eds., CRC Press, Boca Raton, FL, 2003, pp. 1–8.

Kongsoontornkijkul, P., Ekwongsupasam, P., Chiewchan, N., and Devahastin, S., Effects of drying methods and tea preparation temperature on the amount of vitamin C in Indian gooseberry tea, *Dry. Technol.*, 24, 1509–1513, 2006.

Krokida, M.K., Maroulis, Z.B., and Saravacos, G.D., The effects of the method of drying on the colour of dehydrated products, *Int. J. Food Sci. Technol.*, 36, 53–59, 2001.

Kumar, A., Membrane separation technology in processing bioactive components, in *Functional Food Ingredients and Nutraceuticals: Processing Technologies*, Shi, J., Ed., CRC Press, Boca Raton, FL, 2007, pp. 193–208.

Kundu, K.M., Drying of oilseeds, in *Dehydration of Products of Biological Origin*, Mujumdar, A.S., Ed., Science Publishers, Enfield, NH, 2004, pp. 275–295.

Lewicki, P.P., Le, H.V., and Pomarańska-Lazuka, W., Effect of pre-treatment on convective drying of tomatoes, *J. Food Eng.*, 54, 141–146, 2002.

Li, Y., Xu, S.-Y., and Sun, D.-W., Preparation of garlic powder with high allicin content by using combined microwave–vacuum and vacuum drying as well as microencapsulation, *J. Food Eng.*, 83, 76–83, 2007.

Litwinienko, G. and Kasprzycka-Guttman, T., Study on the autoxidation kinetic of fat components by differential scanning calorimetry. 2. Unsaturated fatty acids and their esters, *Ind. Eng. Chem. Res.*, 39, 13–17, 2000.

Madene, A., Jacquot, M., Scher, J., and Desobry, S., Flavour encapsulation and controlled release—A review, *Int. J. Food Sci. Technol.*, 41, 1–21, 2006.

Manthey, F.A., Lee, R.E., and Hall, C.A., III, Processing and cooking effects on lipid content and stability of α-linolenic acid in spaghetti containing ground flaxseed, *J. Agric. Food Chem.*, 50, 1668–1671, 2002.

Márquez, M.C., Fernández, V., and Alonzo, R., Effect of dry heat on the in vitro digestibility and trypsin inhibitor activity of chickpea flour, *Int. J. Food Sci. Technol.*, 33, 527–532, 1998.

Martynenko, A.I., Computer-vision system for control of drying processes, *Dry. Technol.*, 24, 879–888, 2006.

Martynenko, A.I., Brown, R.B., and Davidson, V.J., Physical and physiological factors of ginseng drying, *Appl. Eng. Agric.*, 22, 571–576, 2006.

Mathur, K.B. and Epstein, N., *Spouted Beds*, Academic Press, New York, 1974.

Neto, C.C., Cranberry and its phytochemicals: A review of in vitro anticancer studies, *Am. Soc. Nutr.*, 137, 186S–193S, 2007.

Nindo, C.I., Powers, J.R., and Tang, J., Influence of refractance window evaporation on quality of juices from small fruits, *LWT Food Sci. Technol.*, 40, 1000–1007, 2007.

Nindo, C.I. and Tang, J., Refractance window dehydration technology: A novel contact drying method, *Dry. Technol.*, 25, 37–48, 2007.

Nsonzi, F. and Ramaswamy, H.S., Quality evaluation of osmo-convective dried blueberries, *Dry. Technol.*, 16, 705–723, 1998.

Oliveira, W.P, Bott, R.F., and Souza, C.R.F., Manufacture of standardized dried extracts from medicinal Brazilian plants, *Dry. Technol.*, 24, 523–533, 2006.

Oomah, B.D. and Mazza, G., Functional foods, in *The Wiley Encyclopedia of Food Science and Technology*, Vol. 2, 2nd ed., Francis, F.J., Ed., John Wiley & Sons, New York, 2000, pp. 1176–1182.

Oomah, B.D. and Mazza, G., Optimization of a spray drying process for flaxseed gum, *Int. J. Food Sci. Technol.*, 36, 135–143, 2001.

Orsat, V. and Raghavan, G.S.V., Dehydration technologies to retain bioactive components, in *Functional Food Ingredients and Nutraceuticals: Processing Technologies*, Shi, J., Ed., CRC Press, Boca Raton, FL, 2007, pp. 173–191.

Pan, Y.K., Zhao, L.J., Zhang, Y., Chen, G., and Mujumdar, A.S., Osmotic dehydration pretreatment in drying of fruits and vegetables, *Dry. Technol.*, 21, 1101–1114, 2003.

Passardi, R.L., Schvezov, C.E., Schmalko, M.E., and González, A.D., Drying of *Ilex paraguariensis* Saint Hilaire by microwave radiation, *Dry. Technol.*, 24, 1437–1442, 2006.

Pokorný, J. and Schmidt, Š., The impact of food processing in phytochemicals: The case of antioxidants, in *Phytochemical Functional Foods*, Johnson, I. and Williamson, G., Eds., CRC Press, Boca Raton, FL, 2003, pp. 298–314.

Popovich, D.G., Hu, C., Durance, T.D., and Kitts, D.D., Retention of ginsenosides in dried ginseng root: Comparison of drying methods, *J. Food Sci.*, 70, 355–358, 2005.

Qing-guo, H., Min, Z., Mujumdar, A.S., Wei-hua, D., and Jin-cai, S., Effects of different drying methods on the quality changes of granular edamame, *Dry. Technol.*, 24, 1025–1032, 2006.

Ratti, C., Hot air and freeze-drying of high-value foods: A review, *J. Food Eng.*, 49, 311–319, 2001.

Regier, M., Mayer-Miebach, E., Behsnilian, D., Neff, E., and Schuchmann, H.P., Influences of drying and storage of lycopene-rich carrots on the carotenoid content, *Dry. Technol.*, 23, 989–998, 2005.

Rohloff, J., Draglund, S., Mordal, R., and Iversen, T.-H., Effect of harvest time and drying method on biomass production, essential oil yield, and quality of peppermint (*Mentha x piperita* L.), *J. Agric. Food Chem.*, 53, 4143–4148, 2005.

Roos, Y.H., Phase and state transitions in dehydration of biomaterials and foods, in *Dehydration of Products of Biological Origin*, Mujumdar, A.S., Ed., Science Publishers, Enfield, NH, 2004, pp. 3–22.

Sablani, S.S., Drying of fruits and vegetables: Retention of nutritional/functional quality, *Dry. Technol.*, 24, 123–135, 2006.

Sanguansri L. and Augustin, M.A., Microencapsulation and delivery of omega-3 fatty acids, in *Functional Food Ingredients and Nutraceuticals: Processing Technologies*, Shi, J., Ed., CRC Press, Boca Raton, FL, 2007, pp. 297–327.

Schiffmann, R.F., Microwave and dielectric drying, in *Handbook of Industrial Drying*, Mujumdar, A.S., Ed., Marcel Dekker, New York, 1995, pp. 345–372.

Shahidi, F. and Naczk, M., *Phenolics in Food and Nutraceuticals*, CRC Press, Boca Raton, FL, 2004.

Shi, J., Kassama, L.S., and Kakuda, Y., Supercritical fluid technology for extraction of bioactive components, in *Functional Food Ingredients and Nutraceuticals: Processing Technologies*, Shi, J., Ed., CRC Press, Boca Raton, FL, 2007, pp. 3–43.

Shi, J., LeMaguer, M., and Bryan, M., Lycopene from tomatoes, in *Functional Foods: Biochemical and Processing Aspects*, Vol. 2, Shi, J., Mazza, G., and Le Maguer, M., Eds., CRC Press, Boca Raton, FL, 2002, pp. 135–167.

Skrede, G. and Wrolstad, R.E., Flavonoids from berries and grapes, in *Functional Foods: Biochemical and Processing Aspects*, Vol. 2, Shi, J., Mazza, G., and Le Maguer, M., Eds., CRC Press, Boca Raton, FL, 2002, pp. 71–98.

Small, E. and Catling, P.M., *Canadian Medicinal Crops*, NRC Research Press, Ottawa, 1999.

Sokhansanj, S. and Jayas, D.S., Drying of foodstuffs, in *Handbook of Industrial Drying*, 3rd ed., Mujumdar, A.S., Ed., CRC Press, Boca Raton, FL, 2006, pp. 522–546.

Sunjka, P.S. and Raghavan, G.S.V., Assessment of pretreatment methods and osmotic dehydration for cranberries, *Can. Biosyst. Eng.*, 46, 35–40, 2004.

Tian, S., Kyle, W.S.A., and Small, D.M., Pilot scale isolation of proteins from field peas (*Pisum sativum* L.) for use as food ingredients, *Int. J. Food Sci. Technol.*, 34, 33–39, 1999.

Unadi, A., Fuller, R.J., and Macmillan, R.H., Strategies for drying tomatoes in a tunnel dehydrator, *Dry. Technol.*, 20, 1407–1425, 2002.

Vaidya, S., Bhosale, R., and Singhal, R.S., Microencapsulation of cinnamon oleoresin by spray drying using different wall materials, *Dry. Technol.*, 24, 983–992, 2006.

Van Den Berg, C., Water activity, in *Concentration and Drying of Foods*, MacCarthy, D., Ed., Elsevier Applied Science, New York, 1986, pp. 11–51.

Vega-Mercado, H., Góngora-Nieto, M.M., and Barbosa-Cánovas, G.V., Advances in dehydration of foods, *J. Food Eng.*, 49, 271–289, 2001.

Vvedenskaya, I.O., Rosen, R.T., Guido, J.E., Russell, D.J., Mills, K.A., and Vorsa, N., Characterization of flavonols in cranberry (*Vaccinium macrocarpon*) powder, *J. Agric. Food Chem.*, 52, 188–195, 2004.

Wasowicz, E., Cholesterol and phytosterols, in *Chemical and Functional Properties of Food Lipids*, Sikorski, Z.E. and Kołakowska, A., Eds., CRC Press, Boca Raton, FL, 2003, pp. 93–107.

Webb, G.P., *Dietary Supplements and Functional Foods*, Blackwell Publishing, Oxford, U.K., 2006.

Weiss, J., Takhistov, P., and McClements, D.J., Functional materials in food nanotechnology—Scientific status summary, *J. Food Sci.*, 71, R107–R115, 2006.

WHO, Global strategy on diet, physical activity and health, http://www.who.int/dietphysicalactivity/strategy/eb11344/strategy_english_web.pdf, 2004.

Wildman, R.E.C., Nutraceuticals: A brief review of historical and teleological aspects, in *Handbook of Nutraceuticals and Functional Foods*, Wildman, R.E.C., Ed., CRC Press, Boca Raton, FL, 2001a, pp. 1–12.

Wildman, R.E.C., Classifying nutraceuticals, in *Handbook of Nutraceuticals and Functional Foods*, Wildman, R.E.C., Ed., CRC Press, Boca Raton, FL, 2001b, pp. 13–30.

Wiriyaumpaiwong, S., Soponronnarit, S., and Prachayawarakorn, S., Comparative study of heating processes for full-fat soybeans, *J. Food Eng.*, 65, 371–382, 2004.

Xie, Y.-L., Zhou, H.-M., and Zhang, Z.-R., Effect of relative humidity on retention and stability of vitamin A microencapsulated by spray drying, *J. Food Biochem.*, 31, 68–80, 2007.

Yoshii, H., Yasuda, M., Furuta, T., Kuwahara, H., Ohkawara, M., and Linko, P., Retention of cyclodextrin complexed shiitake (*Lentinus edodes*) flavors with spray drying, *Dry. Technol.*, 23, 1205–1215, 2005.

11 Drying of Microorganisms for Food Applications

Janusz Adamiec

CONTENTS

11.1 INTRODUCTION

Food and pharmaceutical industries have utilized drying technologies as the preferred methods for preserving a multitude of microbiological and drug preparations in bulk quantities. However, desiccation of microorganisms as culture collections of cells for long-term storage and for future propagation has been the preferred preservation method for decades. The term microorganisms encompasses a variety of living organisms of microscopic size that are broadly classified into bacteria, fungi, molds, yeasts, algae, protozoa, and viruses. Microorganisms play key positive and negative roles in food technology. The negative role denotes a risk of the uncontrolled development of microorganisms that may change their positive functions and finally result in deterioration of quality or even in complete degradation of a food product. There is often a risk that poisonous and noxious substances may form.

However, under controlled conditions microorganisms most frequently play a positive role in biotechnological processes, specifically in manufacturing, recombination, and conservation of foods. The spectrum of food products produced with the assistance of microorganisms is very broad and extends from common-life products

315

such as wine, beer, cheese, or bread, which have been produced historically by fermentation, to newer products, such as fungal protein and the technology of a single cell protein (SCP). This new and dynamically developing technology is based on a fermentation process resulting in extraneous single cell microorganisms with high protein content. After separation, and sometimes purification and preservation, the protein concentrate is used as a nutrient additive in many foodstuffs. One of the most important biotechnologies is milk processing that not only produces cream, cheese, yogurt, and so forth, but also includes biopreparations of lactic acid bacteria starters and vitamins as well as the separation of milk components such as casein, lactose, and lactic acid. The cutting edge application of microorganisms exploits their sensitivity to environmental conditions. Thus, such "biosensors" could constitute cell-based devices for environmental monitoring, if long-term preservation of both the viability and activity of the yeast or bacteria are secured by drying, immobilizing, microencapsulation, and similar techniques (Abadias et al., 2001; Bjerketorp et al., 2006).

Note that biotechnological processes affect a broad spectrum of human activities, including the chemical and pharmaceutical industry, agriculture, health, environmental protection, and so forth. The remarkable differences in the definitions of biotechnology used in the scientific and technical literature result from process complexity and a growing number of applications. Thus, representatives of "pure science" in this field will include molecular biology, genetic engineering, and biochemical processes on a microscale, such as those occurring within particular elements constituting a cell. In contrast, a technologist will approach biotechnology more practically as an interdisciplinary science that deals with the technical utilization of biological materials and processes. This is supported by the achievements of numerous, apparently detached areas of science that integrate molecular biology, biochemistry, microbiology, engineering, mathematics, physics, mechanics, economics, and so forth in industrial applications of microorganisms.

In both meanings of biotechnology, two different terms are utilized: "biomaterial" and "bioproduct." It is generally assumed that a biomaterial is a substance that is related in any way to a living organism and its vital functions and activities. Most frequently it is a single cell or a system of cells that form a living organism. However, this can be the structural part of a cell or an organism that constitutes a compact entity, a component built into various structural parts of the organism, or a substance that is formed or subject to transformations as a result of processes in which living organisms are involved. Diversification of biomaterials causes that the same substance to be termed differently and to perform more than one function. Bioproduct is regarded as the equivalent to a biomaterial and denotes a substance that is a product of biotechnological transformation with the use of microorganisms, their active elements, or biochemically active substances. Our definition highlights the features and vital functions of the biomaterial. However, not every bioproduct possesses such properties. Thus, bioproducts do not always fall into the category of a biomaterial, but every biomaterial is a bioproduct by definition. However, in many processes the substance can be considered as both a biomaterial and bioproduct, for example, baker's yeast in the process of biosynthesis. Thus, yeast, vaccines, enzymes, antibiotics, and hormones can be called bioproducts and biomaterials, whereas ethanol, acetone, citric acid, yogurt, and beer can be considered only as bioproducts.

Subsequent terms denoting substances that are within the group of biomaterials include microorganisms, biopolymers, and biomass. The broadest meaning is ascribed to biomass because it is used to cover the two other notions. Hence, biomass is the mixture of reagents subjected to transformations and products formed during the main bioprocess, together with by-products, remnants of substrates, stabilizing agents, and others. It is also correct to call biomass a pure biomaterial separated from the reaction mixture. Biomass essentially consists of microorganisms or their structural components and/or of biopolymers. It can also constitute a mixture of these elements. Microorganisms are biomaterials that usually appear as single cell organisms or as specialized multicell systems that are similar to animal cells in terms of their chemical structure (Figure 11.1). They are able to carry out the same biochemical reactions; that is, a complete metabolic cycle can take place but they are unable to control their own temperature.

Regarding the main profile of this book, these biotechnology and process engineering definitions and classifications appear to be too general, specifically from the microbiology, biochemistry, or genetic engineering points of view. A systematic classification and description of microorganisms can be found in many monographs on microbiology (Ainsworth et al., 1965–1973; Arora, 2004; Atlas, 1984; Barnett et al., 2000; Gunsalus, 1986; Kirst et al., 2001; Rose and Harrison, 1987–1991; Salminen et al., 2004; Schlegel, 1992).

Biomaterials are produced on an industrial scale with the application of fermentation processes that are most often carried out in a liquid or, more exactly, water phase. Taking into account the biomass growth stages, optimal activity, and aging, it is necessary to stop fermentation at the most appropriate stage to separate biomass from the reaction medium (i.e., water solution) and to fix its required qualitative properties. If biomass is left in a fermentation medium, the biochemical activity of the microorganisms decreases dramatically and secondary reactions are triggered that destroy the biomass (e.g., putrefaction). The removal of water from biomass includes mechanical methods of water separation (filtration, centrifugation) and thermal evaporation of water (concentration in evaporators, drying until solid biomass with low moisture content is reached). The fermentation broth is subjected to solvent

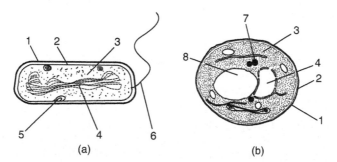

(a)   (b)

FIGURE 11.1 Cross section through (a) bacterial and (b) yeast cells: (1) cell wall, (2) cytoplasmic membrane, (3) cytoplasm, (4) nucleid material (nucleus), (5) mesosomes, (6) flagella, (7) lipid droplets, and (8) vacuoles.

(water) removal. It usually contains from 0.1 to 5–7% mass of dry solids (rarely the solids content exceeds 12 to 15% mass). Most microbiological products should have moisture content from 5 to 12% mass after drying. Hence, in high capacity technologies it is necessary to quickly remove huge amounts of water from the fermentation broth to prevent deterioration of the biomass. A dry microbiological product with biochemical activity preserved for long-term storage is a valuable and often absolutely necessary raw material for the production of various food products.

In addition to the preservation of products of large-scale biotechnological production, in technologies involving microorganisms the preservation of fermentation products on a much smaller scale is equally important, namely, production, preservation, and storage of microorganism strains that are to be collected in the so-called pure culture banks (culture collections of cells). These are collections of particular species and strains of microorganisms that are produced in carefully selected and controlled conditions and with a high degree of microbiological and genetic purity. Such collections can be used either for the preservation and growth of originally produced strains with valuable properties or for distribution as a starting material to initiate fermentation processes on an industrial scale. There are detailed procedures that determine the conditions for production and preservation, storage methods, and time for every strain of bacteria, yeast, mold, and so forth. Although the majority of these materials are produced and stored in a water environment (slurry), in many cases it is advantageous to separate microorganisms from the fermentation broth and to preserve them by drying. Some specific preservation procedures for pure cultures will be presented here later.

Thus, there are many microbiologically different living microorganisms, related substances, and products of their biotechnological transformations. The differentiation of a microbiological material as a drying object (its origin, structure, susceptibility to the environment, individual production conditions, dry product destination, etc.) requires individual treatment and selection of the most appropriate method and drying conditions. Some of the most important issues related to drying of microorganisms (and bioproducts) will be discussed that are crucial to obtain a product with the required stable qualitative properties and to preserve the active features of a living organism in particular.

11.2 BACKGROUND ON MICROORGANISMS

Microorganisms are simple living organisms that are small in size (usually below the limits of normal visibility; Table 11.1) and have a low level of morphological complexity. The main forms of microorganisms and products of microbiological biosynthesis with preserved viability include bacteria, fungi, yeast, molds, protozoa, algae, viruses, vitamins, enzymes, proteins (amino acids), antibiotics, and so forth, the activity of which is strongly related to living microorganisms.

Microorganisms can live in extreme conditions, and living cells can be found almost anywhere where water is in the liquid state. The optimum temperature, pH, and moisture levels vary from one organism to another. Some cells can grow at −20°C, provided that water remains in the liquid state (*psychrophiles*), although optimum is 15°C; others (*mesophiles*) can grow at temperature optima ranging from

TABLE 11.1

Shape and Size of Representative Microorganisms Compared with Protein Molecules

Object	Shape	Size
Bacteria	Spherical	0.5–4 µm diameter
	Cylindrical	2–20 µm length
	Filamentous	
Fungi (molds, yeasts)	Spherical	10–40 to 200 µm diameter
	Ellipsoidal	9–15 × 5–10 µm
	Filamentous	
Algae	Ovoid	28–30 × 8–12 µm
	Filamentous	4–8 µm diameter
Protozoa	Amorphous	<600 µm diameter
Viruses		<1 µm
Protein molecules		<4–6 nm

20 to 50°C. Organisms that grow best at a temperature higher than 50°C (up to 110°C, where the water pressure is high enough to prevent boiling) are classified as *thermophiles*. For each species, and sometimes for each strain, there is either a point or a narrow range of temperatures (2 to 4°C) at which the growth rate of microorganisms is maximal. Many microorganisms have an optimum environmental pH that is far from neutral; for example, some prefer pH 1 or 2 whereas others may grow well at pH 9. Although most microorganisms can grow only when water activity is high (Figure 11.2), others can grow on barely moist solid surfaces or in solutions with high salt concentrations. Many papers and monographs are dedicated to the growth of microorganisms in relation to the water activity of the medium in regard to the preservation and storage of food (Rockland and Beuchat, 1987).

The technological processes of microorganism production (fermentation) can be carried out in batch cultures (batch bioreactors) and in continuous cultures (continuous bioreactors). In batch cultures the conditions of growth change with time: the concentration of microorganisms (bacteria, yeast) increases, whereas the concentration of reaction substrates decreases. When microbial cells are inoculated into a batch reactor containing a fresh culture medium and their increase in concentration is monitored, several distinct phases of growth can be observed (Figure 11.3). There is an initial phase of stagnation called a *lag phase*, which is of variable duration and covers the period of time from inoculation of the fermentation broth until the propagation rate attains the maximum. This is followed by the *exponential growth phase*, where the number of cells (and dry weight) increases *exponentially*. This phase is also called the *logarithmic phase*, the name arising from the common method of plotting the logarithm of the cell number against time. Thereafter, a short phase of *declining growth* exists followed by the *stationary phase*, where the cell numbers are the highest. Finally, the number of cells decline during the *death phase*. It is reasonable to expect the fermentation process to be stopped in the stationary phase for maximum

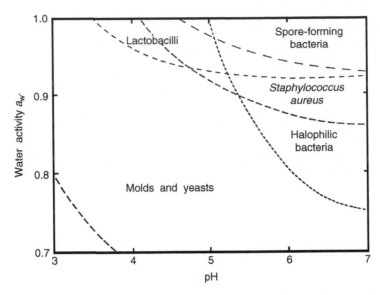

FIGURE 11.2 Schematic diagram of the combined influence of pH and water activity (a_w) on microbial growth. (From Hocking, A.D., in *Proceedings of the ISOPOW/IUFOST Symposium. Food Preservation by Moisture Control*, 1988. With permission.)

productivity and the best quality of microorganisms for preservation and drying (desiccation). However, the specific properties of various types of microorganisms and even particular strains do not support this expectation. For example, the cells of *Lactobacillus rhamnosus* in the stationary phase had the highest stability after drying (31 to 50%), whereas early log phase cells exhibited 14% survival (Corcoran et al., 2004). In contrast, there was a higher survival level of *Sinorhizobium* and *Bradyrhizobium* when sampled in the lag phase of growth (Boumahdi et al., 1999). It

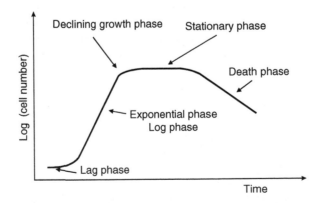

FIGURE 11.3 Typical growth characteristics of microorganisms in a batch reactor.

is generally known that a concentration higher than 1×10^8 cells/mL is sufficient to start the process of preservation or drying, after which the preserved biomaterial will have a sufficiently high concentration of living microorganisms that have the properties of a well-preserved microbiological material.

Another numerical example of batch fermentation is provided in Figure 11.4 according to Hernández et al. (2007). They grew bacteria at subsequent stages (using the pure culture collection in a preliminary inoculation in a 0.5-l bioreactor, fermentation in a 15-l bioreactor, and final fermentation in a 300-l bioreactor) to investigate the stability of *Tsukamurella paurometabola* C-924 powder obtained via spray drying of biomass. The kinetics of the solid-phase (cells) concentration growth in the fermentation broth and an increase of the biomass activity as cell viability were determined (Figure 11.4) to fully estimate the stability of the test microorganisms. The biomass of C-924 bacteria obtained after 20-h fermentation in the 300-l bioreactor was assumed to be an optimal microbiological material to investigate a spray drying series and to study the stability and preservation of the viability of bacterial cells in a dry powder stored under different conditions.

To complete the basic information concerning fermentation, note that in continuous fermentation it is necessary to maintain a long period of stable bioreaction conditions in which, at given culture parameters, a constant concentration of reaction substrates is guaranteed (new substrate is added continuously), microbial cells are constantly kept at the logarithmic biomass growth phase (log phase), and a controlled volume of fermentation product is regularly received. This method of culture growth is termed the equilibrium state.

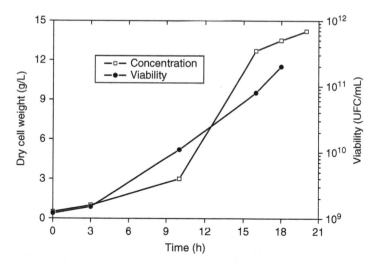

FIGURE 11.4 Kinetic behavior of strain *Tsukamurella paurometabola* C-924 cultured in a 300-l effective volume bioreactor. Experimental conditions: pH 6, 30°C, 300 r/min agitation speed, and 300 l/min air volumetric flow. Fermentation was stopped at 20 h. (From Hernández, A. et al., *Biotechnol. Lett.*, 29, 1723–1728, 2007. With permission.)

11.3   MICROORGANISMS AS DRYING OBJECTS

Living microorganisms and the products of their metabolism are the most complex heterogeneous systems to be dried. In the microorganism structure (a typical bacterial cell), there are numerous important components (cell membrane, nucleus, cytoplasm) built of numerous inorganic and organic compounds with a broad range of molecular weights. The specifics of heat and mass transfer in these systems are due not only to their characteristic structure but also to their effect on the transfer processes of their physicochemical and biological transformations, which follow the drying process. Particular cultures of microorganisms and biosynthesis products are characterized by different physical properties. They behave in different ways during moisture removal, reacting differently to elevated temperature and loss of moisture; and they conduct heat in different ways.

The nature and structure of a variety of microorganisms do not enable a precise specification of the form of material–moisture binding. In general, it is assumed that microorganisms belong to a colloidal–capillary–porous group in which all forms of material–moisture bonds are possible: water as a solvent, free water, and bound water. A significant part of water existing in the cells of microorganisms is free water, whereas that in the macromolecules of biopolymers and products of biochemical transformations is mostly bound water. Water present in the cell as free water is a medium for various reactions and a solvent of many substances (nutritive, enzymes, colloids). Fluctuations in the water level can cause disturbances in biochemical processes that take place in the cell as a living organism (e.g., enzymatic reactions, transformation of proteins). Interactions typical of drying that cause changes in the moisture content and temperature of a product include many transformations affecting its quality, which are related to four types of quality changes: biochemical (or microbiological), enzymatic, chemical, and physical (Table 11.2). Changes in the quality of microorganisms (biomaterials) are usually seen as changes in the viability and death of cells or inactivation and denaturation of the structure. The most drastic change in biomaterial quality is the death of cells, which is regarded as an irreversible change in the material structure that stops the life functions of the organisms (viability), terminating the metabolic transformations that result in growth, reproduction, and active interrelations with the surrounding substances, conditions, and so forth. Inactivation and denaturation or destruction of biomaterials are related to

---

TABLE 11.2

Main Changes of Microorganism Properties during Drying

Biochemical (Microbiological)	Enzymatic	Chemical	Physical
Atrophy of microorganism cells (bacteria, molds, yeast)	Loss of activity of enzymes and vitamins	Decrease of nutritive values and activity of proteins, carbohydrates, fats, antibiotics, amino acids	Decrease of utility properties (solubility, rehydration, shrinkage, loss of aroma)

various phenomena that finally lead to a decrease in biological and microbiological activity. This concerns not only living microorganisms and cells or elements of their structure but also biomaterials (biopolymers) such as enzymes, antibiotics, amino acids, and vitamins, which do not possess the properties of living organisms. The loss of activity or denaturation can be reversible (anabiosis) or irreversible and can be caused by a number of factors.

Anabiosis is a temporary but reversible suspension of life functions induced by external factors, a latency in which all processes take place at a minimum rate and can be resumed when favorable conditions reappear. It is often encountered in nature, for example, as a resting stage in plants and in protozoa, insects, bacteria sporulation, and so forth. Anabiosis can also be regarded also as a factor of evolution that can lead to biological structures capable of survival under extreme conditions. The main physical factors that control the intensity of life processes are temperature and moisture content. Therefore, the following types of anabiosis are distinguished: anhydrobiosis that is induced by water evaporation, cryobiosis caused by low temperature and freezing of water, and osmobiosis when water is extracted from cells because of osmotic pressure. During desiccation, microorganisms lose biologically active free water and retain only bound water, which is biologically inactive. Depending on the dryness, the balance between both forms of water can be disturbed in favor of bound water and the formation of local structures in which moisture is isolated and cannot take part in the transformations. Anabiosis is characterized by the presence of only bound water in the form of hydrated envelopes of polar groups that constitute a skeleton of the protoplasm gel. Microorganisms are subjected to anabiosis mainly to preserve their microbiological features and to store them under safe conditions so that, when necessary, their life functions can be restored after rehydration at an appropriate temperature. However, anabiosis induces some processes related to permanent changes such as a reduced number of cells capable of living and distinct changes on the cell surfaces, nuclei, and membrane structures. Thus, when selecting the methods for carrying out anabiosis and storage and activation of the biomaterial, we should take into account not only what the anabiosis is but also predict and analyze physical and chemical changes that occur during the latency of microorganisms.

Life functions and activity of microorganisms are closely related to two basic forms of bacterial cells: vegetative and sporulating. A bacterial cell is in a vegetative form during the cell's normal (undisturbed) functioning with typical transformations related to its growth, reproduction, and aging. The inhibition of life functions, followed by the transition into anabiosis, is typical behavior of any vegetative cell exposed to conditions unsuitable for its normal functioning. Further, when the external conditions remain unfavorable, irreversible changes in the cell can take place, leading to destruction and death. However, some types of microorganisms (e.g., species of *Bacillus*, *Clostridium*, or *Sporosarcina*) are able to produce a special sporulating form (a spore) that can survive in extreme conditions such as drought, elevated temperature, freezing, and radiation and in the presence of many toxic chemical compounds or enzymes. Note that the bacterial cells' ability to form spores is valuable when it is necessary to preserve a biomaterial during long-term storage under conditions that are different from those during the bioprocess. In contrast, sporulation is undesirable and makes the process of sterilization very difficult

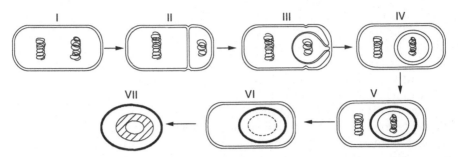

FIGURE 11.5   Schematic of morphological changes during sporulation: (I) vegetative cell, (II–III) asymmetric division and spore enveloping, (IV–V) formation of envelopes and cortex of spores, (VI) spore maturation, and (VII) mature spore.

because even the application of drastic conditions can appear insufficient to kill the spores of bacteria. Hence, in view of the two processes of preservation and sterilization, studies on many aspects of spores have been carried out for a long time (Gould, 1999, 2006; Leuschner and Lillford, 2003).

The ability to form spores is a genetically encoded property, which is induced by activation of the genes responsible for sporulation. This process is not reproductive because one bacterial cell usually produces only one spore. Sporulation is a complex process that encompasses morphological changes of cells, synthesis of new structures, and specific biochemical transformations. There are seven stages of sporulation, covering the formation of prespores, development of external spore envelopes, and maturing of the spore (Figure 11.5).

The survival and stability of specific quality factors of dried microorganisms depend strictly on two basic properties: resistance to temperature changes (both high and low temperature) and sensitivity to changes of moisture content in the biomass or single cells. Thus, biomaterials are classified according to their thermolability or thermostability and xerosensitivity or xeroresistivity. The intensity and time of thermal processing (fermentation) as well as pH and water activity of the environment stimulate the growth of microorganisms, affect their morphology and metabolism, and can induce their thermal death or destruction and denaturation. Note that the thermal stability of microorganisms is determined not only by their type, nature, and genetic variety but also by a number of other factors such as the bioprocess run, type of raw materials, nutrients and additives, stabilizers used in the fermentation process, and so forth. For instance, solutions containing sucrose, glucose, proteins, and ions of some metals (calcium, magnesium) increase the thermal stability of biomaterials.

To estimate the thermal resistance of a given microbiological material, the results of relevant investigations are presented in the form of thermograms, which give the effect of temperature and time of thermal impact on the survival or biochemical activity of the biomaterial. An example of survival curves for *Escherichia coli* for various external conditions (temperature and heating time) is illustrated in Figure 11.6. Another example is the water activity that affects the heat resistance of microorganisms in food powders (cf. Laroche et al., 2005). Two microorganisms, the yeast *Saccharomyces cerevisiae* and the bacterium *Lactobacillus plantarum* (both remain

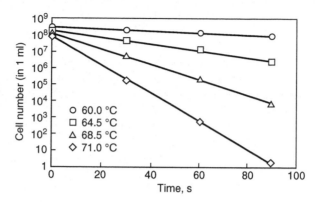

FIGURE 11.6 Survivor curves of *Escherichia coli* cells as a function of temperature and heating time. (From Palaniappan, S. et al., *Biotechnol. Bioeng.*, 39, 225–232, 1992. With permission.)

in a vegetative form), were mixed and then dried alive with milk or wheat flour powders. After heat treatment for 5 to 30 s at 150 and 200°C, the viability of cells in terms of the initial water activity was observed (Figure 11.7). Laroche et al. (2005), among others, concluded that for each strain there was an initial water activity range (0.3 to 0.4) over which the microorganisms were more resistant and that *L. plantarum* was more sensitive to drying and heat treatment on wheat flour than on skim milk; a reverse phenomenon was noted for *S. cerevisiae*.

Unfortunately, the applicability of results from such studies is limited because of specific individual features of microorganisms (cf. the growth curve during fermentation), as well as the effect of the conditions, parameters, and method of given studies. These conditions are not always described precisely in the published studies, and differences in the methodology applied by various researchers do not allow absolute values to be used. However, they provide valuable information on trends and scale of changes or on the behavior of microorganisms in relation to a given strain or type.

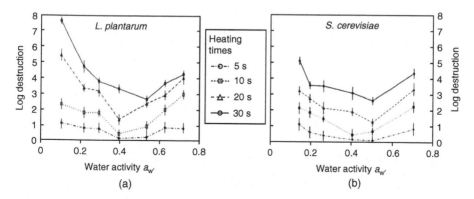

FIGURE 11.7 Destruction of (a) *Lactobacillus plantarum* and (b) *Saccharomyces cerevisiae* dried on wheat flour at 150°C for different heating times. (From Laroche, C. et al., *Int. J. Food Microbiol.*, 97, 307–315, 2005. With permission.)

Zimmermann and Bauer (1986) estimated the biochemical activity of *S. cerevisiae* subjected to heating and drying, based on the determination of the activity (enzyme deactivation kinetics) of one of the yeast enzymes, which is alcohol dehydrogenase (ADH). The effect of temperature was determined by heating whole yeast cells (pellets) and pure isolated enzyme solved in a phosphate buffer for 20 min at 30 to 70°C. The enzymes in the microorganism cells have a higher thermal stability than pure isolated enzymes.

Another method for estimation and numerical comparison of the thermostability of different biomaterials uses the Arrhenius equation [Eq. (11.1)], which describes the thermal death of microorganisms that typically follows a first-order process:

$$\frac{dN_v}{dt} = -kN_v \tag{11.1}$$

where the reaction rate constant k is

$$k = -k_0 \exp\left(-\frac{E_a}{RT(t)}\right) \tag{11.2}$$

or in logarithmic form,

$$\ln k = -\frac{E_a}{R}\left(\frac{1}{T(t)}\right) + \ln k_0 \tag{11.3}$$

where N_v is the number of viable microorganisms per unit volume of medium, t is the time, E_a is the activation energy for the death of the organism, R is the gas constant, $T(t)$ is the absolute temperature, and k_0 is the frequency coefficient.

The activation energy and frequency coefficient are characteristic values for a given substance and depend on process parameters like the biomass moisture content, for example. Equation (11.3) is graphically illustrated in Figure 11.8.

The slope of this plot (calculation of E_a and k_0) provides a measure of the susceptibility of microorganisms to heat. Typical E_a values for bacterial spores range from 250 to 300 kJ/mol, so the spores are typically much more resistant to temperature than heat-sensitive nutrients with medium 70 to 100 kJ/mol E_a values. Enzymes have activation energies intermediate between microbial spores and vitamins (Table 11.3; Blanch and Clark, 1997). Kuts and Tutova (1983–1984) distinguished and characterized the thermal resistance of particular groups of microbiological materials; however, they found lower E_a values. Moreover, these researchers made the kinetic constants dependent on the duration of a thermal impact. Luyben et al. (1982) and Kaminski et al. (1992) used changes in the activity of enzymes during drying and the Arrhenius equation to describe the reaction kinetics of enzyme deactivation by substituting both the temperature and the moisture content of the tested material in the empirical equations. They then used these formulae in a model description and optimization of drying parameters for biomaterials. Note that the E_a, k_0, and decimal reduction time are important and fundamental factors in the determination of the temperature–time profile for sterilization processes.

Biomaterials and microorganisms are especially complex multicomponent objects that require special treatment not only during drying but also after drying. The drying kinetics of raw biomass, included in the group of colloidal–capillary–porous bodies

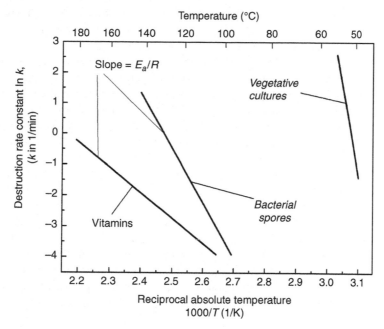

FIGURE 11.8 Dependency of destruction and microbial death on temperature: illustration of Arrhenius correlation as Eq. (11.3).

TABLE 11.3
Selected Values of the Activation Energy for Nutrients, Enzymes, and Cells

Compound or Reaction	E_a (kJ/mol)
Spores	
Bacillus stearothermophilus	287.2
Bacillus subtilis	318.0
Clostridium botulinum	343.1
Nutrients	
Vitamin B_{12}	96.6
Thiamine HCl (B_6)	92.0
Riboflavin (B_2)	98.7
Folic acid	70.2
d-Pantothenyl alcohol	87.8
Enzymes	
Trypsin	170.5
Peroxidase	98.7
Pancreatic lipase	192.3

Source: Blanch, H.W. and Clark, D.S., *Biochemical Engineering*, Marcel Dekker, New York, 1997. With permission.

because of its structure, is characterized by intensive evaporation of free-solving water in the first drying period (process parameters can affect its run) and a prolonged drying time in the second drying period with a particularly low drying rate in the final stage of the process. This is related to the shrinkage of biomass elements and their surface structures that lead to entrapping the water molecules inside the body, which hampers outward transport (diffusion). This characteristic phenomenon is magnified because the inner water is bound by relatively strong adsorption and osmotic forces.

After drying, the biomass is still a complex multicomponent system (proteins, carbohydrates, lipids, biopolymers as enzymes, vitamins, residues from a fermentation broth, and water) with properties reflected by the diversified yet individual character of the equilibrium curves during moisture sorption by a dry product. We emphasize that such systems are most often characterized by a distinct hysteresis of the isothermal equilibrium curves during moisture sorption–desorption (Kapsalis, 1987). These products are frequently characterized by moderate or low hygroscopicity as a mild effect of temperature. In addition, the depth of hysteresis is different and varies with temperature, so there is no regular relation that means that every material requires separate studies. Determination of these characteristic equilibrium properties of a dry product is a precondition for specifying most proper and often very demanding conditions for packing, storage, transportation, and distribution.

11.4   GENERAL RULES ON SELECTION OF DRYING METHOD

Considering only the positive impact, microorganisms play a crucial role in numerous biotechnological processes of food production and preservation, health protection in a broad sense, environmental protection, chemical industry, power engineering, and many other branches of industry. The basic aim, scope, and development of applications of modern biotechnologies in food production, agriculture, and health protection are as follows:

1. Improvement of traditional fermentation processes: fermented dairy and vegetable products, bread baking, production of yeast, and alcoholic beverages
2. Development of new microbiological technologies for production of SCPs, amino acids, vitamins, polysaccharides, and organic acids
3. More efficient application of enzymes as biocatalysts in the production of dairy products, fruit products, vegetable products, fermented beverages, and starch products
4. Production of microbiological preparations (inhibitors, stabilizers) for food preservation
5. Production of valuable fodder supplements (protein, vitamin and antibiotic preparations, growth stimulators) and vegetable ensilage, plant protection products (bioinsecticides, biopesticides, antibiotics), and animal protection remedies (antibiotics, vaccines)
6. Development of modern methods for growing of tissues and cells *in vitro*
7. Development of techniques for microbiological biosynthesis of natural microorganism metabolites (antibiotics, vitamins, enzymes, enzyme inhibitors, organic acids, alkaloids, hormones, antigens) as well as microbiological

and enzymatic biotransformation in the production of a large group of drugs, amino acids, and vitamins

8. Development and improvement of cultivation techniques for microorganisms and animal cells to produce vaccines and antibodies

In these applications of microorganisms, there are different requirements related to the form, physical and biochemical properties, and microbiological activity of individual biomaterials. These requirements should be related to the specific properties of microorganisms such as sensitivity to temperature and dryness or fermentation conditions. As a result of these conditions, preservation or drying of microbiological substances can be considered with respect to three types of actions for three groups of preserved products:

1. Pure culture microorganisms that produce starters (bacteria, yeast) that have to be preserved for long-term storage of a specific type and strain (stored material) kept in the pure culture collections: The processes of fermentation and preservation most often refer to small production on a laboratory scale.

2. Microorganisms and fermentation products that can be preserved to maintain their microbiological activity and biological structure for future application in a variety of biotechnological processes at industrial scale: Examples are the active yeast that promotes fermentation in bread baking and beverage production, active bacteria and enzymes needed in milk processing and production of dairy products and processing of fruits and vegetables, and production of pesticides for agriculture and biopreparations for environmental protection such as those used in wastewater treatment plants.

3. Microorganisms and fermentation products that can be devoid of the properties of a live microorganism but with their chemical and biochemical composition intact: Examples are production of SCPs as a source of protein with high nutritive value or inactive yeast with high protein content used as fodder yeast.

Drying of biomaterials from the third group proceeds according to general rules and parameters applied in the drying of typical food products (Mujumdar, 2004). Thus, the process is carried out in such a way that the nutritive values of a dried product are preserved and will not degrade during storage or rehydration. This means that the most valuable nutrients, proteins, vitamins, fats, enzymes, and aromas will be kept in an unchanged biochemical form and the product's physical form after rehydration will be maintained to be suitable for future applications (no structural changes including excessive and irreversible shrinkage, caramelization or crystallization, and unintended change of color). These topics are discussed in detail elsewhere in this book and hence are only mentioned in this chapter.

With advances in biotechnology, progressive growth of bioproduction, and escalating use of microorganisms, the preservation of properties required for long-term storage and transportation is a subject of both research and publications in which a variety of the process and equipment issues are comprehensively discussed. Worth mentioning are the monographs by Tutova and Kuts (1987) and Kudra and Strumillo

(1998), some chapters in Mujumdar's work (2007), and review articles by Morgan et al. (2006) and Santivarangkna et al. (2007).

The broad spectrum of biotechnological products indicates that it is possible to use various drying and dewatering methods. However, because of thermo- and xero-lability and the risk of many qualitative changes in these products, some process conditions should be strictly monitored. For example, the temperature and moisture content of a product should be controlled, rapid changes of pressure should be eliminated, and mechanical interactions should be avoided because they may damage cells of microorganisms. In addition, many processes must be carried out under sterile conditions using sterile drying agents. In general, the following solutions can be recommended, which consider the properties of biotechnological products:

- Processing at mild conditions
- Vacuum or freeze-drying
- Application of multistage systems
- Special techniques based on the decrease of the initial material moisture content

The phase diagram of thermodynamic water equilibrium in temperature–pressure coordinates is very helpful in the initial selection of drying methods and operating parameters (Figure 11.9). In particular, the equilibrium of the ice–water

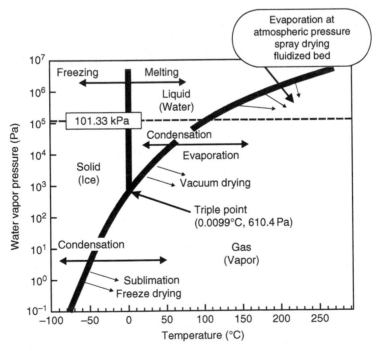

FIGURE 11.9 Phase diagram of water in relation to the important drying processes.

vapor system is one of the important elements in the proper selection of parameters that are required in vacuum and freeze-drying processes.

11.5 DRYING OF STARTER CULTURES

Before the advent of concentrated starter cultures, commercial starter cultures were initially supplied in a liquid form. For many reasons, such as the advancement and development of biotechnological and engineering processes, better equipment and control, and an increasing quantitative demand for starter cultures, the liquid form has been replaced by a much better solid form. Hence, in a growing number of biotechnological processes it is absolutely necessary to preserve high microbiological quality microorganisms (their original physiological activity) in the first group, which is pure culture microorganisms (producing starters) during long-term storage. However, it is difficult to select the optimum preservation method on the basis of experimental data presented in numerous publications. Therefore, each problem calls for individual consideration that may be guided by some general rules that allow the technologies of preservation to be divided into two groups:

1. Dry preservation or storage methods, which are applied to products that are freeze-dried, dried on solid substrates, dried in gel disks, and so forth
2. Wet conservation under slow transfer processes (in distilled water, under a cover of mineral oil, in a protective medium at freezing temperature, etc.)

The first group of dry preservation methods will be characterized in more detail, in accordance with the profile of this book.

In *storage of freeze-dried cultures*, microbial suspension in a protective medium is placed in ampoules, frozen, and vacuum dried with a continuous supply of external energy. After completing the process, the ampoules are sealed under vacuum and stored. A proper protective medium and freeze-drying parameters are very important. The advantages of the method are infections and wetting of the culture during storage are eliminated, a high concentration of inoculant is produced, and small samples reduce storage space. The disadvantages are costly and complex equipment, high energy consumption, and a difficult choice of optimum dehydration conditions. Freeze-drying is the preferred method for culture collections worldwide, including the American Type Culture Collection and the National Collection of Type Cultures.

In *storage on solid carriers*, the culture is stored directly in a carrier–filler that is also a sorbent, protective medium, and in some cases a substrate; organic and inorganic substances (grain, sand, etc.) are used as carriers. Simple sample preparation, application of different (inexpensive) drying methods, and multiple controls are the advantages of this method. The risk of sample infection during preservation and storage and unrepeatable properties of the carriers–fillers are its disadvantages.

Storage in liquid nitrogen is utilized in freeze-drying. During freezing at $-196°C$, the microbial cultures are transferred to the cryobiosis state. This method is recommended for storage of cultures that die during freeze-drying because it offers a high degree of cell viability, no degeneration even during long-term storage, and the production of homogeneous material for further investigations and applications.

Miyamoto-Shinohara et al. (2006) analyzed the survival of various freeze-dried species of microorganisms deposited in the International Patent Organism Depository and stored for up to 20 years. The goals of these studies were (1) to determine the variation of the survival rates for six groups of microbes by periodic testing for up to 20 years [(a) yeast, (b)–(d) Gram-negative bacteria, and (e)–(f) Gram-positive bacteria, see Table 11.4] and the number of strains (from 5 to 144 strains, see Table 11.4), (2) to plot the survival curves for microorganisms stored in ampoules, and (3) to compare the survival patterns of the different species. Strains of the same species were treated as one group, and the survival rates of the same group in the same year were expressed as mean values with standard deviations. The log survival rates from 1 to 20 years (Y_1–Y_{20}) were satisfactorily fitted to a linear function by the least squares method:

$$\log_{10} Y = A + BX \tag{11.4}$$

where Y is the survival rate, X is the number of years of storage, A is the intercept on the y axis, and B is the slope of the line relating the log survival to storage time.

The results of tests and calculated coefficients A and B for each species are listed in Table 11.4, and the survival curves plotted against the storage time are provided in Figure 11.10. Each species showed a slightly different survival curve, depending to some extent on the kind of microorganisms. For example, *S. cerevisiae* showed only Y_D of 8% survival just after the freeze-drying process. The B value of −0.01 indicates that 97.7% of the living cells after freeze-drying will survive the first year of storage

TABLE 11.4
Coefficients of Fitted Survival Eq. (11.4) for Six Species after Freeze-Drying

Genus and Species	Year	Y_D	Y_1	A	B	n
(a) *Saccharomyces cerevisiae*	20	8.0 ± 10.2 (102)	6.7 ± 10.4 (87)	0.8 ± 0.0	−0.010 ± 0.004	9
(b) *Escherichia coli*	20	42.6 ± 23.9 (144)	29.1 ± 24.2 (130)	1.4 ± 0.0	−0.041 ± 0.005	9
(c) *Pseudomonas putida*	20	33.5 ± 26.5 (45)	17.4 ± 18.6 (40)	1.1 ± 0.1	−0.058 ± 0.011	8
(d) *Enterobacter cloacae*	15	50.8 ± 30.2 (11)	26.1 ± 26.0 (11)	1.4 ± 0.2	−0.073 ± 0.027	5
(e) *Lactobacillus acidophilus*	15	62.5 ± 24.5 (8)	36.5 ± 22.5 (8)	1.7 ± 0.1	−0.018 ± 0.009	5
(f) *Enterococcus faecium*	16	85.2 ± 4.8 (5)	86.8 ± 5.3 (4)	2.0 ± 0.1	−0.016 ± 0.007	6

Year indicates storage years; Y_D is the percentage survival rate several days after freeze-drying and Y_1 is the rate after 1 year with the number of strains examined in parentheses; A is the y intercept and B is the slope of the survival equation $\log_{10} Y = A + BX$, where Y is the survival rate and X is the year; and n is the number of data points.

Source: Miyamoto-Shinohara, Y. et al., *Cryobiology*, 52, 27–32, 2006. With permission.

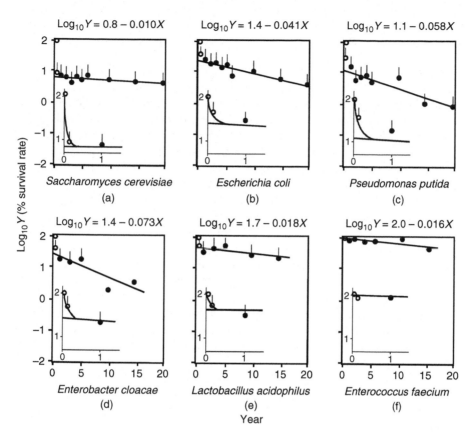

FIGURE 11.10   Survival curves of six species. (○–) Mean values and (●–) standard deviations of the survival rate at each year. Open symbols show the survival rate before freeze-drying (100%) and several days after freeze-drying (Y_D), and closed symbols show the survival rate after more than 1 year. Only the upper bars of the standard deviations are shown. See Table 11.4 for the equation $\log_{10} Y = A + BX$. Each graph includes an enlarged graph for the first year after freeze-drying. (From Miyamoto-Shinohara, Y. et al., *Cryobiology*, 52, 27–32, 2006. With permission.)

($10^{-0.01X} \times 100\% = 10^{-0.01} \times 100\% = 97.7\%$), and the same percentage of living cells from the previous year will survive each consecutive year. In the case of *L. acidophilus* (the Gram-positive bacteria), the survival just after freeze-drying (Y_D) was at 62.5% and B at –0.018, which means that 96.0% of the cells that survived freeze-drying will survive each year of storage ($10^{-0.018X} \times 100\% = 10^{-0.018} \times 100\% = 96.0\%$). The difference in survival curves should reflect structural differences in the cell wall and cell membrane of the microorganisms. The authors (after Antheunisse, 1973; Rudge, 1991) noticed that freeze-drying is not applicable to such microorganisms as *Clostridium* species, *Helicobacter pylori*, and *Microcyclus* strains. Finally, Miyamoto-Shinohara et al. (2006) compared their results with other published data for different drying conditions and postulated that the survival during storage is strongly influenced by the degree of vacuum under which the ampoules were sealed. Thus, the excellent survival after the freeze-drying process might be attributed to perfect sealing under a vacuum.

Santivarangkna et al. (2007) presented a detailed survey of 115 publications on industrial drying methods for preservation of lactic acid starter cultures. Among the group of bacterial starter cultures, lactic acid starter culture is the most important, and only a few non-lactic acid starter cultures are commercially available. For instance, the dairy industry currently uses U.S. $250 million/year of commercial starter cultures. Although freeze-drying is the best and frequently the only method for preservation of starter cultures, the high costs (Table 11.5) drive fundamental and applied research toward new, less expensive techniques. These methods are tested and recommended as suitable even if they sometimes refer to a limited group of microorganisms and biotechnological processes.

Santivarangkna and colleagues (2007) concentrated their survey on the publications that described drying methods other than freeze-drying. In a typical freeze-drying process, cells are initially frozen at −196°C and then dried by sublimation under high vacuum. The inactivation of the cells is mostly attendant on the freezing step, and 60 to 70% of cells that survived the freezing step live through the dehydration step. In contrast to such a very low temperature, the alternative drying processes are divided into two major groups related to the drying temperature: those carried out at a high temperature such as spray drying and those at intermediate temperatures such as fluidized bed drying and vacuum drying. The well-known and historically used spray drying method, with its numerous advantages and disadvantages, is also very useful for the drying of a large amount of starter cultures. The risk of inactivation and dying of microbial cells appears at subsequent stages of the process. This risk is more severe when the raw biomass of starter cells is more concentrated. At the stage of spraying with an atomizer or rotating disk, the cells are already affected by significant, often destroying mechanical forces. In the next stage, the microorganism cells are under the combined influence of two parameters: high air temperature and exposure time to it. At the beginning of drying, the higher temperature is not directly correlated to the inactivation and has only a slight effect, because

TABLE 11.5
Costs of Drying Processes Referenced to That of Freeze-Drying

Drying Process	Fixed Costs (%)	Manufac. Costs (%)
Freeze-drying	100.0	100.0
Vacuum drying	52.2	51.6
Spray drying	12.0	20.0
Drum drying	9.3	24.1
Fluidized bed drying	8.8	17.9
Air drying	5.3	17.9

Source: Santivarangkna, Ch. et al., *Biotechnol. Prog.*, 23, 302–315, 2007. With permission.

the temperature of spray dried particles and thermal inactivation are limited to the wet bulb temperature by the evaporative cooling effect. Finally, the optimum residence time in the dryer is the time for completion of desired moisture removal with a minimum increase in temperature of the dried products. The technological system can be easily kept in aseptic conditions. After leaving the dryer the product is packed and stored in conditions that meet all hygienic requirements.

Another analyzed method for drying of starter cultures (fluidized bed) has a drying time that is longer than that of spray drying (from 1 min to 2 h), but thermal inactivation can be minimized and easily controlled by using relatively low air temperatures. As reported by Santivarangkna et al. (2007), only a few studies have been made on lactic acid starter cultures; more publications refer to studies and applications of this method in the production of commercial dry yeast. A limitation is that only big particles can form a fluid bed, so the biomass can be dried in the agglomerated and granulated forms only. Therefore, the cells must be granulated, entrapped, or encapsulated in support materials such as skim milk, casein, and so forth. Obviously, in this case not only pure culture starter cells but also the support materials are components in the inoculum, together with the cells. Such materials should fit the foods to be fermented, for example, alginate or skim milk for fermented milk, maltodextrin for fermented sausages, and starch or flour for sourdough bread. A disadvantage of this method is reduced viability of cells caused by osmotic shock after mixing of the biomass with extremely desiccated support material characterized by very low water activity. Moreover, the risk exists for easier contamination of the biomass because of contact with the support materials or problems with the preservation of aseptic conditions during additional operations, such as mixing, granulation, and transport of raw materials.

Vacuum drying is recognized as a process suitable for heat-sensitive materials, despite much longer drying times (range = 20 to >100 h) compared to spray or fluidized bed drying. However, this disadvantage can be overcome by the use of a continuous vacuum dryer, which is able to dehydrate the material to 1% moisture (wet basis) at 40°C within 5 to 10 min (Hayashi et al., 1983). Because the dryer operates under vacuum, moisture can be removed from materials at low temperature, because the boiling point of water drops from 100°C at 0.1 MPa to 7°C at 1 kPa. In addition, the oxidation reactions during drying can be minimized for oxygen-sensitive lactic acid bacteria. King and Su (1993) studied freeze-, vacuum, and low-temperature vacuum drying of lactic acid starter cultures (L. acidophilus) and found comparable viability rates of 52.8 to 77.5% for freeze-drying and 50.0 to 74.6% for low-temperature vacuum drying, in contrast to 15.4 to 30.4% for vacuum drying. More details regarding King and Su's (1993) studies are presented in the preservation of probiotic bacteria discussion later in the chapter.

Following technological progress and commercial requests, there is growing interest in preserving the biomass of starter cultures using mixed drying systems to combine spray and fluidized bed drying, spray and vacuum drying, fluidized bed drying, and freeze-drying ("atmospheric pressure freeze-drying"). Using these two-stage drying methods, it is possible to optimally control the parameters of each stage, such as the temperature, moisture content, and residence time of a dried material. As a result, a product of the best possible biochemical properties is obtained.

11.6 CONVENTIONAL AND ADVANCED METHODS IN DRYING AND PRESERVATION OF MICROORGANISMS

Drying and preservation of microorganisms (biomass) from the second group of bio-materials, which are aimed at preserving their microbiological activity and biological structure, are successfully carried out using the described methods. These methods are applied for starter cultures, although a higher process yield and bigger scale are needed for the equipment, which often induces additional problems related to process control. Despite the large dimensions of the spray dryers and ancillary equipment, spray drying is a predominant method in many food technologies (milk and whey in the dairy industry, bacteria, yeast, enzymes), particularly when high process efficiency is required at relatively low costs. In such a case, a liquid feed is atomized at high velocity in the chamber where it contacts hot air (200 to 220°C), so the droplets are dried into powder before they hit the side wall of the dryer chamber. The outlet temperature is close to the temperature of a product leaving the dryer and depends on the product dryness and material residence time. The inlet and outlet temperatures should be lower than that permissible for a microbiological material being dried. High survival rates can be obtained when the temperature is lower than normal, which is ~150°C at the inlet and ~75°C at the outlet from the dryer. However, the drying efficiency is much lower (about 50%) than that specified as a standard for the drying equipment (Adamiec and Strumillo, 1998).

The interesting option of not utilizing the full capacity of a given drying technology is now the acceptable practice, provided that several complementary drying methods are exploited in an entire process. Chávez and Ledeboer (2007) proposed low-temperature spray drying in the presence of protecting carrier materials in order to enhance the survival of probiotics during drying and storage at 30°C. The Food and Agriculture Organization of the United Nations and World Health Organisation define probiotic bacteria as "live microorganisms which, when administered in adequate numbers, confer a health benefit on the host." For instance, some probiotic bacteria are *Lactobacillus* and *Bifidobacterium* species. For many reasons, mainly referring to health issues, probiotics enhance the population of beneficial bacteria in the human gut, suppress pathogens, build up resistance against intestinal diseases, and reduce intestinal side effects associated with antibiotics. In some cases these bacteria lessen lactose intolerance and food allergies. There has been increasing interest recently in food products containing probiotics (not only as fermented milk drinks), which are dried foods with a long-term ambient shelf-live (Gardiner et al., 2000). Probiotics are sensitive to heat, so freeze-drying is the best technology to preserve viability, although spray drying is four- to sevenfold cheaper (Horaczek and Viernstein, 2004). Thus, spray drying must be mild enough to avoid deterioration but sufficiently efficient to yield a powder with moisture content below 4%, which is required for storage stability (Ananta et al., 2004).

It is noteworthy that there is continuous interest in the use of both methods for drying of probiotic bacteria. These studies focus on identification of the best carriers (protective environment) and on establishing a drying strategy (usually by applying variable process conditions) that will offer the highest survival and the longest storage time. A broader overview of these studies is out of the scope of this chapter, but relevant information can be gotten from Bâadi et al. (2000), Fu et al. (1995), Johnson and Etzel (1994), Lian et al. (2002), Rodriguez-Huezo et al. (2007), Teixeira et al. (1995), Wang et al. (2004), and Zamora et al. (2006).

Chávez and Ledeboer (2007) analyzed publications from 2001 to 2005 and two articles published in 1993 and 1995 that deal with spray drying of probiotic bacteria (lactobacilli) in the presence of various carriers, such as proteins, polysaccharides, sugars, gum arabic, maltodextrin, gelatin, starch, lactose, sucrose, trehalose, skim milk powder, and soy milk. They concluded that industrial drying of probiotic foods is still a challenging task to be overcome. The percentage of survival ranged from an acceptable value of ~40% at 25°C after 4 months to the unsatisfactory values, when the mortality was considerable, of 2 to 3 logs in only 5 days. Because one of the important factors that affected the survival of these bacteria is the drying temperature (mortality of bacteria starts to occur during drying and continues during storage), they tested a two-stage drying process consisting of spray drying followed by vacuum drying. The test probiotic was a culture of *Bifidobacterium lactis BB12* that, after appropriate inoculation and separation, was added to a carrier solution immediately before spray drying. The aqueous feed contained 20% (w/w) of various dissolved compounds in addition to the skim milk powder or soy protein isolate to which carbohydrates (gum arabic, maltodextrin or trehalose, lactose, sucrose) were added at a 1:1 ratio. A pilot-scale spray dryer was operated at 80 and 48°C at the dryer inlet and outlet, respectively. Then, the second drying step was performed in a vacuum oven at 45°C under a reduced pressure of 10 mbar over 24h. The bacterial stability and survival in the powder were determined under different conditions (but all at 30°C); the samples were sealed in aluminum bags under vacuum, as well as with no vacuum but under different, controlled relative humidity or in an anaerobic jar (with wet oxygen scavenger). Figure 11.11

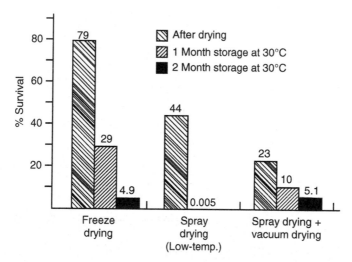

FIGURE 11.11 Survival of bifidobacteria after drying and after 1 and 2 months storage at 30°C (soy protein isolate + maltodextrin DE 20). (From Chávez, B.E. and Ledeboer, A.M., *Int. J. Dry. Technol.*, 25, 1193–1201, 2007. With permission.)

compares the bacteria survival for various drying methods. Clearly, two-step drying (spray drying + vacuum drying) under mild conditions is a realistic alternative to freeze-drying when producing food powders containing viable probiotics. Vacuum drying as the second drying step is needed to bring the moisture content and water activity below 5% and 0.25, respectively, which are safe for long-term storage. Furthermore, such a two-step process is estimated to be threefold cheaper than freeze-drying.

King and Su (1993) studied dehydration of probiotic bacteria *L. acidophilus* in aqueous suspension of the described carriers as well as $CaCO_3$ (1 g/l) using freeze-drying, vacuum drying, and controlled low-temperature vacuum drying. Aside from the positive results described previously, we provide more details regarding vacuum drying at controlled low temperature. The temperature passing the zone of maximum ice crystal formation in tested suspensions was monitored at the beginning using an automatic time–temperature recorder, and the results were between −2 and −3°C. This was applied as the standard for temperature control in ongoing drying experiments. During dehydration the temperature of concentrated cell suspensions was controlled within −2 and +2°C by the combination of shelf heating and vacuum adjustment. A typical run of the material temperature is shown in Figure 11.12. Because of the difficulty of water removal in the falling rate drying period that is due to shrinkage and case hardening, the temperature of the sample rose rapidly to 30°C before a final moisture content of 5% was achieved. This temperature rise (thermal shock) affected the survival of bacteria in the dry product. Note that the controlled low-temperature vacuum drying took only 4 h compared to 24 h for freeze-drying.

Ahmad et al. (2007) studied microwave-assisted vacuum drying of probiotic microorganisms (*L. salivarius* and *L. brevis*) and food yeasts (*S. cerevisiae*) as a drying method competitive to freeze-drying. This hybrid technology, called "traveling radiant energy under vacuum" (t-REV), is a rapid and efficient drying method that

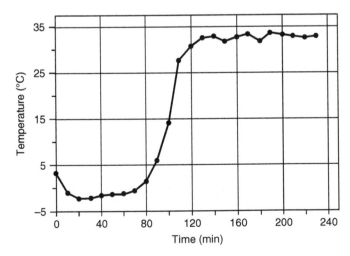

FIGURE 11.12 Temperature variation in controlled low-temperature vacuum dehydration operation of concentrated *Lactobacillus acidophilus* cell suspension. (From King, B.A.E. and Su, J.T., *Process Biochem.*, 28, 47–52, 1993. With permission.)

yields products with unique characteristics while retaining biological functions (Durance et al., 2007; Kudra and Mujumdar, 2001; Schiffmann, 2007). Although microwaves provide volumetric heating, the vacuum not only reduces the boiling point to maintain a low material temperature but also creates a water vapor pressure gradient so the drying rate is increased. As a result, the microbiological material is dried in a shorter time and at a lower temperature compared to other methods. For instance, Ahmad et al. (2007) monitored the material temperature every minute during t-REV drying of L. salivarius 417 using an infrared temperature sensor. They found that down to the final moisture content equivalent to 0.35 ± 0.05 water activity, the maximum temperature was 37°C at an absolute pressure of 30 torr and 24°C at an absolute pressure of 10 torr. Both temperatures were in the optimum range for bacterial growth. Although the viability of microorganisms dehydrated by the t-REV method was equal or superior to that by freeze-drying, the advantages of t-REV are a much shorter drying time (20 to 25 min compared to 72 h) and lower energy consumption. Similar results were reported by Kim et al. (1997), who studied the preservation of bacterial cultures in vacuum–microwave (VM) dried yogurt and found much higher values of survival than in spray drying or freeze-drying.

However, nonconventional energy sources such as infrared, radiofrequency, and microwave irradiation can alter the physiology of the living cells in a different way and to a different extent. Yaghmaee and Durance (2005) studied the effect of 2450-MHz microwave irradiation under vacuum (at microwave power settings of 711 ± 20 and 510 ± 5 W) on the survival and injury of E. coli. The vacuum was used to control the boiling point of water and to maintain the temperature in the bacterial suspension at the specified levels from 49 to 64°C. An analysis of the results showed that the microwave power from 510 to 711 W did not affect the thermal sensitivity of E. coli in this temperature range, but microbes were more sensitive to temperature changes during VM treatment than in the control treatment under a vacuum without microwave heating. For temperatures over 53°C in VM at 711 W and temperatures over 56°C in VM at 510 W, microbial destruction was faster than in the control treatment.

Linders et al. (1996) investigated the influence of the drying rate at low and mild temperatures on inactivation due to dehydration (with negligible thermal inactivation) of L. plantarum. Spray drying was used as a fast drying method because the estimated residence time of spray droplets in the laboratory minispray dryer was 0.45 s. The inlet and outlet drying air temperatures varied from 60 to 80°C and 42 to 53°C, respectively. Such slow drying was performed in a glass vacuum desiccator at 4 and 30°C. Linders and colleagues sought to verify their hypothesis that during drying at low temperature, when thermal inactivation of the cells can be neglected, the microorganisms adapt physiologically to ambient conditions (temperature and humidity) and maintain high activity after dehydration. They concluded that the residence time in this spray dryer was too short whereas the temperature of 4°C in a vacuum desiccator was too low for such an adaptation. The hypothesis on the preservation of the cell activity via such an adaptation was confirmed by vacuum dehydration at 30°C under high water activity.

An interesting option to alleviate the thermal effect on living microorganisms is heat pump drying (Alves-Filho et al., 1998). Experiments were carried out in a fluidized bed of inert particles in the heat pump dryer. The bacterial suspension

(*Streptococcus thermophilus* in a potassium phosphate buffer with and without trehalose protectant) was dried by spraying it over the fluid bed of 3.5-mm polypropylene beads. The inlet air temperature was kept below 35°C by using a classical heat pump circuit. The powdery product, which was dried with the addition of 100 mM of trehalose, showed 86% viability at an inlet air temperature of 10°C. Thus, Alves-Filho and coworkers concluded that this technique could be competitive to heat pump assisted freeze-drying in a fluidized bed that offers 100% viability but when drying at −20°C for 15 h, −10°C for 2 h, −5°C for 0.5 h, and finally 10°C for 0.3 h.

Apart from bacteria, yeasts constitute the second important group of microorganisms because of a wide variety of applications to various food product technologies as well as a considerable yield (productivity) of such processes. The technology of yeast preservation is one of the drying methods aimed at preservation of the microbial activity of cells, that is, active dry yeasts (ADY; the second group of drying technologies), in addition to preservation of high nutrient quality, that is, SCP (a third group). Note that other microorganisms such as bacteria, algae, and molds also constitute an important source of protein such as SCP. Nonetheless, yeasts (e.g., *S. cerevisiae*) possess the relatively highest concentration of intracellular proteins located inside the yeast cell when compared to *E. coli* or bacillus species, for instance. Furthermore, a genetically diversified protein structure affects nutrient quality, availability, and targeted activity. Moreover, proteins from microorganisms utilize their structure–availability relationship the best. Thus, research seeks to identify such a recombination of the protein structure that its suitability, both versatile and oriented, can be exploited in the best way. At present, for instance, only one marketed therapeutic protein (insulin) is produced from yeast. Detailed information on the biochemical phenomena related to the structure and activity of fungi and yeast can be found in the work of Arora (2004) and Ratledge and Kristensen (2006).

Another form of yeast used in the production of food is autolyzed yeast extract. The substrate for producing yeast extracts is a biomass of brewer's yeast and molasses-grown baker's yeast, *S. cerevisiae* and *Kluyveromyces fragilis* grown on whey, and *Candida utilis* grown on carbohydrates wastes. Autolysis is an enzymatic self-digestion of the yeast cell when the yeast cream is heated slowly to 55°C. The process is slow and the yield depends on the temperature, pH, time, and type of yeast. Finally, the slurry of biomass is pasteurized at 70°C, which deactivates the enzymes and kills the vegetative cells of the bacteria. The cell wall residues are removed by centrifugation and spray dried separately to be utilized as fodder. The clear liquid is concentrated in a falling film evaporator to approximately 50% total solids and then spray dried. The obtained powder is hygroscopic, so the use of dehumidified air for cooling and pneumatic transport is recommended. Spray dried autolyzed yeast extracts are rich in soluble nitrogen and vitamin B, and they are used as nutritional supplements in feed and in industrial fermentations. They can be used as additives in human food products for accentuation of flavor and for aroma in soups and meat products.

In practice, yeast in particular are dried in this way as ADY or SCP. That is why the products dried using the methods suitable for the third group should be considered. As previously mentioned, the aim of biomaterials drying in the third group of drying technologies is to obtain products of microbial origin and of high nutrient quality. These are important substances extracted from microorganisms, both from the widely available ones and those originating from other raw foods. Nevertheless, other components that rarely occur or are not encountered in other raw materials are reported as well. At this stage, proteins or, in general, amino acids, vitamins, lipids, and organic dyestuffs may be considered along with enzymes and antibiotics, which require the preservation of their biochemical activity. Through disruption (lysis) of the cell walls (using mechanical or biochemical methods, e.g., enzymatic reactions), certain components of cells are extracted and then separated, concentrated, and preserved using the appropriate methods of drying with the simultaneous preservation of the parameters attributed to a given group of products such as amino acids, vitamins, and so forth. Nonetheless, the profitability of the industrial production of SCP is largely dependent on the costs of the raw materials. In addition, the commercially profitable technologies are based on the use of semifinished products and waste materials. Examples are spent sulfite liquor, methane, carbohydrates, or pentoses and hexoses in waste liquor–cellulose hydrolysates. Here, the strains of *Candida utilis* and *Candida tropicalis* are used as protein (SCP) producers. The appropriate biotechnological process of biomass SCP production after inoculation can be carried out continuously for several months.

Yeast is harvested by subsequent centrifugation. The 1% growth suspension is concentrated to 15 to 22% total solids, but sometimes the yeast cream is concentrated up to 27% in falling film evaporators. As a standard, the spray drying method is applied with inlet and outlet temperatures up to 350 and 100°C, respectively. The temperature and time conditions during spray drying of yeast are such that the living cells of yeast and some pathogenic bacteria such as *Salmonella* are practically killed (up to 8 log), but the nutrition of the product is saved.

Production of an SCP for human consumption was regarded at the beginning as a major method of food production, which eliminates seasonal and variable weather constraints. Along with biotechnology development, because the choice of a microorganism can be based on nutritional value and protein content, large-scale production could help to alleviate the worldwide nutritional deficiency problem today. SCP could be used to supplement natural foods to increase their nutritional value. If it is to be used for human (and animal) consumption, the influence of production techniques or quality factors such as flavor, color, solubility, digestibility, and viable cell count deserve attention.

In their pioneering research on the production of dry SCP using *S. cerevisiae* baker's yeasts, Labuza and Santos (1971) analyzed the soluble solids, soluble proteins, and color score as the indicators of product quality, among others. The products obtained in a double drum dryer equipped with steam heated rolls and in a laboratory spray dryer with a centrifugal atomizer were compared. The results demonstrated the high quality powder after spray drying (vivid color and high protein quality in dough quality tests) and a high dependence of the powder properties on the operating parameters after drum drying (considerable change into a brown color just after 75 s of drying, decreasing content of soluble protein with an increase in pressure of the steam heating drums). Hedenskog and Morgen (1973) examined the production of dry protein concentrates with a low content of nucleic acid (drum drying and mechanical disintegration of the cell). They found that the nutritive value of the protein concentrate was higher than that of the starting yeast material. Ameri and Maa (2006) presented an analysis of the possibilities and advantages of the spray drying process for biopharmaceuticals, for example, the stability of protein- or peptide-based drug formulations for different therapeutic applications.

ADYs constitute an indispensable basic microbial factor in a number of relevant food technologies such as the production of baker's yeasts (yeasts biomass) and beer, wine, and alcohol beverages (ethanol production); in baking (CO_2 production); and the production of various ingredients altering the flavor, smell, or preservation by producing organic acids, unsaturated fatty acids, and so forth. The most frequently used strains are *S. cerevisiae*, *Pichia guillermondi*, and *Kluyveromyces lactis*, as well as yeasts with different morphology and shape (filamentous fungi), such as *Aspergillus niger*, *Agaricus bisporus*, and *Trichoderma viride*. It is noteworthy that the genetic modifications of the yeast cells are frequently applied to the technologies of enzyme and antibiotics production. Not only do such diverse applications require defined activity of the yeast cell in the inoculum, but they also need an appropriate physical structure. Then the yeasts' biomass is preserved using the proper methods for a given application. Although compressed wet yeasts were widely applied in many technologies (baking or beer production), the development of ADY has caused them to be replaced by dry yeasts because dry yeasts are stable for a long time and can be stored and used thereafter without restricted conditions (refrigeration).

Generally, compressed yeasts are dried from a moisture content of about 70% to a moisture level of 3 to 8%. Any dry yeast must survive two drastic changes prior to use by the baker: (1) when drying yeast cells, intracellular water must pass through the cell membrane and cell wall, so the drying process must not damage the cell membrane or decrease the intracellular enzymatic activity (Gervais et al., 1999); (2) before use, water must again pass through the membrane into the cell to activate the yeast. Poirier et al. (1999) demonstrated the importance of rehydration kinetics on the viability of *S. cerevisiae*, specifically the importance of slow rehydration kinetics in the water activity (a_w) range from 0.117 to 0.455. Any substantial degree of damage will dramatically reduce the gas (CO_2) productivity of the dry yeast, and yeast's ability to survive these rigorous conditions will affect its viability as a suitable replacement for compressed yeast. Since the development of ADY in the late 1930s, yeast manufacturers have been unable to effectively dry compressed yeast with a protein content higher than 44% (higher protein yeasts were unable to survive the rigors of drying). Conventional drying is carried out in band, tray and tunnel, or drum dryers, although the drying is relatively long at 4 to 12 h. Spray drying is much faster, but it still exposes the yeast cell to high temperatures with little control (resulting in deterioration of the enzymatic activity).

Due to many reasons (e.g., the availability of a research material, wide application of several different indices to determine the material quality), yeasts still constitute a relevant research and test material to be used in the assessment of new methods and drying parameters or in the experimental verification of mathematical modeling and computer simulation for the design of the preservation process.

There are numerous papers on this subject, but only the key studies will be presented here. Fluidized bed drying of baker's yeast is the most frequently investigated technology because it is easy to form fluidizable granules from commercial compressed yeast. Bayrock and Ingledew (1997a, 1997b) documented that the kinetics of drying and cell death depend not only on the origin of the commercial compressed yeast but also on the quality of the ADY product when dried at 35 to 80°C for up to 400 min. Oszlanyi (1980) demonstrated the advantages of fluidized bed drying of compressed baker's yeast when mixed before extrusion into 0.4-mm granules with <1% liquid sorbitan monostearate as an emulsifier. Because of the precise temperature control of the yeast cells in the fluidized bed at a maximum of 40.5°C, an "instant" product was obtained after 20-min drying. The microbiological activity of the ADY product was almost the same as compressed yeast (CO_2 gasing activity of instant yeast was >87% of the gas produced by compressed yeast). Taeymans and Thursfield (1987) dried immobilized baker's yeast (yeast entrapped in a gel matrix) in a fluidized bed. Alginate at 2% in the slurry with a 30% mass of yeast and a $CaCl_2$ solution were used to solidify the slurry drops pumped through the needles. Zimmermann and Bauer (1986, 1990) studied fluid bed drying of granulated baker's yeast and evaluated its activity through the activity of ADH and glucose-6-dchydrogenase enzymes. The results of these studies and those of Luyben et al. (1982) were used to formulate a mathematical model allowing the material temperature, moisture content, and product activity to be determined as a function of the drying time. Using this model and multiobjective optimization, a procedure was developed to design a fluid bed dryer for granulated baker's yeast (Adamiec et al., 2007; Strumillo et al., 1989).

Luna-Solano et al. (2005) presented the optimal process calculations for spray drying of brewer's yeast (*Saccharomyces* sp.) cream with solids content of 14 to 18% (w/w) in a pilot-scale dryer with a 1.2-m³ chamber at a maximum temperature of 145°C. They established empirical equations for the specific total cost, cell viability, production rate, and product moisture content. Their results were $26.7/kg cost, 10 kg/h productivity, 0.013 kg/kg dry matter product moisture content, and 1×10^6 cfu/g cell viability at an outlet air temperature of 59°C.

Alsina and colleagues (2005) utilized pressed yeast (by-product in the production of ethyl alcohol) to study the possibility of obtaining a powdery *S. cerevisiae* that could be used as a source of protein (raw protein content of 50%), vitamins, and flavoring in animal feed formulas. A relatively novel drying method was employed, namely, drying on a fluid bed of inert particles that is similar to the jet spouted bed (Markowski, 1993). Batches of the wet material (compressed yeast) and some volume of polyethylene particles (3.26-mm mean volume diameter) in different mass ratios were fluidized together and dried in a laboratory fluid bed dryer. The research was aimed at determination of the bed hydrodynamics as a function of the wet material/ inert particles ratio, and a mathematical model was developed to determine the dryer throughput for different operating conditions. Characteristically, stable fluidization was obtained for pressed yeast with the same average moisture content as for commercial pressed yeast, which is on the order of 70% wet basis.

Grabowski et al. (1997) performed an experimental study to obtain high quality ADY in fluidized bed and spouted bed dryers. Wet commercial pressed yeast was extruded through a steel screen with 0.8-, 0.9-, and 1-mm perforations, giving cylindrical particle lengths of around 1.5, 2.5, and 3 mm, respectively. Experiments were carried out in batch laboratory scale dryers. The results from the bed hydrodynamics and drying kinetics for the same batches of material indicated the need for higher (by almost 25%) air velocity in a spouted bed during the initial drying phase down to a yeast moisture content of about 30 to 35% wet basis. However, the drying time to attain the final moisture content of 6 to 8% wet basis was shorter for the spouted bed, and this difference was dependent on the temperature. Based on these results, a two-stage process was arranged in which the spouted bed was turned into a fluidized bed when the moisture content of the yeast was reduced to 35% wet basis. This two-stage drying reduced the energy consumption because of the lower air-flow rate and the viability of the yeast cells was higher (Figure 11.13). With better technology and development of controlling equipment, old methods such as tunnel, band, and drum drying can be successfully replaced by spray drying, fluidized bed, spouted bed, and vibrofluidized bed drying.

Enzymes are simple or complex proteins possessing catalytic activity that allows chemical reactions in a living cell to occur at ambient temperature at a high rate. Every biotechnological process is based on enzymatic reactions. Enzymes are produced by plants, animals, and microorganisms as a result of the fermentation processes. Single enzymes or systems of enzymes are formed in a cell. The structure of enzymes is similar to the structure of amino acids and proteins, with the so-called active center as the central region of the long biopolymer chain that is directly engaged in a biochemical reaction. Based on the biochemical reaction mechanism, a large variety of enzymes are commonly categorized into six basic groups: oxyreductases, transferases, hydrolases, lyases, isomerases, and ligases. The rate of the enzymatic reaction is affected by many factors including the duration of the reaction, temperature, pH, enzyme stability, substrate concentration, and presence of activators. In addition, enzymes are biological catalysts and their action can be targeted and controlled, or it can be self-induced in a biomaterial. Drapron (1985) provided

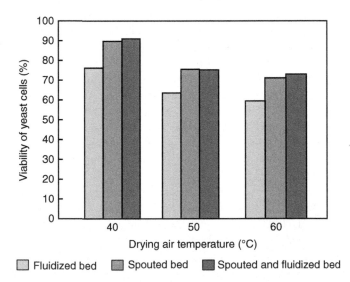

FIGURE 11.13 Viability of ADY at standard moisture content (6 to 8% wet basis). (From Grabowski, S. et al., *Int. J. Dry. Technol.*, 15, 625–634, 1997. With permission.)

numerous examples of enzyme activity as a function of the water activity of a food material. During storage and transformations of foodstuffs, enzymes continuously play a leading role, which is positive or negative (e.g., Maillard reactions).

Because enzymes are proteins, their structures change easily under the action of the mentioned factors, which may reduce even their catalytic ability completely. Because of enzymatic treatment, the source materials are used economically, the processes are more efficient, the product's quality and durability are improved, and the reaction rates can easily be controlled. The specific advantages of enzymes or enzymatic preparations have been utilized in many areas of human activity, causing revolutionary changes in numerous technological processes. Examples of industrial uses of enzymes are provided in Table 11.6.

Because most enzymes are not stable in solution, drying is often used to improve the stability during storage. Depending on the enzyme's characteristics and industrial designation, the dry product should possess certain quality indices, such as high and stable activity and purity, and other physical features such as solubility and dispersability, homogeneity (particle size distribution), bulk density, color, and odor. Enzymes are thermolabile substances, so they are deactivated because of heat-induced structural changes. The methods used for heat-sensitive microorganisms are useful for enzymes as well. Although freeze-drying is usually favored for drying of

TABLE 11.6
Major Food Industry Applications of Microbiologically Produced Enzymes

Application	Enzyme	Microbial Origin
Dairy	Protease	*Rhizomucor*
	Lipase	*Aspergillus*
	Lactase	*Klyveromyces, Aspergillus*
Bakery	Amylase	*Bacillus*
	Protease	*Aspergillus, Bacillus*
	Glucose oxidase	*Aspergillus*
Starch industry	Amylase	*Bacillus*
	Glucoamylase	*Aspergillus*
	Glucose isomerase	*Bacillus, Streptomyces*
Wine and juice	Pectinase	Calf stomach
	Cellulase	*Aspergillus*
	Cellobiase	*Aspergillus, Trichoderma*
Brewery	β-Clucanase	*Aspergillus*
	Acetoacetate decarboxylase	*Aspergillus, Bacillus*
Distilling industry	Amylase	*Trametes*
	Glucoamylase	*Aspergillus*
Animal feed	Phytase	*Aspergillus*
	Cellulase	*Aspergillus*

Source: Adapted from Pilosof, A.M.R. and Sánches, V.E., in *Handbook of Industrial Drying*, Taylor & Francis, Boca Raton, FL, 2007, pp. 881–992. With permission.

high-valued and thermally labile enzymes in small quantities (used as analytical enzymes), spray drying is largely applied in large-scale production of enzymes.

Pilosof and Terebiznik (2000) compared both drying methods based on numerous publications (>100). Some of them are worth mentioning.

1. Yamamoto et al. (1984) studied the kinetics of enzyme inactivation (glucose oxidase, alkaline phosphatase, and alkohol dehydrogenase) during water evaporation from a single droplet of the solution.
2. Etzel et al. (1996) investigated enzyme inactivation (alkaline phosphatase) in a spray-dried condensed skim milk.
3. Samborska et al. (2005) looked at the effect of spray-drying conditions on the activity of a fungal enzyme (α-amylase) dissolved in a maltodextrin.
4. Luyben et al. (1982) used the experimental results on enzyme inactivation kinetics (alkaline phosphatase, lipase, and catalase) to develop a mathematical model and to simulate the kinetics of quality changes, including inactivation of enzymes during drying.

When the high throughput of dry enzymes is needed (in food technology, production of detergents, and the pulp and paper industry), the enzyme powder is produced in a two-stage drying process, which comprises a conventional spray dryer as the first stage and a fluidized bed dryer as the second stage (Masters, 2002). An example is when the isolated and concentrated enzymes solution is mixed with heat-protecting and carrier substances, and the slurry with up to 40% total solids is spray dried in a conical spray dryer at air inlet temperatures from 170 to 200°C. The exhaust temperature is kept as low as possible to permit powder to leave the cone at low temperature and a high moisture content of 10 to 20% wet basis. The still wet enzyme powder is then dried to about 2% wet basis in an external, continuous, plug-flow fluid bed with a fluidizing gas temperature of <90°C. In the last section of the fluid bed, the powder can be cooled with dehumidified air.

Enzymes, which are active polymers with no attributes of living microorganisms but acting as reusable catalysts of biochemical reactions, are frequently preserved by immobilizing (entrapping, encapsulating) them in carrier materials, which preserve and stabilize the enzyme's activity. Note that in living cells most enzymes are bound to the cell membranes or structural elements of the cells, so they are protected in a natural way. Immobilized enzymes are much less sensitive to external factors than free enzymes. For example, immobilization of β-glucosidase on a silty material increases its thermal resistance by about 100 times. Eduardo et al. (1986) studied the effect of dehydration on the thermostability of alcohol oxidase (*Pichia pastoris*) immobilized on diaminoethyl cellulose. The results indicated that dehydration, in contrast to immobilization by wet adsorption, dramatically improved the thermoresistance of the enzyme. For instance, the half-life was increased from <2 min to ~3 h at 60°C, and 25% of the initial enzymatic activity was still present even after 20 h of heating at 80°C. Because of immobilization, the enzyme catalytic properties are often improved, which makes it possible to carry out a wide range of biotechnological operations under restricted conditions. For instance, glucose isomerase can be used continuously for over 1000 h at temperatures from 60 to 65°C. Regarding the

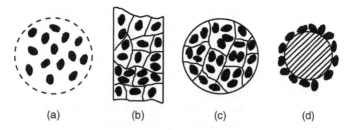

FIGURE 11.14 Examples of methods of production and forms of immobilized biocatalysts: (a) encapsulation, (b,c) closing in membrane and gel structures, and (d) physicochemical bonds of adsorption and adhesion.

mechanisms of immobilization, the typical methods are depicted in Figure 11.14. These are (a) trapping by semipermeable membranes (encapsulation and microencapsulation); (b) and (c) embedding a gel carrier inside the matrix (in natural or synthetic gels such as agar, alginate, or collagen); and (d) adsorption and adhesion in the carrier or on its surface because of the ionic strength, hydrogen bonds, and other weak physical–chemical interactions.

Champagne et al. (1994) conducted an extensive review of immobilized cell technologies in the dairy industry, where lactic acid bacteria play an essential role in the processing of milk into fermented products. For instance, when discussing the process of cheese production, results were quoted on the application of three strains of mesophilic lactic acid bacteria (*Lactococcus lactis* ssp. *lactis*, *Lactococcus lactis* ssp. *cremoris*, and *Lactococcus lactis* ssp. *lactis* var. *diacetylactis*) entrapped separately in calcium alginate beads during continuous prefermentation of milk. Very high productivity was shown for a continuous stirred tank reactor operated at 34°C, a pH of 6.2 that was controlled by the incoming pasteurized milk, and a population of 1.5 to 2.0×10^8 cfu/mL in the prefermented milk. Using immobilized cell technologies, the incubation time required to reach a pH of approximately 4.6 was reduced by 50% compared with milk inoculated with a freeze-dried starter used in industrial production. In this application, immobilized cell technology demonstrated attractive features, including high productivity, better control of the mixed culture with impact on the quality of the product (i.e., final acidity, texture, and sensory attributes), and process economics.

There is continuous and growing interest in immobilized enzymes and cells and a large number of practical applications in various branches of the chemical, pharmaceutical, and food industries as well as in medicine, health care, environmental protection, and the treatment of industrial and municipal wastes. In particular, the encapsulation technology has recently received a great deal of interest mainly because of its suitability in the production of drugs and foods with enhanced nutritive properties. In the food industry this method is most frequently used to protect the active component from the negative influence of the environment (oxygen, light, humidity); to entrap the fragrance components; to change the physical state from liquid to powder because the latter is easier to store and handle; and to preserve the activity of microorganisms, enzymes, vitamins, and other components of interest. The key publications, in the order of their importance, are those by Adamiec and Kalemba (2006), Anal and Singh (2007), Bhandari (2004), Han et al. (2007), Patil et al. (2006), Re (1998), Shahidi and Han (1993), Yoshii et al. (2001), and Yoshii et al. (2006). The procedure leading to microcapsules (most frequently spherical particles having a diameter in the range of microns, typically from 30 to 50 µm, with a maximum of 500 to 700 µm) obtained from a multicomponent system starts with the formation of a suitable emulsion followed by spray drying. The selection of the proper encapsulating medium along with the method and conditions for optimal drying and process control allow the production of powdery microcapsules that offer a high-value substrate that can preserve its properties over long-term storage prior to use or when already incorporated in the food product. A number of preservation methods via dehydration that are described in this chapter can be successfully used to preserve the complex structures of immobilized and microencapsulated products.

11.7   CONCLUSIONS

In summary, a variety of microorganisms can be applied to process raw materials, intermediate foods, and final foods. These microorganisms differ not only in their inherent properties but also in the destination of the final product. Thus, the proper dehydration technique greatly depends on the production rate, the end use of the product, and its market value (cost of the raw material and price of the product). It is reasonable to assume that with the progress in drying and dewatering as well as advanced process control, two-stage or multistage technologies will be favored, especially for large-scale production. It is quite likely that mild and variable operating conditions will be used in addition to additives, which will preserve the vital functions of the product and stabilize it for a long time. Of special interest here is the rapidly growing technique of microencapsulation, which is one of the immobilization methods.

NOMENCLATURE

ADH	alcohol dehydrogenase
A	intercept on the y axis
ADY	active dry yeast
a_w	water activity
B	slope of the line relating log survival to storage time
E_a	activation energy for the death of the organism
k	reaction rate constant
k_0	frequency coefficient
N_v	number of viable microorganisms per unit volume of medium
R	gas constant
SCP	single cell protein
t	time
t-REV	traveling radiant energy under vacuum
$T(t)$	absolute temperature
VM	vacuum–microwave

X	number of years of storage,
Y	survival rate
Y_1-Y_{20}	log survival rates from 1 to 20 years

REFERENCES

Abadias, M., Benabarre, A., Teixidó, N., Usall, J., and Viñas, I., Effect of freeze drying and protectants on viability of the biocontrol yeast *Candida sake*, *Int. J. Food Microbiol.*, 65, 173–182, 2001.

Adamiec, J. and Kalemba, D., Analysis of microencapsulation ability of essential oils during spray drying, *Dry. Technol.*, 24, 1127–1132, 2006.

Adamiec, J., Kaminski, W., Markowski, A.S., and Strumillo, C., Drying of biotechnological products, in *Handbook of Industrial Drying*, 3rd ed., Mujumdar, A.S., Ed., Taylor & Francis, Boca Raton, FL, 2007, pp. 905–929.

Adamiec, J. and Strumillo, C., Attempts of low-temperature spray drying of mixed population of lactic acid bacteria and yeast, in *Drying '98, Proceedings of the 11th International Drying Symposium*, Vol. C, Akritidis, C.B., Marinos-Kouris, D., and Saravakos, G.D., Eds., Ziti Editions, Thessaloniki, Greece, 1998, pp. 1669–1674.

Ahmad, S., Yaghmaee, P., and Durance, T., Survival of probiotic bacteria and food yeast dehydrated with microwave energy under vacuum, in *41st Annual Microwave Symposium Proceedings*, International Microwave Power Institute, Mechanicsville, VA, 2007, pp. 30–34.

Ainsworth, G.C., Sparrow, F.K., and Sussman, A.S., *The Fungi*, Vols. 1–4, Academic Press, New York, 1965–1973.

Alsina, O.L.S., Silva, V.S., Silva, F.L.H., and Rocha, A.P.T., Drying of yeasts in fluidized beds: Powder producing kinetics, in *Proceedings of the 3rd Inter-American Drying Conference*, Orsat, V., Raghavan, G.S.V., and Kudra, T., Eds., Montreal, Canada, 2005, paper XIV-3.

Alves-Filho, O., Strømmen, I., Aasprong, A., Torsveit, A.K., Boman, H.C., and Hovin, W., Heat pump fluidized bed drying for lactic acid suspensions using inert particles and freeze drying, in *Drying '98, Proceedings of the 11th International Drying Symposium*, Vol. C, Akritidis, C.B., Marinos-Kouris, D., and Saravakos, G.D., Eds., Ziti Editions, Thessaloniki, Greece, 1998, pp. 1833–1840.

Ameri, M. and Maa, Y.-F., Spray drying of biopharmaceuticals: Stability and process considerations, *Dry. Technol.*, 24, 763–768, 2006.

Anal, A.K. and Singh, H., Recent advances in microencapsulation of probiotics for industrial applications and targeted delivery, *Trends Food Sci. Technol.*, 18, 240–251, 2007.

Ananta, E., Birkeland, S.E., Corcoran, B.M., Fitzgerald, G.F., Hinz, S., Klijn, A., Mättö, J., Mercernier, A., Nilsson, U., Nyman, M., O'Sullivan, E., Parche, S., Rautonen, N., Ross, R.P., Saarela, M., Stanton, C., Stahl, U., Suomalainen, T., Vincken, J.-P., Virkajärvi, I., Voragen, F., Wesenfeld, J., Wouters, R., and Knorr, D., Processing effects on the nutritional advancement of probiotics and prebiotics, *Microb. Ecol. Health Dis.*, 16, 113–124, 2004.

Antheunisse, J., Viability of lyophilized microorganisms after storage, *Anton. Leeuwen.*, 39, 243–248, 1973.

Arora, D.K., Ed., *Handbook of Fungal Biotechnology*, 2nd ed., Marcel Dekker, New York, 2004.

Atlas, R.M., *Microbiology, Fundamentals and Applications*, Macmillan, New York, 1984.

Bâati, L., Fabre-Gea, C., Auriol, D., and Blanc, P.J., Study of the cryotolerance of *Lactobacillus acidophilus*: effect of culture and freezing conditions on the viability and cellular protein levels. *Int. J. Food Microbiol.*, 59, 241–247, 2000.

Barnett, J.A., Payne, R.W., and Yarrow, D., *Yeasts, Characteristics and Identification*, 3rd ed., Cambridge University Press, New York, 2000.

Bayrock, D. and Ingledew, W.M., Fluidized bed drying of bakers's yeast: Moisture levels, drying rates, and viability changes during drying, *Food Res. Int.*, 30, 407–415, 1997a.

Bayrock, D. and Ingledew, W.M., Mechanism of viability loss during fluidized bed drying of baker's yeast, *Food Res. Int.*, 30, 417–425, 1997b.

Bhandari, B., Spray-drying: An encapsulation technique for food flavors, in *Dehydration of Products of Biological Origin*, Mujumdar, A.S., Ed., Science Publishers, Enfield, NH, 2004, pp. 513–533.

Bjerketorp, J., Håkansson, S., Belkin, S., and Jansson, J.K., Advances in preservation methods: Keeping biosensor microorganisms alive and active, *Curr. Opin. Biotechnol.*, 17, 43–49, 2006.

Blanch, H.W. and Clark, D.S., *Biochemical Engineering*, Marcel Dekker, New York, 1997.

Boumahdi, M., Mary, P., and Hornez, J.-P., Influence of growth phases and desiccation on the degrees of unsaturation of fatty acids and the survival rates of rhizobia, *J. Appl. Microbiol.*, 87, 611–619, 1999.

Champagne, C.P., Lacroix, Ch., and Sodini-Gallot, I., Immobilized cell technologies for the dairy industry, *Crit. Rev. Biotechnol.*, 14, 109–134, 1994.

Chávez, B.E. and Ledeboer, A.M., Drying of probiotics: Optimization of formulation and process to enhance storage survival, *Dry. Technol.*, 25, 1193–1201, 2007.

Corcoran, B.M., Ross, R.P., Fitzgerald, G.F., and Stanton, C., Comparative survival of probiotic lactobacilli spray-dried in the presence of prebiotic substances, *J. Appl. Microbiol.*, 96, 1024–1039, 2004.

Drapron, R., Enzyme activity as a function of water activity, in *Properties of Water in Foods*, Simatos, D. and Multon, J.L., Eds., Martinus Nijhoff Publishers, Dordrecht, 1985, pp. 171–190.

Durance, T., Yaghmaee, P., Ahmad, S., and Zhang, G., Method for dehydrating microorganisms and vaccines, *PCT*, European Patent Application 05772090.6, 2007.

Eduardo, B.-G., Klibanow, A.M., and Karel, M., Enhanced thermoresistance of enzymes by drying, in *Drying '86, Proceedings of the 5th International Drying Symposium*, Mujumdar, A.S., Ed., Hemisphere Publishing, Washington, DC, 1986, pp. 428–431.

Etzel, M.R., Suen, S.-Y., Halverson, S.L., and Budijono, S., Enzyme inactivation in a droplet forming a bubble during drying, *J. Food Eng.*, 27, 17–34, 1996.

Fu, W.-Y., Suen, S.-Y., and Etzel, M.R., Inactivation of *Lactococcus lactis* ssp. *lactis* C2 and alkaline phosphatase during spray drying. *Int. J. Drying Technol.*, 13, 1463–1476, 1995.

Gardiner, G.E., O'Sullivan, E., Kelly, J., Auty, M.A., Fitzgerald, G.F., Collins, J.K., Ross, R.P., and Stanton, C., Comparative survival rates of human-derived probiotics *Lactobacillus paracasei* and *Lactobacillus salivarius* strains during heat treatment and spray drying, *Appl. Environ. Microbiol.*, 66, 2605–2612, 2000.

Gervais, P., Beney, L., Martinez de Maranon, I., and Marechal, P.A., Could high water flow rate damage the membrane of *Saccharomyces cerevisiae*? in *Water Management in the Design and Distribution of Quality Foods*, Roos, Y.H., Leslie, R.B., and Lillford, P.J., Eds., Technomic Publishing, Lancaster, PA, 1999, pp. 353–374.

Gould, G.W., Water and the bacterial spore: Resistance, dormancy and germination, in *Water Management in the Design and Distribution of Quality Foods*, Roos, Y.H., Leslie, R.B., and Lillford, P.J., Eds., Technomic Publishing, Lancaster, PA, 1999, pp. 301–323.

Gould, G.W., History of science—Spores, *J. Appl. Microbiol.*, 101, 507–513, 2006.

Grabowski, S., Mujumdar, A.S., Ramaswamy, H.S., and Strumillo, C., Evaluation of fluidized versus spouted bed drying of baker's yeast, *Dry. Technol.*, 15, 625–634, 1997.

Gunsalus, I.C., *The Bacteria*, Academic Press, Orlando, FL, 1986.

Han, D., Yu, C., and Zhou, Y., Microencapsulation of Gram-negative bacteria by spray-drying, in *Proceedings of the 5th Asia–Pacific Drying Conference*, Chen, G., Ed., The Hong Kong University of Science and Technology, Hong Kong, 2007, pp. 249–254.

Hayashi, H., Kumuzawa, E., Saeki, Y., and Ishioka, Y., Continuous vacuum dryer for energy saving, *Dry. Technol.*, 1, 275–284, 1983.

Hedenskog, G. and Morgen, H., Some methods for processing of single cell protein, *Biotechnol. Bioeng.*, 15, 129–142, 1973.

Hernández, A., Weekers, F., Mena, J., Pimentel, E., Zamora, J., Borroto, C., and Thonart, Ph., Culture and spray-drying of *Tsukamurella paurometabola* C-924: Stability of formulated powders, *Biotechnol. Lett.*, 29, 1723–1728, 2007.

Hocking, A.D., Moulds and yeasts associated with foods of reduced water activity: Ecological interactions, in *Proceedings of the ISOPOW/IUFOST Symposium. Food Preservation by Moisture Control*, Seow, C.C., Teng, T.T., and Quah, C.H., Eds., Malaysian Institute of Food Technology, and Universiti Sains Malaysia, 1988, pp. 57–72.

Horaczek, A. and Viernstein, H., Comparision of three commonly used drying technologies with respect to activity and longevity of aerial conidia of *Beauveria brongniartii* and *Metarhizium anisopliae*, *Biol. Control*, 31, 65–71, 2004.

Johnson, J.A.C. and Etzel, M.R., Inactivation of lactic acid bacteria during spray drying. *AICHE Symposium Series*, 89, No. 297, 98–106, 1994.

Kaminski, W., Adamiec, J., Grabowski, S., Zbicinski, I., Strumillo, C., and Mujumdar, A.S., Application of degradation kinetics to optimization of drying process for yeasts, in *Drying of Solids*, Mujumdar, A.S., Ed., International Science Publishers, New York, 1992, pp. 250–266.

Kapsalis, J.G., Influences of hysteresis and temperature on moisture sorption isotherms, in *Water Activity: Theory and Applications to Food*, Rockland, L.B. and Beuchat, L.R., Eds., Marcel Dekker, New York, 1987, pp. 173–213.

Kim, S.S., Shin, S.S., Chang, S.Y., Kim, S.Y., Noh, B.S., and Bhomik, S.R., Survival of lactic acid bacteria during microwave vacuum drying of plain yoghurt, *Lebensm.-Wiss. Technol.*, 30, 573–577, 1997.

King, V.A.E. and Su, J.T., Dehydration of *Lactobacillus acidophilus*, *Process. Biochem.*, 28, 47–52, 1993.

Kirst, H.A., Yeh, W.K., and Zmijewski, M.J., Jr., Eds., *Enzyme Technologies for Pharmaceutical and Biotechnological Applications*, Marcel Dekker, New York, 2001.

Kudra, T. and Mujumdar, A.S., *Advanced Drying Technologies*, Marcel Dekker, New York, 2001.

Kudra, T. and Strumillo, C., Eds., *Thermal Processing of Bio-Materials*, Gordon and Breach Science Publishers, Amsterdam, 1998.

Kuts, P.S. and Tutova, E.G., Fundamentals of drying of microbiological materials, *Dry. Technol.*, 2, 171–201, 1983–1984.

Labuza, T.P. and Santos, D.B., Concentration and drying of yeast for human food: Effect of evaporation and drying on cell viability and SCP quality, *Trans. ASAE*, 701–705, 1971.

Laroche, C., Fine, F., and Gervais, P., Water activity affects heat resistance of microorganisms in food powders, *Int. J. Food Microbiol.*, 97, 307–315, 2005.

Leuschner, R.G.K. and Lillford, P.J., Thermal properties of bacterial spores and biopolymers, *Int. J. Food Microbiol.*, 80, 131–143, 2003.

Lian, W.-C., Hsiao, H.-C., and Chou, C.-C., Survival of bifidobacteria after spray drying. *Int. J. Food Microbiol.*, 74, 79–86, 2002 .

Linders, L.J.M., Meerdink, G., and van't Riet, K., Influence of temperature and drying rate on the dehydration inactivation of *Lactobacillus plantarum*, *Trans. Inst. Chem. Eng.*, 74C, 110–114, 1996.

Luna-Solano, G., Salgado-Cervantes, M.A., Rodrigues-Jimenes, G.C., and Garcia-Alvarado, M.A., Optimization of brewer's yeast spray drying process, *J. Food Eng.*, 68, 9–18, 2005.

Luyben, K.C.A.M., Liou, J.K., and Bruin, S., Enzyme degradation during drying, *Biotechnol. Bioeng.*, 24, 533–552, 1982.

Markowski, A.S., Quality interaction in a jet spouted bed dryer for bio-products, *Dry. Technol.*, 11, 369–387, 1993.

Masters, K., *Spray Drying in Practice*, SprayDryConsult International ApS, Charottenlund, Denmark, 2002.

Miyamoto-Shinohara, Y., Sukenobe, J., Imaizumi, T., and Nakahara, T., Survival curves for microbial species stored by freeze-drying, *Cryobiology*, 52, 27–32, 2006.

Morgan, C.A., Herman, N., White, P.A., and Vesey, G., Preservation of microorganisms by drying: A review, *J. Microbiol. Methods*, 66, 183–193, 2006.

Mujumdar, A.S., Ed., *Dehydration of Products of Biological Origin*, Science Publishers, Enfield, NH, 2004.

Mujumdar, A.S., Ed., *Handbook of Industrial Drying*, 3th ed., Taylor & Francis, Boca Raton, FL, 2007.

Oszlanyi, A.G., Instant yeast, *Bakers Dig.*, August, 16–19, 1980.

Palaniappan, S., Sastry, S.K., and Richter, E.R., Effect of electroconductive heat treatment and electrical pretreatment on thermal death kinetics of selected microorganisms, *Biotechnol. Bioeng.*, 39, 225–232, 1992.

Patil, V.V., Joshi, V.S., and Thorat, B.N., Effect of various drying techniques on survival of baker's yeast immobilized in Ca-alginate beads and its applications, in *Proceedings of the 15th International Drying Symposium*, Farkas, I., Ed., Szent Istvan University Publisher, Gödöllő, Hungary, 2006, pp. 1209–1215.

Pilosof, A.M.R. and Sánches, V.E., Drying of enzymes, in *Handbook of Industrial Drying*, 3rd ed., Mujumdar, A.S., Ed., Taylor & Francis, Boca Raton, FL, 2007, pp. 981–992.

Pilosof, A.M.R. and Terebiznik, M.R., Spray and freeze drying of enzymes, in *Developments in Dring, Drying of Foods and Agro-Products*, Vol. I, Mujumdar, A.S. and Suvachittanont, S., Eds., Kasetsart University Press, Bangkok, Thailand, 2000, pp. 71–94.

Poirier, I., Maréchal, P.-A., Richard, S., and Gervais, P., *Saccharomyces cerevisiae* viability is strongly dependant on rehydration kinetics and the temperature of dried cells, *J. Appl. Microbiol.*, 86, 87–92, 1999.

Ratledge, C. and Kristiansen, B., Eds., *Basic Biotechnology*, 3th ed., Cambridge University Press, Cambridge, U.K., 2006.

Re, M.I., Microencapsulation by spray drying, *Dry. Technol.*, 16, 1195–1236, 1998.

Rockland, L.B. and Beuchat, L.R., Eds., *Water Activity: Theory and Applications to Food*, Marcel Dekker, New York, 1987.

Rodriguez-Huezo, M.E., Durán-Lugo, R., Prado-Barragán, L.A., Cruz-Sosa, F., Lobato-Calleros, C., Alvarez-Ramírez, J., and Vernon-Carter, E.J., Pre-selection of protective colloids for enhanced viability of *Bifidobacterium bifidum* following spray-drying and storage, and evaluation of aguamiel as thermoprotective prebiotic. *Food Res. Int.*, 40, 1299–1306, 2007.

Rose, A.H. and Harrison, J.S., Eds., *The Yeasts*, Vols. 1–4, Academic Press Limited, London, 1987–1991.

Rudge, R.H., Maintenance of bacteria by freeze-drying, in *Maintenance of Microorganisms*, 2nd ed., Kirsop, B.E. and Doyle, A., Eds., Academic Press, London, 1991, pp. 31–43.

Salminen, S., von Wright, A., and Ouwehand, A., Eds., *Lactic Acid Bacteria Microbiological and Functional Aspects*, Marcel Dekker, New York, 2004.

Samborska, K., Witrowa-Rajchert, D., and Goncalves, A., Spray-drying of α-amylase—The effect of process variables on the enzyme inactivation, *Dry. Technol.*, 23, 941–953, 2005.

Santivarangkna, Ch., Kulozik, U., and Foerst, P., Alternative drying processes for the industrial preservation of lactic acid starter cultures, *Biotechnol. Prog.*, 23, 302–315, 2007.

Schiffmann, R.F., Microwave and dielectric drying, in *Handbook of Industrial Drying*, 3th ed., Mujumdar, A.S., Ed., Taylor & Francis, Boca Raton, FL, 2007, pp. 285–305.

Schlegel, H.G., *Allgemeine Mikrobiologie*, Georg Thieme Verlag, Stuttgart, Germany, 1992.

Shahidi, F. and Han, X., Encapsulation of food ingredients, *Crit. Rev. Food Sci. Nutr.*, 33, 501–547, 1993.

Strumillo, C., Grabowski, S., Kaminski, W., and Zbicinski, I., Simulation of fluidized bed drying of biosynthesis products, *Chem. Eng. Process.*, 26, 139–145, 1989.

Taeymans, D. and Thursfield, J., Fluidbed-drying of immobilized yeasts, in *Drying '87, Proceedings of the 5th International Drying Symposium*, Mujumdar, A.S., Ed., Hemisphere Publishing, Washington, DC, 1987, pp. 160–165.

Teixeira, P.C., Castro, M.H., and Kirby, R.M., Death kinetics of *Lactobacillus bulgaricus* in a spray drying process. *J. Food Prot.*, 57, 934–936, 1995.

Tutova, E.G. and Kuts, P.S., *Drying of Microbiological Products* (in Russian), Agropromizdat, Moscow, 1987.

Wang, Y.-C., Yu, R.-C., and Chou, C.-C., Viability of lactic acid bacteria and bifidobacteria in fermented soymilk after drying, subsequent rehydration and storage. *Int. J. Food Microbiol.*, 93, 209–217, 2004.

Yaghmaee, P. and Durance, T.D., Destruction and injury of Escherichia coli during microwave heating under vacuum. *J. Applied Microbiol.*, 98, 498–506, 2005.

Yamamoto, S., Agawa, M., Nakano, H., and Sano, Y., Enzyme inactivation during drying of a single droplet, in *Proceedings of the 4th International Drying Symposium*, Toei, R. and Mujumdar, A.S., Eds., Hemisphere Publishing, Washington, DC, 1984, pp. 328–335.

Yoshii, H., Buche, F., Takeuchi, N., Ohgawara, M., and Furuta, T., Encapsulation of ADH enzyme in trehalose/protein matrices by spray drying, in *Proceedings of the 15th International Drying Symposium*, Farkas, I., Ed., Szent Istvan University Publisher, Gödöllő, Hungary, 2006, pp. 1139–1142.

Yoshii, H., Shiga, H., Ishikawa, M., Furuta, T., Forssell, P., Poutanen, K., and Linko, P., Enzyme encapsulation with crystal transformation of anhydrous trehalose, in *Proceedings of the 1st Nordic Drying Conference*, Alves-Filho, O., Eikevik, T.M., and Strømmen, I., Eds., Trondheim, Norway, 2001, paper 22.

Zamora, L.M., Carretero, C., and Parés, D., Comparative survival rates of lactic acid bacteria isolated from blood, following spray-drying and freeze-drying. *Food Sci. Technol. Int.*, 12, 77–84, 2006.

Zimmermann, K. and Bauer, W., The influence of drying conditions upon reactivation of baker's yeast, in *Food Engineering and Process Applications. Proceedings of the 4th International Congress on Engineering and Food*, Vol. 1, Le Maguer, M. and Jelen, P., Eds., Elsevier Applied Science, London, 1986, pp. 425–437.

Zimmermann, K. and Bauer, W., Fluidized bed drying of microorganisms on carrier material, in *Engineering and Food. Preservation Processes and Related Techniques. Proceedings of the 5th International Congress on Engineering and Food*, Vol. 2, Spiess, W.E.L. and Schubert, H., Eds., Elsevier Applied Science, London, 1990, pp. 666–678.

12 Dryer Modeling

Catherine Bonazzi, Bertrand Broyart,
and Francis Courtois

CONTENTS

12.1   INTRODUCTION

There are many dryer types and subtypes in the food industry around the world. There is even more diversity in the products to be dried: solids, liquid, gels, and so forth. The objectives and constraints that apply to drying processes are evolving more and more toward high expectations. If the main objective 30 years ago was to achieve a certain degree of dehydration with minimal energy cost, engineers now design, build, and set the performance of industrial dryers with respect to energy, product quality, controllability, and much more. Designing, building, tuning, and controlling industrial dryers clearly becomes more and more complex, requiring highly skilled engineers with efficient simulation tools whenever possible.

Chemical engineers can use commercial software such as ASPEN or PROSIM to help design, set, and optimize their processes. A good question is why food engineers cannot buy similar software.

Dryers are also used for nonfood products (e.g., chemical, wood, or paper industries), but the innovations cannot be simply transferred to food specificities for several reasons:

- Food product properties and moisture contents are generally highly variable, as opposed to manufactured products such as paper or chemical compounds. For example, when looking at the distribution of the grain moisture content of a sample of rice grains from the same harvested field it is common to observe a moisture content distribution from 5% wet basis (w.b.) up to 35% w.b. with a mean value of 20% w.b. Similarly, the geometric characteristics can vary over a wide range, even for homogeneous samples. Currently, there is an active research theme concerning the prediction of food properties based on their composition and the construction of databases.
- Food products are not usually made of homogeneous materials. They partly share this characteristic with wood. Food is often an assembly of different cellular tissues with specific resistance to water transport. Moreover, foods usually shrink considerably during drying and have important visible changes in aspect, texture, color, and so forth. As an illustration, think of the high shrinkage of carrot or apple slices or the change in surface texture of maize kernels (from elastic to rigid).

- The design of food dryers cannot be reduced to a pure energy optimization problem. Apart from the main moisture content target, dried food has to comply with many other "qualities" (e.g., color, firmness, odor, taste, etc.). Although those qualities can be quantified in a laboratory, they are seldom available as models or simple equations.

For these reasons, dryer models and dryer simulators are required. Different situations may arise such as the following:

1. A new dryer must be designed and built. Industrial dryers typically have volumes greater than hundreds of cubic meters or more, so it is necessary to consider laboratory-scale experimental dryers and/or model-based simulation tools. Note that the scaling of results obtained at the laboratory level up to the industrial level is never trivial.
2. There is an existing dryer with a given technology, but a new product or at least new objectives are set. In that case, the engineering work will focus on the setting (design is already done) and possibly the control of the drying system. Because it is practically impossible to test different drying strategies, a model-based simulator is required.
3. There is an existing dryer and the settings are satisfactory, but on-line recordings of dried product moisture content and quality do not match the objectives nor the constraints. This is a classical control problem for a nonclassical (e.g., nonlinear, multiple input–multiple output) system. As shown in several publications, successful controllers for drying systems are model based.

As explained later in this chapter, there are other cases where available dryer models may be helpful.

As stated, no polyvalent dryer simulator is available for the food industry. On the one hand, one can find several calculators relative to moist air or steam properties (e.g., LINRIC or PSYCHRO), sometimes including useful tools for dryer design or tuning, but no true versatile software dedicated to all types of food dryers. On the other hand, there are several big commercial packages (e.g., PROSIM or ASPEN) dedicated to the design, setting, and optimization of chemical processes that may be adapted to fit some food drying problems, possibly with a lot of work. However, no "ready to use" tool is available for food dryer design, setting, control, and optimization. Fortunately, numerous drying models exist for several dryer types (e.g., grain dryers, spray dryers, freeze-dryers, drum dryers, etc.) validated on several products. However, there is a chance that we can find a drying model for the right dryer type but not the right product. In that general case, the model does not have to be invented, just extended (or adapted) to the new product.

It is regrettable that when major industrial simulators (e.g., PROSIM) comply with CAPE-OPEN interoperability requirements, there are no available food dryer models that are compatible. This means that no one did the extra work needed to make the model accessible through a standard interface. Most (if not all) dryer simulators are dryer type specific and product specific, and their graphical user interface is generally hard coded in the model. This may not last indefinitely when current trends in research are seeking more generic approaches and results (e.g., SAFES methodology or molecular modeling approach). If there is no "standard" simulation software for food dryers, some drying models exist and are commonly used (with variants), each of them with a certain level of complexity (i.e., physical realism) and an attached domain of validity. For instance, the Brooker and Bakker Arkema model (1974) can probably be considered to be at the root of the development of several close models such as the Nellist–Bruce model (Bruce, 1983) or the Courtois–Abud model (Courtois et al., 1991, 2001). Alternatively, there are several situations where few bibliographies can be found. When considering a very specific product with specific quality criteria in an unusual (i.e., barely found in industry) kind of dryer, the engineer is left almost helpless. For instance, designing a steam dryer for alfalfa may be difficult for several reasons: the product dimensions and texture change considerably during drying (e.g., due to shrinkage of the product and collapse of the stack), and drying occurs above the boiling point and requires a specific drying model.

The specificity of food dryer models can be analyzed at two levels from the particle to the full dryer scale:

1. The first level is the single particle (e.g., a wheat kernel), the droplet (in spray drying), or an elementary element of volume (e.g., for drum drying). This is the place to describe all of the internal transfers: heat, mass, and possibly quality related transformations. At this level there are many well-known equations to describe the phenomena, but there is little knowledge for predicting heat and mass transfer coefficients or even less for quality transformations. This means that internal heat and mass balances rely heavily on unknown coefficients. These heat and mass transfer equations within the product (classic differential equations) require boundary equations to describe what happens at the interface with the surrounding environment (drying air or steam, contact or radiative fluxes, etc.). This level is classically referred to as a *thin layer* of product (see later).

2. The second level is the integration of the first level model in the three dimensions of the dryer. We should say four dimensions because time is an important variable: even if a dryer can be globally considered in the steady state (time independence of state variables), all of the product particles or parts are in an unsteady state (time dependence). Considering a multipart conveyor belt dryer for pet food as an example, the product is spread over the belt with a certain thickness. This is described as a *deep bed*. Depending on the drying conditions, this dryer can be considered as a combination (in series and/or in parallel) of deep beds. The numerical integration of thin layers into a deep bed is usually complex. Typically, the simulation of a dryer implies the numerical resolution of several coupled deep bed models, involving the simulation of many elementary thin layers. Note that there are many cases in which this logic is not applied exactly (e.g., freeze-dryers, drum dryers, etc.).

Even if there are many categories of dryers that cannot fit exactly with this dichotomy, they are often less different than they seem to be. For instance, a solar dryer with indirect solar heating can be considered as the combination of a solar heater and a classical hot air dryer. In addition, a drum dryer can be approximated as a thin layer drying above the boiling point and so forth.

There are two main steps to model an industrial dryer. First, consider the main phenomena responsible for heat and mass transfer within the product and in interaction with the drying media. At this stage, elementary equations describing the drying of a single particle or elementary volume must be written. Second, integrate elementary systems with respect to space and time. At this stage, a large part of the work is related to computer programming, depending on the graphical user interface that is expected.

The next section discusses different approaches used to describe the internal transfers within the product and the external transfers with the drying media (i.e., boundary conditions). Then, the classical method to integrate thin layer models into a deep bed and then at the full dryer scale is described. In the last section, some important issues are discussed and some applications are presented.

12.2 MODELING INTERNAL AND INTERFACIAL TRANSFERS DURING DRYING

The available approaches for modeling heat and mass transfers within the product and at the interfaces with drying media will be discussed in this section. White box models are considered to comply with strong theoretical foundations. They are supposed to be a combination of physically valid equations and laboratory measurable properties. Conversely, black box models are considered to be totally meaningless and unable to extrapolate any results. Usually, situations are not black or white and many models are grey boxes, for instance, meaningful equations with coefficients and properties obtained via model identification procedures. Neural networks (NNs), when intelligently structured, may also include some knowledge.

It is well known that the whiter the model, the more robust it is. In other words, a first principle model can be considered sufficiently robust for interpolation and perhaps extrapolation.

The general drying curve (also known as drying kinetic) is composed of three successive parts (Figure 12.1), which is a common assumption. The initial step is either warming up or cooling down to reach the wet bulb temperature (if convective drying) at which the product is maintained during drying at a constant rate; during this second step it follows an isenthalpic direction and heat brought to the product is entirely used to evaporate water. The curve is then followed by a slowing down part where the product temperature increases as its surface becomes dry.

This classical theory is not directly transferable to food products. A simple example can illustrate this assertion. Considering the thin layer drying kinetics of maize and carrots, whose moisture content is X and its drying rate $-dX/dt$, Figure 12.2 shows how different the actual curves are compared to Figure 12.1.

Drying models can be written on the scale of a single particle (a grain in a bed, a droplet during spray drying, or an entire solid) in order to predict moisture loss and temperature rise kinetics during processing. There can be multiple objectives because they can be used to obtain optimal settings for the drying operation, to set a design

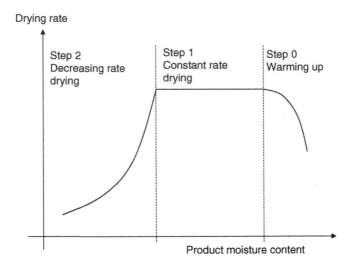

FIGURE 12.1 Generic air drying kinetic with three different steps: warming up, constant rate drying, and decreasing rate drying.

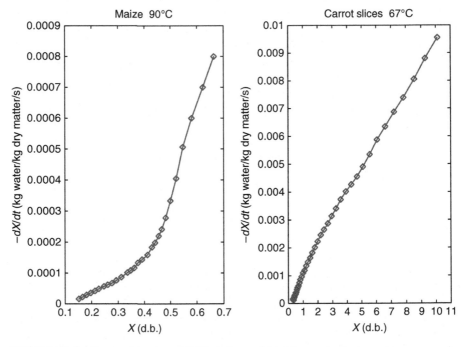

FIGURE 12.2 Drying kinetics of (left) maize and (right) carrots: drying rate versus moisture content (dry basis).

factor for a given dryer, or to understand and quantify interactions between elementary mass transfer modes inside the product during drying. The drying models written on the scale of a single particle can be classified as follows according to the level of empiricism used when formulating the equations:

- The drying models not based on first principle equations where kinetics of moisture loss are described using empirical laws.
- The drying models based on a first principle where an attempt is made to describe internal (inside the product) and external (between the product and its environment) heat and mass transfers: in such models, a variable number of assumptions can be made to simplify the physical (and mathematical) formulation of the transfer phenomena. This gives rise to a large number of model types from oversimplified models, in which internal resistances attibutable to heat and mass transfer are neglected, to a complete mechanistic approach, in which all individual modes of transfer are taken into account when formulating the model's equations.

12.2.1 PHENOMENOLOGICAL AND EMPIRICAL DRYING MODELS

Foods dry mostly in the falling rate period. Experimentally obtained Krischer's curves (drying vs. product water content) as well as the evolution of product temperatures show no constant rate period in most cases. Some products exhibit a very marked initial warming-up period, associated with an increasing drying rate. The falling rate period can sometimes be divided in two or more distinct parts, depending on the product, its dimensions, and the applied temperatures.

The air temperature always has a strong influence on the drying rate constants, which are sometimes "formalized" by an Arrhenius law, without any theoretical ground, but giving an easy basis for comparison. In fact, the drying rate should be more physically correlated to the product temperature, because in the falling rate period the main resistance is due to internal moisture transfer that is strongly dependent on the product temperature. However, this correlation with the air temperature is much more convenient for practical purposes.

Published drying curves for different foods processed under the same conditions show an extreme discrepancy in the magnitude of drying rates. One explanation is certainly that the drying rates are expressed as functions of the decreasing moisture content expressed on a dry matter basis, when surfaces of exchange or specific surfaces (m^2/kg dry matter) would certainly be a better basis of comparison. However, the structures and compositions of food products, especially the sugar and fat concentrations, may also explain variations in the drying rates. For those reasons, the theory is so far unable to predict the drying curves for such products. Moreover, the experimental evaluation of the physical constants (i.e., moisture diffusivity, thermal conductivity, density, specific heat, external heat, and mass transfer coefficients) needed for solving the equations of simultaneous heat and mass transport in a shrinking material (see Section 12.2.2) is a difficult problem, and their dependence on the moisture content and temperature is not well known for biological products.

Although a description of the drying kinetics is important information for dryer design, quite simple models to simulate the drying curves of foods have been frequently proposed. These empirical or semi-empirical models can provide adequate representation of experimental results, although their parameters lack a physical sense.

12.2.1.1 Empirical Models

Different types of empirical models, which are more or less applicable to food materials, have been proposed for fitting the thin layer drying curves. The main similarity between them is the exponential term included in the equations, which gives a decreasing exponential curve that provides a reasonable fit to the continuous decrease of the average moisture content during continuous drying.

The simplest one is based on the description of drying as a first-order kinetics in terms of the moisture ratio:

$$-\frac{d\overline{X}}{dt} = k_d(\overline{X} - X_e) \tag{12.1}$$

where $\overline{X}$ is the average moisture content in the product (dry basis), X_e is the equilibrium moisture content (dry basis), and k_d is the drying constant (/s), which is a lumped parameter used to fit experimental data.

This description was first suggested by Lewis (1921) for drying of thin layers of porous hygroscopic materials in the falling rate period and for constant conditions in the drying air. This equation is analogous to Newton's law for cooling when internal resistance to heat transfer can be neglected.

The main problem is the evaluation of the exact value of the equilibrium moisture content to utilize in this equation. It is sometimes treated as zero to simplify the mathematical solution or extrapolated from the experimental curves representing $-(d\overline{X}/dt)$ as a function of $\overline{X}$; X_e is then considered to be the value of $\overline{X}$ corresponding to $-(d\overline{X}/dt) = 0$ (Jannot et al., 2004). In reality, X_e depends on the temperature and relative humidity of the air and on the material being dried. It should be obtained from the sorption isotherm measured at the temperature of the drying experiment for water activity values corresponding to the relative humidity (RH) of the drying air (Figure 12.3).

In that case, the moisture ratio (MR) is expressed by integrating Eq. (12.1):

$$\text{MR} = \frac{\overline{X} - X_e}{X_0 - X_e} = \exp(-k_d t) \tag{12.2}$$

where X_0 is the initial product moisture content. It shows that the average moisture content of the product decreases exponentially with time (t):

$$\overline{X} = (X_0 - X_e)\exp(-k_d t) + X_e \tag{12.3}$$

As previously discussed, the drying operating conditions have a significant influence on the drying curves. Therefore, k_d can be correlated with the drying

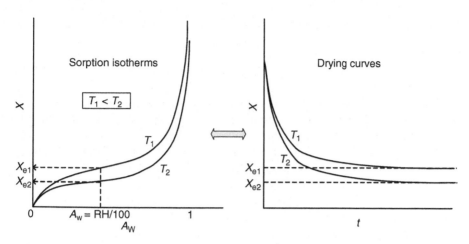

FIGURE 12.3 Equilibrium moisture content measurement from sorption isotherms.

conditions and material dimensions. Henderson and Pabis (1961) proposed the following equation for expressing the dependence on the air temperature (T_a):

$$k_d = k_0 \exp\left(-\frac{k_1}{T}\right) \tag{12.4}$$

However, more complex expressions can be found. For example, Kiranoudis et al. (1997) proposed the following:

$$k_d = k_0 d_p^{k1} T^{k2} v_a^{k3} RH^{k4} \tag{12.5}$$

where d_p is the diameter of the dried material, T the drying temperature, and v_a is the air velocity.

Another similar type of equation is known as Page's equation:

$$MR = \exp(-kt^n) \tag{12.6}$$

This equation has been used extensively for characterizing thin layer drying of cereals, oilseeds, and ear corn (Jayas et al., 1991). It has been utilized to better describe the drying kinetics mainly of starch-based materials when Eq. (12.2) failed to fit the drying data properly.

Sometimes the experimental data show that a single first-order reaction kinetic may not be fitted to the whole drying curve. In such cases, equations expressed as a sum of two or three exponentials have been proposed. Nellist and O'Callaghan (1971) fitted the data for ryegrass seeds using a two-term exponential equation. Karathanos (1999) proposed a three exponential equation [Eq. (12.7)] for drying of fruits rich in sugars, suggesting that each term may correspond to one of the main mechanisms of weight loss, that is, water evaporation and two different reactions of sugar decomposition.

$$MR = c_1 \exp(-k_1 t) - c_2 \exp(-k_2 t) - c_3 \exp(-k_3 t) \qquad (12.7)$$

More than 12 equations of this type are more or less applicable to biological products. A huge number of publications deal with the application of such models to various vegetables, fruits, or agroproducts. Jayas et al. (1991) reviewed and discussed many different equations. Among many others, Akpinar (2006), Celma et al. (2007), and Togrul and Pehlivan (2002) provided descriptions of the most widely used empirical models to describe drying kinetics.

Such empirical models are generally fitted to experimental data by regression analysis using statistical tools. Karathanos (1999) preferred a method of successive residuals described by Mohsenin (1980) for fitting coefficients in Eq. (12.7). The first residual was calculated by the difference between the experimental moisture ratios and those calculated by a first linear regression using only the first exponential term. This residual was then refitted to a second linear regression equation and the same for the third. In this way, Mohsenin proposed to discriminate the weight loss due to the decomposition of sugars from water evaporation.

Empirical models mathematically relate the average moisture content and drying time with a variable level of physical realism. They neglect the fundamentals of the drying process and the controlling transfer mechanisms, and their parameters have no physical meaning.

12.2.1.2 Characteristic Drying Curves (CDC)

The concept of CDC, first proposed by van Meel in 1958, states that there is a unique shape for the falling rate drying curve for a particular material, independent of the external conditions. For certain assumptions, a series of drying curves of the same material that are experimentally obtained under different drying conditions are supposed to be geometrically similar and can be brought down to a unique dimensionless curve called the CDC.

A CDC of a moist material can therefore be drawn from a single or a reduced number of pilot-scale experiments by plotting the *relative drying rate* as a function of the *characteristic moisture content*.

The relative drying rate (f) can be defined as $f = (N/N_c)$, where N is the drying rate and N_c is the drying rate in the constant rate drying period. Function f is supposed to be specific to the product being dried and independent of external conditions.

The dimensionless characteristic moisture content (ϕ) is the ratio of the free moisture content to the difference in moisture content between the critical point (X_{cr}, the end point of the constant drying rate period) and the equilibrium moisture content.

$$\phi = \frac{\overline{X} - X_e}{X_{cr} - X_e} \qquad (12.8)$$

If there is no clear constant drying period, N_c can be taken as the maximal rate and X_{cr} is equal to X_0 (Baini and Langrish, 2007; Jannot et al., 2004).

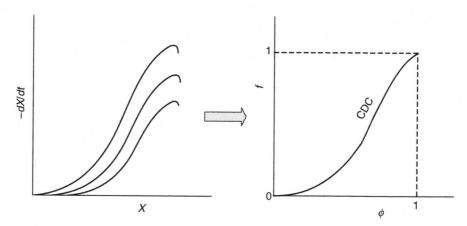

FIGURE 12.4 Achieving a CDC from drying curves measured in different process conditions.

The CDC is normalized to pass through the points (1, 1) at the critical point and (0, 0) at equilibrium, and an empirical function for simulating f can be fitted on this curve (Figure 12.4).

This is an attractive representation because it can lead to a simple lumped-parameter expression for the drying rate:

$$N = f(N_c) = f(S\beta(p_{vs} - p_{vg}))$$ (12.9)

where β is the external mass transfer coefficient (kg/m² Pa s), p_{vs} is the saturated vapor pressure, and p_{vg} is the vapor pressure in the bulk gas. Such a representation allows the separation of the parameters that influence the drying process because f depends on the material itself, β depends on the design of the dryer (value of the air velocity around the product), and p_{vs} and p_{vg} depend on the processing conditions (Langrish and Kockel, 2001).

Fornell (1979) proposed a similar but simplified transformation for food and agricultural products with no constant drying rate period, where $f = N/(T_a - T_{wb})\sqrt{v_a}$ and $\phi = \overline{X}$. This transformation was successfully applied to products with an isotropic and homogeneous physical structure and seemed justified in the usual domain of drying air conditions.

CDCs can be efficient for small (<20 mm), microporous, and nonshrinking materials (Keey and Suzuki, 1974; Langrish and Kockel, 2001). On such products, Fyhr and Kemp (1998) calculated that, in most cases, the CDC predictions were as good as or even sometimes better than those from a diffusion model.

For bioproducts showing a falling rate period with two different decreasing drying rates, it is possible to establish a CDC by segments with two different forms (Jannot et al., 2004; Ng et al., 2006). A transition point has to be identified for calculating two characteristic moisture contents ϕ_1 and ϕ_2, one for each period, and for fitting two expressions for f_1 and f_2.

It is important to outline that empirical models and the CDC method can be useful for describing the drying kinetics of products dried continuously, but they cannot be used to describe intermittent drying and relaxation processes (Baini and Langrish, 2007).

12.2.1.3 Neural Networks (NNs)

Calculating the moisture content using a black box model is quite easy but its application is limited, especially for variable drying conditions that are necessary, for example, for process control purposes. In such cases, the artificial NN is an interesting tool because it allows fast and easy simulation that can be used in predictive control algorithms, when the process model has to be simulated many times in a sampling period.

NNs have the ability to "learn" the process behavior from a set of experimental data and to provide a smooth and reasonable interpolation of new data. When compared to traditional function approximators like polynomials or Fourier series, NNs require generally less adjustable coefficients, especially if the number of variables is high. In order to obtain a dynamic model, the time can be incorporated as an explicit variable. However, it is important to note that NNs usually require more experimental data than physical models because all relevant combinations of inputs have to be present in the training set. The "generalization" concerns only interpolation, and results for combinations of inputs not previously seen are unpredictable.

The inputs of the structure can be variables such as air-flow rates, inlet and outlet temperatures, and relative humidities, which are easily measured. One of the first applications to drying was proposed by Huang and Mujumdar (1993) to predict the temperature and moisture content of tissues during industrial drying. Trelea et al. (1997a) used explicit time and recurrent NNs with three hidden neurons for simulating the moisture content of thin layer corn during drying (Figure 12.5). The inputs of the model were the initial moisture content, the temperature of the drying air, and the drying time. Explicit time NNs allowed the prediction of the final moisture content and recurrent NNs the whole drying kinetics.

These researchers showed that such models can be used for simulations in variable temperature conditions, if the number of temperature levels is not too high. Bala et al. (2005) used a multilayered NN approach to predict the performance of a solar tunnel drier. Martynenko and Yang (2006) compared the accuracy of an NN

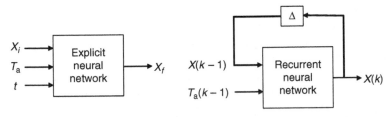

FIGURE 12.5 Explicit and recurrent NNs as used by Trelea et al. (1997a); k, increment in sampling time; Δ, one sampling period delay.

with exponential and Page models for simulating the drying rate of ginseng. The NN gave the best fit to the experimental data, especially for small roots, because it can predict the nonlinear effects of different parameters on the mass transfer.

NNs are a possible methodology for building fast nonlinear dynamic models of drying. No assumptions on the underlying physical mechanisms are made; the models are based on experimental data and can be reliable in interpolation only.

12.2.1.4 Compartmental Models

Physical dynamic models of drying, which are based on the description of underlying diffusion mechanisms (see Section 12.2.2), usually result in coupled partial differential equations. Such models can be simplified using a semi-empirical compartmental approach (Courtois et al., 1991; Toyoda, 1988) that describes the macroscopic behavior of the product. They allow the prediction of the steady-state and dynamic behaviors of any dryer.

In most cases, the product is described as being composed of a certain number of concentric compartments for mass transfer and a uniform one for heat transfer, that is, as a multiple compartment water system and as a single one for heat (Figure 12.6). Mass transfer inside the product is based on diffusion, and vaporization takes place only at the surface. Such a compartmental approach is mainly used to efficiently account for the internal resistance to mass transfer for computing.

The external heat transfer coefficient (h) and the external (β_1) and internal (β_2) mass transfer coefficients must be adjusted with the help of constant condition drying kinetics. Basically, this modeling approach is preferred for its ability to represent many phenomena well (even if containing lumped parameters), its fast speed of computing, its physical backgrounds, and its optimal parameterization.

12.2.2 DRYING MODELS BASED ON FIRST PRINCIPLE EQUATIONS

Before listing the different types of models written on the scale of a single particle, the individual modes of heat and mass transfer are described, making the distinction between internal transfers (inside the drying product) and external transfers (between the product and its environment).

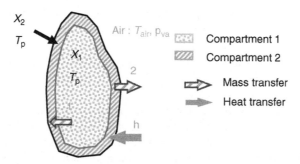

FIGURE 12.6 Principle of the compartmental approach in modeling. The product is seen here as two tanks in series.

12.2.2.1 Importance of Description of Product Internal Physical Structure

It is important to note at this level that the description of the internal mass transfer phenomena is always related to a description of the internal structure of the product that is as precise as possible. Important factors are the presence or absence of pores, the size of pores, and the level of interactions between water and dry matter.

Most food products can be viewed as hygroscopic porous or capillary–porous media or as colloidal (nonporous) media such as gels or liquid solutions. The distinction between porous and capillary–porous is made by considering the size of the pores as discussed by Datta (2007a): porous products have pore diameters $\geq 10^{-7}$ m and capillary–porous products have pore diameters of $<10^{-7}$ m. Analyzing the level of physicochemical interactions between water and dry matter allows us to make the distinction between nonhygroscopic media (negligible amount of bound water in the pores) and hygroscopic media (mixture of bound and unbound water in the pores in a variable ratio as drying proceeds).

Note that many dried food products have an initial porosity that is significantly developed during drying. This is the case for apples whose porosity increases from 0.220 initially (mainly open porosity) to 0.609 when completely dried at 80°C. However, this product ends up with a very low level of open porosity (0.130) because of the collapse of surface pores (Rahman et al., 2005). If this experimental information about the internal structure is available, it must be taken into account when discussing the correct simplifying assumptions used to describe internal mass transfers during drying.

12.2.2.2 Internal Heat and Mass Transfer Mechanisms in Drying of Food Products

According to the discussion in the previous paragraph, drying of food products can be now considered as a combination of heat and mass transport mechanisms occurring in hygroscopic porous media with changing (size and nature) internal porosity. We now list and physically describe all elementary heat and mass transport mechanisms occurring during a food product undergoing drying.

12.2.2.2.1 Internal Mass Transport Mechanism

The main elementary mass transport mechanisms occurring during heat processing of porous media are ordinary diffusion, convection of liquid or gas driven by pressure gradients, capillary diffusion, and the evaporation–condensation phenomenon. All of these modes of heat and mass transfer are precisely described and quantified by Datta (2007a) and Peczalski and Laurent (2000). We give a physical description of each of these elementary modes of transfer.

Ordinary diffusion is the movement of particles from zones of high concentration to zones of lower concentration. It is the statistical result of a random process occurring on a microscopic scale. Ordinary diffusion occurs in the liquid phase for drying of nonporous media and in the gas phase for porous materials if the open porosity of the product is high enough. The ordinary diffusion flux is proportional to the gradient of the mass or molar fraction.

The convection phenomenon (or hydrodynamic flow) is driven by total pressure gradients and is also noted in porous products. This mode of mass transfer is of prime importance during drying at high temperature where internal pressure gradients are expected to occur.

Capillary diffusion (or capillary flow) occurs in a porous medium. It is attributable to the intermolecular forces between the liquid and the solid phase, which are stronger than intermolecular forces inside the liquid phase, and it explains why the liquid phase is migrating in the porous structure. This flux of liquid is proportional to the gradients of capillary pressure. Because the capillary pressure is a function of both the moisture content and temperature, the capillary flow of liquid can be decomposed in the sum of a mass flux proportional to the temperature gradient and a mass flux proportional to the moisture concentration gradient.

Finally, there is a migration of moisture in a porous medium that is initiated by a temperature gradient inside the product. This mechanism is known as evaporation–condensation because liquid water evaporates in the hot zone and condenses in colder zones after diffusion in the gas phase. This mechanism can happen between two menisci in an open pore or between different zones of the cell walls surrounding a gas cell subjected to a temperature gradient. By modifying the liquid moisture concentration when condensing, this mechanism can be amplified by capillary diffusion operating in the return from zones of high moisture concentration to zones of lower moisture concentration.

12.2.2.2.2 Internal Heat Transport Mechanism

In continuous media, heat is driven by thermal conduction from zones of high temperature to zones of lower temperature. The proportionality factor between the heat flux and temperature gradient is by definition the product thermal conductivity. In porous media during drying, this elementary mechanism is completed by energy inflow and outflow due to different mass transfers. In the evaporation–condensation phenomenon, a significant transfer of latent heat of vaporization accompanies the mass transport from hot zones (where evaporation occurs) to cold zones (where condensation occurs). This effect can appear explicitly in the heat balance or be taken into account in a modified thermal conductivity term, the so-called *apparent thermal conductivity*. This apparent coefficient globalizes the different mechanisms contributing to the overall heat transport due to temperature gradients.

12.2.2.3 External Heat and Mass Transfer Mechanisms in Drying of Food Products

The mechanisms of heat and mass transfer occurring between the product and its environment during drying (external transfers) are those classically encountered in all unit operations involving heat transfer (heating, drying, baking, frying, etc.). External heat transfer mechanisms are convection, thermal radiation, and heat transfer occurring at the interface between two solids (heat transfer by contact).

Thermal convection is a mode of heat transfer resulting of the bulk motion of a fluid relative to the surface of a solid when the two phases have different temperatures. Thermal and hydrodynamic boundary layers develop at the immediate vicinity of the solid surface. The patterns of variation of the temperature and velocity distributions in these boundary layers determine the efficiency of energy exchange between the fluid and the solid. This efficiency is quantified by a convective heat transfer coefficient.

The same phenomenon explains the convective drying of the product. In this case, the vapor flux emitted by the surface of the product is proportional to the difference of the vapor concentration (or partial pressure) between the immediate vicinity of the product surface and the drying air. The proportionality factor is physically related to the convective heat transfer coefficient. The more efficient the thermal convection is, the more important the vapor flux emitted by the surface of the product. At the product surface, the composition of humid air in the immediate vicinity of the product surface is estimated by assuming an instantaneous thermodynamic equilibrium between liquid moisture at the surface of the product and vapor just above the product. It is thus calculated using moisture equilibrium isotherms, as discussed in Section 12.2.2.1.

Thermal radiation is the result of energy exchanges between solid surfaces at different temperatures, and this energy is transported by electromagnetic waves (alternatively, photons). The net energy emitted by the environment and absorbed by the surface of a solid during drying is influenced by the emissivity of each of the emitting surfaces (which is a measure of the ability of a surface to emit thermal energy relative to an ideal solid called a black body), by the relative position between the emitting surfaces (quantified by a geometric factor called view factor), and by the temperature difference (ΔT^4) between the different emitting surfaces.

Finally, external heat transfer occurs at the interface between two solids put into contact. As a result of the surface roughness effects, air filled gaps always appear at mesoscopic levels when two solids are put into contact. Heat transfer across these gaps occurs by conduction, which is known to be a nonefficient phenomenon in a gas phase. Macroscopically, thermal contact resistance appears at the interface between the two solids and assuming the equality of temperature for the two solids at the interface can be an incorrect hypothesis. The heat flux between the two solids is hence proportional to the temperature difference between the two solids at the interface, and the inverse of the proportionality factor is called thermal contact resistance. This thermal contact resistance is always difficult to evaluate because surface temperatures and contact flux are difficult to measure experimentally without affecting this mechanism of transfer.

12.2.2.4 Models Neglecting Resistance to Internal Heat and Mass Transfers

The influence of the product internal structure on the nature of internal heat and mass transfers was discussed and the different mechanisms of internal and external heat and mass transfers were listed and briefly described. Thus, the first category of drying models to be presented concerns the first principle based models in which resistance to internal heat and mass transfer is neglected.

According to this simplifying hypothesis, the temperature and moisture content are assumed to be uniform (in the space volume occupied by the product) at any drying time. In this case, the product drying rate is controlled by the nature and intensity of the external modes of heat transfer (thermal convection, radiation, heat transfer by contact, convective drying) that thus must be measured or estimated as precisely as possible. This assumption can be satisfactory for small products (millimetric size or lower such as in a liquid droplet during spray drying) and for drying in the constant drying rate period or at low levels of external heat flux.

Using this hypothesis, the model for drying takes the mathematical form of simple ordinary differential equations expressing the overall heat and mass conservation equations for the product. The major drawback of this type of approach is that an eventual (always present, even if weak) internal resistance to heat and mass transfer is likely to rely numerically on the value of a parameter related to external transfer (e.g., the convective heat transfer coefficient). In this case, this parameter, which is initially of clear physical significance, becomes a pseudo-empirical parameter with an initial order of magnitude that can be assessed experimentally but the precise final value will be fixed by a comparison between predicted and experimental drying curves. Moreover, this type of model is difficult to extrapolate to other sizes of products (when increasing the size of the product, the internal resistance to heat and mass transfer increases) and to other dryer types (levels of external heat and mass flux can change when using another type of dryer).

12.2.2.5 Models Using an Apparent Diffusivity Approach

In this category of models, we make the assumption that, whatever the migrating phase (liquid or gas), all of the elementary internal mass transport mechanisms can be mathematically described by Fick's law of diffusion that was initially developed for ordinary binary diffusion. In this case, the resulting moisture flux (indifferently in the liquid and gas phases) is assumed to be proportional to the moisture concentration or moisture content gradient, and the proportionality factor is defined as the apparent diffusivity.

The apparent diffusivity can be assumed constant in a range of moisture contents. Nevertheless, because the apparent diffusivity is a property used to mathematically describe a wide range of mass transport mechanisms, it is more generally expressed as an empirical function of product composition factors (moisture, fat and salt contents, etc.), physical structure factors (level and type of porosity, etc.), or environmental conditions (mainly temperature). From its definition, the apparent diffusivity is a pseudo-empirical parameter and its experimental determination becomes very difficult. When trying to experimentally measure the apparent diffusivity, we must ensure that the mass transfer conditions (nature and intensity of heat and mass flux, level of deformation of the product, etc.) are of the same order of magnitude on the apparatus used for measurement as the ones experienced by the product during the drying process. A large catalogue of values for the apparent moisture diffusivity and a clear discussion on the methodology and precautions needed to measure this transport property can be found in Saravacos and Maroulis (2001).

The same methodology can be used to describe all internal heat transfer mechanisms by defining an apparent thermal conductivity and using the classical Fourier's law. The experimental determination of this apparent heat transport property globalizing a wide range of elementary mechanisms poses the same problem as the one encountered when measuring the apparent diffusivity. Fortunately, the internal heat resistance (in products up to centimeter size) is often neglected for reasonable levels of external heat flux.

The benefit of this type of approach is that it has been widely used for numerous products and dryers. Hence, there is profuse literature about this subject and the validity of all of the simplifying assumptions made when formulating the model has been already discussed for a lot of cases. Moreover, if the internal heat and mass transfer phenomena can be assumed as unidirectional and if the product can be assumed as a typical shape (slab, cylinder, or sphere), the obtained equations are in the form of coupled partial differential equations (assuming internal resistance to heat and mass transfer), which can still be numerically solved using classical software for technical computing such as Matlab®.

The major drawback of this approach is the pseudo-empirical nature of the apparent heat and mass transport properties and the difficulty of measure it on a laboratory scale. When comparing the drying model to experimental results, we must simultaneously assume the validity of the hypothesis made when formulating the model and identify the optimal values for parameters appearing in expressions of the apparent diffusivity and conductivity by minimizing the sum square of errors between the measured and predicted values. This strategy is always dangerous when dealing with uncertain model principles and uncertain parameter values. Particular care must also be taken when characterizing external heat and mass transfer (often by experimental measurement) in order to not report uncertainties in the values of the internal transport properties and hence lowering their level of physical realism.

12.2.2.6 Fully Mechanistic Models

In this approach each individual mode of internal heat and mass transfer is clearly made explicit and formulated mathematically. Note that these individual modes of transfer are highly coupled, which gives rise to serious complications when solving the obtained equations. The most complete works on modeling coupled heat and mass transfer during heating and drying of porous media are certainly those of Berger and Pei (1973), Perré and Degiovanni (1990), Philip and De Vries (1957), and Whitaker (1977).

In this category of models, food or nonfood products are always considered as polyphasic materials in which the internal porosity is partially filled with a mixture of bound water, unbound water, and a gas phase that is assumed to be a mixture of dry air and vapor. We must be particularly careful to describe the internal structure of the product as precisely as possible. The proportion and nature of the porosity between completely closed porosity (gas cells entrapped in a continuous solid of the liquid phase) and open or flow-through pores allowing hydrodynamic flow of liquid or gas is sometimes taken into account. The link between internal structure parameters and the significant elementary mass transfer mechanisms is always made. The respective influence between the mechanical properties of the product and mass transfer mechanisms is sometimes discussed when considering product deformation during drying. The levels of local deformation are described in one or more directions in space in this case.

Conservation equations are written on the microscopic scale at the level of a representative elementary volume including water, vapor, air, and solid matrix phases and are then volume averaged to obtain final equations expressing continuity and flux equations in a fictitious equivalent continuous medium (Datta, 2007a). As discussed in detail by Datta (2007a), two complementary approaches are distinguished at this level: a first approach considering a uniformly distributed evaporation in liquid, vapor, and heat conservation equations (the evaporation term in mass balance equations corresponds to a heat sink term in heat conservation equations) and a second approach in which evaporation is supposed to occur at a moving interface between a dry phase and a phase still containing liquid water.

The obvious benefit of this type of model is its ability to be extrapolated to other product geometries and other types of dryers. This is possible because the physical basis used to develop the model is well established and very clear. Moreover, this type of approach allows evaluating and then organizing into a hierarchy all of the individual modes of internal heat and mass transfer. Limiting phenomena to overall mass transport can also be identified.

By contrast, the major drawbacks of these approaches lie in the complexity of the mathematical formulation of internal heat and mass transfer and the complexity of solving the resulting coupled equations. Moreover, the hypotheses made when developing the drying model are critical and their validity is sometimes difficult to judge because no experimental technique is available on-line to separately evaluate the influence of each of the individual transfer modes on the overall product moisture loss. Finally, this type of model needs the knowledge of an important number of properties related to the product structure (nature and level of porosity and its evolution as drying proceeds) and their impact on transport properties such as capillarity or molecular diffusivity, vapor, and liquid permeability. Because these values are sometimes difficult to assess on the laboratory scale and are not available in the literature, this gives rise to another complication when comparing the predictions of the drying model to experimentally measured values. Datta (2007b) recently discussed the problem of the availability of precise property data needed when developing a drying model for a porous product.

12.2.2.7 Modeling of Product Volume Change

The volume change of food products submitted to drying occurs for a large family of products such as fruits or vegetables. Generally, overall shrinkage of the product occurs as drying progresses. Some cases of swelling can also be noted when rewetting dried products or when internal vaporization of water occurs inside individual gas cells separated by a still deformable liquid phase. This deformation has a great influence on the overall product moisture loss kinetics because of the following:

- The product surface area can significantly decrease in a product submitted to shrinkage. This phenomenon can lead to conflicting results about the presence or absence of a first drying stage. May and Perré (2002) investigated this in the drying of various fruits and vegetables, where the external mass flux per surface area was precisely measured by an original experimental setup.
- The internal structure of the product is significantly modified when shrinkage or swelling occurs during drying. In mechanistic models, the nature

and proportion of individual transfer modes can be drastically affected by this change of the internal structure.

Faced with this supplementary complexity, some models do not take this volume change into account or simply consider a dimension change that is directly proportional to the moisture loss or gain in each zone of the product. The model is hence written in a referential attached to the dry matter of the product (Lagrangian coordinates) and the Eulerian coordinates are then recalculated at the end of the simulation by considering that, whatever the drying time, the volume of the product is always equal to the sum of the partial volumes of the liquid and dry matter phases. At this level of description of deformation phenomena, the shrinkage is often considered as unidirectional and seldom bi- or tridirectional. Another alternative consists of imposing on the material a deformation law measured experimentally or simply predicted from the evolution of the apparent density of the product (making the assumption that the volume of the product is equal to the sum of the partial volumes of water, dry matter, and gas phases). The latter point is discussed by Mayor and Sereno (2004). Some recent studies presented drying models that were completely coupled to small or large deformation models, taking into account the mechanical properties of the product and their change during drying. One example of this type of approach can be found in Zhang et al. (2005).

12.3   SPACE AND TIME INTEGRATION OF ELEMENTARY DRYING MODELS

The previous section listed the phenomena describing heat and mass transfers within the product and at the interface with the drying medium. This level of analysis is usually called a "thin layer" of product. This terminology refers to the fact that air (or steam) is considered as constant and uniform around the product particle. A more practical definition is given for the air drying of a static bed: the output temperature and moisture content of used air are not significantly different from the input values.

In fact, the modeling approaches provided in Section 12.2 may be sufficient to describe some industrial dryers. For instance, when considering a fluidized bed dryer, the air flow is "strong" enough to maintain product particles in suspension with perfect agitation. In other words, all particles are submitted to the same average drying air and dry the same way. In that case, the model is a mathematical equivalent to an average particle and there is no need for a time-consuming numerical integration over the bed volume.

In a more generic approach, every time the particles are well agitated and/or their drying medium (air most likely) is not dependent upon the drying of other particles (with whom air would already have some transfers), the dryer can be modeled as average particle drying. To be precise, one may calculate the following index of saturation (IoS) empirical ratio:

$$\text{IoS} = 100 \times \frac{\varphi_w}{\varphi_a} \times \frac{1}{\Delta Y_{max}} \ (\%) \qquad (12.10)$$

where φ_a and φ_w are the air-flow and evaporation rates, respectively; ΔY_{max} is the maximum saturation capacity. Then, the maximum evaporation rate (max φ_w) is

$$\max \varphi_w = \max \left(-\frac{dX}{dt} \right) \times S \times H \times \rho_{dm}$$

and

$$\varphi_a = \rho_{da} \times S \times v_a$$

where S is the section area; H is the bed height; and ρ_{da} and ρ_{dm} are the dry air and dry matter densities, respectively.

The previous equation can be simplified to

$$IoS \approx 100 \times \frac{\max \left(-\frac{dX}{dt} \right) \times H \times \rho_{dm}}{v_a} \times \frac{1}{\Delta Y_{max}}$$

by neglecting variations of the air density with the temperature.

The ΔY_{max} of drying air is usually taken as the (positive) difference between its actual absolute humidity ratio (kg water/kg dry air) and the same variable at saturation assuming isenthalpic saturation. For instance, 90°C drying air and 0.010 kg water/kg dry air will saturate in an isenthalpic manner at 33.6°C (wet bulb temperature) for an absolute humidity ratio of 0.034 kg water/kg dry air. In that case, the ΔY_{max} of this drying air is 0.024 kg water/kg dry air.

When the index of saturation is low (e.g., <20%), it is probable that the deep bed can be considered as an elementary particle of the product of the same weight. In that case, there is no need to model and integrate the air moisture and temperature gradients within the bed because it behaves mostly like a thin layer of product.

In any other case, the progressive saturation of the drying air throughout the product deep bed, with respect to both time and space, can no longer be neglected. Furthermore, the air gradients within the bed must be computed to correctly simulate successive product layers. Taking into consideration the energy costs, it is generally a good idea to design a dryer in such a way that the drying air finally outputs close to (but below) saturation in order to use its maximum drying capacity and minimize the energy spent at the burners.

Note that the index of saturation can reach values of over 100% in local condensations at the air outlet ducts that are quite common in industrial dryers.

12.3.1 FROM THIN LAYER TO DEEP BED

As previously stated, most deep bed models for dryers assume a deep bed of product to be equivalent to a group of thin layers in series (Figure 12.7). Many industrial dryers are simulated that way, such as conveyor belt dryers, mixed-flow dryers, and

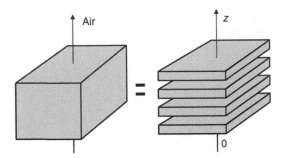

FIGURE 12.7 Virtual decomposition of a deep bed of product into a series of successive thin layers.

so forth. For that purpose, in such an approach, there are several assumptions (on air drying) to make:

- The air flow is plug type and unidirectional [one-dimensional (1-D) models, see discussion in Section 12.3.2].
- Direct heat and mass transfers between product particles are negligible. All transfers are assumed to take place either within the particles or at the interface with the surrounding air.
- There is no significant change in air properties within a thin layer of product, only between two successive layers.
- The air pressure drop within the bed is usually neglected.
- The overall shrinkage of the bed is usually neglected. The same is true for the changes in bed porosity.
- The heat losses throughout the walls are usually considered to be negligible.

In this approach the air moisture content and temperatures are no longer considered to be constant and uniform: they are truly new variables to include specifically in the equation set. Therefore, it is necessary to extend the heat and mass balances from the product to the drying air. The direction axis over which the air flows is usually noted oz, with z being the position in the bed (0 at the input and H at the output of the bed).

The mass balances for the air at level z are the result of the difference between the water taken from the product at the same level z and the renewing of the air moisture attributable to its gradient over the bed height:

$$\frac{\partial(\rho_a \times Y)}{\partial t} = \frac{\partial(v_a \times \rho_a \times Y)}{\partial z} + \Phi_w \times a \times \frac{1-\varepsilon}{\varepsilon} \tag{12.11}$$

where v_a is the actual air velocity within the bed, depending mainly on the bed porosity (ε), and Φ_w is the water flow density between the air and the product.

In a similar way, the heat balance for this air is defined as

$$\frac{\partial(\rho_a \times c_{pa} \times T)}{\partial t} = \frac{\partial(v_a \times \rho_a \times c_{pa} \times T)}{\partial z} + (\Phi_w \times L_v + \Phi_h) \times a \times \frac{1-\varepsilon}{\varepsilon} \tag{12.12}$$

Note that the $\rho_a \times c_{pa}$ (volumetric heat capacity of the air) term is not constant because it highly depends on Y. One common way of dealing with this term is to consider that it is the sum of two terms: one corresponding to the dry air fraction, and the other to the vapor fraction. Here, Φ_w is the water exchanged between the air and product at level z and Φ_h is the heat exchanged between the air and product at level z. They are both calculated using the equations described in Section 12.2 (i.e., thin layer equations). They constitute the hard link between air related balances and product related balances.

Solving the resulting system results in the ability to predict both air and product temperatures and moisture contents at any location within the bed. Unfortunately, there have been few research attempts to find analytical solutions of such systems. There is an interesting application of Adomian's method (1996) for coupled heat and mass transfers that is far less complex than what is found in food dryers: no complicated food properties or phase changes. There is also a nice work from Sebastian (Aregba et al., 2006; Nadeau et Puiggali, 1995) to analytically integrate some simple thin layer models over a deep bed with strong assumptions concerning the product temperature behavior. It is therefore common to find numerical solutions of this type of models. For that purpose, continuous equations are usually discretized over space (Δz) and time (Δt) using either an explicit or implicit finite difference scheme.

There are basically two methods to solve the discrete version of this model type, depending on the finite difference scheme (for further explanations, see Press et al., 2007):

1. If the scheme is time explicit, Runge–Kutta methods or similar ones can be used to calculate, time step after time step, the space evolution of all air and product temperatures and moisture contents. Because of the stiffness of the equation set, the integration procedure can diverge very easily.
2. If the scheme is time implicit, the resolution implies, at each time step, the inversion of a (variable) matrix. The extra cost of computation required by this inversion is usually fully compensated by the gain in stability and thus the gain in integration speed (at least a factor 100).

In any case, the choice of the space increment Δz and the time increment Δt constitutes a major issue. Many methods are available to adjust Δt in order to ensure stability, convergence, and precision of the numerical integration as performed by Matlab®, Scilab®, Octave®, or any other platform for numerical computation implementing ordinary differential equation solvers. In contrast, Δz is usually a constant fixed at the beginning of the algorithm. However, Δz should be reduced as the integration progresses in order to take the bed shrinkage into account.

In actuality, it is possible to extend this method to two-dimensional (2-D), even three-dimensional (3-D), models with a considerable amount of complexity added into the numerical integration procedure. It is barely used in works found in the literature, probably because the resulting complexity is not always justified to attain a precise simulator for industrial dryers.

After having written the equations and solved them with an appropriate numerical method on a computer, the user can obtain a lot of valuable information concerning the process: moisture contents and temperatures for air and product at each $z_i = i\Delta z$. Therefore, it is possible to compare some of the simulated variables with corresponding measured values from air temperature or relative humidity sensors located at the outlet or within the deep bed (e.g., see Techasena et al., 1992). Some problems usually arise at this point. On the one hand, thin layer models are usually strongly validated but only with regard to the overall grain moisture content, which is usually the only reliable on-line measurement available. On the other hand, more reliable information is available at the deep bed level (on the air, possibly at different locations within the deep bed). Air temperature measurements within the bed are arguable because they may correspond to the temperature of the mixture of air and solid. Similarly, on-line measurements of the relative humidity of the air are often biased by the probe system inertia. At least the outlet (used) air temperature just outside of the deep bed may be considered as a reliable on-line measurement that can be used for comparison, and thus validation, purposes. Techasena et al. (1992) and many other authors (e.g., Courtois, 1991) state that it is common to have good agreement between simulation and experimental measurements only for the product moisture content but not for the air temperature. As shown in Figure 12.8, even if the simulation correctly renders

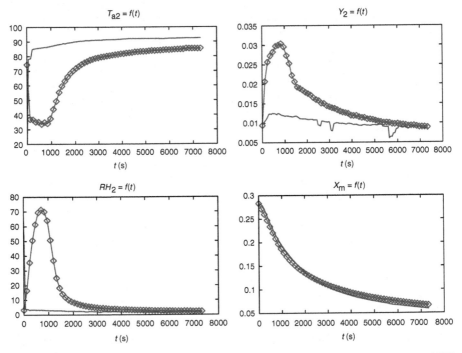

FIGURE 12.8 Comparison of simulation and experiment for parboiled rice dried at 95°C, ambient moisture content, and 2 m/s. For a small error in the simulation of the mean grain moisture content (X_m), there are significant differences in the outlet air temperature (T_{a2}) and moisture content (Y_2), and even more in the relative humidity (RH_2) predicted trajectories.

the overall shape of the registered data curves, the fast dynamics of the air temperature change is not accounted correctly. There may be several reasons for that:

- The heat and mass transfer coefficients obtained for the thin layer may no longer be valid when the particles are stacked with very different aerodynamics.
- The heat exchanges directly between particles or with the walls may not be as negligible as expected.
- The ratio of air to solid densities shows that a very small error in the predic-tion of the heat and mass transfers leads to large errors in the change of air temperature and moisture content.

12.3.2 FROM DEEP BED TO FULL DRYER

There are many dryers worldwide that can be considered to be a static deep bed. In developing countries, for instance, many dryers are simply composed of a fixed bed of grains and a fan, sometimes coupled with a burner. Every time the weather conditions allow it, the drying air can be solar heated.

Large-scale industrial dryers, especially grain dryers, are usually composed of several stages, each of them an assembly of elementary sections. Each of these elements is generally assumed to be composed of several deep beds. It is therefore common to consider an industrial dryer as a network of deep bed dryers in which the connections depend on the air trajectory, especially if there is some air-recycling scheme.

For example, a multiple conveyor belt dryer (i.e., a succession of several single conveyor belts with air recycling from one to another) can be considered as a collec-tion of several deep beds with drying conditions depending on each other. Similarly, grain dryers are huge vertical towers combining several deep beds with a complex air trajectory designed to minimize energy loss.

There are basically five types of elementary deep beds:

1. Cross-current deep beds where air and product trajectories have perpendi-cular axes
2. Cocurrent deep beds where air and product trajectories have the same axes and directions
3. Countercurrent deep beds where air and product trajectories have the same axes but reverse directions
4. Tempering deep beds where there is no air flow and the product is main-tained at the same temperature without drying
5. Mixing deep beds where all product particles are mixed to improve the bed homogeneity

Three other types of beds corresponding to cooling should be added: cocurrent, countercurrent, and cross-flow cooling beds. This distinction between cooling and drying beds depends on the domain of validity of the thin layer drying model that is used. In fact, many thin layer models are validated only for some particular drying conditions, for example, when the air temperature is higher than the product temperature. Using such a model for simulating the behavior in the cooling sections is generally not a good idea because the complementary drying happening in these stages will be significantly underestimated.

Tempering sections are common in mixed-flow dryers for cereal grains. They are supposed to allow, at no energy loss, a decrease in the internal moisture gradient within the grain, leading to a subsequent higher drying rate in the next drying stage (see Figure 12.9). There are several publications about the intermittent drying of rice. From the product quality viewpoint, if considering the breakage of rice grains after drying, tempering periods are of interest for reducing the strengths caused by moisture gradients and involved in cracking and breakage. Conversely, when considering the biochemical properties of a dried product, maintaining a product at high temperature without drying is generally a bad idea (e.g., wet milling quality of corn in Courtois, 1991). Furthermore, because a resting period has to be long enough to significantly decrease moisture gradients, the building of an extra tempering zone is synonymous with extra investment costs and loss in compactness.

Countercurrent sections are more difficult to simulate because calculating the drying kinetics of a top-first thin layer relies on the calculation of the air used for drying the bottom-last thin layers. The usual method is to simulate such a deep bed as if it was a cocurrent one and then to reinject simulated air temperatures and moisture gradients within the bed to reiterate until reaching a steady state.

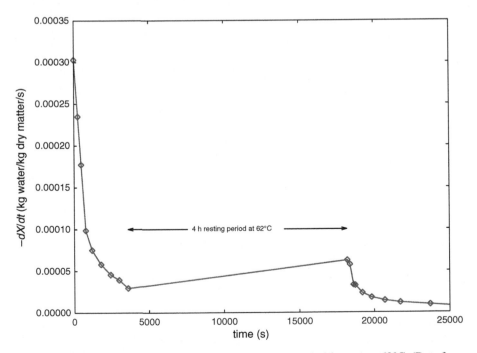

FIGURE 12.9 Increase of drying rate after a 4-h resting period for corn at 62°C. (Data from Courtois, F., Ph.D. thesis, AgroParisTech, Massy, France, 1991. With permission.)

There are mainly two types of grain dryers in the industry: cross-flow and mixed flow. The former is comparable to a big cross-current deep bed. The latter is usually considered as a vertical succession of alternative cocurrent and countercurrent deep beds. Olesen (1987) studied the air flow between the air ducts in a mixed-flow grain dryer and evaluated the air and grains exchanges as 32% cocurrents, 32% counter-currents, and 36% cross-currents (in volumes of grains). These proportions may vary according to the geometry of these air ducts and their distribution in the dryer volume. In practice, a vast majority of researchers consider that each physical deep bed is of only one type. Giner et al. (1998) and Giner and Bruce (1998) considered that the classical 1-D decomposition of the drier in cocurrent and countercurrent deep beds underestimates the grain temperature near the air ducts and hence overestimates the final grain quality (here seed viability). They proposed a 2-D model where co-, counter-, and cross-currents are coexisting. This conclusion is in contradiction with results from Courtois et al. (1991) on corn wet milling quality and Courtois et al. (2001) on the rice breakage ratio. This may mean that, depending on the quality criterion used, the dynamics of the quality decrease may require a more precise prediction of local grain temperature.

In some cases the volumes of the air ducts are subtracted from the total volume of the bed, as demonstrated in Figure 12.10. The full-scale dryer can then be considered as a pile of blocks of elementary deep beds linked by the specific air circulation (and recycling) between each stage. Behind this scheme there are strong assumptions concerning the evenness of air flows in each duct. Furthermore, the aerodynamics within the dryer is supposed to distribute the air flows equally within each section, which is almost never true in practice. The positive point is that, even if it is not true for local air-flow rates, the global air-flow rate is generally known with good precision so that we can expect automatic compensations between overestimates and underestimates. Actually, Courtois et al. (2001) and others observed less than 10% error between simulated (1-D model) and measured grain moisture content and quality in industrial dryers.

Continuing further with grain dryers, the decomposition into deep beds should be completed by the right air trajectories connecting all beds. Actually, air recycling patterns are common in mixed-flow dryers for obvious energy saving considerations. The most common scheme is presented in Figure 12.11. Ambient air is used at the

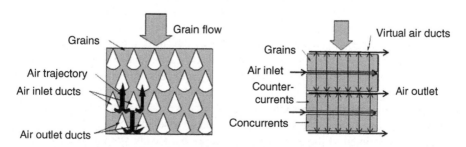

FIGURE 12.10 Side view of an elementary drying section in a mixed-flow dryer. Virtualization (from left to right) of air ducts within the grain bed. Here, the section is assumed to be composed of either cocurrent or countercurrent deep beds.

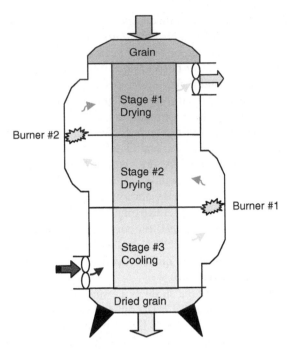

FIGURE 12.11 Typical schematic of a three-stage mixed-flow dryer used for grain drying. Air is recycled from bottom to top, and grain flows from top to bottom. Two burners are used to control temperature set points at stages 1 and 2.

very bottom of the tower to cool the grain for safe storage. The removed sensible heat is then used to reduce the energy needed for heating the air to the temperature set point of the second drying stage (numbering starts from the top, i.e., the grain entrance). Because the grain has already been partially dried in stage 1, the exhaust air from stage 2 is not saturated and is still warm enough to be reheated to the temperature set point of stage 1. At the end, exhaust air from stage 1, which is close to saturation, is rejected outside. From a very superficial point of view, the global exchange is of countercurrents. As explained for countercurrent deep beds, the calculation is more complex because the drying kinetics at the top-first beds depend on the air outlet from bottom-last (at the bottom of the dryer) beds. An iterative procedure is usually the best option to avoid an excessively long calculation.

There is an additional difficulty in simulating the full-scale dryer: if burners are used for air heating, we must pay attention to the additional water vapor coming from the combustion of gas or fuel. However, in practice burners do not work continuously. They are connected to temperature controllers so that the additional air moisture content created varies with time.

Although there is no steady state at the product level in such a dryer, simulators are usually restricted to the simulation of the global steady state of the operation. The unsteady-state simulation will be discussed later in this chapter. Mixed-flow dryers usually have a discontinuous extraction of grains at the bottom to ensure no static dead volume in order to avoid fire risks. This is due to the presence of the numerous air ducts that break the natural easy flow of grains by gravity. Each grain discharge and extraction forces a nearly plug-type grain flow. A typical setting is 1-s extractions every 3 min. For some other types of grain dryers (e.g., cross-flow dryers), where no air ducts are on the way of the grain flow, the discharge rate is a continuous function. For the former case, at each extraction, the whole dryer content (i.e., numerous thin layers) is shifted downward and the upper part is filled with thin layers of grain at the initial state. In the latter case, this must be done at a higher frequency calculated from the time needed to extract one full elementary thin layer. Therefore, for the same equations (thin layer and deep bed), there are many options when computing the algorithm for linking them with respect to the discharge rate.

Commercial software for simulating such dryers is available (e.g., DRYER3000, http://www.drying.org). A (specific) dryer simulator can be very helpful to calculate some interesting ratios:

- Final product temperature (should be compatible with safe storage)
- Final grain moisture gradients within each discharged quantity (the less the better for safe storage)
- Final grain quality (see discussion later in this chapter)
- Total heat power used for temperature control (e.g., necessary for the dimensioning of the burners)
- Energy cost of each kilogram of water evaporated from the product (to be compared to the thermodynamic constant L_v)
- Volume efficiency of the dryer in kilograms of evaporated water per second and per cubic meter of dryer (the higher the better for a more compact dryer)
- Percentage of unused dryer volume due to air having reached saturation (the higher the better)
- Average level of saturation of exhaust air rejected by the dryer (the higher the better for more efficient energy savings)

12.3.3 Other Approaches

12.3.3.1 Use of Computational Fluid Dynamics (CFD) in Drying Simulation

The CFD approach has enabled progress in understanding the circulation of gazes around products, in predicting changes in the product, and therefore in improving the design of dryers.

Modeling of drying can also be used for analyzing flows at the dryer level. Thanks to the rapid increase in the calculation power of computers and to the development of commercial codes, CFD techniques are increasingly used in the field of food drying. Such codes can solve problems of the flow of uncompressible and compressible fluids via heat and mass transfer and turbulence phenomena. Many research groups used CFD techniques for spray dryers (Ducept et al., 2001; Straatsma et al., 1999a). They allow the simulations of 2-D or 3-D gas flows, calculating the particle trajectories throughout the dryer and calculating the drying kinetics of the atomized droplets, which are influenced by both internal and external transport phenomena. The direct results of a CFD calculation are particle size distributions (from interparticle collision and agglomeration calculations), drying states, and related material properties. Blei and Sommerfeld (2007) provide the results of a good study on the use of CFD for simulating spray dryers. They are effective tools in giving indications on how to adapt industrial dryers to attain better product quality or reduce fouling problems (change in process conditions, change in shape of the drying chamber, choice of another atomization position or nozzle). However, the degree of physical detail that has to be described and incorporated in such a model is rather high, as is the computational effort.

CFD techniques can also be used for calculating the fields of air velocities and controlling the homogeneity of drying. Many applications can be found in the literature. Anglerot et al. (2001) used them for calculating air flows in a convective rack dryer and for eliminating air recirculation. Marchal (2001) employed CFD for piloting an industrial dryer combining convective and radiative heat transfers and for adapting radiative flux in order to consider the constraints from the nature of the product. Mirade (2001) utilized this technique for piloting air flows in sausage cabinet dryers and for obtaining good surface drying of the product, which is known for piloting the internal process of fermentation. Pajonk (2001) used CFD for calculating air flows in a refrigerated cheese maturing cabinet.

12.3.3.2 Simulation of Drum Dryers

Drum dryers have high drying rates and high energy efficiencies, and they are suitable for slurries in which the particles are too large for spray drying. Hence, drum drying is used to produce potato flakes, precooked cereals, molasses, some dried soups and fruit purees, and so forth. Actually, it is the only process to produce flakes in the industry. However, the high investment cost of the machined drums and threat for heat damage to sensitive foods caused by high drum temperatures have caused a move to spray drying for many bulk dried foods.

Slowly rotating hollow steel drums are heated internally by pressurized steam at 120 to 170°C. A thin layer of food is spread uniformly over the outer surface by dipping, spraying, or spreading or by auxiliary feed rollers. Before the drum has completed one revolution (usually within 20 s to 3 min, sometimes <20 s), the dried food is scraped off by a cutting blade contacting the drum surface uniformly along its length (Figure 12.12).

To simulate such a drying process, most groups (Abchir, 1988; Kozempel et al., 1986, 1995; Trystram, 1985; Vasseur, 1983) use some common simplifying assumptions:

- A drum rotation cycle is composed of two elementary phases (Figure 12.13): deposit, drying, and removal of product, and then heat accumulation.
- The product behaves as an infinite slab, receiving heat by conduction from internal condensation of vapor.
- Convection and radiation exchanges between the product or drum wall and air are neglected compared to conduction with the drum wall.
- The product is thin enough to neglect any moisture and temperature gradients along the axis perpendicular to the product layer.

FIGURE 12.12 A detailed view of the "bourbier" between the drum and its feed roller (pilot plant, AgroParisTech, Massy).

- The product temperature is equal to the drum surface temperature.
- No border effects along the cylinder axis are taken into consideration.
- The product is placed on the drum surface at approximately 100°C (confirmed by measurements) and warming of the product can be neglected.
- The product is supposed to dry at the pure water boiling temperature (this can be relaxed) and hence evaporative flow is proportional to the heat flow between the drum wall and product.

Finally, the heat flow from the drum wall to the product is modeled with an empirical equation related to its temperature, water load (kg water/m² drum wall), and dry matter load. The equation contains three to six unknown parameters to identify from experimental runs. As a recent example, Karapantsios (2006) used such equations derived from the work of Trystram and Vasseur

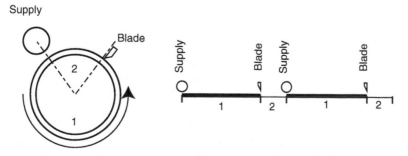

FIGURE 12.13 The two main phases in the drum rotation cycle.

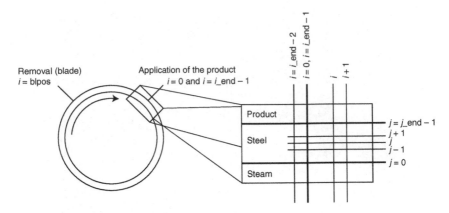

FIGURE 12.14 Discretization scheme of the drum.

(1992) to describe the amount of heat (kW/m^2) used for drum drying of thin films of pregelatinized starch:

$$\Phi_h(t) = b1 \times (T_s(t) - 100)^{b2} \times \left(\frac{1}{C}\right)^{b3} \times (1 - e^{X(t)/b4})^{b5} \qquad (12.13)$$

where T_s is the temperature at the outer surface of the drum wall (i.e., it is also the product temperature); C is the so-called "specific load" (kg dry matter/m^2, a quantity remaining constant during drying); $X(t)$ is the instantaneous moisture content of the film (kg water/kg dry matter); and $b1, b2, \ldots, b5$ are parameters that must be identified in order to best-fit the experimental data. Such an equation can be understood as the heat flow formulation of the classical thin layer equations described in Section 12.1.1.

Most simulators described in the literature solve this model numerically with finite difference schemes as described in Figure 12.14. There are basically two axes: the x axis along the path followed by the product film during its rotation (steps i in Figure 12.14) and the y axis perpendicular to the x axis (steps j in Figure 12.14) where the temperature gradient within the steel wall is discretized.

To our knowledge, there is no commercial simulator available for food drum dryers. The available software remains confined to very few research laboratories.

12.4 IMPORTANT ISSUES

12.4.1 DYNAMIC VERSUS STEADY STATE

Most process simulators (including dryer simulators) use steady-state assumptions. As far as we know, they do no dynamic simulations. There are few articles available on the process control of dryers, even less for food dryers.

At a higher scale, dryer simulators, probably like most process simulators, predict only the steady states of dryers. This means that they are unusable for process control, for instance (see later in this chapter). However, when getting into detail, down to the elementary particle level, the core of the simulator model may not be compatible with any transient period of drying. For example, the moisture ratio as defined in Eq. (12.2) is incompatible with dynamic modeling because it includes references to both the initial moisture content and an equilibrium value. Fohr et al. (1988) demonstrated that a drying model relying on the moisture ratio, although validated for drying kinetics under constant drying conditions, will not work on kinetics with sharp changes in the drying conditions (e.g., see Figure 12.9). As stated earlier, most industrial dryers cannot be considered as drying under constant conditions. Hence, most empirical drying models (at least those relying on the moisture ratio) are incompatible with a valid simulation of a full-scale dryer.

12.4.2 IDENTIFICATION AND VALIDATION

There are insufficient available data on food properties to allow direct simulation of the drying kinetics. In the process of modeling, apart from the equations and some specific properties of the material, some unknown parameters always remain that have to be fitted to the experimental data.

In the common sense, "modeling" is understood as writing the equations describing all of the thermodynamics of the system undergoing drying. In practice, it is only half of the job. The subsequent "identification" of all necessary parameters constitutes the other half.

Concerning the interfacial transfer coefficients, note that they always appear in balance equations coupled with the product external surface area (in contact with the drying medium). If the latter is not constant during drying, there is a chance that it is not known with enough precision to enable a precise simulation. Every time the product shrinks, its surface geometry may change in a complex way. For instance, macroscopic estimations of the product surface are not reliable enough to describe complex roughness changes. In such cases, it is useful to group the more or less known transfer coefficient with the unknown surface area to build an apparent coefficient to fit the experimental data.

Carrying this reasoning to its limits, we would transfer the heavy task of the laboratory measurement of product properties to a higher level of numerical identification of too many unknown coefficients. However, we highly recommend limiting the identification only to those parameters that really cannot be determined with laboratory instruments with a satisfactory level of confidence. For example, product heat capacity, density, volume, water activity, and so forth should be measured whereas internal diffusion coefficients should be identified throughout some curve-fitting algorithm.

Numerical identification of unknown parameters is a well-known, yet difficult, problem. There are numerous articles and books concerning the identification of linear parameter models, especially in control sciences. Much less information is available concerning dynamic models that have nonlinear parameters. Prior to any identification, we should check the theoretical feasibility of such a computation. For instance, it is impossible to distinctly identify the heat transfer coefficient and the product surface area because they are always coupled in the model equations. Precise checking of the structural identifiability of nonlinear models requires fastidious manipulations of the symbolic expressions in the model. Symbolic mathematical packages (e.g., Maple®, Mathematica®, Maxima®, etc.) may be of some help, although they do not guarantee convergence in all cases. Under some simplifying assumptions, it is possible to find approximate answers using linearization techniques or a Taylor series expansion (Rodriguez-Fernandez et al., 2007). Because of the high level of complexity of the exact structural identifiability test and the limited validity of the approximate approaches, most published works skip this step and go directly to the identification itself.

Because of nonlinearities in the parameters, the identification procedure implies the use of a nonlinear optimization algorithm. Common choices are simplex (also known as Nelder and Mead), Levenberg–Marquardt, Broyden–Fletcher–Goldfarb–Shanno, and so forth. Using software like Matlab, Scilab, and Octave can simplify the extra programming needed. As opposed to linear curve fitting, obtaining confidence intervals for the parameters is not straightforward. This can explain why so many articles do not give this information. This is somehow unfortunate because large confidence intervals will inform of potential identifiability (structural or practical) problems. Moreover, the identification procedure is less limited by mathematical or software difficulties than by poor experimental data.

In a classical approach, unknown transfer coefficients are fitted using thin layer drying kinetics in constant conditions over an expected domain of validity. Depending on the industry needs, drying conditions (temperature mostly, plus eventually relative humidity and velocity) are varied between each experiment. The resulting experimental data obtained are then split in two subsets: one for the identification itself, and the rest for the validation. Fohr et al. (1988) discussed that it is generally a good idea to add kinetics obtained with varying drying conditions in the second subset to improve the validation. In the field of food drying, there is no practical explanation of how the drying experiments could be designed to minimize both the confidence intervals and the total experimental time.

We emphasize that an interesting consequence of avoiding black box models and limiting the number of parameters to be identified is the ability to keep the same model for several product varieties. For instance, Courtois et al. (2004) demonstrated that the same model could be used for two totally different kinds of rice grains just by updating the product properties; there was no need for a second identification.

Model validation is the final step of modeling. Depending on the choice of model equations and experimental data in the training and validation subsets, the model may or may not be robust in a wide range of conditions. In fact, it is very common to see models validated over a narrow range of temperatures (e.g., from 20 to 50°C). In a similar manner, most models cannot correctly simulate the drying rate during cooling. It is also quite uncommon to discuss the residuals (i.e., the part of the data that remains unexplained by the simulation), although they contain some information about the model validity. If the model is perfect, the residuals should be distributed as a normal law (white noise). In practice, only visual agreement between the simulated and experimental kinetics are used, in conjunction with some accounting of mean errors between simulations and experimental data. As an illustration, DRYER3000 claims to simulate the final grain moisture content in an industrial mixed-flow dryer with an error of $<5\%$.

Thus, the question is, what would be a good criterion to measure the goodness of fit of the model over the data? The common answer is the absolute or relative error observed in the prediction of the product moisture content. It is the only criterion that is used in practice. The error concerning the product temperature is not used because of the numerous biases and problems concerning the measurement. Reverse engineering might be of interest to try to identify the transfer coefficients directly from deep bed experiments using, in addition to the classical moisture content, the errors between simulations and experiments for the output air temperature and relative humidity.

12.5   APPLICATIONS OF DRYING MODELS

12.5.1   Coupling with a Quality Model

The quality of dried products is of prime importance for consumers and food manufacturers. This notion of quality covers various concepts for food products: appearance (color, gloss, shape, and natural aspect), taste (aroma, texture), convenience of use (rapidity of rehydration or dissolution, consistency of apparent density for constant delivery quantities, no stickiness), composition and nutritional characteristics (vitamin content in fruits, furosine content in pasta, protein degradation, etc.), microbial innocuousness, and so forth. Quality is also very important for many agroproducts like grains or seeds, for which the mechanical resistance (e.g., head grain yield), biochemical composition (wheat gluten content for breadmaking, ability of starch or protein separation in corn), or germination potential have to be preserved by drying for further use.

The physicochemical phenomena responsible for the quality changes depend mainly on the dryer design and on control of the drying operating conditions (process parameters and product residence time distribution). In fact, reactions rates are time–temperature dependent and very sensitive to water activity. Depending on the nature, structure, and size of the product (solid, paste, liquid), different drying processes are used with various drying times (seconds to hours) and consequences in the evolution of the product during drying. The environmental drying conditions (temperature, relative humidity of the air) also result in creating more or less intense moisture and temperature gradients during drying. The gradients of water concentration provoke differential shrinkage and induce mechanical stresses that can lead to cracks or breakage, and therefore often a loss of quality and economic value. Case hardening is also a common problem in dried bioproducts that is attributable to rapid drying, and phenomena like the glass transition or crystallization can be observed.

It is therefore important to develop combined modeling approaches of drying that integrate a simulation of the chemical and biochemical reactions and/or the stress-induced cracks to assess the quality parameters. Such simulation tools are necessary to obtain acceptable, even improved, quality and a high production capacity. However, the difficulty is that the definition of the desired final quality is often a multidimensional problem, involving complicated reactions and transformations, for which a target function must be defined. Most of the time the initial state of the product is variable, and this variability must also be taken into account.

The optimization of processing conditions can be helped by such mathematical modeling and lead to a dynamic prediction of the quality of the product being dried. Different kinds of models have been developed:

- nth-Order reaction equations (Abud-Archila et al., 2000; Beke et al., 1993; Bruce, 1992; Courtois et al., 1991; Migliori et al., 2005; Nellist, 1981; Straatsma et al., 1999b)
- NNs (Rocha Mier, 1993; Trelea et al., 1997a)
- Fuzzy logic (Perrot et al., 1998; Zhang and Lichtfield, 1990)
- Mechanical description of tensile stress profiles using finite elements (Lichtfield and Okos, 1998; Peczalski et al., 1996; Ponsart, 2002)

For the three first types of models, the purpose is to express the time derivative of a quality index as a function of the moisture content (average moisture content or gradient of moisture content for mechanical properties) and temperature in order to calculate it during drying by integrating through the changing product moisture contents and temperatures. The quality model is used as a module of the drying simulation model.

Reaction equations of the nth order have been widely used to give the variations of the concentration of a sensitive component or the variations of a quality index (Q) with time:

$$\frac{dQ}{dt} = -k_Q \times Q^n \tag{12.14}$$

with a constant of reaction k_Q usually depending on X and n generally equal to 0, 1, or 2. An Arrhenius law is frequently used to describe the temperature (T) dependence:

$$k_Q = k_0 [X,\ldots,] \exp\left(-\frac{E_a}{RT}\right) \tag{12.15}$$

where E_a is the pseudoactivation energy and R is the perfect gas constant.

In most cases, the rate of variation of the quality index can be assessed in experimental tests under constant conditions at different temperatures and for different water activities (*sealed heating experiments*) and then used for calculating the cumulative damage in a drying model. It has been used for very different types of quality indexes. Courtois et al. (1991), Migliori et al. (2005), Nellist (1981), and Straatsma et al. (1999b) applied this procedure for predicting the loss of wheat seed viability, loss of corn wet milling quality, and the insolubility index of milk powder or furosine production in pasta, respectively.

When quality is not dependent on the temperature and moisture content only, such tests cannot be applied. Bruce (1992) and Abud-Archila et al. (2000) measured the variations of a quality index directly by sampling and analyzing products during thin layer drying experiments. In the first study, the relative loaf volumes were measured for loaves baked from wheat samples dried in thin layers for various times and they were described using a first-order process and a constant of the reaction function of the moisture content of the kernels. In the second study, the head rice yield after milling was modeled by an analogy with a second-order chemical reaction. Qualitative analysis of the experimental results as a function of the drying conditions suggested that there was a very close relationship between the moisture content gradient and the processing quality. An Arrhenius-type expression was used to describe the effect of the kernel temperature (T_k):

$$\frac{dQ}{dt} = -kQ^2 \quad \text{with} \quad k = 5.71 \times 10^{26}\Delta X^5 \exp\left(\frac{-1.657 \times 10^5}{R(T_k + 273.1)}\right) \quad (12.16)$$

There are numerous studies on the impact of drying parameters on quality attributes, but few published articles deal with the validation of such models on the industrial scale, imposing variable conditions in separated zones during the treatment (Courtois, 1995; Courtois et al., 2001; Migliori et al., 2005). However, such models can be used in dryer design for optimization of process parameters or for control. For example, Courtois (1991) designed a control algorithm taking the milling quality of dried corn into account according to the potentialities of the dynamic model used to simulate the unsteady-state behavior of the dryer. This control algorithm was validated on the industrial scale.

The prediction of the mechanical behavior of a solid during drying is more complex. Heat and moisture gradients induce differential shrinkage, so strain and stress in the product and stress cracking can appear. However, the rheological behavior of the solid in turn impacts the mass transport inside the product. To predict this it is necessary to *simultaneously* model the evolution of the local variables inside the product during the drying process, such as the water content, temperature, deformation, and drying-induced stress. For elastic products, the influence of the rheology on heat and mass transport can be taken into account (Jomaa and Puiggali, 1991; Mrani et al., 1995). For viscoelastic products, solving the coupling is very complex and the influence of strains and stresses on internal transport must generally be neglected in order to describe the interactions between the process conditions and crack formation in the product.

In all cases, the stress must be calculated by using an experimental behavior law linking stress to strain, depending on both the moisture and temperature. However, at the temperatures reached by the products during drying, very few data on the Young's modulus and tensile and compressive strengths as a function of the strain rate and moisture content are available in the literature for food materials.

12.5.2 ON-LINE USES OF DRYER MODELS

As mentioned previously, there are very few articles describing drying models, including a quality equation and validated on the industrial scale. In fact, there are even fewer articles describing (a) how to set up a real-time automatic control of an industrial dryer or (b) how to get an estimate of the product moisture content at the outlet from indirect measurements such as exhaust air temperatures. Dufour (2006) provided some statistics about publications concerning the control of dryers. He reported that recently there were about 4 to 10 new articles/year in that area with approximately 66% of them concerning food, but there were only 0 to 2 new articles 20 years ago. Obviously, this is not an active research topic. Most of these works made limited use of the dryer model to help design and test the control strategy (Eltigani and Bakker-Arkema, 1987; McFarlane and Bruce, 1991). In many cases, the control law was a variant of a classical proportional–integral–derivative controller. Among the improvements over the generic control law, we emphasize the simple linearization technique detailed in Whitfield (1988a, b), which was used in several works: using the logarithm of the product moisture content and a sampling time proportional to the residence time is practically a necessity for linear control.

Trelea et al. (1997b) developed optimization techniques combined with a prediction model to attain nonlinear model based predictive control. Two models, one for the prediction of the product moisture content and the other for the product wet milling quality, are used in an optimization scheme to compute the best air temperature profile satisfying one objective and three constraints (Figure 12.15):

- Minimal energy or drying time or any combination
- Air temperature within specified boundaries
- Final product moisture content below the related set point
- Final product quality above the related set point

The interesting point in this work is in the generic approach. The authors successfully applied a similar method to a batch refrigeration process.

Olmos et al. (2002) performed a similar study on the batch drying of rice and its breakage ratio as the quality index.

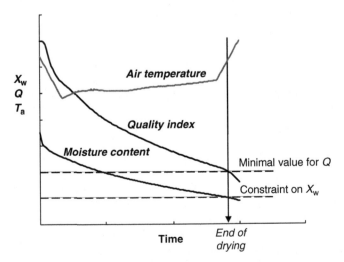

FIGURE 12.15 Optimization of maize drying with respect to a minimal value for the quality index and a maximal value for the moisture content. (From Trelea, I.C. et al., *J. Food Eng.*, 31, 403–421, 1997b. With permission.)

Concerning how to get an estimate of the product moisture content at the outlet from indirect measurements such as exhaust air temperatures, a well-known problem in automatic control is building up a state observer for a highly nonlinear system. Unfortunately, generic solutions are not available and it remains an open field for research. Some authors have tried, sometimes with success, to control the product moisture content indirectly by measuring the final exhaust air temperature (at the very end of the dryer). In most cases, depending on the drying scheme, the relation between the product moisture content and exhaust air temperature is not monotonous. Hence, there is no guarantee that good control of the temperature implies good control of the product moisture content.

12.6 CONCLUSION

There are many models available to simulate and hence predict the performance of food dryers. Unfortunately, each model is available only for a limited number of products and a limited domain of validity in the best case. Furthermore, a model is not exactly a simulator. There are only a few simulators available for very specific problems. None of them can be compared to what is found in the chemical engineering area. There are a few works in progress that are trying to make the modeling of the drying methodology more generic. In contrast, with the increasing price of energy, the optimization of dryers again becomes an important issue. Hence, we can probably expect numerous studies to come from the research laboratories, which will soon be transferred into useful tools for food engineers.

NOMENCLATURE

VARIABLES

a	surface to volume ratio (/m)
$b1, b2, \ldots, b5$	drum drier equation parameters
C	specific load on the drum (kg dry matter/m^2)
c_{pa}	heat capacity of product (J/kg/K)
d_p	diameter of the dried material (m)
E_a	pseudoactivation energy in Arrhenius law (J/mol)
f	relative drying rate
H	bed total height (m)
IoS	index of saturation (empirical ratio)
k_d	drying constant (/s)
L_v	latent heat of water evaporation (J/kg)
MR	reduced moisture content
$n, k, k_Q, k_0, k_1, k_2, k_3, c_1, c_2, c_3$	empirical coefficients
N	drying rate (kg water/kg dry matter/s)
N_c	drying rate in constant rate period
p_{vs}	saturated vapor pressure (Pa)
p_{vg}	vapor pressure in bulk gas (Pa)
Q	product quality

R perfect gas constant (J/mol/K)
RH   air relative humidity (%)
S section surface (m^2)
t time (s)
T temperature (°C)
T_a air temperature (°C)
T_k kernel temperature (°C)
T_s temperature at the outer surface of the drum wall (°C)
T_{wb} wet bulb temperature (°C)
v_a air velocity (m/s)
X product moisture content (kg water/kg dry matter)
X_0 initial product moisture content (kg water/kg dry matter)
X_{cr} critical moisture content (kg water/kg dry matter)
$\overline{X}, X_m$ product average moisture content (kg water/kg dry matter)
X_e product equilibrium moisture content (kg water/kg dry matter)
Y air moisture content (kg vapor/kg dry air)
z bed height (m)

GREEK LETTERS

β external mass transfer coefficient (kg/m^2/Pa/s)
ε deep bed porosity
ϕ dimensionless characteristic moisture content
Φ_h thermal flow density between air and product (W/m^2)
Φ_w water flow density between air and product (kg water/s/m^2)
φ air-flow rate (kg dry air/s)
φ_w evaporation rate (kg water/s)
ρ_a air density (kg/m^3)
ρ_{da} dry air density (kg/m^3)
ρ_{dm} dry matter density (kg/m^3)

ABBREVIATIONS

CDC   characteristic drying curve
CFD   computational fluid dynamics
NN   neural network
w.b.   wet basis
d.b.   dry basis

REFERENCES

Abchir, F., Modélisation du séchage sur cylindre, Ph.D. thesis, AgroParisTech, Massy, France, 1988.

Abud-Archila, M., Courtois, F., Bonazzi, C., and Bimbenet, J.J., Processing quality of rough rice during drying—Modeling of head rice yield versus moisture gradients and kernel temperature, *J. Food Technol.*, 45, 161–169, 2000.

Adomian, G., *Fundamental Theories of Physics. Solving Frontier Problems of Physics: The Decomposition Method*, Vol. 60, Kluwer Academic, Dordrecht, 1996.

Akpinar, E.K., Mathematical modeling of thin layer drying process under open sun of some aromatic plants, *J. Food Eng.*, 77, 864–870, 2006.

Anglerot, D., Couture, F., and Roques, M., Optimisation des écoulements dans un séchoir à claies à l'aide de FLUENT, in *Cahiers de l'AFSIA 19—Compte Rendu des 19èmes Journées de l'AFSIA*, 22–23 mars, Poitiers, France, AFSIA-EFCPE, Villeurbanne, 2001, pp. 69–79.

Aregba, A.W., Sebastian, P., and Nadeau, J.-P., Stationary deep-bed drying: A comparative study between a logarithmic model and a non-equilibrium model, *J. Food Eng.*, 77, 27–40, 2006.

Baini, R. and Langrish, T.A.G., Choosing an appropriate drying model for intermittent and continuous drying of bananas, *J. Food Eng.*, 79, 330–343, 2007.

Bala, B.K., Ashraf, M.A., Uddin, M.A., and Janjai, S., Experimental and neural network prediction of the performance of a solar tunnel drier for drying jackfruit bulbs and leather, *J. Food Proc. Eng.*, 28, 552–566, 2005.

Beke, J., Vas, A., and Mujumdar, A., Impact of process parameters on the nutritional value of convectively dried grains, *Dry. Technol.*, 11, 1415–1428, 1993.

Berger, D. and Pei, D.C.T., Drying of hygroscopic capillary porous solids, a theoretical approach, *Int. J. Heat Mass Transfer*, 16, 293–302, 1973.

Blei, S. and Sommerfeld, M., CFD in drying technology—Spray-dryer simulation, in *Modern Drying Technology. Computational Tools at Different Scales*, Tsotsas, E. and Mujumdar, A.S., Eds., Wiley–VCH Verlag GmbH & Co. KGaA, Weinheim, 2007, pp. 155–208.

Brooker, D.B., Bakker-Arkema, F.W., and Hall, C.W., *Drying Cereal Grains*, AVI Publishing Company, Westport, CT, 1974.

Bruce, D.M., A simulation of multi-bed, concurrent-flow and counter-flow grain driers. Part 1: The model, National Institute of Agricultural Engineering, Silsoe, U.K., 1983, paper DN.1171.

Bruce, D.M., A model of the effect of heated-air drying on the bread baking quality of wheat, *J. Agric. Eng. Res.*, 52, 53–76, 1992.

Celma, A.R., Rojas, S., Lopez, F., Montero, I., and Miranda, T., Thin-layer drying behaviour of sludge of olive oil extraction, *J. Food Eng.*, 80, 1261–1271, 2007.

Courtois, F., Dynamic modeling of drying to improve processing quality of grain, Ph.D. thesis, AgroParisTech, Massy, France, http://francis.courtois.free.fr, 1991.

Courtois, F., Computer-aided design of corn dryers with quality prediction, *Dry. Technol.*, 13, 147–164, 1995.

Courtois, F., Abud Archila, M., Bonazzi, C., Meot, J.M., and Trystram, G., Modeling and control of a mixed-flow rice dryer with emphasis on breakage quality, *J. Food Eng.*, 49, 303–309, 2001.

Courtois, F., Khoshhal, M., Matthieu, O., and Lalanne, V., Simulation of deep bed dryers: Application to the drying of rice in Iran, in *ICEF9, International Congress on Engineering and Food*, Montpellier, France, March 7–11, 2004 (CD).

Courtois, F., Lebert, A., Duquenoy, A., Lasseran, J.C., and Bimbenet, J.J., Modeling of drying in order to improve processing quality of maize, *Dry. Technol.*, 9, 927–945, 1991.

Datta, A.K., Porous media approaches to studying simultaneous heat and mass transfer in food processes. I: Process formulations, *J. Food Eng.*, 80, 80–95, 2007a.

Datta, A.K., Porous media approaches to studying simultaneous heat and mass transfer in food processes. II: Property data and representative results, *J. Food Eng.*, 80, 96–110, 2007b.

Ducept, F., Sionneau, M., and Vasseur, J., Séchage par pulvérisation en vapeur d'eau surchauffée: Modélisation CFD et validation expérimentale, in *Cahiers de l'AFSIA 19—Compte Rendu des 19èmes Journées de l'AFSIA*, 22–23 mars, Poitiers, France, AFSIA-EFCPE, Villeurbanne, 2001, pp. 81–90.

Dufour, P., Control engineering in drying technology: Review and trends, *Dry. Technol.*, 24, 889–904, 2006.

Eltigani, A.Y. and Bakker-Arkema, F.W., Automatic control of commercial crossflow grain dryers, *Dry. Technol.*, 5, 561–575, 1987.

Fohr, J.P., Arnaud, G., Ali Mohamed, A., and Ben Moussa, H., Validity of drying kinetics, in *Proceedings of the Sixth International Drying Symposium (IDS'88)*, Vol. 1, Versailles, France, September 5–8, 1988, OP151–OP157.

Fornell, A., Le séchage de produits biologiques par l'air chaud—Calcul de séchoirs, Ph.D. thesis, AgroParisTech, Massy, France, 1979.

Fyhr, C. and Kemp, I.C., Comparison of different drying kinetics models for single particles, *Dry. Technol.*, 16, 1339–1369, 1998.

Giner, S.A. and Bruce, D.M., Two-dimensional simulation model of steady-state mixed-flow grain drying. Part 2: Experimental validation, *J. Agric. Eng. Res.*, 71, 55–66, 1998.

Giner, S.A., Bruce, D.M., and Mortimore, S., Two-dimensional simulation model of steady-state mixed-flow grain drying. Part 1: The model, *J. Agric. Eng. Res.*, 71, 37–50, 1998.

Henderson, S.M. and Pabis, S., Grain drying theory. I: Temperature effect on drying coefficient, *J. Agric. Eng. Res.*, 6, 169–174, 1961.

Huang, B. and Mujumdar, A.S., Use of neural network to predict industrial dryer performance, *Dry. Technol.*, 11, 525–541, 1993.

Jannot, Y., Talla, A., Nganhou, J., and Puiggali, J.R., Modeling of banana convective drying by the drying characteristic curve (DCC) method, *Dry. Technol.*, 22, 1949–1956, 2004.

Jayas, D.S., Cenkowski, S., Pabis, S., and Muir, W.E., Review of thin-layer drying and wetting equations, *Dry. Technol.*, 9, 551–588, 1991.

Jomaa, W. and Puiggali, J.R., Drying of shrinking materials: Modelling with shrinkage velocity, *Dry. Technol.*, 9, 1271–1293, 1991.

Karapantsios, T.D., Conductive drying kinetics of pregelatinized starch thin films, *J. Food Eng.*, 76, 477–489, 2006.

Karathanos, V.T., Determination of water content of dried fruits by drying kinetics, *J. Food Eng.*, 39, 337–344, 1999.

Keey, R.B. and Suzuki, M., On the characteristic drying curve, *Int. J. Heat Mass Transfer*, 17, 1455–1464, 1974.

Kiranoudis, C.T., Tsami, E., Maroulis, Z.B., and Marinos-Kouris, D., Drying kinetics of some fruits, *Dry. Technol.*, 15, 1399–1418, 1997.

Kozempel, M.F., Sullivan, J.F., Craig, J.C., Heiland, W.K., Jr., and Heiland, W.K., Drum drying potato flakes—A predictive model, *Lebensm.-Wiss. Technol.*, 19, 193–197, 1986.

Kozempel, M.F., Tomasula, P., and Craig, J.C., Jr., The development of the ERRC food process simulator, *Simul. Practice Theory*, 2, 221–236, 1995.

Langrish, T.A.G. and Kockel, T.K., The assessment of a characteristic drying curve for milk powder for use in computational fluid dynamics modeling, *Chem. Eng. J.*, 84, 69–74, 2001.

Lewis, W.K., The rate of drying of solid materials, *J. Ind. Eng. Chem.*, 13, 427–432, 1921.

Lichtfield, J.B. and Okos, M.R., Prediction of corn kernel stress and breakage induced by drying, tempering and cooling, *Trans. ASAE*, 31, 585–594, 1998.

Marchal, D., Simulation du rayonnement dans les fours infrarouge au moyen de codes CFD, in *Cahiers de l'AFSIA 19—Compte Rendu des 19èmes Journées de l'AFSIA*, 22–23 mars, Poitiers, France, AFSIA-EFCPE, Villeurbanne, 2001, pp. 107–116.

Martynenko, A.I. and Yang, S.X., Biologically inspired neural computation for ginseng drying rate, *Biosyst. Eng.*, 95, 385–396, 2006.

May, B.K. and Perré, P., The importance of considering exchange surface area reduction to exhibit a constant drying flux period in foodstuffs, *J. Food Eng.*, 54, 271–282, 2002.

Mayor, L. and Sereno, A.M., Modeling shrinkage during convective drying of food materials: A review, *J. Food Eng.*, 16, 373–386, 2004.

McFarlane, N.J.B. and Bruce, D.M., Control of mixed-flow grain-drier: Development of a feedback-plus-feedforward algorithm, *J. Agric Eng. Res.*, 49, 243–258, 1991.

Migliori, M., Gabriele, D., de Cindio, B., and Pollini, C.M., Modeling of high quality pasta drying: Quality indices and industrial application, *J. Food Eng.*, 71, 242–251, 2005.

Mirade, P.S., Simulation du fonctionnement aéraulique instationnaire d'un séchoir à charcuterie au moyen d'un outil de CFD, in *Cahiers de l'AFSIA 19—Compte Rendu des 19èmes Journées de l'AFSIA*, 22–23 mars, Poitiers, France, AFSIA-EFCPE, Villeurbanne, 2001, pp. 125–133.

Mohsenin, N.N., *Physical Properties of Plant and Animal Materials*, 3rd ed., Gordon and Breach Science Publishers, New York, 1980.

Mrani, I., Benet, J.C., and Fras, G., Transport of water in a biconstituent elastic medium, *Appl. Mech. Rev.*, 48, 717–721, 1995.

Nadeau, J.-P. and Puiggali, J.-R., *Séchage: Des Processus Physiques aux Procédés Industriels*, Lavoisier, Techniques et Documentation, Paris, 1995.

Nellist, M.E., Predicting the viability of seeds dried with heated air, *Seed Sci. Technol.*, 9, 438–455, 1981.

Nellist, M.E. and O'Callaghan, J.R., The measurement of drying rates in thin layers of ryegrass seed, *J. Agric. Eng. Res.*, 16, 192–212, 1971.

Ng, P.P., Tasirin, S.M., and Law, C.L., Thin layer method analysis of spouted bed dried Malaysian paddy—Characteristic drying curves, *J. Food Proc. Eng.*, 29, 414–428, 2006.

Olesen, H.T., *Grain Drying*, Innovation Development Engineering, Thisted, Denmark, 1987.

Olmos, A., Trelea, I.C., Courtois, F., Bonazzi, C., and Trystram, G., Dynamic optimal control of batch rice drying process, *Dry. Technol.*, 20, 1319–1345, 2002.

Pajonk, A., Simulation numérique et étude expérimentale de la circulation de l'air et des échanges de matière dans les enceinte d'affinage de l'emmental, in *Cahiers de l'AFSIA 19—Compte Rendu des 19èmes Journées de l'AFSIA*, 22–23 mars, Poitiers, France, AFSIA-EFCPE, Villeurbanne, 2001, pp. 117–124.

Peczalski, R. and Laurent, M., Transferts dans les aliments solides. Modèles physiques et mathématiques, in *Techniques de l'Ingénieur, Génie des Procédés—Agroalimentaire*, ETI Sciences & Techniques, Paris, vol. F-2000, 2000, pp. 1–18.

Peczalski, R., Laurent, P., Andrieu, J., Boyer, J.C., and Boivin, M., Drying-induced stress build-up within spaghetti, in *Drying '96, Proceedings of the 10th International Drying Symposium*, Vol. B, Strumillo, C. and Pakowski, Z., Eds., Lodz Technical University, Lodz, Poland, 1996, pp. 805–816.

Perré, P. and Degiovanni, A., Simulation par volumes finis des transferts couples en milieux poreux anisotropes: Séchage du bois à basse et haute température, *Int. J. Heat Mass Transfer*, 33, 2463–2478, 1990.

Perrot, N., Bonazzi, C., and Trystram, G., Application of fuzzy rules-based models to prediction of quality degradation of rice and maize during hot air drying, *Dry. Technol.*, 16, 1533–1565, 1998.

Philip, J.R. and De Vries, D.A., Moisture movement in porous materials under temperature gradients, *Trans. Am. Geophys. Union*, 38, 222–232, 1957.

Ponsart, G., Caractérisation expérimentale d'un matériau alimentaire et modélisation des contraintes au cours du séchage. Application aux pâtes spaghetti, Ph.D. thesis, AgroParisTech, Massy, France, 2002.

Press, W.H., Teukolski, S.A., Vetterling, W.T., and Flannery, B.P., *Numerical Recipes: The Art of Scientific Computing*, 3rd ed., Cambridge University Press, Cambridge, U.K., 2007.

Rahman, M.S., Al-Zakwani, I., and Guizani, N., Pore formation in apple during air-drying as a function of temperature: Porosity and pore-distribution size distribution, *J. Sci. Food Agric.*, 85, 979–985, 2005.

Rocha Mier, T., Influence des prétraitements et des conditions de séchage sur la couleur et l'arôme de la menthe (*Mentha spicata* Huds.) et du basilic (*Ocimum basilicum*), Ph.D. thesis, AgroParisTech, Massy, France, 1993.

Rodriguez-Fernandez, M., Balsa-Canto, E., Egea, J.A., and Banga, J.R., Identifiability and robust parameter estimation in food process modeling: Application to a drying model, *J. Food Eng.*, 83, 374–383, 2007.

Saravacos, G.D. and Maroulis, Z.B., *Transport Properties of Foods*, New York, Marcel Dekker, 2001.

Straatsma, J., Van Houwelingen, G., Steenberg, A.E., and De Jong, P., Spray drying of food products. 1. Simulation model, *J. Food Eng.*, 42, 67–72, 1999a.

Straatsma, J., Van Houwelingen, G., Steenberg, A.E., and De Jong, P., Spray drying of food products. 2. Prediction of insolubility index, *J. Food Eng.*, 42, 73–77, 1999b.

Techasena, O., Lebert, A., and Bimbenet, J.-J., Simulation of deep bed drying of carrots, *J. Food Eng.*, 16, 267–281, 1992.

Togrul, I.T. and Pehlivan, D., Mathematical modeling of solar drying of apricots in thin layers, *J. Food Eng.*, 55, 209–216, 2002.

Toyoda, H., Study on intermittent drying of rough rice in a recirculation dryer, in *Sixth International Drying Symposium (IDS'88)*, Vol. 2, Versailles, France, September 5–8, 1988, OP171–OP178.

Trelea, I.C., Courtois, F., and Trystram, G., Dynamic models for drying and wet-milling quality degradation of corn using neural networks, *Dry. Technol.*, 15, 1095–1102, 1997a.

Trelea, I.C., Courtois, F., and Trystram, G., Optimal constrained non linear control of batch processes: Application to corn drying, *J. Food Eng.*, 31, 403–421, 1997b.

Trystram, G., Contribution à l'automatisation des procédés en industrie alimentaire; Cas du séchage sur cylindre, Ph.D. thesis, Institut National Polytechnique de Lorraine, Nancy, France, 1985.

Trystram, G. and Vasseur, J., The modeling and simulation of a drum drying process, *Int. Chem. Eng.*, 32, 689–705, 1992.

van Meel, D.A., Adiabatic convection batch drying with recirculation of air, *Chem. Eng. Sci.*, 9, 36–44, 1958.

Vasseur, J., Etude du séchage d'un produit visqueux, en couche mince sur une paroi chaude, permettant de définir un modèle de séchoir cylindre, Ph.D. thesis, AgroParisTech, Massy, France, 1983.

Whitaker, S., Simultaneous heat, mass and momentum transfer in porous media: A theory of drying, in *Advances in Heat Transfer*, Vol. 13, Academic Press, New York, 1977, pp. 119–203.

Whitfield, R.D., Control of a mixed-flow drier, Part 1: Design of the control algorithm, *J. Agric. Eng. Res.*, 41, 275–287, 1988a.

Whitfield, R.D., Control of a mixed-flow drier, Part 2: Test of the control algorithm, *J. Agric. Eng. Res.*, 41, 289–299, 1988b.

Zhang, J., Datta, A.K., and Mukherjee, S., Transport processes and large deformation during baking of bread, *AIChE J.*, 51, 2569–2580, 2005.

Zhang, Q. and Lichtfield, J.B., Fuzzy expert systems: A prototype for control of corn breakage during drying, *J. Food Proc. Eng.*, 12, 259–273, 1990.

APPENDIX

WEB LINKS TO SOFTWARE RELATED TO FOOD DRYERS

Psychrometric calculations software: http://www.linric.com
Psychrometric calculator: http://www.psychro.com
Commercial mixed-flow grain dryer simulator: http://www.drying.org

WEB LINKS TO SOFTWARE RELATED TO THE CHEMICAL INDUSTRY

Commercial software for design, tuning, and optimization of chemical processes (no food, no drying): http://www.prosim.net

Commercial software for design, tuning, and optimization of chemical processes (some tools for drying): http://www.aspentech.com

Network promoting the CAPE-OPEN initiative for software interoperability standards: http://www.colan.org

13 Nonconventional Heating Sources during Drying

Valérie Orsat and G.S. Vijaya Raghavan

CONTENTS

13.1 INTRODUCTION

This chapter focuses on drying processes that have been developed using nonconventional sources of energy for drying agricultural and food products. Nonconventional heating sources such as microwaves (MWs), radiofrequency (RF), infrared (IR), and heat pumps have been increasingly utilized in drying in recent years because of the improved drying kinetics, energy savings, and quality dried products that they can provide. The benefit of using these heating sources lies in the short processing time that is the most advantageous for product quality when compared to other drying techniques, especially when dealing with fruits and vegetables that contain high moisture and are sensitive to heat. To ensure good physical appearance and maintain the nutrient value of dried products, combination techniques with new sources of energy need to be considered for the successful drying of biological materials.

Population growth, the world economy, and climate change are placing severe pressures on natural resources and food production systems. Appropriate management of resources will require innovative measures and practices that are researched and developed by interdisciplinary teams of engineers and scientists. Innovation contributes to the development of more value-added chains in the agrifood economy while providing the foundation to ensure food security and safety and maintain the health of our environment.

Greenhouse gas emissions and global warming are motivating the development of technologies using electrical energy to replace fossil fuels in a wide spectrum of activities, with one obvious restriction: the electrical energy must come from renewable sources. These include hydro, solar, tidal, and wind energies and exclude combustion of coal, oil, and natural gas. To a great extent, electricity produced from renewable sources is considered clean and relatively friendly to the environment. Electricity can also be transformed into radiant energy over practically the entire useful electromagnetic spectrum (Figure 13.1), thus providing processing options that are not available through combustion-based technologies, other than through the electrical intermediate.

Evolution and competition in the market place drive processors to continuously look for new processing methods to differentiate their products through improved quality, lower costs, or value-added features. Dielectric heating (MW and RF) has a number of advantages over other drying processes such as the volumetric dissipation of energy throughout a product and the ability to automatically level any moisture variation within it (Decareau, 1985). However, it also has some limitations, which are occasionally technical, but principally economic. MW technology in industrial processes is no doubt expensive, but it is rapid and efficient. In general, dielectric heating can be justified in processes involving material of high value that is being processed in modest quantities or in combination with other technologies. The benefits of the development of combination processes with nonconventional heating sources are the decrease in energy consumption for the production of high-quality dried products, a reduction in chemicals used for pretreatment of the products prior to drying, and a better understanding of the kinetics involved in drying heat-sensitive materials. Reductions in processing time and substantially more efficient energy utilization should make nonconventional hybrid processes very cost effective and attractive to all levels of the agrifood processing chain.

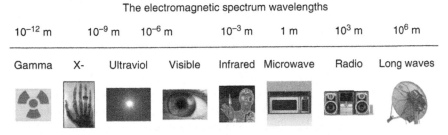

FIGURE 13.1 Electromagnetic spectrum wavelengths for gamma rays, x-rays, ultraviolet, visible, IR, MW, radio waves, and long waves. Are all electromagnetic energy, and the differences among them are their wavelengths. (From Langley Science Center, NASA, http://www.nasa.gov/centers/langley/science/FIRST.html. With permission.)

The benefit of drying is not only to preserve crops as food but also to reduce package and transport cost in terms of weight and volume while offering the possibility to add value to harvested commodities. The challenge in food drying, especially for high moisture fruits and vegetables, is to reduce the moisture content to a certain level where microbiological growth will not occur while maintaining the high nutrient value, especially the functional micronutrients. For the majority of existing drying processes, excluding freeze-drying, applying heat during drying through conduction, convection, and radiation is the basic technique to force water to vaporize, whereas forced air or a vacuum is applied to encourage the removal of the moisture vapor away from the drying material.

The operating conditions in drying and thermal processing have a considerable influence on the quality of processed biological materials. Browning reactions occur in the presence of high heat with lower moisture, causing a decrease in nutritional value and overall quality. Color loss, shrinkage, denaturation, loss of aroma and flavor, vitamin loss, and textural changes are also important quality changes affected by thermal processing. In order to optimize practices for thermal processing and drying of biological materials, extensive research has been carried out worldwide on various drying methods such as fluidized bed drying, spouted bed drying, particulate medium conduction drying, flash drying, freeze-drying, spray drying, drum drying, tunnel convective drying, tray drying, MW drying, RF drying, IR drying, electric field drying, superheated steam, osmotic dehydration, conveyor dryers, impingement drying, solar drying, rotary drying, and so forth (Mujumdar, 2007). All of the drying methods have been developed to optimize the drying of target materials, because not all drying processes are suitable for all types of biological materials.

High frequency electromagnetic energy transfer in the MW and RF ranges can be considered as heat arising from the oscillation of molecular dipoles and the movement of ionic constituents in response to an alternating electric field. Unlike conventional energy, this energy is absorbed throughout the volume of the wet material. The evaporation of moisture within the material results in an increase in the internal pressure that can rapidly drive the moisture from the interior to the exterior of the solid. Care must be taken to limit this pressure build-up because it can cause the material to burst (Venkatachalapathy and Raghavan, 1999).

Resistive or ohmic heating is the term used to describe the release of heat in food products by direct electrical resistance. The process allows rapid, uniform heating for high-quality processing of liquid food products through the direct dissipation of electrical energy. Metallic materials, insulators, and foodstuffs are highly conductive at main frequencies, so that the source energy is dissipated ohmically in direct resistive heating. In such applications, the material comes in direct contact with the electrodes. Ohmic heating has great application for thermal treatments, potentially for enhancing water evaporation, rather than in direct drying applications (Goullieux and Pain, 2005).

In contrast, induction heating has had few applications in biological materials but has been used in the metal industry. The majority of applicators for induction are based on coils wrapped around the material. The electric current flowing in the coil establishes an axial magnetic flux that in turn sets up a circumferential current in the material, the depth of which depends on the frequency of operation.

The flowing current is responsible for imparting electrical energy to the material (Metaxas, 1996).

IR radiation falls in the short wavelength electromagnetic spectrum and offers great potential for surface heat treatments in thin layer drying applications (Ratti and Mujumdar, 2007).

Heat pumps are used to reduce the energy consumption of existing dryers. Their usefulness lies in their dehumidification process for the recovery of the latent heat of evaporation of water being lost in the exhaust air from the dryer (Chou and Chua, 2007; Sosle, 2006).

This chapter concentrates on drying applications with MW and RF drying, IR heating, and heat pump dehumidification.

13.2   MW DRYING

The electromagnetic spectrum covers a large span of wavelengths from low frequency telecommunications, MWs, and IR radiation, up to gamma rays. Dielectric and MW heating lie in the electromagnetic spectrum in the range of frequencies from 300 kHz to 300 GHz. RFs range from 300 kHz to 300 MHz, and MWs range from 300 MHz to 300 GHz.

The difference between RF and MW is in the means of transferring the energy to the processed material. In RF, an electric field is developed between electrodes; in MW, a wave is propagated and reflected under the laws of optics. RF works well with bulky materials having high ionic conductivity. At 27.12 MHz the wavelength is on the order of 10 m. MW works well with smaller quantities of a dipolar nature. At 2450 MHz the wavelength is on the order of 10 cm.

Research on mutual interactions between food products and dielectric processing equipment is still needed to provide a practical basis for process control and minimization of process energy costs and to ensure microbial safety and product quality. Processes for high moisture solid foods have been less successful in extensive research conducted at 2450 MHz, unless combined with convection heating methods. This results to some extent from low penetration depths that could be increased at frequencies below 2450 MHz, such as 915 and 27.12 MHz.

The degree to which a given material responds to an electromagnetic field depends on the frequency and intensity of the field and on the characteristics of the material itself. The electrical basis of interaction is described in terms of the dielectric properties: the dielectric constant (ε') and dielectric loss factor (ε''). These represent the proportion of energy that can penetrate the material and the amount of energy that can be absorbed by the material, respectively. The loss tangent (tan δ) is the ratio of $\varepsilon'/\varepsilon''$, which is a combination of the material's ability to be penetrated by an electric field and to dissipate that energy as heat (Mudgett, 1986). The dielectric properties of materials are of great importance in the consideration of high frequency or MW applications with biological materials.

Fundamental studies aimed at characterizing the dielectric responses of complex biological materials over a wide range of frequencies, temperatures, and moisture contents still need to be undertaken. Although a great deal of progress has been achieved in the development of instrumentation and in the understanding of the functional relationship between dielectric properties and other parameters, the growing trend toward electromagnetic processing has created a need for more flexible and accessible equipment to quantify the dielectric properties of products under a wide variety of conditions. Existing measurement technologies such as the cavity perturbation technique (Kudra et al., 1992) and the open ended coaxial transmission line technique (Tulasidas et al., 1995; Venkatesh and Raghavan, 2005) are expensive and require the use of a network analyzer.

13.2.1 MATERIAL PROPERTIES

The overall process of heating by MWs and RF is defined by the dissipation of electrical energy in "lossy media." The polarization effect is a function of the electromagnetic frequency, the dielectric and electric properties of the material, the viscosity of the medium, and the size of the polar molecules.

Water is the major absorber of electromagnetic waves in foods; consequently, the higher the moisture content is, the better the heating. The organic constituents of foods are dielectrically inert ($\varepsilon' < 3$ and $\varepsilon'' < 0.1$) and, compared to aqueous ionic fluids or water, may be considered transparent to electromagnetic waves. It is only at very low moisture levels, when the remaining traces of water are bound and unaffected by the rapidly alternating field, that the components of low specific heat become the major factor in dielectric heating.

The dielectric constant varies significantly with the temperature and frequency for many typical workload substances. In many MW and dielectric heating applications, it is actually necessary to account for that change during processing.

The dielectric behavior and heating characteristics of foods vary with the frequency and temperature and are significantly affected by moisture and salt contents. Ionic losses for a particular product are much higher and dipole losses are much lower at 915 MHz than at 2450 MHz and vice versa. At higher MW frequencies (5800 MHz) the dipole losses for most products are much greater and the ionic losses become negligible. In contrast, ionic losses are increasingly greater as the frequency decreases at sub-MW frequencies, and dipole losses for free water become negligible. The effects of frequency variations in the RF region on the dielectric constant are negligible because the dielectric constant of water in this region is close to its static value. Moisture and dissolved salts are the major determinants of dielectric activity in the liquid phase of such products as modified by the volumetric exclusion effects of an inert solids phase containing colloidal or undissolved lipid, proteins, carbohydrate, ash, or bound water.

Many reports have been made on the measurement of the dielectric properties of a variety of products (Mudgett, 1985; Nelson, 1973; Wang, Tang, et al., 2003; Wang Wig, et al., 2003). Recent reviews have been prepared by Piyasena et al. (2003) and Venkatesh and Raghavan (2004, 2005).

An excessively high ε'' value will result in a small skin depth, which annuls the desirable volumetric heating effect. In contrast, an ε'' value that is too low renders the material practically transparent to the incoming energy. Experience has shown that materials with an effective loss factor in the range $10^2 < \varepsilon'' < 2$ will be suitable for processing with dielectric heating (Metaxas, 1996). Selective polar or ionic additions to a low loss host material can enhance its effective loss and render it suitable for dielectric processing. It is sometimes possible to modify a low loss factor material without significantly altering its other properties via a small amount of high loss factor additive, such as carbon added to natural rubber or sodium chloride added to urea formaldehyde glues (Ikediala et al., 2002; Orsat and Raghavan, 2005).

Two major mechanisms are involved in the MW–matter heating interaction: dipolar rotation and ionic conduction. In ionic conduction, ions are accelerated by electric fields causing them to move toward the direction opposite to their own polarity. The movement of the ions provokes collisions with the molecules of the material. A disordered kinetic energy is created and as a consequence heat is generated. For example, polar molecules subjected to MW radiation at 2450 MHz will rotate 2.45×10^9 times/s. The friction between the fast rotating molecules generates heat throughout the material instead of being transferred from the surface to the inner part as is the case in conventional hot air drying.

Equation (13.1) is used to calculate the energy absorption by products during dielectric heating:

$$P = 2\pi f \varepsilon_0 \, \varepsilon'' \, |E|^2 \qquad\qquad (13.1)$$

where P is the energy developed per unit volume (W/m^3), f is the frequency (Hz), ε_0 is the absolute permittivity of a vacuum (8.854188×10^{-12} F/m), and $|E|$ is the electric field strength (V/m). Uniformity of drying is made possible with the control of the MW power density and duty cycle.

13.2.2 MW–Convective Drying

The heat and mass transfer through conduction, convection, and electroheating each have their own advantages and disadvantages. Process equipment can benefit from the combination of multiple processes (Orsat et al., 2007). Utilizing MW energy in drying offers reduced drying times and complements conventional drying in later stages by specifically targeting the internal residual moisture (Osepchuk, 2002). Because the dielectric properties of the material being processed vary as a function of the frequency along with temperature and moisture content as variables, they should point to a specific frequency range for a target application.

The ability of MWs to selectively heat areas with higher dielectric loss factors is very important in this application. Unlike other conventional systems, dielectric drying has an inherent ability to "moisture level" a given product, with more power being applied to regions with higher moisture content than to regions with lower moisture content. If the drying process is considered as a whole, it becomes apparent that, as long as the physical surface of a material can be maintained wet, conventional drying can still be economical. The argument for combining volumetric (dielectric) heating and surface (conventional) heating is based on sufficient heat being provided by the dielectric heating within the body of the material to cause water to move to the surface. This minimizes the retreating wet front, maintains the evaporating surface, and perhaps permits entrainment–evaporation. If MW energy can be used to bring moisture continually to the surface, then only a portion of the total drying heat needs to be provided by the dielectric heating system; the bulk of the heat is required for evaporation, and it can be provided at the surface with convective air drying.

MW drying uses electrical energy in the frequency range from 300 MHz to 300 GHz with 2450 MHz being the most commonly used frequency, especially for heating applications with the popular domestic MW oven. MWs are generated inside a cavity by stepping up the 60-Hz frequency of alternating current from domestic power lines to 2450 MHz. A magnetron device accomplishes this (Orsat et al., 2005). The step-up transformer that powers the magnetron accounts for more than half of the weight of the domestic MW ovens. A waveguide channels the MWs into the cavity that holds samples for heating. Domestic ovens have reflecting cavity walls that produce several modes of MWs, maximizing the efficiency of heating with a variety of materials. In industrial applications, however, it may be desirable to use single mode ovens that homogenously distribute the MWs into the cavity.

The basic requirements in MW power applications are efficiency of power transfer, low cost, and reliability of system operation. The most successful device for power application is the magnetron (Osepchuk, 2002). Magnetrons in MW ovens operate from 300 to 3000 W and in high power applications operate from 5 to 100 kW. The supplied power is fed into a cavity known as the applicator. The most common applicators include the multimode cavity, the single mode cavity, and the traveling wave device.

Because the heat is generated within the material, mass transfer occurs from the pressure gradient created by vapor generation inside the material, rapidly forcing water vapor to the product's surface creating a pumping action. This mechanism pushes water out of the product with great efficiency as the moisture content of the product decreases. The rapidity of the process can thus yield energy savings (Schiffmann, 2007).

For large-scale heating applications, RF's longer wavelength makes it less prone to standing waves and the resulting nonuniform heating. Moisture leveling is more effective at RF for wet planar materials in drying applications. In contrast, if drying needs to be conducted under a vacuum to reduce the boiling point, MW energy is better suited in drying cases with temperature-sensitive materials because the likelihood of arcing and corona discharge is much smaller.

From an environmentally friendly perspective, more efficient utilization of energy is required in the food drying and processing sector. In this respect, dielectric techniques may play an important role in the future. Selection of dielectric energy and its combination with conventional technologies will lead to products that meet, and often improve, the quality requirements of existing products and open the field for new product development (Demeczky, 1985; Jones, 1987).

The use of MW energy for drying requires moderately low energy consumption (Tulasidas et al., 1995). The volumetric heating and reduced processing time make MWs an attractive source of thermal energy. Because MWs alone cannot complete a drying process, combining techniques, such as forced air or vacuum, is recommended in order to further improve the MW process efficiency (Chou and Chua, 2001).

In MW drying, when the material couples with the MW energy, heat is generated within the product through molecular excitation. The critical next step is to immediately remove the water vapor. The simplest method to remove the released water is to pass air over the surface of the material, hence combining processes to create "MW–convective drying." In many cases when MW drying is mentioned, MW–convective drying is implied. The air temperature passing through the product can be varied to speed up the drying process by increasing its moisture carrying capacity (Chua et al., 2000). The selected air temperature is dependent on the products' characteristics and their sensitivity to heat. In order to control the overall drying process temperature, it is necessary to control the MW power density (watts/grams of material) or the duty cycle (time of power on or off; Changrue et al., 2004). Drying banana slices with MW–convective drying demonstrated that good quality dried products can be achieved by optimizing the power density and duty cycle time (Nemes et al., 2005). Similar results were reported in dried carrots, in which there was excellent quality in terms of color, shrinkage, and rehydration properties (Orsat et al., 2007; Wang and Xi 2005). Tulasidas et al. (1995) conducted a comparison of the specific energy consumption and drying time and showed the advantage of MW over convective drying for grapes (Table 13.1). The quality of raisins dried by MW drying was superior to hot air dried samples in terms of color, damage, darkness, crystallized sugar, stickiness, and uniformity (Tulasidas et al., 1995a; Tulasidas et al., 1995).

In studying MW drying kinetics, it is clearly during the falling rate period of the drying process that the use of MWs can prove most beneficial. As the material absorbs the MW energy, a temperature gradient occurs where the center temperature is greater, effectively forcing the residual moisture out (Erle, 2005; Soysal et al., 2006). Combination drying with an initial conventional drying process followed by a finish MW or MW–vacuum process reduces the drying time while improving product quality and minimizing the overall energy requirements (Erle, 2005; Raghavan et al., 2005; Soysal et al., 2006; Yanyang et al., 2004).

A significant industrial application is in MW–air tunnel drying that is used principally as a finish drying process to level-off moisture content in pasta, cracker, or chip drying (Osepchuk, 2002). However, when the heat generation within the product is too rapid, it can cause the generation of great internal steam pressure,

TABLE 13.1

Comparison of Convective and MW Drying of Grapes

Drying Method	Air Temp. (°C)	Air-Flow Rate (kg/s)	Drying Time (h)	Specific Energy Consumption (MJ/kg Water)
Convective	50	0.0210	23.66	90.35
	60	0.0204	16.75	81.15
MW	50	0.0210	5.58	21.86
	60	0.0204	3.86	19.08

Source: From Tulasidas, T.N. et al., *Dry. Technol.*, 13, 1973–1992, 1995a. With permission.

resulting in expansion that can lead to product collapse or material explosion as experienced by Venkatachalapathy and Raghavan (2000) during the MW drying of whole strawberries.

MW–hot air tunnel dryers are commercially available from various sources (Microdry Incorporated, 2007; Sairem Microwave & Radio Frequency, 2007), and the industrial equipment manufacturers have to work with their customer to adjust the equipment design to meet target product requirements and their specific limitations.

13.2.3 MW–VACUUM DRYING

Further drying improvements can be obtained by using subatmospheric pressures. Water evaporation takes place at lower temperatures under a vacuum; hence, the product processing temperature can be significantly lower, offering higher product quality. Many comparisons have been made between MW–vacuum drying and other systems, mainly hot air and freeze-drying.

MW–vacuum dehydration was first used for concentration of citrus juice. In the food industry, MW–vacuum drying is used for drying of pastas, powders, and many porous solids. The main purpose of vacuum drying is to enable the removal of moisture at a much lower temperature than the boiling point under ambient conditions. For example, the boiling point of water is reduced to 29°C at 40 mbar. A low temperature is important for many products that are heat sensitive. Furthermore, the absence of air, especially oxygen, helps preserve many of the components that are sensitive to oxidation (Regier et al. 2005).

Drouzas and Schubert (1996) investigated vacuum–MW drying of banana slices. They found the product quality was excellent in terms of taste, aroma, and rehydration. Yongsawatddigul and Gunasekaran (1996) used MW–vacuum drying to dry cranberries. Their experiment was done in pulsed and continuous mode duty cycles. The dried products were redder and had a softer texture than those dried by the conventional hot air method.

Lin et al. (1998) compared dried carrot slices using vacuum–MW drying with air and freeze-drying on the basis of rehydration potential, color, density, nutritional value, and textural properties. MW–vacuum dried carrot slices had higher rehydration potential, higher β-carotene and vitamin C content, lower density, and softer texture than those prepared by air drying. Although freeze-drying of carrot slices yielded a product with improved rehydration potential, appearance, and nutrient retention, MW–vacuum drying was rated as equal to freeze-dried. Similar results were reported by Regier et al. (2005); MW–vacuum dried carrots had the highest carotenoid retention compared to freeze-drying and convection drying. Giri and Prasad (2007) studied the drying kinetics of MW–vacuum and convective dried mushrooms. MW–vacuum was much faster than hot air drying. The drying rate was controlled by the level of applied MW power.

Adjusting the MW power or selecting an intermittent mode of application is recommended to control the product temperature during the MW–vacuum drying process. Orsat et al. (2007) found that at the vacuum level of 5.1-kPa absolute pressure, 1.5 W/g power density provides more suitable drying conditions in continuous mode for MW drying of carrots. An intermittent MW mode, operating at 90s on and

30 s off with 2.0 W/g MW at an absolute pressure of 5.1 kPa, provided the best drying conditions. These results also showed that the combination of continuous and intermittent modes is an alternate MW operating sequence that improves energy utilization and accelerates the drying rate (Orsat et al., 2007).

Böhm et al. (2006) studied the MW–vacuum drying of strawberries with the objective of maximizing their nutritional quality. Control of the process temperature at low levels (<60°C) was critical to maintain the stability of ascorbic acid and the total content of phenolic compounds.

Cui et al. (2003) conducted a comparative study for garlic drying. Freeze-drying, MW–vacuum drying, and hot air drying effects on loss of pyruvate were compared. The best dried garlic quality was obtained with freeze-drying and MW–vacuum drying as a close second. In contrast, there was a great loss in garlic pungency with the hot air dried samples. Sharma and Prasad (2006) came to a similar conclusion when comparing hot air drying with MW–convective drying of garlic. They obtained a drying time reduction of 80% with superior quality dried garlic when combining MWs at 0.4 W/g with hot air at 60 to 70°C.

In a comparative study conducted by Sunjka et al. (2004), MW–vacuum drying of cranberries produced enhanced quality characteristics when compared to MW–convective drying. The drying performance results (defined as the mass of evaporated water per unit of supplied energy) showed that MW–vacuum drying is more energy efficient than MW–convective drying.

Vacuum–MW drying was compared with freeze-drying, air drying, and a combination drying method of air drying and MW–vacuum drying for preserving the antioxidant functional properties of Saskatoon berries (Kwok et al., 2004). Air drying yielded a considerable loss of anthocyanin. The best retention was obtained with freeze-drying followed by vacuum–MW drying, combination drying, and air drying (75°C).

Commercially available MW–vacuum dryers are custom made to specific product requirements (Enwave Corporation, 2007). GEA Niro Inc. (2007) and Püschner Microwave Power Systems (2007) offer various industrial models of vacuum–MW dryers (Figures 13.2 and 13.3). Their use is predominant when oxidation of the product requires the drying to be carried out under an inert atmosphere or if the volatiles that are produced need to be recovered for economic and environmental reasons. When considering vacuum drying alone, there is a significant disadvantage in the long drying periods that require adding thermal energy, where MW is selected to push the moisture out with increased drying rates. The demands for producing higher product quality with shorter drying times for temperature-sensitive products can be addressed using MW heat treatment in combination with vacuum driers.

13.2.4 MW–Osmotic Dehydration

Osmosis is known as a partial dehydration process. Although it does not remove enough moisture from the product to be considered as a dried product, the process has the advantage of requiring little energy because the driving force pulling water out resides in the osmotic gradient created between the product and the osmotic agent.

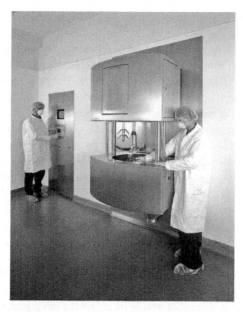

FIGURE 13.2 UltimaPro™ one-pot stirring–MW–vacuum dryer. (From GEA Niro Pharma Systems, http://www.niropharma.com. With permission.)

FIGURE 13.3 Batch MW–vacuum dryer. (From Püschner Microwave Power Systems, http://www.pueschner.com, 2007. With permission.)

It works well as a pretreatment prior to other drying processes (Barbosa-Canovas and Vega-Mercado, 1996). The application of osmotic dehydration to fruits and to a lesser extent to vegetables has received attention in recent years to reduce energy consumption and heat damage, thus allowing greater retention of nutrients (Beaudry et al., 2004; Jayaraman and Gupta, 1992; Shi et al., 1997). The protective skin of some food products impedes water transport through the surface. In these cases, pretreating the skin to increase its permeability is required before osmosis, which Sunjka and Raghavan (2004) described for the pretreatment of cranberries. Nsonzi and Ramaswamy (1998) modeled the mass transfer process with respect to moisture loss and solids gain for the osmotic concentration of blueberries. They stated that, even though the moisture loss and solids gain occurred at the same time, the rate of moisture loss was much higher than the rate of solids gain. Lenart (1996) described the main advantages of using osmotic dehydration as the reduction of the process temperature, sweeter taste of the dehydrated product, a 20 to 30% reduction in energy consumption when carbohydrates were used as the osmotic agent, and a shorter drying time following osmotic dehydration.

Beaudry et al. (2004) studied the effect of MW–convective (0.7 W/g and 62°C), hot air (62°C), freeze-drying, and vacuum drying (94.6 kPa) methods on the quality of osmotically dehydrated cranberries. With all of the drying methods, they reported that the constant drying rate period is no longer present following osmotic dehydration. Piotrowski et al. (2004) also reported this effect for strawberries. The fastest drying method was MW–convective, and the longest was hot air drying at 62°C. All dried samples were judged acceptable by sensory evaluation, and the texture of the MW dried ones scored the closest to commercially available dried cranberries (Beaudry et al., 2004). Venkatachalapathy and Raghavan (1999) reported that combined osmotic–MW dried strawberries were close to the freeze-dried product in terms of rehydration characteristics and overall sensory evaluation.

13.3   RF DRYING

In an RF heating system, the RF generator creates an alternating electric field between two electrodes. The electrode configuration is designed according to the characteristics of the product being processed (Orsat and Raghavan, 2004). The material is placed between the electrodes where the alternating energy causes polarization, and the molecules in the material continuously reorient themselves to face opposite poles. When the electric field is alternating at RFs, for example, 27.12 MHz, the electric field alternates 27,120,000 cycles/s. The friction resulting from the rotational movement of the molecules and the space charge displacement causes the material to be rapidly heated throughout its mass. The amount of heat generated in the product is determined by the frequency, the square of the applied voltage, the dimensions of the product, and the dielectric loss factor of the material [Eq. (13.1)].

Industrial development that has been undertaken thus far in RF drying applications is limited because of the small size of the corresponding equipment manufacturing industry and the limited market demand, causing the investment costs to remain high. Nonetheless, there are some well-known applications of RF energy in drying, especially in wood (Avramidis and Zwick, 1996), textiles (Pai et al., 1989), and postbaking drying (Mermelstein, 1998). The most successful applications often combine two or more drying techniques (RF–heat pump, forced air with RF, etc.).

Murphy et al. (1992) conducted experiments on RF drying (27 MHz) of alfalfa. To dry alfalfa from 80 to 12% moisture content requires the removal of 3.5 kg water/1.0 kg dried alfalfa. Assuming that the water removed is free water, the alfalfa is initially at 20°C, and the water is vaporized at 100°C, the energy required is 8.8 MJ or 2.4 kW h. Typically, the efficiency of RF ovens is about 50 to 60%, so 4.8 kW h of input energy would be required to produce 1kg of dried alfalfa. At a rate of $0.04/kW h, the energy costs alone would amount to $200/t dried alfalfa, which is more than the current market value. Hence, the only economically feasible application of RF power to dry alfalfa is as a supplementary source rather than a major power source, perhaps in the final stages of drying for moisture leveling.

RF drying has mainly been used for postbake drying of cookies, crackers, and pasta (Mermelstein, 1998; Union Internationale d'Électrothermie, 1992). Cookies and crackers, fresh out of the oven, have a nonuniform moisture distribution that may yield to cracking during handling. RF heating can help even out the moisture distribution after baking by targeting the remaining moisture pockets. Recent development trends are in investigating hybrid drying systems involving RF to cater to the special needs of heat-sensitive foodstuffs (Chou and Chua, 2001; Vega-Mercado et al., 2001; Zhao et al., 2000).

The Radio Frequency Macrowave Company produces a line of RF dryers operating at 40 MHz (Figure 13.4). Their dryers are designed with programmable logic controllers to provide complete diagnostics and centralized control of all dryer functions. The macrowave heat reclamation system uses waste heat from the RF power generator to scavenge moisture and increase drying efficiency.

13.4 IR DRYING

IR radiation falls in the wavelength range of 0.75 to 100 μm; hence, solid bodies will generally absorb the radiation in a narrow layer at the surface. Therefore, for a successful IR drying application, the radiation properties of both the radiator and the material must be matched (Ratti and Mujumdar, 2007). The emissivity, absorptivity, reflectivity, and transmissivity are the properties of interest. IR heating is used in industry principally for surface drying or thin layer drying. IR drying offers high efficiency of conversion of the electrical energy into heat (average 80% efficiency) in addition to being a simple, inexpensive and easy to control technology. The limitation is that applications are more or less limited to thin layers of material. IR dryers are commonly designed to be integrated with conventional dryers (conveyor hot air dryers), where standard IR radiators are used in order to ensure that the IR radiation is directed on the product being dried. The design must take into account the product's radiation properties and the efficiency of conversion from electric to IR energy for the emitters and radiators to optimize the process. The cavity must be equipped with reflectors and adequate air movement and exhaust to remove moisture, gases, and so forth. The heating and drying chambers require some particular design components. The inner surface of the cavity should have a mirrored polished finish to ensure maximum reflection of the radiation onto the bed of material. The IR emitters

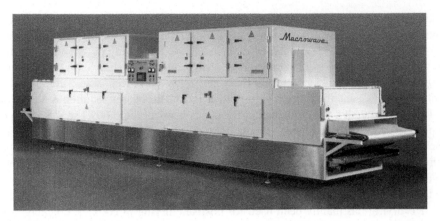

FIGURE 13.4 The 7000 Series Macrowave dryers. (From Radio Frequency Macrowave Company, http://www.radiofrequency.com/products/mac7000f.html, 2007. With permission.)

can either be electrical or gas fired. The efficiency of the IR heating system depends on the coupling of the heat source with the material because the heat conversion only occurs if the material absorbs the radiation. The absorption range for food materials is generally between 2.5 and 3.0 μm.

The combination of IR and hot air provides a synergistic effect, resulting in greater efficiency with a higher rate of mass transfer. When a material is exposed to IR radiation, there is increased molecular vibration in the material at the inner surface layers. The rate of moisture movement in the material toward the surface is increased and the convective air removes this moisture from the surface, reducing the product's temperature for improved product quality (Hebbar et al., 2004). Combination IR and hot air drying performed better than either IR (17 kW/m², 2.4 to 3 μm) or hot air alone (80°C, 1 m/s) for drying carrots and potatoes (Hebbar et al., 2004). Table 13.2 presents the specific energy consumption for these types of drying as reported by Hebbar et al. (2004).

TABLE 13.2
Specific Energy Consumption during Hot Air, IR, and Combined IR and Hot Air Drying of Potatoes and Carrots

	Specific Energy (MJ/kg) of Evaporated Water	
Drying Mode	Potato Drying	Carrot Drying
Hot air	17.17	16.15
IR	7.60	7.15
IR and hot air	6.43	6.04

Source: From Hebbar, H.U. et al., *J. Food Eng.*, 65, 557–563, 2004. With permission.

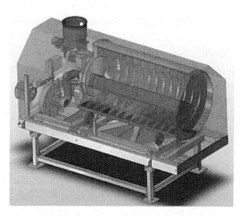

FIGURE 13.5 Kreyenborg Infrared Rapid Dryer. (From Kreyenborg GmbH, http://www.kreyenborg.com/en/kreyenborg/index.php, 2007. With permission.)

Pathare and Sharma (2006) conducted an experiment on IR–convective drying of onion slices. The operating parameters were IR intensities of 26 to 44 kW/m², air temperatures ranging between 30 and 60°C, and air velocities from 0 to 1.5 m/s. The drying occurred entirely in the falling rate period, and the drying rate increased with an increase in process intensity with a dominant effect from an increase in IR intensity.

Tireki et al. (2006) studied the combined near-IR–MW drying of breadcrumbs. MW inputs of 210, 350, and 490 W/20 g samples and near-IR heating in the 0.7 to 1.1 µm wavelength were tested. Halogen lamps (1350, 2250, and 3150 W) were used. Combining the processes shortened the drying time while ensuring product quality. Wang and Sheng (2006) investigated sequential far-IR (>25 µm) and MW drying of peaches. For MW drying, power intensities of 0.5, 0.7, and 1 kW/kg were studied with a 1 m/s flow of air. The far-IR heating operated at 0.5, 0.7, and 1 kW/kg with a 1 m/s flow of air. The drying rate increased with increasing intensity, but MW drying achieved greater drying rates at lower energy consumption than IR drying (Wang and Sheng, 2006).

The Kreyenborg GmbH (2007) offers a line of IR dryers. The drying chamber consists of a screw coil that conveys the drying material lengthwise through the system (Figure 13.5). The residence time is determined by the speed of the screw conveyor. The product temperature is permanently controlled, and the radiators' capacity is readjusted if necessary. The material is stirred by the rotation of the drum and at the same time permanently irradiated by the IR radiation, ensuring uniformity of the process. The wavelength of the IR rays is adjusted according to the process parameters. The radiation is directly absorbed by the bulk material and vaporizes the water to the surface of the material.

13.5 HEAT PUMP DRYING

Heat pumps have been studied for use on farms since the early 1950s and have found some applications in various sectors like dairy, grain drying, timber drying, and so forth. Heat pump dryers have many advantages compared to conventional systems such as improved product quality due to mild drying conditions, reduced energy consumption, and reduced environmental incidence. An electrically driven heat pump shares the advantage of low temperature drying and manifests the same desirable characteristics of resistance heat, yet is considerably more energy efficient (Strommen and Jonassen, 1996).

In drying, there are two distinctly different types of moisture removal: one in which the problem is to eliminate airborne water vapor and another in which liquid water is to be extracted from some material. In the first, the heat has already been supplied from somewhere to evaporate the water and increase the air humidity; in the second, the water is still present as liquid, and part of the drying process is to supply the heat to vaporize it (Strommen and Kramer, 1994). The simplest drying method is to blow heated air over the moist material and vent the moist air to the atmosphere. All of the latent heat carried in the water vapor is lost to the atmosphere. In a heat pump, the moisture in the exhaust is condensed at the evaporator and the latent heat is recovered to the circulating air stream via the condenser (Heap, 1979). The heat pump, in a closed cycle, reclaims heat and uses it at a cost equal to that of operating the compressor motor.

A heat pump dehumidifier works on the principle of refrigeration to cool an air stream and condense the water contained in it. This not only renders the air dry but also recovers the latent heat of evaporation represented by the water vapor removed, which can be routed back to the air, thus increasing its capacity to pick up more moisture. This technology is ideal for application in the drying sector where the purpose is removal of water at a minimal energy cost. The introduction of the dehumidifier into the process enhances the energy efficiency by recovering the latent heat that is exhausted to air in a conventional process. The use of MW heating provides an efficient supply of energy required to speed up the process of moisture rejection from the material undergoing drying. This component overcomes some of the heat and mass transfer drawbacks of convective air drying.

The schematic presented in Figure 13.6 shows the arrangement of a typical heat pump system as described by Sosle et al. (2003). The drying material is placed on perforated trays stacked on a platform in the drying chamber. The platform rests on load cells that monitor the mass of the load. There is an electrical heater that is used to control the temperature of the drying air. The heat pump itself has a centrifugal fan that can be run independently of the compressor, and this fan drives the air in the loop. The heat pump unit is equipped with a secondary water-cooled condenser. There is a butterfly valve in the circuit and two dampers to switch between closed and semiclosed modes of air circulation.

The air circuit can be operated in a completely closed, partially closed, or completely open mode. In the closed mode, an initial quantity of air is introduced into the system and recirculated throughout the process. In the other two modes, there is partial or complete discharge–intake of the air that acts as the working fluid for the drying system. The primary condenser is located in the path of the air stream, which implies that any heat rejection over this retains the energy within the system, whereas the secondary water-cooled coil channels the energy out of the system to the environment.

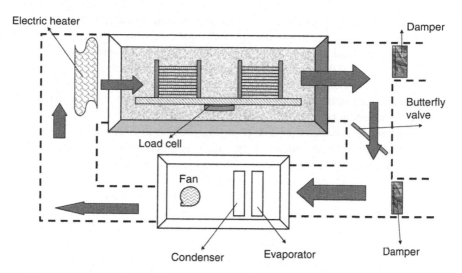

FIGURE 13.6 Schematic of a heat pump dryer. (From Sosle, V. et al., *Dry. Technol.*, 21, 539–554, 2003. With permission.)

Various agriproducts such as apple, banana, onion, and potato rings were heat pump dried by Sosle et al. (2003). In most cases, the products were dried from an initial moisture content of 85 to 90% wet basis to a final 10 to 12% wet basis. The products obtained by the process were of exceptionally high quality. The color degradation was negligible and the flavor was free from any process-induced aberrations (Sosle et al., 2003). Their observations showed a distinct advantage of the novel drying process over conventional methods in preserving the integrity of the cellular structure of the plant material. The low temperature drying that was achieved was less destructive toward the cell walls and led to a gradual, gentle collapse of the cells, which recovered their turgor remarkably upon rehydration (Sosle et al., 2003).

Cardona et al. (2002) studied heat pump drying of lactic acid bacteria. They obtained high-quality dried products with the use of dehydroprotectants and nitrogen purging.

Hawlader et al. (2006) compared the loss of gingerol when ginger was dried by air drying, modified atmosphere heat pump drying, freeze-drying, and vacuum drying. Using inert gases such as N_2 and CO_2, the modified atmosphere heat pump drying process improved the quality and gingerol retention of dried ginger better than freeze-drying.

13.6 CONCLUSION

The trend in recent years has been concentrated on the incorporation of multiple transfer mechanisms and energy sources. Developments are reported for heat pump, IR, RF, and MW dryers.

Because of their selective and volumetric heating effects, MWs bring new characteristics to various bioprocessing techniques such as an increased rate of drying, enhanced final product quality, and improved energy consumption.

The quality of MW dried commodities is often between air dried and freeze-dried products. The rapidity of the process yields better color and retention of aroma. Quality is further improved when a vacuum is used because the thermal and oxidative stresses are reduced.

The operating conditions of a MW drying process are dependent on the commodity. In some cases, an osmotic pretreatment is recommended (as with cranberries) or the process should be combined with either hot air (for moisture removal) or a vacuum (to keep the temperature to a minimum).

Because electrotechnologies are based on electric power sources, which are free of chemicals, they provide a more ecological and stable environment cost than a fuel-based conventional processing method. Past research successfully demonstrated that electromagnetic energy can be used efficiently at a fraction of the energy consumption of conventional equipment presently found in the agrifood sector.

The development of new applications for MW, RF, IR, and electroheating of materials has been steadily increasing over the years. This is a high tech growth industry with much untapped potential, and it has recently been adapted to biomedical needs. We can expect this area to become more important in parallel with the increased development of sources of electrical energy, whether they are hydro, solar, wind, or nuclear driven. Furthermore, the increased pressure to reduce combustion of fossil fuels is forcing the private sector to investigate alternate sources of heating for a wide variety of processes that require thermal input.

NOMENCLATURE

P	energy per unit volume (W/m^3)
E	electric field strength (V/m)
ε_0	absolute permittivity of vacuum (8.854188 × 10^{-12} F/m)
ε'	dielectric constant (relative real permittivity, F/m)
ε''	dielectric loss factor (relative dielectric loss factor, F/m)
f	frequency (Hz)
$\tan \delta = \varepsilon''/\varepsilon'$	dielectric loss tangent

REFERENCES

Avramidis, S. and Zwick, R.L., Commercial-scale RF/V drying of softwood lumber. 2: Drying characteristics and lumber quality, *Forest Prod. J.*, 46, 27–36, 1996.

Barbosa-Canovas, S.G. and Vega-Mercado, H., *Dehydration of Foods*, Chapman & Hall, New York, 1996.

Beaudry, C., Raghavan, G.S.V., Ratti, C., and Rennie, T.J., Effect of four drying methods on the quality of osmotically dehydrated cranberries, *Dry. Technol.*, 22, 521–539, 2004.

Böhm V., Kühnert, S., Rohm, H., and Scholze, G., Improving the nutritional quality of microwave–vacuum dried strawberries: A preliminary study, *Food Sci. Technol. Int.*, 12, 67–75, 2006.

Cardona, T.D., Driscoll, R.H., Paterson, J.L., Srzednicki, G.S., and Kim, W.S., Optimizing conditions for heat pump dehydration of lactic acid bacteria, *Dry. Technol.*, 20, 1611–1632, 2002.

Changrue, V., Sunjka, P.S., Gariepy, Y., Raghavan, G.S.V., and Wang, N., Real-time control of microwave drying process, in *Drying 2004, Proceedings of the 14th International Drying Symposium*, Silv, M.A. and Rocha, S.C.S., Eds., State University of Campinas, Campinas, Brazil, 2004, pp. 941–948.

Chou, S.K. and Chua, K.J., New hybrid drying technologies for heat sensitive foodstuffs, *Trends Food Sci. Technol.*, 12, 359–369, 2001.

Chou, S.K. and Chua, K.J., Heat pump drying systems, in *Handbook of Industrial Drying*, 3rd ed., Mujumdar, A.S., Ed., CRC Press, Boca Raton, FL, 2007, pp. 1103–1131.

Chua, K.J., Mujumdar, A.S., Chou, S.K., Hawlader, M.N.A., and Ho, J.C., Convective drying of banana, guava and potato pieces: Effect of cyclical variations of air temperature on drying kinetics and color change, *Dry. Technol.*, 18, 907–936, 2000.

Cui, Z.-W., Xu, S.-Y., and Sun, D.-W., Dehydration of garlic slices by combined microwave vacuum and air drying, *Dry. Technol.*, 21, 1173–1184, 2003.

Decareau, R.V., *Microwaves in the Food Processing Industry*, Academic Press, Orlando, FL, 1985.

Demeczky, M., Application of dielectric techniques in food production and preservation, in *Developments in Food Preservation*, Thorne, S., Ed., Elsevier, London, 1985, pp. 127–181.

Drouzas, A.E. and Schubert, H., Microwave application in vacuum drying of fruit, *J. Food Eng.*, 28, 203–209, 1996.

EnWave Corporation, http://www.enwave.net, 2007.

Erle, U., Drying using microwave processing, in *The Microwave Processing of Foods*, Schubert, H. and Regier, M., Eds., Woodhead Publishing, Cambridge, U.K., 2005, pp. 142–152.

GEA Niro Pharma Systems, http://www.niropharma.com, 2007.

Giri, S.K. and Prasad, S., Drying kinetics and rehydration characteristics of microwave–vacuum and convective hot air dried mushrooms, *J. Food Eng.*, 78, 512–521, 2007.

Goullieux, A. and Pain, J.-P., Ohmic heating, in *Emerging Technologies for Food Processing*, Sun, D.W., Ed., Elsevier Ltd., London, 2005, pp. 469–505.

Hawlader, M.N.A., Perera, C.O., and Tian, M., Comparison of the retention of 6-gingerol in drying of ginger under modified atmosphere heat pump drying and other drying methods, *Dry. Technol.*, 24, 51–56, 2006.

Heap, R.D., *Heat Pumps*, E. & F.N. Spon, London, 1979.

Hebbar, H.U., Vishwanathan, K.H., and Ramesh, M.N., Development of combined infrared and hot air dryer for vegetables, *J. Food Eng.*, 65, 557–563, 2004.

Ikediala, J.N., Hansen, J.D., Tang, J., Drake, S.R., and Wang, S., Development of a saline water immersion technique with RF energy as a postharvest treatment against codling moth in cherries, *Postharv. Biol. Technol.*, 24, 25–37, 2002.

Jayaraman, K.S. and Gupta, D.K., Dehydration of fruit and vegetable—Recent developments in principles and techniques, *Dry. Technol.*, 10, 1–50, 1992.

Jones, P.L., Radio-frequency processing in Europe, *J. Microwave Power Electromagn. Energy*, 22, 143–153, 1987.

Kreyenborg GmbH, http://www.kreyenborg.com/en/kreyenborg/index.php, 2007.

Kudra, T., Raghavan, G.S.V., Akyel, C., Bosisio, R., and van de Voort, F., Electromagnetic properties of milk and its constituents at 2.45 GHz, *J. Microwave Power Electromagn. Energy*, 27, 199–203, 1992.

Kwok, B.H.L., Hu, C., Durance, T., and Kitts, D.D., Dehydration techniques affect phytochemical contents and free radical scavenging activities of Saskatoon berries (*Amelanchier alnifolia* Nutt.), *J. Food Sci.*, 69, S122–S126, 2004.

Lenart, A., Osmo-convective drying of fruits and vegetables: Technology and application, *Dry. Technol.*, 14, 391–413, 1996.

Lin, T.M., Durance, T.D., and Scaman, C.H., Characterization of vacuum microwave, air and freeze dried carrot slices, *Food Res. Int.*, 31, 111–117, 1998.

Mermelstein, N.H., Microwave and radio frequency drying, *Food Technol.*, 52, 84–85, 1998.

Metaxas, A.C., *Foundations of Electroheat: A Unified Approach*, John Wiley & Sons Ltd., Chichester, U.K., 1996.

Microdry Incorporated, http://www.microdry.com, 2007.

Mudgett, R.E., Electrical properties of foods, in *Microwave in the Food Processing Industry*, Decareau, R., Ed., Academic Press, London, 1985.

Mudgett, R.E., Microwave properties and heating characteristics of foods, *Food Technol.*, 40, 84–93, 1986.

Mujumdar, A.S., Ed., *Handbook of Industrial Drying*, 3d ed., CRC Press, Boca Raton, FL, 2007, p. 1280.

Murphy, A., Morrow, R., and Besley, L., Combined radio-frequency and forced-air drying of alfalfa, *J. Microwave Power Electromagn. Energy*, 27, 223–232, 1992.

Nelson, S.O., Electrical properties of agricultural products—A critical review, *Trans. ASAE*, 16, 384–400, 1973.

Nemes, S.M., Sivakumar, S.S., Gariepy, Y., and Raghavan, G.S.V., Microwave vacuum drying of banana slices under different power levels, in *Proceedings of the 3rd Inter-American Drying Conference (IADC 2005)*, Orsat, V., Raghavan, G.S.V., and Kudra, T., Eds., McGill University, Montreal, 2005.

Nsonzi, F. and Ramaswamy, H.S., Osmotic dehydration kinetics of blueberries, *Dry. Technol.*, 16, 725–741, 1998.

Orsat, V. and Raghavan, G.S.V., Review of the design aspects of radio-frequency heating for development of agri-food applications, in *Dehydration of Products of Biological Origin*, Mujumdar, A.S., Ed., Science Publishers, Enfield, NH, 2004, pp. 467–492.

Orsat, V. and Raghavan, G.S.V., Radio frequency processing, in *Emerging Technologies for Food Processing*, Sun, D.W., Ed., Elsevier Ltd., London, 2005, pp. 445–468.

Orsat, V., Raghavan, G.S.V., and Meda, V., Microwave technology for food processing: An overview, in *The Microwave Processing of Foods*, Schubert, H. and Regier, M.M., Eds., Woodhead Publishing, Cambridge, U.K., 2005, pp. 105–118.

Orsat, V., Yang, W., Changrue, V., and Raghavan, G.S.V., Microwave-assisted drying of biomaterials, *Food Bioprod. Process.*, 85, 1–9, 2007.

Osepchuk, J.M., Microwave power applications, *IEEE Trans. Microwave Theory Tech.*, 50, 975–985, 2002.

Pai, G.A., Mock, G.N., Grady, P.L., Graham, R.W., Crabtree, K.K., and Moore, E.J., Radio frequency drying in the textile industry, paper presented at the IEEE 1989 Annual Textile Industry Technical Conference, paper 89CH2697-1 P9/1-9/3, 1989.

Pathare, P.B. and Sharma, G.P., Effective moisture diffusivity of onion slices undergoing infrared convective drying, *Biosyst. Eng.*, 93, 285–291, 2006.

Piotrowski, D., Lenart, A., and Wardzynski, A., Influence of osmotic dehydration on micro-wave–convective drying of frozen strawberries, *J. Food Eng.*, 65, 519–525, 2004.

Piyasena, P., Dussault, C., Koutchma, T., Ramaswamy, H.S., and Awuah, G.B., Radio-frequency heating of foods: Principles, applications and related properties—A review, *Crit. Rev. Food Sci. Nutr.*, 43, 587–606, 2003.

Püschner Microwave Power Systems, http://www.pueschner.com, 2007.

Raghavan, G.S.V., Rennie, T.J., Sunjka, P.S., Orsat, V., Phaphuangwittayakul, W., and Terdtoon, P., Overview of new techniques for drying biological materials with emphasis on energy aspects, *Brazil. J. Chem. Eng.*, 22, 195–201, 2005.

Ratti, C. and Mujumdar, A.S., Infrared drying, in *Handbook of Industrial Drying*, 3rd ed., Mujumdar, A.S., Ed., CRC Press, Boca Raton, FL, 2007, pp. 423–438.

Regier, M., Mayer-Miebach, E., Behsnilian, D., Neff, E., and Schuchmann, H.P., Influences of drying and storage of lycopene-rich carrots on the carotenoid content, *Dry. Technol.*, 23, 989–998, 2005.

Sairem Microwave & Radio Frequency, http://www.sairem.com, 2007.

Schiffmann, R.F., Microwave and dielectric drying, in *Handbook of Industrial Drying*, 3rd ed., Mujumdar, A.S., Ed., CRC Press, Boca Raton, FL, 2007, pp. 285–305.

Sharma, G.P. and Prasad, S., Optimization of process parameters for microwave drying of garlic cloves, *J. Food Eng.*, 75, 441–446, 2006.

Shi, J.X., Le Maguer, M., Wang, S.L., and Liptay, A., Application of osmotic treatment in tomato processing—Effect of skin treatments on mass transfer in osmotic dehydration of tomatoes, *Food Res. Int.*, 30, 669–674, 1997.

Sosle, V., Heat pump drying, *Stewart Postharv. Rev.*, 2, 3, 2006.

Sosle, V., Raghavan, G.S.V., and Kittler, R., Low-temperature drying using a versatile heat pump dehumidifier, *Dry. Technol.*, 21, 539–554, 2003.

Soysal, Y., Oztekin, S., and Eren, O., Microwave drying of parsley modeling, kinetics and energy aspects, *Biosyst. Eng.*, 93, 403–413, 2006.

Strommen, I. and Jonassen, O., Performance tests of a new 2-stage counter-current heat pump fluidized bed dryer, in *Drying '96, Proceedings of IDS 96*, Vol. A, Strumillo, C. and Pakowski, Z., Eds., Lodz Technical University Publisher, Lodz, Poland, 1996, pp. 563–568.

Strommen, I. and Kramer, K., New applications of heat pumps in drying processes, *Dry. Technol.*, 12, 889–901, 1994.

Sunjka, P.S. and Raghavan, G.S.V., Assessment of pretreatment methods and osmotic dehydration of cranberries, *Can. Biosyst. Eng.*, 4, 3.35–3.40, 2004.

Sunjka, P.S., Rennie, T.J., Beaudry, C., and Raghavan, G.S.V., Microwave/convective and microwave/vacuum drying of cranberries: A comparative study, *Dry. Technol.*, 22, 1217–1231, 2004.

Tireki, S., Sumnu, G., and Esin, A., Production of breadcrumb by infrared-assisted microwave drying, *Eur. Food Res. Technol.*, 222, 8–14, 2006.

Tulasidas, T.N., Raghavan, G.S.V., and Mujumdar, A.S., Microwave drying of grapes in a single mode cavity at 2450 MHz—II: Quality and energy aspects, *Dry. Technol.*, 13, 1973–1992, 1995a.

Tulasidas, T.N., Raghavan, G.S.V., and Mujumdar, A.S., Microwave drying of grapes in a single mode cavity at 2450MHz. I. Drying kinetics, *Dry. Technol.*, 13, 1949–1971, 1995b.

Tulasidas, T.N., Raghavan, G.S.V., van de Voort, F., and Girard, R., Dielectric properties of grapes and sugar solutions at 2.45 GHz, *J. Microwave Power Electromagn. Energy*, 30, 117–123, 1995.

Union Internationale d'Électrothermie (International Union of Electroheat), Dielectric heating for industrial processes, UIE Working Group paper, La Defense, Paris, 1992.

Vega-Mercado, H., Gongora-Nieto, M.M., and Barbosa-Canovas, G.V., Advances in dehydration of foods, *J. Food Eng.*, 49, 271–289, 2001.

Venkatachalapathy, K. and Raghavan, G.S.V., Combined osmotic and microwave drying of strawberry, *Dry. Technol.*, 17, 837–853, 1999.

Venkatachalapathy, K. and Raghavan, G.S.V., Microwave drying of whole, sliced and pureed strawberries, *Agric. Eng. J.*, 9, 29–39, 2000.

Venkatesh M.S. and Raghavan, G.S.V., An overview of microwave processing and dielectric properties of agri food materials, *Biosyst. Eng.*, 88, 1–18, 2004.

Venkatesh, M.S. and Raghavan, G.S.V., An overview of dielectric properties measuring techniques, *Can. Biosyst. Eng.*, 47, 15–30, 2005.

Wang, J. and Sheng, K., Far-infrared and microwave drying of peach, *Lebensm.-Wiss. Technol.*, 39, 247–255, 2006.

Wang, J. and Xi, Y.S., Drying characteristics and drying quality of carrot using a two-stage microwave process, *J. Food Eng.*, 68, 505–511, 2005.

Wang, S., Tang, J., Johnson, J.A., Mitcham, E., Hansen, J.D., Hallman, G., Drake, S.R., and Wang, Y., Dielectric properties of fruits and insect pests as related to radio-frequency and microwave treatments, *Biosyst. Eng.*, 85, 201–212, 2003.

Wang, Y., Wig, T.D., Tang, J., and Hallberg, L.M., Dielectric properties of foods relevant to RF and microwave pasteurization and sterilization, *J. Food Eng.*, 57, 257–268, 2003.

Yanyang, X., Min, Z., Mujumdar, A.S., Le-qun, Z., and Jian-cai, S., Studies on hot air and microwave vacuum drying of wild cabbage, *Dry. Technol.*, 22, 2201–2209, 2004.

Yongsawatddigul, J. and Gunasekaran, S., Microwave–vacuum-drying of cranberries: Part I: Energy use and efficiency. Part II: Quality evaluation, *J. Food Process. Preserv.*, 20, 121–156, 1996.

Zhao, Y., Flugstad, B., Kolbe, E., Park, J.W., and Wells, J.H., Using capacitive (radio frequency) dielectric heating in food processing and preservation—A review, *J. Food Process Eng.*, 23, 25–55, 2000.

14 Energy Aspects in Food Dehydration

Tadeusz Kudra

CONTENTS

14.1 INTRODUCTION

Thermal drying has long been recognized as one of the most energy-intensive operations in solids–liquid separation. The share of drying energy in industrial energy use ranges from 10 to 25%, although in some industrial sectors it is much higher, reaching 35% in paper production and even 50% in finishing of textile fabrics (Kudra, 2004; Mujumdar, 2007). It is difficult to estimate drying energy use in the agrifood sector because of the diversity of food products, multiplicity of food producers, and frequent use of process coproducts as supplementary fuel. Nevertheless, 12% was quoted for England and 27% for France over two decades ago (Strumillo et al., 2007). At present, higher numbers can be anticipated because the percentage of drying energy is progressively increasing (Baker, 2005). For example, Kemp

423

(1998) indicates that the drying proportion of the U.K. industrial energy consumption has risen by 4% over 12 years. Because almost 99% of applications involve removal of water (Mujumdar, 2007), such high energy consumption is primarily due to the latent heat of water vaporization, which is 2500 kJ/kg at 0°C. The fraction of thermal energy can be favorably reduced by upstream mechanical separation or nonthermal dewatering such as membrane processing and osmotic dehydration. An interesting option for drying certain fruits such as cranberries is osmotic dehydration prior to thermal drying because such a hybrid technology not only reduces energy consumption but also offers a product of enhanced quality (Grabowski et al., 2002). Basically, inefficient convective dryers, which account for about 85% of all industrial dryers (Mujumdar, 2007), are the second contributor to the high energy demands for drying. Thus, concentrating the liquid feed by evaporation prior to spray drying can notably reduce the heat load to the dryer because the energy demand for evaporation ranges from 360 to 3600 kJ/kg versus 2900 to 5400 kJ/kg for a single spray dryer, even though higher values were also reported, such as 9040 kJ/kg for spray drying of sweet whey (Trägårdh, 1986).

Except for certain technologies such as contact drying, dielectric drying, sorption drying, freeze-drying, or drying on inert particles (Kudra and Mujumdar, 2002), the majority of foodstuffs are convection dried either by dispersing liquid materials into a hot air stream as in spray drying or by blowing hot air through a wet granular material as in rotary, fluidized bed, or belt dryers. Hot air drying requires high energy input because of inefficient air–material heat transfer and the significant amount of energy lost with the exhaust air, even if its temperature approaches the wet bulb temperature. Various measures, such as partial recycling of exhaust gas or heat recovery, can increase the overall energy efficiency of the dryer. However, the energy cost in terms of heat represents a significant percentage of the total drying costs of food and agriproducts, especially when using inefficient methods such as indirect heating of the drying air.

Energy consumption can vary considerably, depending on the physical and thermal characteristics of the material, product quality requirements, and drying technology. An example of a fairly energy efficient process is spray drying of milk, in which raw milk concentrated by evaporation to approximately 50% solids is then dried to a final moisture content of 3 to 4% wet basis (w.b.). The specific energy consumption amounts to 5300 kJ/kg water evaporated in a single-stage concurrent spray dryer (Tang et al., 1999). This yields an energy efficiency (defined as the ratio of energy needed for moisture evaporation to the energy supplied to the dryer) of about 47%. The use of two-stage and three-stage drying systems reduces specific energy consumption to 4500 and 4000 kJ/kg H_2O, respectively. Thus, replacing the most frequently used two-stage dryers with three-stage ones could reduce the energy demand by 465 kJ/kg milk powder. In Canada, which produced 72,420 tonnes of skim milk powder in 2006 (Canadian Dairy Information Centre, 2007), such technology upgrades would result in appreciable energy savings.

An example of a highly energy intensive process is the drying of ginseng root. The initial moisture content of ~70% w.b. should be reduced to <10% w.b. for safe storage. Typically, the whole roots are placed on perforated trays stacked in a heated chamber and dried by hot air flowing through the trays. At 38°C, which is the optimum temperature for quality product, the drying time is on the order of 240 to 280 h. Under such conditions the specific energy consumption is close to 40,000 kJ/kg evaporated water (Reynolds, 1998), although twofold higher values were calculated based on the energy use by ginseng farmers. Accepting an activation energy of 3650 kJ/kg water at 38°C (Martynenko, 2007), the energy efficiency ranges from 5 to 9%. The annual ginseng production in Canada amounts to 2510 tonnes, so the energy spent for drying is on the order of 2×10^{11} kJ/year.

These two extreme examples clearly indicate that energy consumption is an important issue in the drying of constantly increasing food production, even for materials that contain significant amounts of easily removed water and that are dried in relatively efficient dryers. The attempts to improve energy efficiency start with an energy audit for a dryer in use followed by a thorough analysis of the existing literature and technical information to identify options to reduce energy consumption. However, the reported data should be considered with caution: inconsistent terminology and calculation methods are found that, although starting with apparently identical indices, can lead to drastically different results (Kudra, 2004). For example, if a given index is not explicitly defined, care should be taken to not confuse drying efficiency with energy efficiency, because the latter generally yields lower values.

This chapter provides a concise overview of the most frequently used energy indices in order to establish a common platform for various measures of energy use, permitting the objective comparison of various dryers and a credible assessment of options for reducing energy consumption.

14.2 ENERGY MEASURES

14.2.1 ENERGY AND THERMAL EFFICIENCY

The performance of a drying system is characterized by various indices, including energy efficiency, thermal efficiency, volumetric evaporation rate, specific heat consumption, surface heat losses, unit steam consumption, and so forth, that are defined to reflect the particularities of various drying processes such as intermittent drying, microwave-assisted drying, or heat pump drying. Of these indices, energy efficiency and specific heat consumption are most frequently used to assess dryer performance from the energy viewpoint.

The energy efficiency (η) relates the energy used for moisture evaporation at the feed temperature (E_{ev}) to the total energy supplied to the dryer (E_t):

$$\eta = \frac{E_{ev}}{E_t} \tag{14.1}$$

The energy needed for moisture evaporation is commonly calculated from the mass of evaporated water and the latent heat of vaporization (ΔH). The calculation of the mass of the evaporated water can be based on either the initial and final material moisture content or the inlet and outlet air (drying agent) humidity. Because the material temperature is rarely known, the latent heat of vaporization is taken as the feed temperature, although some calculations of energy efficiency are based on the latent heat of vaporization at 100°C. However, utilizing the feed temperature is more accurate because during convective drying the material temperature is frequently equal or close to the wet bulb temperature. Thus, the difference between both temperatures is much smaller than between 100°C and the wet bulb temperature, even at higher drying temperatures.

The assumption on evaporation from a free water surface penalizes a majority of foods because they dry mostly in the falling drying rate period. Therefore, it is worthwhile to consider another measure of energy in the numerator of Eq. (14.1) such as the sorption energy.

Thermal energy supplied to the external or internal heater or combustion heat released in the burner is a major part of the total energy supplied to the dryer. For some dryers, however, additional energy inputs cannot be neglected. This is the case in the electrical energy needed to run a microwave generator in microwave-assisted drying; additional energy is necessary to vibrate or rotate the dryer, disperse the feed, and stir the product. A notable fraction of energy can be needed to run the blowers that force an air stream through the drying system not only to remove the evaporated moisture but also to provide the required hydrodynamic conditions, especially in dryers with active hydrodynamics (progressive dryers) such as fluidized bed dryers, spouted bed dryers, pneumatic dryers, or dryers with impinging streams.

Regarding only the energy used to heat drying air that enters a single pass, theoretical convective dryer, Eq. (14.1) can be rewritten as

$$\eta^t = \frac{\Delta H}{W^*(I_1 - I_a)} = \frac{\Delta H \, (Y_2 - Y_a)}{I_1 - I_a} = \frac{\Delta H (Y_2 - Y_a)}{I_2 - I_a} \tag{14.2}$$

where Y_2 and Y_a are the outlet and ambient air humidities, respectively; I_1 and I_a are the inlet and ambient enthalpies, respectively; and W^* is the specific air consumption given as the mass of air required to evaporate 1 kg of water (Strumillo and Kudra, 1986):

$$W^* = \frac{W_g}{W_{ev}} = \frac{1}{Y_2 - Y_a} \tag{14.3}$$

where W_g and W_{ev} are the gas and evaporation flow rates, respectively.

A theoretical dryer is defined as a dryer with no heat spent for heating the material and transportation equipment, heat is not supplied to the internal heater, heat is not lost to the ambient atmosphere, and the inlet material temperature is 0°C. This means that energy supplied to the external heater is used solely to increase the enthalpy of the exit air (Strumillo and Kudra, 1986). A difference in the heat consumption in the external heater for the real and theoretical dryers quantifies the heat losses (ΣQ_l) per kilogram of evaporated water. Thus, the energy efficiency of a real dryer can be determined from the following relation:

$$\eta = \frac{\Delta H}{W^*(I_2 - I_a) + \Sigma Q_l} \tag{14.4}$$

or expressed in terms of the theoretical dryer (Nevenkin, 1985):

$$\eta = \frac{1}{(1/\eta^t) + (\Sigma Q_l / \Delta H)} \tag{14.5}$$

For low humidity and low temperature convective drying, the energy efficiency can be approximated by the thermal efficiency (η_T) that is based on the inlet air temperature (T_1), the outlet air temperature (T_2) and the ambient temperature (T_a) (Strumillo et al., 2007):

$$\eta_T = \frac{T_1 - T_2}{T_1 - T_a} \tag{14.6}$$

Equation (14.6) shows that η_T can vary from 0 (for $T_2 = T_1$) to 1 (for $T_2 = T_a$) if $T_1 > T_a$.

Because of water evaporation the outlet air temperature is lower than the inlet temperature. Theoretically, its minimum value will attain either the adiabatic saturation temperature (T_{AS}) or the wet bulb temperature (T_{WB}), depending on the air–water ratio. Thus, the maximum thermal efficiency ($\eta_{T,max}$) can be determined from the following relations:

$$\eta_{T,max} = \frac{T_1 - T_{WB}}{T_1 - T_a} \tag{14.7}$$

or

$$\eta_{T,max} = \frac{T_1 - T_{AS}}{T_1 - T_a} \tag{14.8}$$

Similar to Eq. (14.6), both equations for maximum thermal efficiency do not hold true when T_1 approaches T_a.

At the low temperatures and humidities of air–water systems, the wet bulb temperature is negligibly higher than the adiabatic saturation temperature (Bond, 2002; Pakowski, 2001). Thus, either equation can be used to determine by how much a given thermal efficiency departs from the maximum one:

$$\chi = \frac{\eta_T}{\eta_{T,\,max}} = \frac{T_1 - T_2}{T_1 - T_{WB}} = \frac{T_1 - T_2}{T_1 - T_{AS}} \tag{14.9}$$

Parameter χ is an equivalent of the evaporative efficiency frequently used in spray drying (Masters, 2002) because it reflects the ratio of the actual evaporative capacity of the dryer to the capacity obtained in the ideal case of exhausting air at saturation.

As shown in Eqs. (14.7) and (14.8), the maximum thermal efficiency is based on the wet bulb temperature or the adiabatic saturation temperature. This implies that unbound water in a drying material is either in adiabatic contact with a limited amount of air to attain equilibrium or in contact with a large amount of unsaturated air. Further, the material is either nonhygroscopic or contains large amounts of surface water to exhibit a constant rate drying period under constant drying conditions. However, most materials, especially foods, dry in the constant and falling rate periods or only in the falling rate period, so the material temperature rises progressively from the wet bulb temperature once unbound water is removed. For such materials, it is more realistic to determine the maximum energy efficiency based on a relationship between the material moisture content and air humidity.

From Eq. (14.2) it follows that maximum energy efficiency exists when the exhaust air is saturated with water vapor, although this never occurs in practice because of operational constraints such as condensation problems. For a given evaporation rate, saturation of the exhaust air (Y_2^{sat}) can be attained when the air flow is minimal yet securing both heat and hydrodynamic requirements. Thus,

$$\eta_{max} = \frac{\Delta H(Y_2^{sat} - Y_a)}{I_2 - I_a} \tag{14.10}$$

Equation (14.10) can be easily transformed into Eqs. (14.7) and (14.8) based on fundamental relationships for humid air. However, all of these equations are valid for convective drying during the constant rate period when unbound water is evaporated. When hygroscopic materials are being processed, the relative humidity of the air stream at any point in the dryer should be less than the equilibrium relative humidity at this point. Thus, for such materials the maximum energy efficiency is restricted not by saturation but by the equilibrium air humidity (Y_2^{eq}) determined from sorption isotherms:

$$\eta_{max} = \frac{\Delta H(Y_2^{eq} - Y_a)}{I_2 - I_a} \tag{14.11}$$

Figure 14.1 presents the maximum energy efficiency plotted against the inlet air temperature for a nonhygroscopic material [Eq. (14.10)] and a hygroscopic material

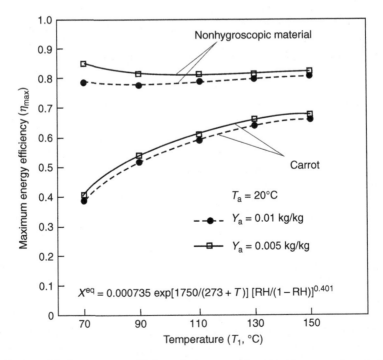

FIGURE 14.1   Maximum energy efficiency for hygroscopic and nonhygroscopic materials. The equation for the equilibrium moisture content was taken from Maroulis and Saravacos (2003).

[Eq. (14.11)] represented by carrots, the sorption isotherm $X^{eq} = f(\text{RH}, T)$ for which was taken from Maroulis and Saravacos (2003), where X^{eq} is the equilibrium material moisture content, T is the temperature, and RH is the relative humidity. To determine the Y_2^{eq}, the typical moisture content of dry carrots of 0.1kg/kg dry basis (d.b.) was assumed to be the same as the X^{eq}. Then, the air relative humidity was calculated from the equilibrium equation for several temperatures ranging from 25 to 75°C, which covers the foreseen range of outlet air temperatures during adiabatic drying at inlet air temperatures varying from 70 to 150°C. The line drawn on the Mollier chart across the points (RH, T) depicts the variation of the Y_2^{eq} with temperature for a carrot moisture content of 0.1kg/kg d.b. The particular value of Y_2^{eq} for a given inlet air temperature is determined by the intersection of the isenthalpic and equilibrium lines. Similarly, the intersection of the isenthalpic and saturation lines specifies the value of Y_2^{sat}.

The relatively high and almost constant energy efficiency for nonhygroscopic materials can be anticipated in view of unbound water evaporation. Although a particular variation of the maximum energy efficiency for hygroscopic materials is dictated by the sorption isotherms, the curves are located well below their counterparts for nonhygroscopic materials. Thus, evaluation of the energy performance of a given dryer may result in overestimated values if the sorption isotherms are not taken into account.

According to Eqs. (14.2) to (14.11), the thermal efficiency (and indirectly the energy efficiency) depends on the inlet and outlet temperatures of a drying agent, the ambient temperature, and the air humidity. Because the energy performance of the dryer is affected by ambient conditions, it will vary from day to day, depending on climatic conditions. For given ambient parameters, high energy efficiencies are attained by using high inlet temperatures and by arranging for outlet air conditions that are close to saturation. However, certain constraints should be considered when playing with operating parameters to maximize energy efficiency. For instance, high humidity implies reduced driving force that could result in slower drying. Condensation may occur on the dryer wall and downstream equipment, so the already dry product will stick to the walls and eventually clog the system. Although higher efficiencies can be obtained at higher inlet air temperatures, the maximum temperature is limited by quality loss that is attributable to thermal degradation of the product. In heat-sensitive materials, the temperature at any point in the dryer should be lower than the permissible material temperature. In general, the inlet air temperature in drying of foods does not exceed 200°C, and 300°C is the upper limit in spray drying of some materials such as coffee extract and fodder yeast (Masters, 2002). An excessive temperature may change the temperature profiles across the dryers and bring about an increase in the outlet air temperature. This can lead to overdrying and provoke adverse effects such as deactivation of enzymes, deterioration of proteins, and generation of poisonous (carcinogenic) substances. Moreover, the particle size and bulk density can be greatly reduced because of the expansion of the vapor and air bubbles occluded in the droplets, which may also lead to increased volatile retention. Furthermore, if the outlet air temperature is higher than the glass-transition temperature, it may cause stickiness of certain products, which leads to lower product yields and operation problems such as clogging of the dryer. In addition to amorphous products, the materials prone to stickiness are those containing low molecular weight carbohydrates, fat, and proteins. Typical examples are tomato paste, milk, and orange juice (Brennan et al., 1971; Goula et al., 2004; Kota and Langrish, 2006). The temperature level can also be constrained by economic reasons such as the additional costs of air heating and more expensive materials of construction that can withstand higher temperatures in the heater and the dryer.

The effect of the air-flow rate on the thermal efficiency is not explicit because this parameter is not incorporated into Eqs. (14.6) through (14.9). However, if the air-flow rate is not limited by the heat demand, then at lower flow rates and the same evaporative capacity the exhaust air stream will be more humid and give higher energy efficiency, as can be concluded from Eq. (14.2). Note that a significant amount of energy is lost with exhaust air even if its temperature approaches the wet bulb temperature. Moreover, heat dissipated with the exhaust air increases with the wet bulb temperature because of higher water content. According to Cook and DuMont (1991), the reject heat ranges typically from 20 to 40% of the drying energy but extreme values up to 60% are also possible. Thus, aside from higher energy efficiency, the air-flow rate should be kept minimal to reduce the absolute values of heat carried away with the exhaust air.

In addition to these factors, the energy efficiency also depends on the heat fluxes supplied and exhausted, number of internal heating zones, material preheating, air recycle ratio, fractional air saturation, and so on. Ashworth (1982) presents an extensive review of various expressions for energy efficiency along with a parametric analysis of various drying conditions.

14.2.2 Specific Energy Consumption

An alternative measure of dryer efficiency is the specific energy consumption (E_s), which is defined as the heat input to the dryer (Q_h) per unit mass of water evaporated. For convective dryers the heat input is given as the power supplied to the heater, so

$$E_s = \frac{Q_h}{W_{ev}} = \frac{Q_h}{F(X_1 - X_2)} \tag{14.12}$$

where F is the dry basis material feed rate and X_1 and X_2 are the moisture contents at the inlet (feed) and the outlet (product), respectively. (To facilitate future analysis of source publications by interested readers, the original nomenclature used by Baker and coworkers has been retained here.)

Baker and McKenzie (2005) derived the following expression for the specific energy consumption of a theoretical dryer that indicates that the specific energy consumption of an unspecified indirectly heated convective dryer depends on the temperature and humidity of the outlet air and the heat loss:

$$E_s = c_g \left(\frac{T_2/(1 - \eta_1) - T_a}{Y_2 - Y_a} \right) + \Delta H_{ref} \left(\frac{Y_2/(1 - \eta_1) - Y_a}{Y_2 - Y_a} \right) \tag{14.13}$$

where c_g is the heat capacity of air (gas), ΔH_{ref} is the latent heat of evaporation at 0°C taken as the reference temperature, and η_1 is the thermal loss factor of the dryer.

Parameter η_l in Eq. (14.13) is defined as

$$\eta_l = \frac{Q_l}{W_g I_1} \tag{14.14}$$

where Q_l represents the heat loss and W_g is the dry basis air-flow rate. If the dryer is adiabatic, then $\eta_l = 0$ and Eq. (14.13) reduces to

$$E_{s,a} = c_g \left(\frac{T_2 - T_a}{Y_2 - Y_a}\right) + \Delta H_{ref} \tag{14.15}$$

where $E_{s,a}$ is the specific energy consumption of the adiabatic dryer.

From Eq. (14.15) it follows that a plot of $E_{s,a}$ against $(T_2 - T_a)/(Y_2 - Y_a)$ is linear with a slope of c_g and an intercept of ΔH_{ref}. Thus, this plot can be used as a benchmark against which the performance of the real dryer can be assessed.

Figure 14.2 demonstrates the performance of several industrial spray dryers with different capacities plotted as the specific energy consumption against $(T_2 - T_a)/(Y_2 - Y_a)$, where T_a was taken as 25°C and Y_a as 0.005 kg/kg (Baker and McKenzie, 2005). As expected, most of the points are located above the straight line that depicts the performance of a theoretical dryer. Clearly, more efficient dryers are characterized by lower specific energy consumption values.

The difference between the specific energy consumption of the real and theoretical dryers operated at the same exhaust air temperature and humidity gives the excess specific energy consumption ($E_{s,x}$), which is a measure of the wasted energy due to heat losses and other inefficiencies (Baker and McKenzie, 2005):

$$E_{s,x} = E_s - E_{s,a} \tag{14.16}$$

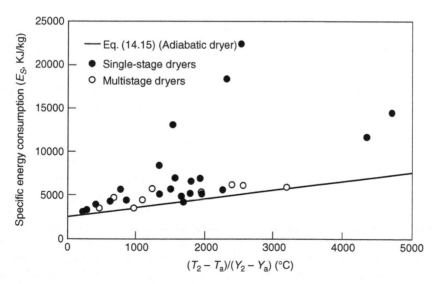

FIGURE 14.2 Specific energy consumption for several industrial spray dryers. (From Baker, C.G.J. and McKenzie, K.A., *Dry. Technol.*, 23, 365–386, 2005. With permission.)

The concept of using specific energy consumption to evaluate the performance of indirect heated spray dryers (Baker and McKenzie, 2005) and extended to fluidized bed dryers (Baker, 2005; Baker and Al-Adwani, 2007; Baker et al., 2006) was successfully validated here for other single-stage convective dryers such as pneumatic and rotary dryers and various combined dryers including a filtermat dryer and a spray dryer with an integrated fluidized bed. This concept is also valid for gas-fired dryers, if combustion air is accounted for.

14.2.3 ENERGY INDICES FOR PARTICULAR TECHNOLOGIES

An example of a particular energy index that is often used for heat pump dryers is the specific moisture extraction rate (SMER), which indicates how much water can be removed from the system by a unit of input energy:

$$\text{SMER} = \frac{\text{mass of water evaporated}}{\text{energy input}} \text{ (kg/kW h)} \qquad (14.17)$$

For conventional drying, the theoretical maximum value for SMER is 1.55 kg/kW h, which equals the latent heat of water evaporation at 100°C (Strumillo et al., 2007). The SMER value ranges from 1 to 4 kg/kW h for heat pump drying (typically from 2 to 3 kg/kW h), which is much higher than 0.12 to 1.28 kg/kW h for hot air drying (Chou and Chua, 2007).

Another example of an energy index developed for specific applications is the one used to calculate the energy consumption rate (DE) per unit mass of water during microwave-assisted drying with an intermittent power supply (Beaudry et al., 2003; Yongsawatdigul and Gunasekaran, 1996):

$$\text{DE} = \frac{t_{on} P(1 - X_f') \times 10^{-3}}{m_i (X_i' - X_f')} \text{ (kJ/kg)} \qquad (14.18)$$

where t_{on} is the total time that the microwave power is on (s); P is the input microwave power (W); m_i is the initial mass of the material (kg); and X_i' and X_f' are the initial and final material moisture contents, respectively (fraction, wet basis). Note that Eq. (14.18) reflects the energy demand attributable to intermittent microwave irradiation, but it does not account for the energy demand from the vacuum (Yongsawatdigul and Gunasekaran, 1996) or energy supplied by hot air (Beaudry et al., 2003).

14.2.4 INSTANTANEOUS ENERGY INDICES

The energy indices determined from these specified equations hold true if all of the operating parameters remain constant. With a few exceptions such as spray drying of liquids, where the constant rate period dominates the drying kinetics, most of the materials, including the majority of food products, dry in the falling rate period where energy is used for capillary-bound water removal, heating of the wet material as its temperature changes in the course of drying, local superheating of the vapor, and overheating of the already dry layers of the material otherwise necessary to maintain the required temperature gradient. Thus, even at a constant inlet air temperature, the outlet air temperature and humidity vary as drying progresses. Clearly, the energy indices calculated from the initial-final or inlet-outlet data are averaged over time for batch dryers and over the dryer length for continuous dryers. The averaged indices are useful when comparing different dryers for the same product or different products dried in the same dryer, but they have limited application when analyzing options to reduce energy consumption by modifications to the dryer design and operating parameters. Such an analysis can be performed when using the instantaneous energy efficiency (η') defined as (Kudra, 1998)

$$\eta'(t) = \frac{\text{energy used for evaporation at time } t}{\text{input energy at time } t} \tag{14.19}$$

The cumulative energy efficiency can be calculated by integrating the instantaneous energy efficiency with respect to time (t):

$$\eta = \frac{1}{t} \int_0^t \eta'(t) dt \tag{14.20}$$

By definition, the cumulative energy efficiency is the same as that calculated from Eq. (14.1). It is also the same as the one calculated from Eq. (14.6) if the outlet air temperature is constant, the time interval is short and the moisture content variation is not significant, or the outlet temperature is calculated as an integral average over the same drying time.

The energy and thermal efficiency indicates the degree of energy utilization, but it fails when the ability of the heat to remove moisture from the product has to be quantified. An example is drying of grains in a spouted bed where the air-flow rate needed for spouting is significantly higher than required to supply heat for the drying. Another example is a drying tunnel used to study drying kinetics in which high air velocity is needed to simulate conditions in an industrial dryer. However, a relatively small sample does not change the air temperature to the same degree as in an industrial dryer, so the energy efficiency calculated for such conditions gives very low and therefore meaningless values (Figure 14.3). Better insight into the "drying quality" provides instantaneous drying efficiency (η_D') because it accounts for the enthalpy of the exhaust air:

$$\eta_D'(t) = \frac{\text{energy used for evaporation at time } t}{(\text{input energy} - \text{output energy with outlet air}) \text{ at time } t} \tag{14.21}$$

Similar to the instantaneous energy efficiency, this equation can also be integrated to give the cumulative drying efficiency (η_D):

$$\eta_D = \frac{1}{t} \int_0^t \eta_D'(t) dt \tag{14.22}$$

Figure 14.3 shows the relationship between instantaneous and cumulative drying and energy indices for batch drying of corn in a fluidized bed, which represents a hygroscopic solid with considerable internal mass transfer resistance. It is evident that the energy and drying efficiencies both depend on the solid moisture content, although their absolute values differ significantly. A distinct maximum for the instantaneous drying efficiency can be attributed to the control of the drying process by internal moisture diffusion. In the initial drying period, a significant fraction of the input energy is utilized for material heating. As drying proceeds, the sensible

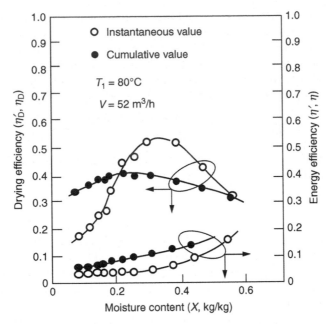

FIGURE 14.3 Instantaneous and cumulative drying and energy indices for fluid bed drying of corn. (From Kudra, T., *Chem. Process. Eng.*, 19, 163–172, 1998. With permission.)

heat in a drying agent is used for evaporation of nonhygroscopic water near the material surface. When this water is evaporated, the instantaneous drying efficiency falls rapidly because most of the heat is now utilized for overheating of the already dry surface layers and removal of the microcapillary water (the bound water moisture content for grains = ~22% w.b., but the hygroscopic moisture content = ~36.5% w.b.). Another example of the variation of drying indices with moisture content is shown later.

Considering the sensitivity of instantaneous and cumulative indices to moisture content variation, which also holds for alumina beads and potato cubes (Kudra, 1998), the drying efficiency index appears to be more suitable for energy performance analysis than the energy efficiency index, and the instantaneous indices are more useful than the cumulative ones. An example of energy performance analysis for a cross-flow pilot-scale vibrated fluid bed dryer is given below based on experimental data by Nilsson and Wimmerstedt (1987). Details of this analysis as well as performance analysis for a concurrent rotary dryer and batch flow-through dryer can be found elsewhere (Kudra, 1998).

Figure 14.4 demonstrates that instantaneous drying and energy efficiencies attain maximum values near the dryer inlet, which results from high drying rates due to the evaporation of surface moisture. In this section of a dryer the outlet air leaving the bed is saturated, which practically limits the drying rate. Further along the dryer, the energy and drying efficiencies diminish dramatically as a result of the decreasing drying rate. Although the temperature of the outlet air remains nearly constant along

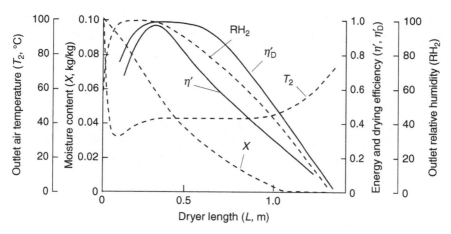

FIGURE 14.4 Performance characteristics of a vibrated fluid bed dryer. (From Kudra, T., *Chem. Process. Eng.*, 19, 163–172, 1998. With permission.).

the dryer length from about 0.4 to 1.0 m, the relative humidity falls sharply because of the reduced evaporation rate. At the dryer outlet, where the material is practically dry, the air temperature starts to increase so the relative humidity drops to 5%. Considering the pattern of energy and drying efficiencies in the dryer, we can conclude that better energy performance can be achieved by reducing the air-flow rate at the dryer end. In practice, starting at 0.4 m from the dryer inlet, the internal mass transfer controls the drying rate, so reducing the gas velocity may lower the energy consumption without significantly decreasing the drying rate. Because at the dryer inlet the outlet air is saturated, higher air-flow rates are recommended to supply additional heat for water evaporation and to increase the mass transfer driving force. The higher inlet air temperature in this part of the dryer can also be taken into account. Thus, the improved performance of this particular dryer in regard to the drying rate and energy consumption can be achieved by varying the air-flow rate and air temperature along the dryer. This can be done by dividing the gas plenum into separate compartments that are fed with individual air streams. The temperatures and superficial air velocities of each air stream can be set to maximize the instantaneous drying efficiency throughout the dryer.

Note that the concept of instantaneous energy and drying efficiency, which is an explicit function of the material moisture content, applies to other energy indices. For example, specific energy consumption in a plug flow fluidized bed dryer increased with the distance from the feed point as the evaporation rate declined over the length of the dryer (Baker et al., 2006).

14.3 SELECTED ENERGY-SAVING TECHNOLOGIES

Various options to reduce energy consumption including methods of heat recovery, heat pump drying, mechanical vapor recompression, hybrid technologies, process control, and so forth are well described in easily accessible studies (Baker, 2005;

Cook and DuMont, 1991; Kemp, 2005; Marcotte and Grabowski, 2008; Masters, 2002; Raghavan et al., 2005; Strumillo et al., 2007). Therefore, we provide only a concise overview of three selected technologies that are particularly suited for drying of food products.

14.3.1 Foam-Mat Drying

Foaming of liquid and semiliquid materials is recognized as one of the methods to shorten the drying time. Recently, this relatively old technology, known as foam-mat drying, received renewed attention because of its added ability to process hard to dry materials, obtain products of desired properties (e.g., favorable rehydration, controlled density), and retain volatiles that otherwise would be lost during the drying of nonfoamed materials. Thus, current research is directed not only toward convective drying of purposely foamed materials in spray dryers, plate dryers, and band dryers but also toward conventional freeze-drying, as well as microwave drying of frozen foams with and without dielectric inserts as complementary heat sources (Ratti and Kudra, 2006).

Foam-mat drying is a process by which a liquid is whipped by various means to form a stable foam and then dehydrated by evaporation of water. Wet foams are usually dried by convection, via hot air flowing over or through a relatively thin (3 to 10 mm) layer of the foamed material. Craters can be made in the foam layer by impinging the air jets to accelerate the drying rate. Foams can also be extruded into "spaghetti" up to several millimeters in diameter.

The main advantages of foam-mat drying are lower temperatures and shorter drying times, compared to the nonfoamed material dried in the same type of dryer, which should lead to energy savings when drying foamed materials. Figure 14.5

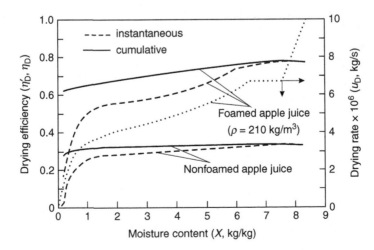

FIGURE 14.5 Drying rate and drying efficiency versus moisture content for foamed and nonfoamed apple juice air dried as a 19-mm layer at 55°C. Data are compiled from Kudra and Ratti (2006).

displays the variation of the drying rate and energy indices as defined by Eqs. (14.21) and (14.22) with the moisture content during drying of nonfoamed and foamed apple juice dried in a drying tunnel at a superficial air velocity of 0.7 m/s (Kudra and Ratti, 2006). Characteristically, both indices for foamed and nonfoamed juice exhibit very similar runs. Namely, after reaching a weak maximum at the beginning of drying the instantaneous drying efficiency gradually decreases as water evaporates. An abrupt drop occurs at the end of drying, and such a run conforms to the drying rate curve. The run of cumulative indices is more tempered than the respective run of instantaneous indices because the cumulative efficiency represents an integral average value from the beginning of drying to a given instant.

As expected, drying of foamed juice is more energy efficient, so the curves for the instantaneous and cumulative drying indices for foamed juice are both located above the ones for nonfoamed juice. Higher energy efficiency can be ascribed to (1) the foam particularities facilitating the removal of moisture at the beginning of drying, such as extended interfacial area and enhanced transport of liquid to the evaporation front; and (2) the porous structure of the foamed material developed at a certain instant and retained throughout drying. Thus, the foamed materials dry similarly to the capillary–porous solid, which opens several possibilities for energy savings. For example, once the rate of drying starts to be controlled by internal heat and mass transfer conditions, there is no reason to maintain the same level of heat input because a greater fraction of it is used to raise the sensible heat of the solid rather than that needed for evaporation. It is thus logical to reduce the intensity of the heat input once the surface moisture is removed, especially for heat-sensitive materials. This will not only improve product quality but also increase the overall energy performance of the dryer.

The energy consumption for drying of foamed apple juice was 20% of the one for drying of nonfoamed juice, which results from a shorter drying time (220 vs. 600 min) and higher drying efficiency, as shown in Figure 14.5.

14.3.2 INTERMITTENT DRYING

Drying of most solid foods occurs in the falling drying rate period because the resistance to mass transfer inside the product dominates the process. Depending on the operating conditions and characteristics of the drying material, a temporal moisture gradient is established inside the dried solid. When the evaporation front recedes and leaves behind a dry near-surface region, only a fraction of the thermal energy from the drying air is utilized for evaporation, leading to lower energy efficiency and higher specific energy consumption as pointed out earlier. Thus, from the energy standpoint, optimal drying takes place when the energy supplied to the dryer matches the energy for water evaporation at any instant. One of the technical solutions to optimized drying is "intermittent drying," which is based on a controlled supply of thermal energy that varies periodically with time according to the dryer design or its operation (Chua et al., 2003; Mujumdar, 2004). Intermittent drying offers better energy efficiency because of reduced heat input, a shorter effective drying time, and lower air consumption, in addition to improved product quality as a result of the lower material temperature.

An example of intermittent drying is the rotating jet spouted bed (Figure 14.6). In this bed the granular material is periodically exposed to hot air over the period resulting from the rotational speed of one or several air jets that move circumferentially in the annular region between the dryer wall and the central spout (Kudra and Mujumdar, 2002). Such a configuration reduces air consumption by up to 10% without any significant increase in drying time, which translates into proportional energy savings due to lower heat input (Devahastin et al., 1998).

Another mode of intermittent drying consists of a continuous heat supply over a predetermined period (usually to the critical moisture content), a tempering period over which redistribution of temperature and moisture content takes place with essentially no heat supply, and continuous drying to the final moisture content (Pan et al., 1999).

The reason for intermittent drying, in addition to a straightforward reduction of energy consumption, is the so-called tempering effect. Here, during the "rest" period (no heat supply), the moisture diffuses from the wet core of the dried material toward its surface, so the already dry near-surface region becomes wet. During subsequent

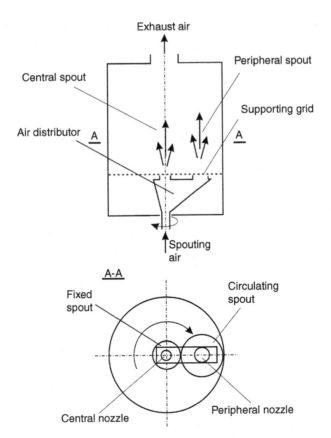

FIGURE 14.6 Rotating jet spouted bed with central and peripheral spouts.

heat supply, water evaporates from the material surface, which accelerates the drying rate. It is apparent that the intermittency (heating–drying time ratio) has to be adjusted to the material properties and drying characteristics; that is, the subsequent heating period should start when the moisture reaches the material surface. However, the start of the heating period can be delayed to accumulate more water in the near-surface layers. Clearly, the process needs optimization to take full advantage of intermittent operation. The results for the simulation of intermittent drying to minimize energy consumption reveal the following advantages of intermittent drying over continuous drying (Bon and Kudra, 2007):

- Reduced energy consumption attributable to shorter total heating time (from 2.3 to 6.3%) at constant air temperature in all heating periods or lower average temperature of the drying air when the total heating time is constrained
- Lower average enthalpy of the product by up to 23%, which results in quality advantages

Another mode of intermittent drying is based on a pulsed supply of microwave energy by alternating between microwave power-on and power-off. This cycling schedule is usually preset according to the power density given as the microwave power per unit mass of the wet material and according to the properties of a dried material (Beaudry et al., 2003; Sunjka et al., 2004; Yongsawatdigul and Gunasekaran, 1996). Overall, pulsed application of microwave energy is more efficient than continuous application, even though the total drying time is longer. Beaudry et al. (2003) demonstrated that the specific energy consumption for pulsed microwave drying of osmotically dehydrated cranberries depends on the microwave power density and cycling period; the minimum value was 8700 kJ/kg for a power density of 1.00 W/g and a cycling period of 30 s on and 30 s off whereas the maximum specific energy consumption was 9900 kJ/kg for a power density of 1.25 W/g and a cycling period of 30 s on and 30 s off.

Further improvement of the energy performance in pulsed microwave drying can be achieved by drying under a vacuum. Yongsawatdigul and Gunasekaran (1996) found the best energy performance for microwave–vacuum drying of osmotically dehydrated cranberries when a 100-g sample was exposed intermittently to 250-W microwave power that was on for 30 s and off for 150 s. Under such conditions, the minimum specific energy consumption was 2470 kJ/kg water at 10.67 kPa and an initial moisture content of 76% w.b., which represents a 47% improvement over continuous microwave–vacuum drying. The maximum specific energy consumption of 4490 kJ/kg was attained at a power level of 250 W, a cycling schedule of 60 s on and 60 s off, a vacuum level of 10.67 kPa, and an initial moisture content of 62% w.b. These values match the results of Sunjka et al. (2004) who found the specific energy consumption confined to 2850 to 7690 kJ/kg for microwave–vacuum drying of osmotically dehydrated cranberries at 3.4-kPa absolute pressure, 1.00 to 1.25 W/g power density, and two cycling schedules of 30 s on and 30 s off and 30 s on and 45 s off.

14.3.3 COMBINED OSMOTIC DEHYDRATION AND THERMAL DRYING

Osmotic dehydration is regarded as the partial removal of water by immersing wet materials such as fruits and vegetables in a hypertonic solution. The mass transfer driving force originates from the water and the solute activity gradient across the material–solution interface. Although various solutes can be used such as glucose, high fructose corn syrup, common salt, sorbitol, and starch, the most popular is sucrose syrup because the sugar-infused material may form a new market product for intermediate moisture foods with enhanced nutritional, sensory, and functional properties (Grabowski et al., 2002, 2007). Osmotic dehydration is frequently followed by complementary processing such as thermal drying to attain a product with reduced moisture content and thus a long shelf life. Because osmotic dehydration is carried out at moderate temperatures, usually from 30 to 40°C, it is regarded as an energy-saving technology when used prior to thermal drying.

In osmotic dehydration the energy is needed for (1) heating of the material and hypertonic solution to the required temperature and compensating for heat losses and (2) reconstitution of the diluted hypertonic solution by evaporation of excess water that is equal to the mass of water removed from the material during dehydration. The input energy for solution pumping and circulation (~10 kJ/kg evaporated water) and for dissolution of a hypertonic substance in a diluted solution (~1 kJ/kg water removed) can be neglected as being much smaller than for heating and evaporation. As pointed out by Lewicki and Lenart (1992), energy for water evaporation constitutes a basic component of energy consumption. For example, in osmotic dehydration of apples from 7.6 to 4.0 kg/kg d.b. carried out at 20 to 40°C, dehydration consumes 100 to 500 kJ/kg water removed whereas concentration of the diluted solution by evaporation increases energy consumption up to 2500 kJ/kg unless multiple effect evaporators are used. Therefore, an interesting option is to design osmotic dehydration in such a way that the spent syrup from the osmotic dehydration can be sold as a market product. An example is sucrose syrup from osmotic dehydration of cranberries and blueberries, which is sold as a table syrup that contains anthocyanins and other valuable compounds that were partially leached from the berries (Grabowski et al., 2007).

Regarding osmoconvective drying as a hybrid technology, osmotic dehydration prior to thermal drying results in reduced energy consumption, although to different degrees, depending on the drying kinetics. For example, during convective drying of apples nearly 40% of the water is evaporated during the constant rate period. Convective drying of osmotically dehydrated apples shows no constant rate period because the water content is below its critical value. Even though the drying rate of osmotically dehydrated apples is slightly lower because of the surface layer made up of a concentrated syrup, the initial osmotic dehydration shortens the drying time by 10 to 15% compared with the drying time of raw apples. Washing out the surface sugar reduces the resistance to mass transfer to such an extent that the drying time is shortened by ~20%, even though extra water is added to the material during washing (Lewicki and Lenart, 1992). Convective drying of apples consumes ~5000 kJ/kg evaporated water. Application of osmotic dehydration preceding convective drying reduces the total energy consumption by 24 to 75%, depending on the process conditions and the method to reconstitute the hypertonic solution. As anticipated, reduction of the water content in apples by 76% instead of 49% decreases the specific energy consumption by 46 to 51%, depending on the temperature of the osmotic dehydration (Lewicki and Lenart, 1992).

The same slow-down effect of the infused sugar on the air drying rate for cranberries was noted by Grabowski et al. (2002). However, this negative effect was completely offset by lower energy consumption due to a shorter drying time and nonthermal water removal. With respect to the total processing time, the osmotic dehydration of cranberries adds 5 h to about 50 to 60 min of the convective drying whereas the drying time of raw berries is about 3 h. However, drying of raw cranberries starts at relatively high moisture content (e.g., 87.4% w.b.) but finish drying of osmotically dehydrated cranberries begins at a moisture content of about 50% w.b. This means that osmotic dehydration allows nearly 75% of the water to be removed nonthermally, which translates into energy savings of about 2150 kJ/kg fresh berries when convective drying in the vibrated fluid bed or pulsed fluid bed follows osmotic dehydration.

14.4   SUMMARY

An evaluation of the actual energy performance of a given dryer for an objective comparison with alternative dryers and a credible assessment of the options for reducing energy consumption require a common platform to be established for various measures of energy performance. Of a variety of indices, the energy efficiency and specific energy consumption are most frequently used to assess the dryer performance from the energy viewpoint. Because the energy performance for a given dryer depends on the drying material and operating parameters, the absolute value of the energy efficiency is not informative because it does not indicate whether the dryer operates at or close to its maximum efficiency.

A departure from such a maximum can be quantified when calculating the maximum energy efficiency in hygroscopic materials such as foods that should be based on thermodynamic equilibrium (sorption isotherms). Similarly, the departure from the maximum performance of convective dryers can be quantified when using the specific energy consumption.

Instantaneous and cumulative energy and drying indices allow the analysis of the dryer performance that varies in the course of drying because of decreasing moisture content. Therefore, optional modifications are highlighted to improve temporal or spatial inefficiencies by altering the operating parameters through implementing temperature regimes or varying the air-flow rate, for example, or modifying the dryer design to a multistage dryer or turning to hybrid technologies.

NOMENCLATURE

ABBREVIATIONS AND VARIABLES

c	heat capacity [kJ/(kg K)]
d.b.	dry basis
DE	energy consumption rate [Eq. (14.18), kJ/kg H_2O]

E	energy (J)
E_s	specific energy consumption (kJ/kg H_2O)
$E_{s,a}$	specific energy consumption of adiabatic dryer (kJ/kg H_2O)
$E_{s,x}$	excess specific energy consumption (kJ/kg H_2O)
F	feed rate (dry basis, kg/s)
ΔH	latent heat of vaporization (kJ/kg)
I	enthalpy (kJ/kg)
L	length (m)
m	mass (kg)
P	microwave power (W)
ΣQ_1	heat losses (kJ/kg H_2O)
Q	heat rate (kJ/s)
RH	relative air humidity
SMER	specific moisture extraction rate (kg/kW h)
t	time (s)
t_{on}	time of microwave power on (s)
T	temperature (K, °C)
u_D	drying rate (kg H_2O/s)
V	volumetric air-flow rate (m^3/h)
W	flow rate (kg/s)
W^*	specific air consumption (kg air/kg H_2O)
w.b.	wet basis
X	material moisture content (dry basis, kg H_2O/kg dry material)
X'	material moisture content (wet basis, kg H_2O/kg wet material)
Y	air humidity (kg H_2O/kg dry air)

GREEK SYMBOLS

η	efficiency
η	cumulative energy efficiency
η_D	cumulative drying efficiency
η_1	thermal loss factor
η'	instantaneous energy efficiency
η'_D	instantaneous drying efficiency
ρ	density (kg/m^3)
χ	efficiency ratio or evaporative efficiency

SUBSCRIPTS

a	ambient
AS	adiabatic saturation
ev	evaporation
f	final
g	gas (air)
h	heater
i	initial
l	losses

max	maximum
ref	reference
t	total
T	thermal
WB	wet bulb
1	inlet
2	outlet

SUPERSCRIPTS

eq	equilibrium
t	theoretical
sat	saturation

REFERENCES

Ashworth, J.C., Energy performance of drying and application of heat recovery devices, in *The Scientific Approach to Solids Drying Problems*, Ashworth, J.C., Ed., Drying Research Ltd., Wolverhampton, U.K., 1982.

Baker, C.G.J., Energy efficient dryer operation—An update on developments, *Dry. Technol.*, 23, 2071–2087, 2005.

Baker, C.G.J. and Al-Adwani, H.H., An evaluation of factors influencing the energy-efficient operation of well-mixed fluidized bed dryers, *Dry. Technol.*, 25, 311–318, 2007.

Baker, C.G.J., Khan, A.R., Ali, Y.I., and Damyar, K., Simulation of plug flow fluidized bed dryers, *Chem. Eng. Process.*, 45, 641–651, 2006.

Baker, C.G.J. and McKenzie, K.A., Energy consumption of industrial spray dryers, *Dry. Technol.*, 23, 365–386, 2005.

Beaudry, C., Raghavan, G.S.V., and Rennie, T.J., Microwave finish drying of osmotically dehydrated cranberries, *Dry. Technol.*, 21, 1797–1810, 2003.

Bon, J. and Kudra, T., Enthalpy-driven optimization of intermittent drying, *Dry. Technol.*, 25, 523–532, 2007.

Bond, J.-F., *Visual Metrix*, Drying Doctor, Inc., Verdun, QC, Canada, 2002.

Brennan, J.G., Herrera, J., and Jowitt, R., A study of some of the factors affecting the spray drying of concentrated orange juice on a laboratory scale, *J. Food Technol.*, 6, 295–307, 1971.

Canadian Dairy Information Centre, Milk powders production, http://www.dairyinfo.gc.ca, 2007.

Chou, S.K. and Chua, K.J., Heat pump drying systems, in *Handbook of Industrial Drying*, 3rd ed., Mujumdar, A.S., Ed., Taylor & Francis, Boca Raton, FL, 2007, pp. 1103–1131.

Chua, K.J., Mujumdar, A.S., and Chou, S.K., Intermittent drying of bioproducts: An overview, *Biores. Technol.*, 90, 285–295, 2003.

Cook, E.M. and DuMont, H.D., *Process Drying Practice*, McGraw-Hill, New York, 1991.

Devahastin, S., Mujumdar, A.S., and Raghavan, G.S.V., Diffusion-controlled batch drying of particles in a novel rotating jet annular spouted bed, *Dry. Technol.*, 16, 525–543, 1998.

Goula, A.M., Adamopoulos, K.G., and Kazakis, N.A., Influence of spray drying conditions on tomato powder properties, *Dry. Technol.*, 22, 1129–1151, 2004.

Grabowski, S., Marcotte, M., Poirier, M., and Kudra, T., Drying characteristics of osmotically pretreated cranberries—Energy and quality aspects, *Dry. Technol.*, 20, 1989–2004, 2002.

Grabowski, S., Marcotte, M., Quan, D., Taherian, A.R., Zareifard, M.R., Poirier, M., and Kudra, T., Kinetics and quality aspects of Canadian blueberries and cranberries dried by osmo-convective method, *Dry. Technol.*, 25, 367–374, 2007.

Kemp, I.C., Reducing dryer energy use by process integration and pinch analysis, *Dry. Technol.*, 23, 2089–2104, 2005.

Kemp, J.C., Thermal drying: Dryer models, energy statistics and conclusions, paper presented at the Expert Workshop on Energy Efficiency in Separation Technologies, Meeting of the International Energy Agency, Mainz, Germany, May 18–19, 1998.

Kota, K. and Langrish, T.G.A., Fluxes and patterns of wall deposits from skim milk in a pilot-scale spray dryer, *Dry. Technol.*, 24, 993–1001, 2006.

Kudra, T., Instantaneous dryer indices for energy performance analysis, *Chem. Process. Eng.*, 19, 163–172, 1998.

Kudra, T., Energy aspects in drying, *Dry. Technol.*, 22, 917–932, 2004.

Kudra, T. and Mujumdar, A.S., *Advanced Drying Technologies*, Marcel Dekker, New York, 2002.

Kudra, T. and Ratti, C., Foam-mat drying: Energy and cost analyses, *Can. Biosyst., Eng.*, 48, 3.27–3.32, 2006.

Lewicki, P. and Lenart, A., Energy consumption during osmo-convection drying of fruits and vegetables, in *Drying of Solids*, Mujumdar, A.S., Ed., International Science Publishers, New York, 1992, pp. 354–366.

Marcotte, M. and Grabowski, S., Minimising energy consumption associated with drying, baking and evaporation, in *Improving Water and Energy Management in Food Processing*, Smith, R., Klemes, J., and Kim, J.-K., Eds., Woodhead Publishers Ltd., Cambridge, U.K., 2008, pp. 481–522.

Maroulis, Z.B. and Saravacos, G.D., *Food Process Design*, Marcel Dekker, New York, 2003.

Martynenko, A., Nova Scotia Agricultural College, Truro, NS, Canada, unpublished data, 2007.

Masters, K., *Spray Drying in Practice*, SprayDryConsult International ApS, Charlottenlund, Denmark, 2002.

Mujumdar, A.S., Research and developments in drying: Recent trends and future prospects, *Dry. Technol.*, 22, 1–26, 2004.

Mujumdar, A.S., Principles, classification and selection of dryers, in *Handbook of Industrial Drying*, 3rd ed., Mujumdar, A.S., Ed., Taylor & Francis, Boca Raton, FL, 2007, pp. 3–32.

Nevenkin, S., *Drying and Drying Technique*, Tekhnika, Sofia, Bulgaria, 1985 (in Bulgarian).

Nilsson, L. and Wimmerstedt, R., Drying in longitudinal-flow vibrating fluid beds—Pilot plant experiments and model simulations, *Dry. Technol.*, 5, 337–362, 1987.

Pakowski, Z., *DryPak 2000LE: Psychrometric and Drying Computations*, OMNIKON Ltd., Lodz, Poland, 2001.

Pan, Y.K., Zhao, L.J., Dong, Z.X., Mujumdar, A.S., and Kudra, T., Intermittent drying of carrot in a vibrated fluid bed: Effect on product quality, *Dry. Technol.*, 17, 2323–2340, 1999.

Raghavan, G.S.V., Rennie, T.J., Sunjka, P.S., Orsat, V., Phaphuangwittayakul, W., and Terdtoon, P., Overview of new techniques for drying biological materials with emphasis on energy aspects, *Brazil. J. Chem. Eng.*, 22, 195–201, 2005.

Ratti, C. and Kudra, T., Drying of foamed biological materials: Opportunities and challenges, *Dry. Technol.*, 24, 1101–1108, 2006.

Reynolds, L.B., Effects of drying on chemical and physical characteristics of American ginseng (*Panax quinquefolius* L.), *J. Herbs Spices Med. Plants*, 6(2), 9–21, 1998.

Strumillo, C., Jones, P.L., and Zylla, R., Energy aspects in drying, in *Handbook of Industrial Drying*, 3rd ed., Mujumdar, A.S., Ed., Taylor & Francis, Boca Raton, FL, 2007, pp. 1075–1101.

Strumillo, C. and Kudra, T., *Drying: Principles, Applications and Design*, Gordon and Breach Science Publishers, New York, 1986.

Sunjka, P.S., Rennie, T.J., Beaudry, C., and Raghavan, G.S.V., Microwave–convective and microwave–vacuum drying of cranberries: A comparative study, *Dry. Technol.*, 22, 1217–1231, 2004.

Tang, J.X., Wang, Z.G., and Huang, L.X., Recent progress of spray drying in China, *Dry. Technol.*, 17, 1747–1759, 1999.

Trägårdh, C., Energy analysis for energy conservation in the food processing industry, in *Energy in Food Processing*, Singh, R.P., Ed., Elsevier, Amsterdam, 1986, pp. 89–100.

Yongsawatdigul, J. and Gunasekaran, S., Microwave–vacuum drying of cranberries: Part 1. Energy use and efficiency, *J. Food Process. Preserv.*, 20, 121–143, 1996.

15 Novel Food Dryers and Future Perspectives

Cristina Ratti

CONTENTS

15.1 INTRODUCTION

The drying process has been known for centuries. However, there is still a clear desire in the scientific community to improve and perfect this unit operation. As times changes, the need for dry foodstuffs is based not only on the extension of their shelf life and better microbiological preservation but also on the development of dry foods and ingredients with added value in terms of nutritional and physicochemical quality. In addition, energy and the environment are important issues that presently worry researchers and the general public worldwide. Cleaner and more energy efficient technologies are thus the focal point of current scientific challenges. Thus, the most recent research tendencies in dryer developments are for the enhancement of the nutritional quality of dry products, to improve dryer energy efficiency, and to integrate "greener" technologies into the process.

The heating utilized in drying has been historically provided by thermal means, commonly by convection. However, nonconventional heat sources are new areas of increased interest these days. The use of microwaves, infrared, or radiofrequency in drying was discussed in Chapter 13 and will not be repeated here.

This last chapter presents new pretreatments and drying technologies that have been developed or substantially improved in the last 10 years. Note that, unfortunately, some of the new technologies have not been thoroughly studied "independently" from commercially oriented tests, and thus some of the information found in the literature regarding these technologies does not completely reflect their real potential. More independent scientific evaluation is therefore needed in some cases. Among the new technologies that will be demonstrated, superheated steam and ultrasound drying technologies have the highest potential for success in industry.

15.2 NOVEL PRETREATMENTS

15.2.1 POULTICE UP PROCESS (PUP)

Case hardening, a problem sometimes encountered during drying, occurs when the surface of the product dehydrates too rapidly and forms a dry layer that is "hard." Because of the presence of this layer, mass transfer and shrinkage are reduced (Ratti, 1994) and the process duration increases. This phenomenon causes an undesirable decrease in quality in the final product, because moisture and the porous structure are not distributed uniformly. Konishi and colleagues propose a PUP to reduce case hardening (Konishi et al., 2001a, 2001b; Konishi and Kobayashi, 2003). In this process, the product is subjected to the first stage of drying, followed by storage (several hours) in a low temperature controlled environment. After this "relaxation time" the moisture profiles are again uniform, and then the product is subjected to the second stage of drying. Figure 15.1a contains a schematic of this process. After this treatment, the final product has a uniform distribution of moisture that contributes to reductions in the drying time and total cost of the drying process (Konishi and Kobayashi, 2003). The treatment has been successfully applied to fish paste and sausages (Konishi et al., 2001a, 2001b).

15.2.2 CONTROLLED PRESSURE DROP

A controlled instantaneous pressure drop or DIC, from the French designation *Décompression Instantanée Controllée*, is a hydrothermomechanical treatment that is mainly used with biomaterials (Louka and Allaf, 2004). High pressure followed by an instantaneous release to vacuum conditions is applied to the product. Thus, the texture of the product changes to a less compact porous structure that facilitates mass transfer during many processes (extraction, drying, etc.).

During drying, the DIC treatment is applied only after the first stage of drying followed by an equilibration of moisture profiles. A second drying stage takes place after the DIC treatment to reduce the product moisture content to low values (see Figure 15.1b). Volume expansion accelerates drying, which has been known for a long time; it has already been used in a process called "explosion puffing," in which the product is also pressurized by heating the air to high temperatures in a hermetic container followed by a sudden decompression to atmospheric conditions (Saca and Lozano, 1992). In the DIC method, however, steam is used to increase the pressure

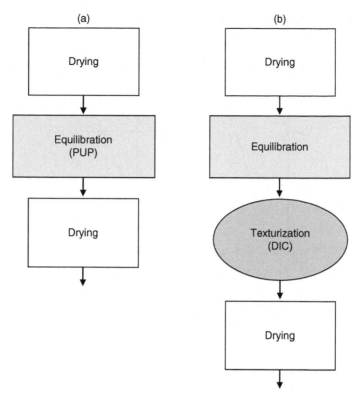

FIGURE 15.1 Schematic representation of two pretreatment processes: (a) PUP and (b) DIC.

in the chamber, and the release of pressure is performed up to vacuum conditions. These two differences (steam and vacuum) make the DIC method a better and gentler texturization process that can be safely used for heat-sensitive foodstuffs.

Figure 15.2 provides a comparison of the pressure profiles that can be encountered during DIC and explosion puffing processes. There are many variables that should be controlled to optimize the increase in the expansion ratio of the product: the steam pressure in the vessel (P_{DIC}), the duration of the pressurization (τ) and decompression, and the vacuum level (P_{vacuum}). A good discussion on the effect of these variables in the process of vegetable dehydration can be found in the work of Louka et al. (2004). Adding an equilibration step to the process increases the drying rate of the second dehydration stage through increased porosity in the sample (Louka and Allaf, 2002).

15.2.3 FOAMING

Some researchers successfully used foaming prior to processing to decrease operating times during conventional hot-air drying of liquids, semi-liquids, and pastes (Kudra and Ratti, 2006). In this technique called "foam-mat drying," a liquid is whipped into a stable foam, and then dehydrated by thermal means. Many studies concerning physical and organoleptic characteristics have been done on foam-mat dried foods

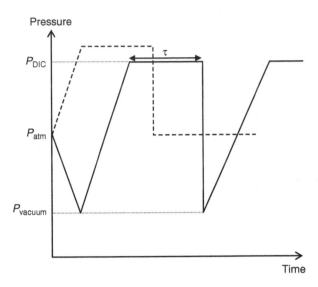

FIGURE 15.2 Pressure scheme (—) during DIC texturization treatment and (- - -) in explosion puffing; P_{DIC}, steam pressure in the vessel; P_{atm}, atmospheric pressure; P_{vacuum}, vacuum level. (Adapted from Haddad, J. and Allaf, K., *J. Food Eng.*, 79, 353–357, 2007. With permission.)

such as mango, tamarind juice, starfruit puree, whole egg, or ripe bananas. These studies showed better final quality characteristics after foam-mat drying than using simply the air drying process. Current research is directed toward convective drying of foamed materials in spray dryers, plate dryers, and band dryers but also toward conventional freeze-drying, as well as microwave drying of frozen foams (Ratti and Kudra, 2006).

Air drying of foamed materials is generally faster than air drying of nonfoamed ones, although certain foams such soymilk (Akintoye and Oguntunde, 1991) or starfruit (Karim and Wai, 1999) exhibit higher drying rates in the beginning of foam-mat drying. For other materials such as tomato paste (Lewicki, 1975), bananas (Sankat and Castaigne, 2004), and mango (Cooke et al., 1976) the drying rates are greatly accelerated only at the end of drying.

Freeze-drying of foamed and nonfoamed apple juice was studied to assess whether there is a reduction in process time attributable to foaming (Raharitsifa and Ratti, 2008). Foams were prepared by whipping apple juice with methylcellulose or egg albumin at different concentrations. Foaming reduced the process time when the comparison was done at equal sample thickness. However, a lower density of foamed materials decreased the weight load in the dryer. The optimization of the process did not provide a minimal sample thickness necessary to increase both the drying rate and dryer throughput for egg albumin or methylcellulose foams (Raharitsifa and Ratti, 2008).

In addition to accelerated transport of liquid water to the evaporation front, drying experts have repeatedly isolated the increased interfacial area of foamed materials as the factor responsible for reduced drying time. Because the density of foamed materials (300 to 600kg/m^3) is lower than that of nonfoamed ones, the mass load of the foam-mat dryer is also lower. In some cases, a shorter drying time not only offsets the reduced dryer load but also increases the dryer through-put. For example, the dryer throughput can be 32% higher when air drying foamed apple juice with pulp (unpublished data) and 22% higher when drying foamed mango pulp, which was calculated in this study from the experimental data gener-ated by Rajkumar et al. (2007). In foam-mat freeze-drying, the possibility of increasing the drying kinetics and dryer throughput is dependent on the composi-tion of the foam as well as on the sample thickness (Raharitsifa and Ratti, 2008). Furthermore, shorter drying times per unit mass of foamed materials may not always bring about lower energy consumption and better process economics (Kudra and Ratti, 2006).

Food foams can be considered as biphasic systems in which a gas bubble phase is dispersed in a continuous liquid phase (Herzhaft, 1999; Pernell et al., 2000; Thakur et al., 2003; Vernon-Carter et al., 2001). Foams are inherently unstable (Dickinson and Stainsby 1987; Karim and Wai 1999), resulting in the drainage of the continuous phase through the thin films between the bubbles. As a result, the foam lamellae are thinned, favoring the mass transfer of gas across it, which leads to the growth of large bubbles at the expense of smaller ones (disproportionation or Ostwald ripening). This process continues with bubble coalescence (the thin film between the two bubbles collapses, and the two bubbles merge to form one larger bubble) and ends in phase separation. Foam stability is influenced by the physical and rheological properties of the interface and the continuous phase. Foaming agents contribute to the improvement of the structure and to the control of the tex-ture and stability of food foams by modifying the rheological properties of the continuous phase and the interfacial regions where they are adsorbed. A thorough study of the stability should be performed before applying this pretreatment to accelerate drying in order to determine the optimal foaming agent and its concen-tration (Raharitsifa et al., 2006).

15.3   NOVEL DRYERS

15.3.1   SUPERHEATED STEAM DRYING

Superheated steam vacuum drying is a technique that has recently received increased attention (Defost et al., 2004; Ratti, 2008). This technique uses superheated steam as a heating medium for water evaporation from a solid that is at a temperature above its boiling point. The evaporated moisture becomes part of the drying medium and does not need to be exhausted (Cenkowski et al., 2005) and, because the steam is superheated and unsaturated, a drop in its temperature because of heat transfer to the solid will not cause condensation. In addition, no oxygen is present in the super-heated steam medium, which can make it a choice heating method for oxygen-sensitive materials. Superheated steam vacuum drying has been successfully applied to foods (Devahastin and Suvarnakuta, 2004; Elustondo et al., 2001; Methakhup et al., 2005; Nathakaranakule et al., 2007; Suvarnakuta et al., 2005).

TABLE 15.1
Diffusion Coefficients during Drying of Foods with Hot Air or Superheated Steam at Various Temperatures

Food	Drying Medium	Drying Temp. (°C)	$D_{eff} \times 10^8$ (m²/s)	Ref.
Durian chips	Hot air	130	1.49	Jamradloedluk et al. (2007)
		140	1.69	
		150	2.15	
	Superheated steam	130	0.97	
		140	1.10	
		150	1.44	
Chicken meat	Superheated steam	120	2.71	Nathakaranakule et al. (2007)
		140	4.64	
		160	7.74	

Studies comparing the product quality and drying kinetics of hot air or vacuum drying versus superheated steam drying are inconclusive. Some works report that the final product quality is superior with superheated steam dehydration, but the drying time is longer than that for air or vacuum drying at the same temperatures (Jamradloedluk et al., 2007; Methakhup et al., 2005; Suvarnakuta et al., 2005). Table 15.1 lists the diffusion coefficients for hot air and superheated steam drying of durian chips (Jamradloedluk et al., 2007). The diffusion coefficients for superheated steam are 35% lower than air drying. The steam temperature, however, has a strong effect on the diffusion coefficients of some foodstuffs, as can be seen in Table 15.1 for chicken meat (Nathakaranakule et al., 2007). Thus, above a specific temperature, superheated steam could provide a higher drying rate than hot air (Jamradloedluk et al., 2007). However, the quality of the product processed at an excessive temperature would deteriorate. This approach is therefore not recommended to dry highly sensitive foods.

Models for mass transfer during superheated steam drying have been based mainly on Fick's law (Jamradloedluk et al., 2007; Nathakaranakule et al., 2007). More original semiempirical kinetics models have been also tested against actual foods and gave excellent results (Elustondo et al., 2001).

15.3.2 Refractance Window® (RW)

RW is a relatively new film drying method from which flakes and powders can be obtained. As explained by Nindo and Tang (2007) in an extensive comparative review, the energy in this method is supplied from a hot water bath (95 to 97°C) to the product deposited as a thin film (1 to 2 mm) over a plastic transparent conveyor under atmospheric pressure. Figure 15.3 is a schematic representation of the RW dryer (adapted from Nindo et al., 2007). There are three heat transfer modes in this dehydration method

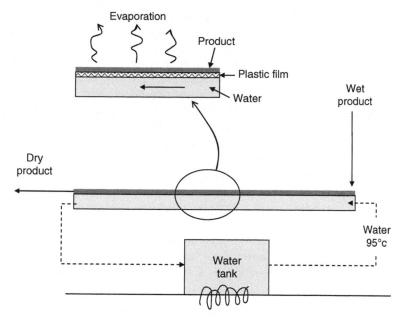

FIGURE 15.3 Refractance Window dryer. (Adapted from Nindo, C.I. et al., *Lebensm.-Wiss. Technol.*, 40, 1000–1007, 2007. With permission.)

(Nindo and Tang, 2007): conduction (from the water bath to the product through the plastic film), radiation (from the water bath to the product through the transparent plastic film that is selected to maximize the heat absorption), and convection (from the product to the atmospheric air conditions). Nindo and Tang (2007) claimed that the use of process water at temperatures just below boiling and the thin plastic conveyor with infrared transmission in the wavelength range that matches the absorption spectrum for water all work together to facilitate rapid drying in <5min.

Some of the advantages of this type of drying technology are (Nindo and Tang, 2007) the following:

- Indirect method that has no chance of cross-contamination
- Rapid heating
- Demands less energy than freeze-drying (equipment costs estimated as one third to one half that of a freeze-dryer and operation costs approximately one half that of freeze-drying, which is not negligible)

By contrast, this dehydration method is only applicable for liquids or pastes. (It is not applicable for solids because of the low penetration depth of infrared radiation in solids and because of the slow down in heat conduction transfer.) In addition, the product should have the right consistency so that it can be spread in very thin films on the conveyor, which implicates some kind of pretreatment and optimization of the product properties prior to the drying process.

The drying kinetics and final physical and bioactive product quality with RW has been compared to other drying methods such as freeze-drying, spray drying, and drum drying for several food pastes and purees (asparagus, carrots, strawberries, etc.). The quality and drying kinetics obtained with RW is clearly superior to spray and drum drying, as shown in various studies (Abonyi et al., 2002; Nindo et al., 2003; Nindo and Tang, 2007). RW provided products with similar or higher quality than freeze-drying in very short times (Abonyi et al., 2002; Nindo et al., 2003). However, these comparisons were made on different bases. Thickness is an extremely important parameter in freeze-drying, which directly determines the dehydration time. If the product is left inside the freeze-drier too long after the completion of dehydration, the product will be overheated and quality losses may increase. Food pastes or purees (i.e., strawberry, carrot or asparagus) with a thickness of 1 to 2mm certainly need <2h to complete freeze-drying at a heating plate temperature of 20°C. Nevertheless, in the comparison between freeze-drying and RW in the work of Abonyi et al. (2002) and Nindo et al. (2003) the freeze-drying was carried out for 24h, although the thickness of the product being freeze-dried was not provided (products dehydrated by RW had a thickness of 1 to 2mm).

15.3.3 Low Temperature with Adsorbent Materials

Adsorption freeze-drying uses a desiccant (e.g., silica gel) to create a high vapor drive at low temperatures (Bell and Mellor, 1990a). Figure 15.4a provides a schematic representation of this type of freeze-dryer. The adsorbent material replaces the condenser, and it reduces the total costs by 50% compared to traditional freeze-drying. Despite its many advantages compared to regular freeze-drying (Bell and Mellor, 1990b), the quality of adsorption freeze-dried foods is slightly reduced and sometimes poor compared to that obtained by traditional freeze-drying. Another problem is related to the enormous amounts of desiccant needed for the operation, which has to be regenerated when saturation occurs.

Condensing the water vapor on adsorbent materials has also been applied in the "zeodration" technique, in which drying takes place at low temperatures in a vacuum cabinet and the water vapor is adsorbed in a zeolite reactor (Figure 15.4a). The difference between the adsorption freeze-dryer and the zeodrator is the temperatures of the product: in the former process the product is frozen and in the latter the temperature is above freezing. Table 15.2 provides a comparison of the three types of vacuum dryers. Zeolites have a higher water adsorption capacity than other materials during wheat dehydration (Revilla et al., 2006). Because the adsorption reaction is exothermic, the heat generated during water adsorption can increase the air temperature in the zeodrator to reduce the energy required for dehydration (Djaeni et al., 2007a). However, zeolite regeneration needs significant energy input, which decreases the energy saved for drying (Djaeni et al., 2007a). Very few articles can be found on a dehydration technique aided by an adsorbent reactor (Djaeni et al., 2007a, 2007b; Rane et al., 2005), and scientific studies on the effect of the method on quality attributes are practically nonexistent. However, drying with adsorbents has been advertized (Bucher Zeodration, 2007) as a gentle and "green" dehydration technique, which is appropriate to maintain a high standard of quality in food products.

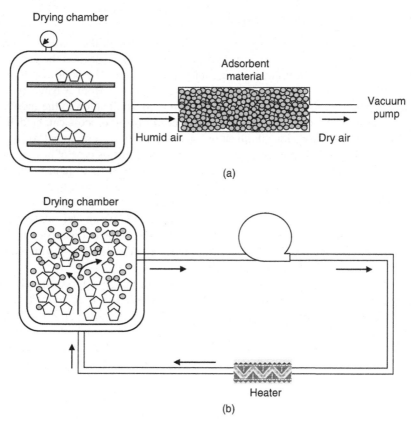

Drying chamber

Adsorbent
material

Humid air

Dry air

Vacuum
pump

(a)

Drying chamber

Heater

(b)

FIGURE 15.4 (a) Freeze-dryer with adsorbents or zeodrator and (b) atmospheric fluidized freeze-dryer.

TABLE 15.2
Comparison of Vacuum Drying Techniques

Dryer Type	Vacuum Level	Temp. Range (°C)	Condenser Type	Characteristics
Vacuum drying	Medium	30–100	Water trap	Inefficient heat transfer
Freeze-drying	Very high	−40 to −15 (product) −20 to 50 (heating plate)	Low temperature	Frozen samples
Zeodrator	High	15–65	Adsorbent material	No use of refrigerants

15.3.4 ATMOSPHERIC FREEZE-DRYING

Atmospheric freeze-drying was developed in the beginning of the 1990s because of the numerous difficulties that a vacuum causes during freeze-drying: the expensive energy cost to maintain a vacuum during long drying times; the poor heat transfer under a vacuum; and the problem of converting a process from a batch type to continuous when a vacuum is applied, which limits its application to the industrial scale. The most common type of atmospheric freeze-drying process can be defined with three words: adsorption, fluidization, and atmospheric pressure (Wolff and Gibert, 1987). Figure 15.4b is a schematic representation of this type of process. The product particles are mixed with an adsorbent medium in order to replace the condenser. Fluidization increases the heat transfer coefficients by more than 1 order of magnitude compared to fixed bed operations (Donsì et al., 1998; Donsì and Ferrari, 1995). The bed temperature and size of the product are key parameters affecting the process dehydration rates (Di Matteo et al., 2003). An excellent discussion on heat and mass transfer phenomena during atmospheric fluidized freeze-drying can be found in the work by Di Matteo et al. (2003).

The energy savings offered by atmospheric freeze-drying are approximately 34% compared to vacuum freeze-drying (Wolff and Gibert, 1990). However, drying times are increased 1 to 3 times because the use of atmospheric pressure turns the control of the process from heat to mass transfer, which makes the kinetics extremely slow. In a recent review on the subject, Claussen et al. (2007) stated that atmospheric freeze-dried foods had higher quality than vacuum freeze-dried foods. However, Lombraña and Villarán

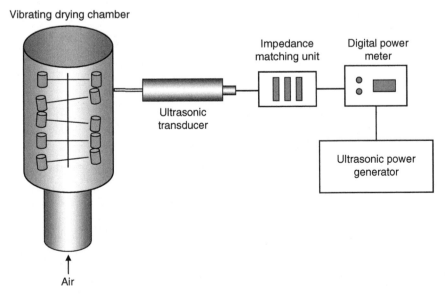

FIGURE 15.5 Schematic of an ultrasound dryer. (Adapted from Cárcel, J.A. et al., *Dry. Technol.*, 25, 185–193, 2007. With permission.)

(1997) showed that the quality of the products is not excellent when atmospheric pressure is used instead of a vacuum, because of the increased risk of product collapse.

15.3.5 ACOUSTIC DRYING

Among the novel dehydration methods, acoustic drying together with superheated steam drying has the highest potential for commercial application. Ultrasound can be applied to the fluid itself (airborne ultrasound drying) or to the solid directly by vibrating the solid surface. Figure 15.5 shows a schematic of an ultrasonic assisted air dryer in which ultrasound is applied directly to food samples. As explained by Cárcel et al. (2007), high-intensity ultrasonic waves during air drying produce a rapid series of alternative contractions and expansions of the solid material that makes the moisture transport within the solid easier. In addition, bound water can be unattached effortlessly in the presence of cavitation produced by ultrasound. Cavitation could also be the cause for enhanced heat and mass transfer at the solid–air interface. Heat transfer in the presence of ultrasonic waves can be increased by approximately 30 to 60% (Mason et al., 1996). However, high-intensity airborne ultrasound is more efficient in the drying constant rate period when the surface of the material is completely covered with moisture, as well as when external transfer is predominant, such as during drying with low air velocities (Gallego-Juárez et al., 1999). Under all other conditions, ultrasonic waves applied directly to the solid material are recommended for enhancing the drying rates. Ultrasound was recently tested during fluidized bed drying of foodstuffs (Garcia-Perez et al., 2006).

Because ultrasound is not an invasive technique, drying can be achieved in reduced times under lower air temperatures but without dramatic changes in the sample structure. This may have an important application in dehydration of high-value foods, such as those that are heat sensitive or contain fragile bioactive compounds.

Acoustic drying has been successfully applied to air drying of carrots (de la Fuente-Blanco et al., 2006), persimmons (Cárcel et al., 2007), and so forth. In addition, it has been utilized as a pretreatment before air drying of mushrooms, Brussels sprouts, and cauliflower (Jambrak et al., 2007). Ultrasound has also been useful in enhancing osmotic dehydration of blueberries (Stojanovic and Silva, 2006), bananas (Fernandes and Rodrigues, 2007), and papaya (Fernandes et al., 2007).

15.3.6 CORONA WIND

Corona wind drying ("electrohydrodynamic drying") is one of the latest nonthermal dehydration technologies applied to foodstuffs. In this method, the food is placed between two electrodes where high-intensity electric fields at 50- to 60-Hz frequencies are applied. As a result, the electric wind created by the ionic injection enhances the heat and mass transfers to and from the particle (Goodenough et al., 2007). Evaporation rates are increased significantly, but energy consumption is equivalent to conventional drying (Goodenough et al., 2007). The greater the voltage that is applied in this type of dehydration method, the greater the drying rate (Bajgai et al., 2006). This drying method was successfully applied to potato slabs (Chen et al., 1994), apple slices (Hashinaga et al., 1999), Japanese radishes (Bajgai and Hashinaga, 2001a), spinach (Bajgai and Hashinaga, 2001b), and biscuits (Goodenough et al., 2007).

15.4 FUTURE PERSPECTIVES

Although progress has been made recently in developing new dehydration technologies and pretreatments to reduce the energy and costs associated with the process and to enhance quality aspects of dried foods, there is no single dehydration method that is suitable for any type of product and application while operating at reduced energy consumption. The available dehydration processes are specific to a particular situation requiring pilot-scale optimization in order to attain the best results. Hybrid dehydration technologies have good potential to cope with present trends, but they still require numerous research tests to generalize their results.

In this sense, modeling and simulation should be regarded with increased attention because they allow extrapolation to different drying situations in terms of handled products, energy requirements, and optimal dried food quality, thus minimizing the experiments done on lab or pilot scales. In addition, realistic models are certainly required in order to control the main variables during dehydration. Dryer control is one of the main areas, if not THE main area, of food engineering research that should be targeted for development in the next decade.

REFERENCES

Abonyi, B.I., Feng, H., Tang, J., Edwards, C.G., Mattinson, D.S., and Fellman, J.K., Quality retention in strawberry and carrot purees dried with Refractance Window system, *J. Food Sci.*, 67, 1051–1056, 2002.

Akintoye, O.A. and Oguntunde, A.O., Preliminary investigation on the effect of foam stabilizers on the physical characteristics and reconstitution properties of foam-mat dried soymilk, *Dry. Technol.*, 9, 245–262, 1991.

Bajgai, T.R. and Hashinaga, F., High electric field drying of Japanese radish, *Dry. Technol.*, 19, 2291–2301, 2001a.

Bajgai, T.R. and Hashinaga, F., Drying of spinach with a high electric field, *Dry. Technol.*, 19, 2331–2341, 2001b.

Bajgai, T.R., Raghavan, G.S.V., Hashinaga, F., and Ngadi, M.O., Electro-hydrodynamic drying—A concise review, *Dry. Technol.*, 24, 905–910, 2006.

Bell, G.A. and Mellor, J.D., Adsorption freeze-drying, *Food Austral.*, 42, 226–227, 1990a.

Bell, G.A. and Mellor, J.D., Further developments in adsorption freeze-drying, *Food Res. Quart.*, 50, 48–53, 1990b.

Bucher Zeodration, http://www.bucherdrytech.com/html/en/5233.html (accessed December 6, 2007).

Cárcel, J.A., García-Pérez, J.V., Riera, E., and Mulet, A., Influence of high-intensity ultrasound on drying kinetics of persimmon, *Dry. Technol.*, 25, 185–193, 2007.

Cenkowski, S., Pronyk, C., and Muir, W.E., Current advances in superheated-steam drying and processing, *Stewart Postharv. Rev.*, 4, 4, 2005.

Chen, Y., Barthakur, N.N., and Arnold, N.P., Electrohydrodynamic (EHD) drying of potato slabs, *J. Food Eng.*, 23, 107–119, 1994.

Claussen, I.C., Ustad, T.S., Strømmen, I., and Walde, P.M., Atmospheric freeze drying—A review, *Dry. Technol.*, 25, 957–967, 2007.

Cooke, R.D., Breag, G.R., Ferber, C.E.M., Best, P.R., and Jones, J., Studies of mango processing. 1. The foam-mat drying of mango (Alphonso cultivar) puree, *J. Food Technol.*, 11, 463–473, 1976.

Defost, M., Fortin, Y., and Cloutier, A., Modeling superheated steam vacuum drying of wood, *Dry. Technol.*, 22, 2231–2253, 2004.

de la Fuente-Blanco, S., Riera-Franco de Saravia, E., Acosta-Aparicio, V.M., Blanco-Blanco, A., and Gallego-Juárez, J.A., Food drying process by power ultrasound, *Ultrasonics*, 44, e523–e527, 2006.

Devahastin, S. and Suvarnakuta, P., Superheated steam drying of food products, in *Dehydration of Products of Biological Origin*, Mujumdar, A.S., Ed., Science Publishers, Enfield, NH, 2004.

Dickinson, E. and Stainsby, G., Progress in the formulation of food emulsions and foams, *Food Technol.*, 41, 74–81, 116, 1987.

Di Matteo, P., Donsì, G., and Ferrari, G., The role of heat and mass transfer phenomena in atmospheric freeze-drying of foods in a fluidised bed, *J. Food Eng.*, 59, 267–275, 2003.

Djaeni, M., Bartels, P., Sanders, J., van Straten, G., and van Boxtel, A.J.B., Process integration for food drying with air dehumidified by zeolites, *Dry. Technol.*, 25, 225–239, 2007a.

Djaeni, M., Bartels, P., Sanders, J., van Straten, G., and van Boxtel, A.J.B., Multistage zeolite drying for energy-efficient drying, *Dry. Technol.*, 25, 1063–1077, 2007b.

Donsì, G. and Ferrari, G., Heat transfer coefficients between gas fluidised beds and immersed spheres: Dependence on the sphere size, *Powder Technol.*, 82, 293–299, 1995.

Donsì, G., Ferrari, G., Nigro, R., and Di Matteo, P., Combination of mild dehydration and freeze-drying processes to obtain high quality dried vegetables and fruits, *Trans. IChemE.*, 76, 181–187, 1998.

Elustondo, D., Elustondo, M.P., and Urbicain, M.J., Mathematical modeling of moisture evaporation from foodstuffs exposed to subatmospheric pressure superheated steam, *J. Food Eng.*, 49, 15–24, 2001.

Fernandes, F.A.N., Oliveira, F.I.P., and Rodrigues, S., Use of ultrasound for dehydration of papayas, *Food Bioprocess. Technol.*, DOI: 10.1007/s11947-007-0019-9, 2007.

Fernandes, F.A.N. and Rodrigues, S., Ultrasound as pre-treatment for drying of fruits: Dehydration of banana, *J. Food Eng.*, 82, 261–267, 2007.

Gallego-Juárez, J.A., Rodríguez-Corral, G., Galvez Moraleda, J.C., and Yang, T.S., A new high-intensity ultrasonic technology for food dehydration, *Dry. Technol.*, 17, 597–608, 1999.

Garcia-Perez, J.V., Cárcel, J.A., de la Fuente-Blanco, S., and Riera-Franco de Sarabia, E., Ultrasonic drying of foodstuff in a fluidized bed: Parametric study, *Ultrasonics*, 44, e539–e543, 2006.

Goodenough, T.I.J., Goodenough, P.W., and Goodenough, S.M., The efficiency of corona wind drying and its application to the food industry, *J. Food Eng.*, 80, 1233–1238, 2007.

Haddad, J. and Allaf, K., A study of the impact of instantaneous controlled pressure drop on the trypsin inhibitors of soybean, *J. Food Eng.*, 79, 353–357, 2007.

Hashinaga, F., Bajgai, T.R., Isobe, S., and Barthakur, N.N., Electrohydrodynamic (EHD) drying of apple slices, *Dry. Technol.*, 17, 479–495, 1999.

Herzhaft, B., Rheology of aqueous foams: A literature review of some experimental works, *Oil Gas Sci. Technol.*, 54, 587–596, 1999.

Jambrak, A.R., Mason, T.J., Paniwnyk, L., and Vesna, L., Accelerated drying of button mushrooms, Brussels sprouts and cauliflower by applying power ultrasound and its rehydration properties, *J. Food Eng.*, 81, 88–97, 2007.

Jamradloedluk, J., Nathakaranakule, A., Soponronnarit, S., and Prachayawarakorn, S., Influences of drying medium and temperature on drying kinetics and quality attributes of durian chip, *J. Food Eng.*, 78, 198–205, 2007.

Karim, A.A. and Wai, C.C., Characteristics of foam prepared from starfruit (*Averrhoa carambola* L.) puree by using methylcellulose, *Food Hydrocolloids*, 13, 203–210, 1999.

Konishi, Y., Horiuchi, Y., and Kobayashi, M., Dynamic evaluation of the dehydration response curves of foods characterized by a poultice-up process using a fish-paste sausage. I. Determination of the mechanism for moisture transfer, *Dry. Technol.*, 19, 1253–1270, 2001a.

Konishi, Y., Horiuchi, Y., and Kobayashi, M., Dynamic evaluation of the dehydration response curves of foods characterized by a poultice-up process using a fish-paste sausage. II. A new tank model for a computer simulation, *Dry. Technol.*, 19, 1271–1285, 2001b.

Konishi Y. and Kobayashi, M., Characteristic innovation of a food drying process revealed by the physicochemical analysis of dehydration dynamics, *J. Food Eng.*, 59, 277–283, 2003.

Kudra, T. and Ratti, C., Foam-mat drying: Energy and cost analyses, *Can. Biosyst. Eng.*, 48, 3.27–3.32, 2006.

Lewicki, P.P., Mechanisms concerned in foam-mat drying of tomato paste, *Trans. Agric. Acad. Warsaw*, 55, 1–67, 1975 (in Polish).

Lombraña, J.I. and Villarán, M., The influence of pressure and temperature on freeze-drying in an adsorbent medium and establishment of drying strategies, *Food Res. Int.*, 30, 213–222, 1997.

Louka, N. and Allaf, K., New process for texturizing partially dehydrated biological products using controlled sudden decompression to the vacuum: Application on potatoes, *J. Food Sci.*, 67, 3033–3038, 2002.

Louka, N. and Allaf, K., Expansion ratio and color improvement of dried vegetables texturized by a new process "controlled sudden decompression to the vacuum". Application to potatoes, carrots and onions, *J. Food Eng.*, 65, 233–243, 2004.

Louka, N., Juhel, F., and Allaf, K., Quality studies on various types of partially dried vegetables texturized by controlled sudden decompression. General patterns for the variation of the expansion ratio, *J. Food Eng.*, 65, 245–253, 2004.

Mason, T.J., Paniwnyk, L., and Lorimer, J.P., The uses of ultrasound in food technology, *Ultrason. Sonochem.*, 3, s253–s260, 1996.

Methakhup, S., Chiewchan, N., and Devahastin, S., Effects of drying methods and conditions on drying kinetics and quality of Indian gooseberry flake, *Lebensm.-Wiss. Technol.*, 38, 579–587, 2005.

Nathakaranakule, A., Kraiwanichkul, W., and Soponronnarit, S., Comparative study of different combined superheated-steam drying techniques for chicken meat, *J. Food Eng.*, 80, 1023–1030, 2007.

Nindo, C.I., Powers, J.R., and Tang, J., Influence of Refractance Window evaporation on quality of juices from small fruits, *Lebensm.-Wiss. Technol.*, 40, 1000–1007, 2007.

Nindo, C.I., Sun, T., Wang, S.W., Tang, J., and Powers, J.R., Evaluation of drying technologies for retention of physical quality and atioxidants in asparagus (*Asparagus officinalis* L.), *Lebensm.-Wiss. Technol.*, 36, 507–516, 2003.

Nindo, C.I. and Tang, J., Refractance Window dehydration technology: A novel contact drying method, *Dry. Technol.*, 25, 37–48, 2007.

Pernell, C.W., Foegeding, E.A., and Daubert, C.R., Measurement of the yield stress of protein foams by vane rheometry, *J. Food Sci.*, 65, 110–114, 2000.

Raharitsifa, N., Genovese, D.B., and Ratti, C., Characterization of apple juice foams for foam-mat drying prepared with egg white protein and methylcellulose, *J. Food Sci.*, 71, E142–E151, 2006.

Raharitsifa, N. and Ratti, C., Foam-mat freeze-drying of apple juice: Experimental data and ANN simulations, *J. Food Process Eng.*, submitted.

Rajkumar, P., Kailappan, R., Viswanathan, R., Raghavan, G.S.V., and Ratti, C., Studies on foam mat drying of Alphonso mango pulp, *Dry. Technol.*, 25, 357–366, 2007.

Rane, M.V., Kota Reddy, S.V., and Easow, R.R., Energy efficient liquid desiccant-based dryer, *Appl. Therm. Eng.*, 25, 769–781, 2005.

Ratti, C., Shrinkage during drying of foodstuffs, *J. Food Eng.*, 23, 91–105, 1994.

Ratti, C., Freeze and vacuum drying of foods, in *Drying Technologies for Food Processing*, Mujumdar, A.S., Ed., Blackwell Publishing, New York, 2008.

Ratti, C. and Kudra, T., Drying of foamed materials: Opportunities and challenges, *Dry. Technol.*, 24, 1101–1108, 2006.

Revilla, G.O., Velázquez, T.G., Cortéz, S.L., and Cárdenas, S.A., Immersion drying of wheat using Al-PILC, zeolite, clay and sand as particulate media, *Dry. Technol.*, 24, 1033–1038, 2006.

Saca, S.A. and Lozano, J.E., Explosion puffing of bananas, *Int. J. Food Sci. Technol.*, 27, 419–426, 1992.

Sankat, C.K. and Castaigne, F., Foaming and drying behaviour of ripe bananas, *Lebensm.-Wiss. Technol.*, 37, 517–525, 2004.

Stojanovic, J. and Silva, J.L., Influence of osmoconcentration, continuous high-frequency ultrasound and dehydration on properties and microstructure of rabbiteye blueberries, *Dry. Technol.*, 24, 165–171, 2006.

Suvarnakuta, P., Devahastin, S., and Mujumdar, A.S., Drying kinetics and β-carotene degradation in carrot undergoing different drying processes, *J. Food Sci.*, 70, S520–S526, 2005.

Thakur, R.K., Vial, C., and Djelveh, G., Influence of operating conditions and impeller design on the continuous manufacturing of food foams, *J. Food Eng.*, 60, 9–20, 2003.

Vernon-Carter, E.J., Espinosa-Paredes, G., Beristain, C.I., and Romero-Tehuitzil, H., Effect of foaming agents on the stability, rheological properties, drying kinetics and flavour retention of tamarind foam-mats, *Food Res. Int.*, 34, 587–598, 2001.

Wolff, E. and Gibert, H., Atmospheric freeze-drying part 2: Modelling drying kinetics using adsorption isotherms, *Dry. Technol.*, 8, 405–428, 1990.

Wolff, E. and Gibert, H., Lyophilisation sous pression atmosphérique, in *Collection Récents Progrès en Génie des Procédés*, Lavoisier, Paris, 1987.

Index